VOLUME ONE HUNDRED AND SEVEN

Current Topics in DEVELOPMENTAL BIOLOGY

Stem Cells in Development and Disease

CURRENT TOPICS IN DEVELOPMENTAL BIOLOGY

"A meeting-ground for critical review and discussion of developmental processes"
A.A. Moscona and Alberto Monroy (Volume 1, 1966)

VOLUME ONE HUNDRED AND SEVEN

CURRENT TOPICS IN DEVELOPMENTAL BIOLOGY

Stem Cells in Development and Disease

Edited by

MICHAEL RENDL
Black Family Stem Cell Institute,
Department of Developmental and Regenerative Biology
Icahn School of Medicine at Mount Sinai, NY, USA

ELSEVIER

AMSTERDAM • BOSTON • HEIDELBERG • LONDON
NEW YORK • OXFORD • PARIS • SAN DIEGO
SAN FRANCISCO • SINGAPORE • SYDNEY • TOKYO
Academic Press is an imprint of Elsevier

Academic Press is an imprint of Elsevier
525 B Street, Suite 1800, San Diego, CA 92101-4495, USA
225 Wyman Street, Waltham, MA 02451, USA
The Boulevard, Langford Lane, Kidlington, Oxford, OX5 1GB, UK
32 Jamestown Road, London NW1 7BY, UK
Radarweg 29, PO Box 211, 1000 AE Amsterdam, The Netherlands

First edition 2014

ISBN: 978-0-12-416022-4
ISSN: 0070-2153

Printed and bound by CPI Group (UK) Ltd, Croydon, CR0 4YY

CONTENTS

CONTRIBUTORS

Yaser Atlasi
Department of Pathology, Josephine Nefkens Institute, Erasmus MC, Rotterdam, The Netherlands

Nick Barker
Institute of Medical Biology, Singapore, and MRC Centre for Regenerative Medicine, The University of Edinburgh, Edinburgh, United Kingdom

Andrew S. Brack
Center for Regenerative Medicine, Massachusetts General Hospital; Harvard Medical School, and Harvard Stem Cell Institute, Boston, Massachusetts, USA

Natasha C. Chang
The Sprott Centre for Stem Cell Research, Ottawa Hospital Research Institute, and Department of Cellular and Molecular Medicine, Faculty of Medicine, University of Ottawa, Ottawa, Ontario, Canada

Shuang Chen
Department of Developmental Biology, Washington University, St. Louis, Missouri, USA

Sean M. Cullen
Stem Cells and Regenerative Medicine Center, and Program in Developmental Biology, Baylor College of Medicine, Houston, Texas, USA

Christine M. Fillmore
Stem Cell Program, Boston Children's Hospital; The Harvard Stem Cell Institute, Cambridge, and Department of Genetics, Harvard Medical School, Boston, Massachusetts, USA

Riccardo Fodde
Department of Pathology, Josephine Nefkens Institute, Erasmus MC, Rotterdam, The Netherlands

Naiyang Fu
Stem Cells and Cancer Division, The Walter and Eliza Hall Institute of Medical Research, and Department of Medical Biology, University of Melbourne, Parkville, Victoria, Australia

Margaret A. Goodell
Stem Cells and Regenerative Medicine Center; Program in Developmental Biology, and Department of Molecular and Human Genetics, Baylor College of Medicine, Houston, Texas, USA

Valerie Horsley
Department of Molecular, Cell and Developmental Biology, Yale University, New Haven, Connecticut, USA

Yunjoon Jung
Center for Regenerative Medicine, Massachusetts General Hospital, Boston, Massachusetts, USA

Klaus H. Kaestner
Department of Genetics and Institute for Diabetes, Obesity, and Metabolism, University of Pennsylvania School of Medicine, Philadelphia, Pennsylvania, USA

Carla F. Kim
Stem Cell Program, Boston Children's Hospital; The Harvard Stem Cell Institute, Cambridge, and Department of Genetics, Harvard Medical School, Boston, Massachusetts, USA

Raphael Kopan
Department of Developmental Biology, Washington University, St. Louis, Missouri, and Division of Developmental Biology, Department of Pediatrics, Cincinnati Children's Hospital, Cincinnati, Ohio, USA

Kristen T. Leeman
Division of Newborn Medicine, Boston Children's Hospital, Harvard Medical School; Stem Cell Program, Boston Children's Hospital; The Harvard Stem Cell Institute, Cambridge, and Department of Genetics, Harvard Medical School, Boston, Massachusetts, USA

Geoffrey J. Lindeman
Stem Cells and Cancer Division, The Walter and Eliza Hall Institute of Medical Research; Department of Medical Oncology, Royal Melbourne Hospital, and Department of Medicine, University of Melbourne, Parkville, Victoria, Australia

Melissa Little
Department of Developmental Biology, Washington University, St. Louis, Missouri, USA, and Institute for Molecular Bioscience, The University of Queensland, St. Lucia, Queensland, Australia

Leendert Looijenga
Department of Pathology, Josephine Nefkens Institute, Erasmus MC, Rotterdam, The Netherlands

Allison Mayle
Stem Cells and Regenerative Medicine Center, and Department of Molecular and Human Genetics, Baylor College of Medicine, Houston, Texas, USA

Jon M. Oatley
Center for Reproductive Biology, School of Molecular Biosciences, College of Veterinary Medicine, Washington State University, Pullman, Washington, USA

Eszter Posfai
Developmental and Stem Cell Biology, The Hospital for Sick Children Research Institute, Toronto, Ontario, Canada

Michael Rendl
Black Family Stem Cell Institute; Department of Developmental and Regenerative Biology; Department of Dermatology, and Graduate School of Biomedical Sciences, Icahn School of Medicine at Mount Sinai, New York, USA

Amélie Rezza
Black Family Stem Cell Institute, and Department of Developmental and Regenerative Biology, Icahn School of Medicine at Mount Sinai, New York, USA

Chiara Rolando
Department of Biomedicine, University of Basel, Basel, Switzerland

Janet Rossant
Developmental and Stem Cell Biology, The Hospital for Sick Children Research Institute, and Department of Molecular Genetics, University of Toronto, Toronto, Ontario, Canada

Lara Rossi
Department of Experimental, Diagnostic, and Specialty Medicine (DIMES), Institute of Hematology "L. and A. Seràgnoli", University of Bologna, Bologna, Italy

Michael A. Rudnicki
The Sprott Centre for Stem Cell Research, Ottawa Hospital Research Institute, and Department of Cellular and Molecular Medicine, Faculty of Medicine, University of Ottawa, Ottawa, Ontario, Canada

Rachel Sennett
Black Family Stem Cell Institute, and Department of Developmental and Regenerative Biology, Icahn School of Medicine at Mount Sinai, New York, USA

Soona Shin
Department of Genetics and Institute for Diabetes, Obesity, and Metabolism, University of Pennsylvania School of Medicine, Philadelphia, Pennsylvania, USA

Ana Mafalda Baptista Tadeu
Department of Molecular, Cell and Developmental Biology, Yale University, New Haven, Connecticut, USA

Oliver H. Tam
Developmental and Stem Cell Biology, The Hospital for Sick Children Research Institute, Toronto, Ontario, Canada

David Wei-Min Tan
Institute of Medical Biology, Singapore

Verdon Taylor
Department of Biomedicine, University of Basel, Basel, Switzerland

Jane E. Visvader
Stem Cells and Cancer Division, The Walter and Eliza Hall Institute of Medical Research, and Department of Medical Biology, University of Melbourne, Parkville, Victoria, Australia

Qi-En Yang
Center for Reproductive Biology, School of Molecular Biosciences, College of Veterinary Medicine, Washington State University, Pullman, Washington, USA

PREFACE

Stem cell research is one of the most fascinating and exciting research areas considering the therapeutical potential of stem cells. Manipulation of endogenous stem cells or the application of exogenous stem cells has the great promise to rejuvenate diseased and aged tissues in regenerative therapies. To fulfill this promise, we first need to identify tissue stem cells, define their physiological lineage relationship to differentiated cells, and characterize their functional control by intrinsic regulators and extrinsic factors from the microenvironment. In the past three decades, tremendous progress has been made: from the successful culture of embryonic stem cells and the early discovery of hematopoietic stem cells and stem cells in rapidly renewing skin and intestinal tissues to the more recent identification of neural, mammary, muscle, lung, and germ stem cells as well as hepatic and kidney progenitors. Recent studies have also shed light on the molecular make-up of stem cells and the complexly controlled balance between stem cell self-renewal and differentiation by endogenous transcriptional regulators, microRNAs and their overall epigenetic state, as well as by signaling and cellular communication from and to the microenvironment.

These insights have led to the establishment of stable stem cell cultures and culminated in recent efforts to engineer three-dimensional intestinal, skin, and other tissues. They further guide attempts at differentiation of embryonic and induced pluripotent stem cells and at direct reprogramming of differentiated somatic cells. Finally, our greatly improved understanding of stem cell control and function directly overlaps with studies of cancer biology and aging, as recently discovered cancer stem cells share many regulatory mechanisms with normal tissue stem cells that go awry in cancers and that also become dysregulated in aging tissues.

In this volume of *Current Topics in Developmental Biology* on *Stem Cells in Development and Disease*, 14 contributions from leading stem cell researchers provide a comprehensive overview about our current knowledge of physiological stem cell functions in 11 different mammalian organ systems and of the mechanisms that contribute to aging and cancer. In Chapter 1, Posfai et al. review embryonic and other pluripotent stem cells in the pre- and postimplantation embryo. They comprehensively discuss the latest insights on the sequential lineage specifications in the early developing embryo. Furthermore, they dissect the similarities and differences of *in vivo* pluripotency development with their stem cell population counterparts propagated *in vitro*. All these developmental aspects are then placed in the context of the major signaling pathways known to play critical roles in the regulation of pluripotency.

Chapters 2–4 are focused on the discussion of well-characterized stem cell systems in the rapidly renewing tissues of the bone marrow, intestine, and skin. In Chapter 2, Goodell and colleagues summarize the main results of many important recent studies on hematopoietic stem cell regulation with a focus on epigenetic control. They thoroughly discuss the latest insights on general principles of epigenetic regulation and the specific epigenetic mechanisms that impact embryonic hematopoietic development and that control the balance between hematopoietic stem cell maintenance and lineage commitment. This review further highlights epigenetic controls in the context of aging and hematopoietic malignancies. In Chapter 3, Tan and Barker review intestinal stem cells and discuss the controversial relationships and hierarchy of different stem cell populations that have been identified within the crypts. They summarize the main competing theories and conflicting results of many important recent studies and offer an integrative discussion of a unifying theory. In Chapter 4, Tadeu and Horsley review the latest insights on epithelial skin stem cells of the epidermis, hair follicles, and sweat glands. They detail the intrinsic and extrinsic controls of adult skin stem cell niches and discuss their relevance for epithelial skin cancer progression and pathology.

In subsequent chapters, other well-established and more recently discovered stem cells and progenitors of several organ systems are discussed. In Chapter 5, Fu et al. review the lineage relationship of different stem and progenitor cell populations of the mammary gland. With an insightful discussion, they provide reconciliatory explanations to new competing theories regarding lineage relationships within the epithelial mammary gland. It continues with a thorough overview of the molecular control of mammary stem and progenitor cell regulation by steroid hormones and other molecular players and finally ties together the latest insights regarding implications for breast cancer. In Chapter 6, several established and novel features of satellite stem cells are reviewed by Chang and Rudnicki in a comprehensive discussion of many recent insights, including: the emerging concept of satellite stem and progenitor cell heterogeneity, novel findings regarding the embryonic origin of muscle stem cells, and the requirement of Pax7 in satellite cells. Finally, this chapter discusses age- and disease-related decline of muscle stem cell function and provides an outlook on potential stem cell transplantation therapies. In Chapter 7, Rolando and Taylor summarize the latest insights on the developmental origin of adult hippocampal neuronal stem cells with a focus on controlling signaling pathways and transcription factors. Then, they dissect the heterogeneous nature and function of adult neurogenic stem cells and niches and, in this context, the role of important signaling pathways. Finally, this chapter discusses recent insights into the role of adult hippocampal neuronal stem cells in epilepsy, aging, and depression. In Chapter 8, Leeman et al. review recently discovered lung stem and progenitor cells. They summarize the identification of several different lung epithelial stem cells, the role of lung stromal and endothelial cells, and their relevance for lung cancers. Discussed are further recent advances toward cell-based therapies by applying bone marrow-derived mesenchymal stem cells and directed differentiation of induced pluripotent stem cells. In Chapter 9, Yang and Oatley review spermatogonial stem cells in physiological and pathological conditions. They first discuss the origin of spermatogonial stem cells, then the molecular mechanisms regulating their maintenance, and the latest insights into germ cell tumor formation. In Chapter 10, Shin and Kaestner describe recently discovered hepatic progenitor cells. They discuss competing and controversial theories regarding the cellular origin and precise identity of these stem cell-like cells. In Chapter 11, Kopan et al. review the role of nephrogenic progenitors in kidney morphogenesis and dissect extrinsic regulatory influences by stromal signals, in addition to intrinsic transcriptional controls. They then discuss the impact of these latest insights on renal hypoplasia and the general control of cessation of nephrogenesis.

In Chapter 12, Rezza et al. provide a niche-centric view on the general intrinsic regulation of stem cells and their extrinsic control by the microenvironment. A dissection of the cellular organization of adult stem cell niches is followed by an extensive discussion of established and novel molecular regulators in adult stem cell niches. It closes with a brief discussion of stem cell niche dysfunction in aging and cancer. The final Chapters 13 and 14 establish in greater detail the stem cell and niche-related processes and molecular mechanisms that are dysregulated in cancers and during aging, respectively. Atlasi et al. in Chapter 13 first provide a general discussion of the historical perspective, the definitions, and the emerging concept of cancer stem cells. In addition, they analyze the control of cancer stem cells by signaling pathways with a focus on Wnt/β-catenin signaling, and discuss the similarities of germ cell tumors with embryonic pluripotent stem cells. Finally, Jung and Brack review in Chapter 14 the dynamic changes in stem cells during aging that are only recently beginning to be uncovered. They discuss the molecular players in cellular aging and the extrinsic regulation mediating aged cellular and tissue homeostasis. They finally provide important examples of aging mechanisms in major well-known stem cell systems.

This volume of *Stem Cells in Development and Disease* is a collection of 14 timely and comprehensive reviews on the current knowledge and recent insights of stem cell functions in several different organ systems. New biotechnological breakthroughs in the past decade have brought tremendous advances that have greatly increased our understanding of stem cell functions, which ultimately brings us closer to the prospect of applying stem cell therapies to human disease. It is my hope that reading the diverse chapters in this volume will provide the different stem cell fields with new ideas and fresh perspectives.

Michael Rendl

CHAPTER ONE

Mechanisms of Pluripotency *In Vivo* and *In Vitro*

Eszter Posfai*, Oliver H. Tam*, Janet Rossant*,†,1
*Developmental and Stem Cell Biology, The Hospital for Sick Children Research Institute, Toronto, Ontario, Canada
†Department of Molecular Genetics, University of Toronto, Toronto, Ontario, Canada
[1]Corresponding author: e-mail address: janet.rossant@sickkids.ca

Contents

Abstract

During the course of preimplantation development, the mammalian embryo develops from a single totipotent cell into a blastocyst that is composed of three distinct cell types. Two waves of lineage specification events take place, setting aside a pluripotent cell population, the epiblast, from extraembryonic tissues. The epiblast that will form the somatic cells and germ line of the adult organism remains pluripotent until gastrulation, which commences shortly after the embryo implants. The epiblast's remarkable property of pluripotency has been harnessed by researchers for decades through derivation of embryonic stem cells and epiblast stem cells. Both types of cells can self-renew indefinitely and still retain the ability of germ layer differentiation. However, a central conundrum to the field of stem cell biology is the extent to which these *in vitro* cultured cells represent their *in vivo* tissue of origin. In this review we discuss the development of

Current Topics in Developmental Biology, Volume 107
ISSN 0070-2153
http://dx.doi.org/10.1016/B978-0-12-416022-4.00001-9

in vivo pluripotency, and compare and contrast the role of signaling pathways and downstream transcription factors in embryo-derived stem cell types and their *in vivo* equivalent lineage counterparts.

1. INTRODUCTION

At the onset of mouse development, fertilization of an oocyte by a sperm generates a 1-cell embryo, also known as a zygote. The zygote possesses totipotency, the ability to generate an entire organism, including its extraembryonic supporting tissues. As the zygote travels through the oviduct, it undergoes cleavage to form an 8-cell embryo around embryonic day 2.5 (E2.5) (Fig. 1.1). Individual 8-cell blastomeres are still totipotent (Kelly, 1977; Tarkowski, 1959; Tarkowski & Wróblewska, 1967). However, this period of totipotency does not persist, with the blastomeres gradually restricting their developmental fate. The first lineage restriction (and the subsequent loss of totipotency) begins with the compaction of the 8-cell embryo and ends in the formation of the inner cell mass (ICM) and the trophectoderm (TE) of the blastocyst at E3.5. The ICM will undergo another lineage segregation event prior to uterine implantation, whereby cells become restricted to either the primitive endoderm (PE) or epiblast (EPI) lineages (Arnold & Robertson, 2009; Schrode et al., 2013; Stephenson, Rossant, & Tam, 2012). These lineages become clearly distinguishable by the late blastocyst stage (~E4.5), with the TE forming an outer layer of epithelium surrounding a fluid-filled cavity (blastocoel), the EPI forming a ball of cells on one side of the blastocyst, and the PE migrating to generate an epithelial layer of cells separating the EPI from the blastocoel cavity. From fertilization to the time of implantation, a period also known as preimplantation development, the single-cell totipotent zygote has segregated into three developmentally distinct lineages, each of which will contribute to discrete cell populations essential for embryonic growth and development.

Although there are clear molecular and morphological differences between the three cell populations (EPI, TE, PE) of the preimplantation embryo, their divergent developmental potential is more apparent after implantation (Nowotschin & Hadjantonakis, 2010). In mouse, the cluster of EPI cells in the blastocyst rapidly expands to form a cup-shaped single-layer epithelium known as the postimplantation EPI. These cells

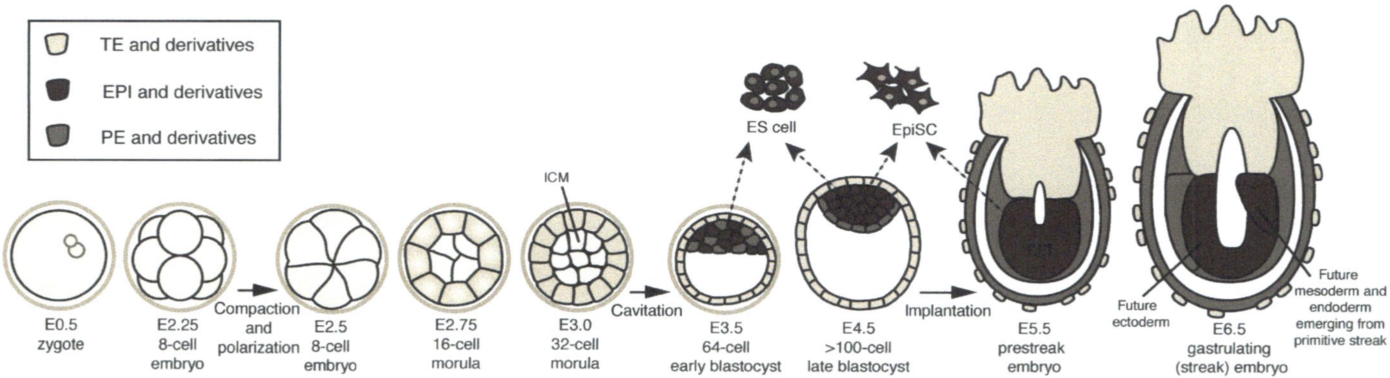

Figure 1.1 Schematic overview of early mouse development and origins of pluripotent stem cell types. The totipotent zygote undergoes three rounds of cleavage, producing an 8-cell embryo, which then undergoes compaction and polarization. From the 8-cell stage onward, cell divisions produce two populations of cells: outside cells, which will become the trophectoderm (TE), and inside cells, which will form the inner cell mass (ICM). The ICM further segregates into the primitive endoderm (PE) and the pluripotent epiblast (EPI) by the blastocyst stage. The EPI matures after implantation and gradually loses pluripotency when gastrulation commences at E6.5 (E, embryonic day). During gastrulation, the three germ layers are formed: mesoderm and endoderm precursors migrate through the primitive streak, while nonmigrating EPI cells form ectodermal tissues. Two types of pluripotent stem cells can be derived from the EPI at indicated stages: embryonic stem (ES) cells and epiblast stem cells (EpiSCs). (See color plate.)

subsequently generate embryonic ectoderm, mesoderm, endoderm, and germ cells of the developing organism, and contribute extraembryonic mesoderm to structures such as the allantois, amnion, yolk sac, and chorion (Lawson, Meneses, & Pedersen, 1991). The ability of the EPI population to generate the three major embryonic germ layers and the germ line has led to its classification as the pluripotent population of the developing embryo. In contrast, the TE and PE populations contribute almost exclusively to the extraembryonic tissues, forming structures such as the placenta (Cross, 2005; Rossant & Cross, 2001) and the parietal yolk sacs (Arnold & Robertson, 2009; Rossant & Tam, 2009).

2. PHASES OF LINEAGE RESTRICTION

A complex cellular process such as lineage restriction and determination is unlikely to be a single-step event. Molecular markers, some of which are lineage-determining transcription factors (TFs), allow us to separate lineage restriction into three phases: initiation, commitment, and maintenance. During the initiation phase, differences in cellular states arise, either "stochastically" or through subtle external cues, in the previously homogenous cell population (Johnston & Desplan, 2010). Despite these differences, these cells remain "plastic" and are able to interconvert between different cellular states depending on external stimuli. Early-response TFs are then expressed in cells that become biased toward a specific developmental fate and are often used to identify populations that progress toward lineage commitment. In the commitment phase, the cellular state is no longer "plastic," and becomes refractory to the initial inducing signal. In turn, specific developmental programs are activated that become increasingly dependent on cell-intrinsic cues, as seen in germ cell differentiation (Magnúsdóttir, Gillich, Grabole, & Surani, 2012). This is typically achieved by the activity of the early-response TFs, which can act by (a) forming positive feedback loops to strengthen their expression and (b) activating downstream (late response) TFs that are hallmarks of the cellular state (Zernicka-Goetz, Morris, & Bruce, 2009). In the final maintenance phase, the cell faces the task of stably maintaining expression of its gene set and inhibiting reversion or conversion into inappropriate cell fates (Pietersen & van Lohuizen, 2008). In many cases (such as in stem or progenitor cells), a cell must also be responsive to cues that direct appropriate downstream differentiation. Therefore, many genes that are involved in the formation of differentiated cell types are likely to be poised for activation, awaiting the right signals to produce downstream lineages (Dillon, 2012).

Given the difficulties in studying lineage specification *in vivo*, there have been significant efforts to establish culture systems to study these events *in vitro*.

We are now able to derive and maintain several stem cell populations from the mouse embryo that recapitulate the three early developmental lineages *in vitro*. These include embryonic stem (ES) (Brook & Gardner, 1997; Evans & Kaufman, 1981; Martin, 1981) and epiblast stem cells (EpiSC) (Brons et al., 2007; Tesar et al., 2007), trophoblast stem (TS) cells (Tanaka, 1998), and extraembryonic endoderm cells (Kunath et al., 2005), which represent the EPI (ES and EpiSC), TE, and PE lineages, respectively. These cell lines serve as potentially useful models for studying the maintenance phase of their respective embryonic cell type as well as their downstream differentiation capabilities.

However, we must be cautious when interpreting observations of *in vitro* cellular states, as they have been artificially stabilized by culture conditions in the absence of various extracellular and positional cues that exist in an intact mouse embryo. Therefore, these cell lines, although invaluable, might be overly simplistic snapshots of *in vivo* states, and their adaptation to culture conditions might have resulted in *de novo* regulative properties.

In this review, we examine and identify gaps in our current knowledge on TF networks and signaling pathways regulating the first lineage specification events in the preimplantation embryo. In particular, we focus on the molecular mechanisms that drive initiation, commitment, and maintenance of the pluripotent lineage *in vivo* and compare it with the pluripotent stem cell state *in vitro*.

3. FIRST LINEAGE SPECIFICATION—ICM VERSUS TE CELL FATE

Although there is growing evidence that developmental differences, or biases, exist in the early mouse embryo (Hartshorn, Eckert, Hartung, & Wangh, 2007; Tabansky et al., 2013; Torres-Padilla, Parfitt, Kouzarides, & Zernicka-Goetz, 2007), each blastomere up to the 8-cell stage (E2.5) is morphologically indistinguishable. Upon reaching the 8-cell stage, the embryo undergoes compaction (E2.75), a process in which blastomeres that were previously loosely associated increase their intracellular adhesion by forming E-cadherin-mediated adherens junctions (AJs) at regions of cell–cell contact. Concurrently, blastomeres undergo apical–basal polarization, characterized by an enrichment of glycoproteins and microvilli on the outward-facing cell surface and of F-actin, myosin, and clathrin in the underlying apical cytoplasm. The apical domain (AD), which is formed on the outside surface of blastomeres, is marked by the presence of the Par (partitioning defective) complex proteins, while other proteins are localized to basolateral surfaces, thereby consolidating the apico-basal polarity of cells (Fig. 1.2A) (Ducibella, Ukena, Karnovsky, & Anderson, 1977; Handyside,

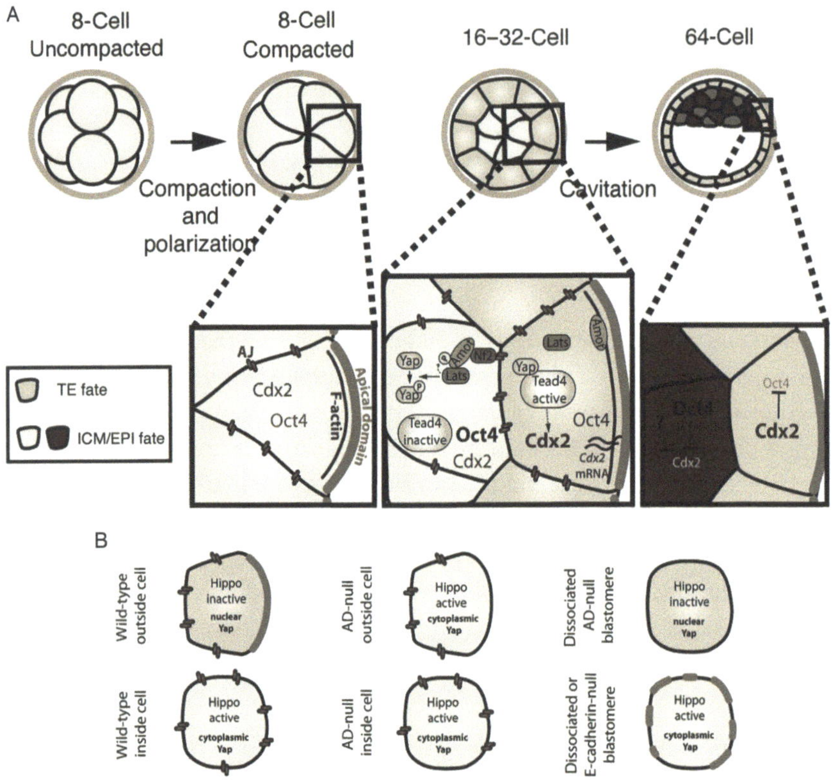

Figure 1.2 Model for ICM–TE specification in the mouse embryo. (A) At the 8-cell stage, the embryo undergoes compaction and polarization, forming an apical domain (AD) on outside surfaces and adherens junctions (AJs) on basolateral surfaces. At this stage, ICM and TE markers *Oct4* and *Cdx2* are coexpressed. Further divisions will yield two cell populations: apolar inside and polar outside cells of the embryo. In inside cells, the Hippo pathway is activated through *Amot* binding to AJs via *Nf2* and activating *Lats1/2* kinases. Active *Lats1/2* phosphorylate Yap, preventing its movement into the nucleus. Without Yap, *Tead4* cannot induce the expression of *Cdx2*. In outside cells, *Amot* is sequestered by the AD and does not bind AJs via *Nf2*; thus, it does not activate Hippo signaling. Yap is not phosphorylated and is free to move into the nucleus, activating *Cdx2*. At the blastocyst stage, *Cdx2* inhibits *Oct4* expression in outside cells. *Oct4* may contribute indirectly to *Cdx2* repression in inside cells. (B) Different genetic and/or mechanistic manipulations of blastomeres of the morula stage embryo result in the presence or absence of AJs and AD proteins. Consequently, the Hippo pathway is either active or inactive, resulting in the cytoplasmic or nuclear localization of Yap and the blastomere adopting an ICM or TE fate, respectively. (See color plate.)

1980; Maro, Johnson, Pickering, & Louvard, 1985; Pratt, Ziomek, Reeve, & Johnson, 1982; Reeve & Ziomek, 1981; Stephenson et al., 2012; Yamanaka, Ralston, Stephenson, & Rossant, 2006).

Compaction and polarization are followed by two rounds of cell division that can be classified as symmetric or asymmetric, depending on the plane of cleavage. In a symmetric division, the cleavage plane is aligned to the apical–basal polarity of the cell, resulting in two polarized daughter cells. If the cleavage plane is orthogonal or oblique to the axis of polarity, the resulting daughter cells will inherit differing amounts of AD markers, generating one polarized cell that remains on the outside of the embryo and one apolar cell that integrates into the inside of the embryo. Subsequent divisions of these inside cells will generate apolar progenies (Johnson & Ziomek, 1981, 1983). Thus, after the 8-cell stage, the embryo contains two populations of cells: inside cells that are apolar and experience a high degree of cell–cell contact and outside cells that are polarized and have an AD on the surface facing the outer environment.

The segregation of inside and outside cell populations coincides with a key event in the ICM–TE specification: differential localization of the transcriptional activator Yes-associated protein (Yap) (Nishioka et al., 2009). In outside cells, as in all cells of the compacted 8-cell embryo prior to differential cell division, Yap is present in the nucleus and is able to interact with its transcriptional partner *Tead4* (Yagi et al., 2007). In contrast, in newly formed inside cells, Yap is phosporylated by the *Lats1/2* kinases and sequestered in the cytoplasm, preventing its transcriptional activity. These differences in Yap localization are correlated with the expression patterns of Yap target genes such as the TE-specific TF *Cdx2*, which becomes upregulated in outside cells and lost from inside cells well before the onset of blastocyst formation (Dietrich & Hiiragi, 2007; Ralston & Rossant, 2008). These observations alone strongly suggest that differences in polarity and/or cell–cell contact between inside and outside cells kick-start the process of ICM–TE segregation. However, because they are tightly linked during normal development, distinguishing which of these two aspects is important for ICM–TE segregation is somewhat more challenging. For instance, elimination of E-cadherin prevents compaction and proper segregation of apical and basal domains in blastomeres. In these embryos, AD proteins are localized in an apolar fashion, and internalized blastomeres continue to express AD proteins on their surface (Stephenson, Yamanaka, & Rossant, 2010; Fig. 1.2B). Interestingly, Yap is nuclear and *Cdx2* is expressed in the majority of E-cadherin-deficient blastomeres, both hallmarks of future TE cells. Similarly, separation of early blastomeres, which disrupts cell–cell contact, also leads to AD protein expression in an apolar fashion in all blastomeres. This leads to nuclear accumulation of Yap (Hirate et al., 2013) and drives the blastomeres to adopt a gene expression profile reminiscent of TE cells (Fig. 1.2B; Lorthongpanich, Doris, Limviphuvadh, Knowles, & Solter, 2012).

Further studies have shown that decreasing AD protein expression (without interfering with cell–cell adhesion) through knockdown of *Pard6b*, *Par3*, or a dominant negative or genetic null aPKC results in more cytoplasmic Yap (Hirate et al., 2013), reduced Cdx2 expression (Alarcon, 2010), and preferential ICM fate adaptation (Plusa et al., 2005), suggesting that the presence of an AD is sufficient to prevent Yap phosphorylation and TE specification. However, in contrast to results obtained in intact embryos, dissociated blastomeres deficient in *Par6b* have largely nuclear Yap, indicating that in addition to an absence of polarity, cell contact is required for Yap phosphorylation (Fig. 1.2B).

Lats1/2, *Yap1*, and *Tead4* are members of the mammalian Hippo signaling pathway, and two recent publications have further identified the Hippo pathway members *Nf2/Merlin* and Angiomotin (*Amot*) as important components of ICM–TE specification (Cockburn, Biechele, Garner, & Rossant, 2013; Hirate et al., 2013). In an elegant study, Hirate and colleagues showed that E-cadherin-mediated AJs are the sites of Hippo activation in inside cells and that AD proteins dictate the cellular localization and binding properties of *Amot*. They propose a model in which *Amot* proteins act as an important scaffold for Hippo activation in the apolar inside cells, where *Amot* binds to AJs via *Nf2* to allow phosphorylation of its N-terminal domain by *Lats1/2*. Phospho-*Amot* in turn strongly binds *Lats1/2*, which is needed for the Hippo-activating function of the kinases. On the other hand, AD proteins in the polar outside cells sequester *Amot* from AJs and promote its binding to cortical F-actin, somehow preventing Hippo pathway activation. Thus, though questions still remain, a picture is emerging in which the presence of an AD prevents cell–cell contact-dependent Hippo pathway activation in outside cells (Fig. 1.2A). Such a mechanism would allow cells to read out their position in the embryo and activate differential downstream pathways, leading to the adoption of ICM or TE-specific transcriptional programs.

With the segregation of inside and outside cell populations and subsequent differential Hippo activation, *Cdx2* becomes preferentially upregulated in outer cells and gradually lost from inside cells (Dietrich & Hiiragi, 2007; Ralston & Rossant, 2008). This upregulation in outside cells is likely largely attributable to Yap activity, but asymmetric localization of

Cdx2 transcripts to the AD from the 8-cell stage onward may also play a role (Skamagki, Wicher, Jedrusik, Ganguly, & Zernicka-Goetz, 2013). The presence of *Cdx2* in outside cells is essential for TE development. Although they are able to initiate polarization of the outer cells and even form a blastocoel cavity, *Cdx2* mutants fail to maintain the TE, resulting in blastocoel collapse and a failure to implant (Strumpf et al., 2005; Wu et al., 2010). Moreover, the pluripotency factors *Oct4* and *Nanog* are ectopically expressed in outside cells of these mutants. Thus, while *Cdx2* is not required for the initial specification of the TE lineage, it has a prominent role in TE commitment and is crucial for the segregation of the TE from the ICM.

Once it has been confined in its expression to inside cells, *Oct4* has been proposed to act reciprocally on *Cdx2*, preventing its expression in the ICM. This hypothesis has been supported largely by experiments done *in vitro*, in which loss of *Oct4* in ES cells results in morphological and molecular changes reminiscent of TS cells (Hay, Sutherland, Clark, & Burdon, 2004; Niwa et al., 2005); the same changes can be observed when *Cdx2* is overexpressed in ES cells. Surprisingly however, *Oct4*-null embryos still express the pluripotent EPI marker *Nanog* and instead show defects in the specification of the PE lineage (see later) (Frum et al., 2013). They also show ectopic *Cdx2* expression in some, but not all, inside cells. Therefore, unlike its central role in preventing TE conversion in ES cells, *Oct4* may be required only indirectly for TE fate repression through upregulating the pluripotency program in the ICM. It will be interesting to see which other pluripotency factors are responsible for *Cdx2* repression in inner cells.

The formation of the ICM and the TE at the blastocyst stage is the first event that separates cells with pluripotent capacity from restricted extraembryonic cell types. Accordingly, ES cells, generated from the ICM, normally lack the capacity to form TE *in vivo* or *in vitro* (Beddington & Robertson, 1989). The TF networks that have been extensively characterized in ES cells can be detected *in vivo* in the ICM, but the full pluripotency network has yet to be stabilized. Therefore, the ES cell state is more akin to the EPI progenitor state that appears with the segregation of the PE and EPI progenitors within the ICM (see later).

In the embryo, Hippo signaling must be activated in the developing inside cells to prevent them from expressing TE markers (Cockburn et al., 2013; Hirate et al., 2013; Lorthongpanich et al., 2013). In contrast, inhibition of Hippo signaling in ES cells, while resulting in a loss of pluripotency, does not appear to convert ES cells into TE fate, but rather promotes differentiation into an endodermal fate (Lian et al., 2010;

Tamm, Böwer, & Annerén, 2011). This suggests that Hippo signaling no longer governs ICM–TE fate specification in ES cells, which is consistent with the model that ES cells represent a stabilized EPI lineage that is now refractory to specification cues present during ICM–TE lineage specification.

4. SECOND LINEAGE SPECIFICATION—EPI VERSUS PE CELL FATE

The second lineage segregation event in the embryo directs the cells of the ICM to become either EPI or PE cells by the end of preimplantation development (Fig. 1.3A). The asymmetric cell divisions that occur after compaction produce inside cells in two separate waves: some cells are produced earlier (in the 8- to 16-cell division wave) and some cells are produced slightly later (in the 16- to 32-cell wave). There has been some debate about whether cells produced by these two different waves vary in their contribution to the EPI or PE lineages. Although one study reported that cells internalized during the first wave were more likely to become EPI, with the second wave biased toward PE fate (Morris et al., 2010), this was not observed in a second study (Yamanaka, Lanner, & Rossant, 2010). Despite the unresolved link between cell fate bias and timing of internalization, it has been demonstrated that the fates of inner cells are not irreversibly determined at these early divisions, and that they are able to switch between EPI–PE identities when exposed to a different environment (Nichols, Silva, Roode, & Smith, 2009; Plusa, Piliszek, Frankenberg, Artus, & Hadjantonakis, 2008; Yamanaka et al., 2010).

Fibroblast growth factor (FGF) signaling has long been implicated in PE development, as embryos with mutations in *Fgf4* (Feldman, Poueymirou, Papaioannou, DeChiara, & Goldfarb, 1995; Goldin & Papaioannou, 2003), *Fgfr2* (Arman, Haffner-Krausz, Chen, Heath, & Lonai, 1998) or a downstream signaling component *Grb2* (Chazaud, Yamanaka, Pawson, & Rossant, 2006; Cheng et al., 1998) fail to generate PE, and die shortly after implantation.

The FGF pathway is a receptor tyrosine kinase signaling pathway that is activated by FGF ligand binding to specific cell surface FGF receptors (FGFRs). The binding of the ligand leads to dimerization of the receptors, and triggers autophosphorylation of their intracellular receptor domains. These phosphorylated residues serve as docking sites for signaling proteins and adaptors that initiate signal transduction through multiple downstream signaling cascades: They include (a) the Janus kinase/signal transducer and activator of transcription

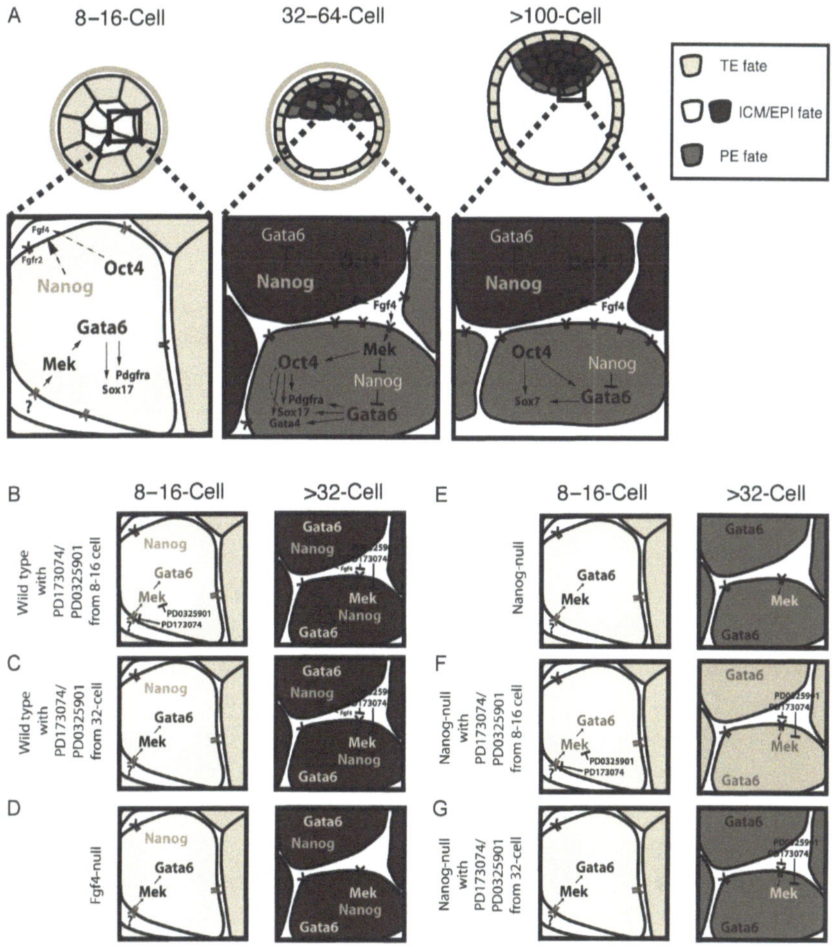

Figure 1.3 Model for EPI–PE specification in the mouse embryo. (A) Prior to the 32-cell stage, the EPI marker *Nanog* and the PE marker *Gata6* are coexpressed. *Gata6* expression is FGF4-independent but Mek-dependent at this stage. Segregation of *Nanog*-expressing EPI and *Gata6*-expressing PE precursors occurs from the 32-cell stage onwards: In EPI cells, *Nanog* inhibits *Gata6* expression, while in *Fgfr2*-expressing PE cells, FGF4 (supplied by the EPI cells) via Mek-mediated signaling inhibits *Nanog*. *Oct4* is activated in an Mek-dependent manner in PE cells. In the absence of *Nanog* in PE precursors, *Gata6* is maintained. *Oct4* and *Gata6* are needed for downstream PE-specific TF (*Pdgfra*, *Sox17*, and *Gata4*) expression. In the late blastocyst, PE cells are sorted into a layer lining the blastocoel cavity. PE cells are no longer sensitive to FGF-signaling and express *Sox7* in a *Gata6*- and *Oct4*-dependent manner. (B–G) Different genetic and/or different timing of treatment with small molecule inhibitors (PD173074 or PD0325901) of Mek-dependent signaling results in altered *Nanog* and/or *Gata6* expression and in cells adopting altered EPI or PE fates. (See color plate.)

(JAK/STAT) pathway, (b) the phosphoinositide phospholipase C (PLCg) pathway, (c) the Ras-Raf–Mek–MAPK/Erk (mitogen-activated protein kinase/extracellular signal-regulated kinase) pathway, and (d) the phosphatidylinositol 3-kinase (PI3K) pathway (Dailey, Ambrosetti, Mansukhani, & Basilico, 2005; Lanner & Rossant, 2010). Some of these pathways share common adaptor proteins, such as *Frs2a* (FGF receptor substrate 2 alpha) and *Grb2* (growth factor receptor-bound protein 2), which could lead to cross talk between them upon FGF binding (Fig. 1.4).

Although there are numerous FGF ligands and receptors in mammals, *Fgf4* and *Fgfr2* are the predominant ligand–receptor pair expressed during preimplantation development (Arman et al., 1998; Niswander & Martin, 1992; Rappolee, Basilico, Patel, & Werb, 1994). The expression of *Fgf4* and *Fgfr2* is initially ubiquitous in the early embryo, but becomes asymmetric at the 32-cell stage, with ligand and receptor higher in future EPI and PE/TE cells, respectively (Guo et al., 2010; Kurimoto et al., 2006). The asymmetric expression within the early ICM becomes mutually exclusive by the 64-cell blastocyst stage, with *Fgf4* expressed only in the EPI cells, and *Fgfr2* in the PE cells (Frankenberg et al., 2011; Messerschmidt & Kemler, 2010).

Modulation of FGF signaling is critical in EPI–PE lineage specification. Treatment with small molecule inhibitors against FGFRs or *Mek1/2* (MAPK/Erk pathway) results in embryos where all ICM cells express the EPI marker *Nanog* and lack the PE marker *Gata6*. Conversely, embryos cultured with exogenous *Fgf4* completely eliminate *Nanog*-expressing cells, with all ICM cells expressing *Gata6* (Nichols et al., 2009; Yamanaka et al., 2010). FGF-signaling levels appear to be tightly regulated in order to generate balanced numbers of EPI and PE progenitors within the ICM, as partial reduction of *Fgf4* (heterozygous *Fgf4* mutant embryos) reduces the number of PE cells in the embryo (Kang, Piliszek, Artus, & Hadjantonakis, 2013). Based on these studies, we can conclude that FGF signaling through the MAPK/Erk pathway plays a crucial role in EPI–PE cell fate specification.

Genetic studies in the mouse have identified two TFs as key determinants of the EPI and PE lineages. The homeodomain TF *Nanog* is associated with ground-state or naive pluripotency and is required cell-autonomously for EPI fate. Mutants lacking Nanog show peri-implantation lethality and a failure to generate EPI cells in the embryo or ES cells *in vitro*. Intriguingly, Nanog is also required non-cell-autonomously for PE formation, resulting from the absence of FGF signals normally generated by *Nanog*-positive cells (Fig. 1.3E) (Frankenberg et al., 2011). *In vitro Nanog*-null ES cells or ICM outgrowths produce only endoderm-like cells reminiscent of PE (Messerschmidt & Kemler, 2010; Mitsui et al., 2003), which can be eliminated by treatment with FGF inhibitors during culture (Silva et al., 2009). The zinc finger TF *Gata6* marks the PE lineage in the late blastocyst (Chazaud et al., 2006), and its absence results in defects in PE (and derivatives) and peri-implantation lethality (Cai, Capo-Chichi, Rula, Yang, & Xu, 2008; Koutsourakis, Langeveld, Patient, Beddington, & Grosveld, 1999; Morrisey et al., 1998). Overexpression of *Gata6* in ES cells is sufficient to induce multiple PE markers and differentiation into PE-like cells (Fujikura et al., 2002; Wang, Smedberg, Cai, Capo-Chichi, & Xu, 2011).

Nanog and *Gata6* are initially coexpressed in all inner cells of the morula (up to E3.0, 32-cell) stage (Fig. 1.3A) (Kang et al., 2013; Plusa et al., 2008). This initial period of EPI–PE marker coexpression seems to be independent of FGF signaling, as maternal/zygotic *Fgf4*-null embryos correctly establish initial expression of *Gata6* and *Nanog* in all ICM cells (Fig. 1.3D; Kang et al., 2013). Intriguingly, Frankenberg and colleagues were able to inhibit early *Gata6* expression with two commonly used inhibitors of the FGF-signaling pathway, PD173074 and PD0325901 (which only becomes evident in *Nanog*-null embryos, see Fig. 1.3 for detailed explanation) (Frankenberg et al., 2011). This can be easily reconciled because of potential off-target effects of PD173074 (an FGFR inhibitor that can also inhibit other pathways, such as insulin, VEGF, EGF, and PDGF) and the broad impacts of *Mek1/2* inhibition by PD0325901 on multiple signaling pathways. These results suggest that the initial induction of *Gata6* expression (and thus the PE program) is dependent on an exogenous signal that acts through *Mek1/2*, but it is unlikely to be *Fgf4*. However, we cannot exclude the possibility that the alternative signaling molecule plays a compensatory (rather than inductive) role only in the absence of *Fgf4* expression.

Concomitant with initial *Fgf4/Fgfr2* differences at the 32-cell stage, individual ICM cells also start to show differences in expression of *Nanog* and *Gata6*. Like *Fgf4/Fgfr2*, these differences in *Nanog* and *Gata6* also resolve into a mutually exclusive expression pattern by the 64-cell stage in a seemingly mixed "salt-and-pepper" fashion (Chazaud et al., 2006; Guo et al., 2010; Kurimoto et al., 2006).

The segregation of EPI and PE expression profiles is dependent on *Fgf4*: although *Gata6* expression is initiated at the 8- to 16-cell stage, it is not maintained in the absence of *Fgf4* beyond the 32-cell stage (Kang et al., 2013), which can be recapitulated through chemical inhibition of *Mek1* or FGFR (Frankenberg et al., 2011). Thus, while initiation of the PE program appears to be *Fgf4*-independent, its maintenance and restriction to future PE cells seems to require *Fgf4*-mediated input.

While there is a strong correlation between FGF-signaling activity and EPI–PE specification, the exact molecular mechanisms linking the two phenomena remain elusive. It has been speculated that *Fgf4* signaling in the PE-precursors (at the "salt-and-pepper" stage) alleviates *Nanog*-dependent suppression of *Gata6*, which is supported by *in vitro* models that show *Nanog* inhibition occurring downstream of FGF/MAPK/Erk signaling (see later in ES section). Furthermore, *Nanog* was shown to bind and repress the *Gata6* promoter in ES cells, suggesting a potential direct link between its activity in regulating PE fate (Mitsui et al., 2003; Singh, Hamazaki, Hankowski, & Terada, 2007; Fig. 1.3).

Gata6 is required to activate additional PE-specific TFs, which is supported by the identification of GATA response elements *in cis*-regulatory regions of *Gata4*, *Sox17*, and *Pdgfra* (Niakan et al., 2010; Wang & Song, 1996). Although *Gata6* overexpression is able to induce expression of multiple PE markers in ES cells (Fujikura et al., 2002; Wang et al., 2011), it is unable to upregulate downstream PE markers in *Nanog*-null embryos, where all inside cells express *Gata6* (Frankenberg et al., 2011; Messerschmidt & Kemler, 2010; Silva et al., 2009). This discrepancy may stem from the requirement of a factor produced by EPI cells (such as Fgf4) to maintain downstream PE marker expression. Indeed, it has been demonstrated that *Gata6* can initiate, but not sustain, *Pdgfra* and *Sox17* expression beyond the 32-cell stage of *Fgf4*-null embryos (Kang et al., 2013). It has also been shown that *Sox17* and *Gata4* expression can be restored in *Nanog*-null embryos by addition of exogenous *Fgf4* (Frankenberg et al., 2011). These data suggest that FGF signaling is also required for maintaining the PE program in parallel with alleviating *Nanog* repression of *Gata6*.

A recent study has uncovered a surprising role for *Oct4* in PE specification (Frum et al., 2013). Frum and colleagues found that *Oct4* is dispensable for initiating the pluripotency program of the EPI lineage (as shown by markers such as *Nanog*), but is required cell-non-autonomously for PE specification through *Fgf4* production. Surprisingly, *Oct4* is also required cell-autonomously for PE fate specification in an FGF/MAPK/Erk signaling-dependent manner, where it mediates expression of *Pdgfra*, *Sox17*, and, to some extent, *Gata4*. Therefore, *Oct4* is a novel component of the FGF/MAPK/Erk-dependent pathway required for activation of the PE program.

Among inner cells of the "salt-and-pepper" stage embryo, *Oct4* is expressed at similar levels in EPI and PE progenitors (Guo et al., 2010; Kurimoto et al., 2006; Palmieri, Peter, Hess, & Schöler, 1994). Results from *in vitro* assays suggest that *Oct4* can switch its interacting partner from *Sox2* to *Sox17*, which converts the Oct4 complex from pluripotency regulation to endodermal gene activation (Aksoy et al., 2013; Jauch et al., 2011). *Sox17* also functions to outcompete and displace Nanog from its binding sites

(Niakan et al., 2010), further suppressing the EPI pluripotency network. Although not proven yet, it is quite plausible that *Sox17* plays similar roles *in vivo* and contributes to consolidating PE fate.

In contrast to inner cells of the embryo, fine-tuned levels of *Oct4* have been recently shown to govern the pluripotent state in ES cells. Somewhat surprisingly, a low level of *Oct4* supports ground-state pluripotency, while elevated or stabilized *Oct4* promotes germ layer differentiation (Karwacki-Neisius et al., 2013; Radzisheuskaya et al., 2013). Thus, the function of *Oct4* in ES cells may be related to regulatory maintenance, rather than initiating gene programs as in the blastocyst.

The "salt-and-pepper" distribution within the ICM is rearranged during the late blastocyst stage, with the PE cells forming an epithelial layer adjacent to the blastocoelic cavity. This cell sorting has been demonstrated through live imaging studies to involve active migration of PE cells to their correct location, coupled with apoptosis of cells that fail to do so (Gerbe, Cox, Rossant, & Chazaud, 2008; Meilhac et al., 2009; Plusa et al., 2008). At this later stage, expression of *Gata6* in PE cells is Oct4-dependent, and together they drive the expression of *Sox7* (Frum et al., 2013; Kang et al., 2013). The appearance of *Sox7* also coincides with EPI and PE cell fates becoming resistant to modulation by addition of exogenous *Fgf4* or FGF inhibitors (Nichols et al., 2009; Plusa et al., 2008; Yamanaka et al., 2010), suggesting that the window of lineage commitment had closed.

5. MAINTENANCE OF EARLY CELL LINEAGES—RELEVANCE TO PLURIPOTENT STEM CELLS

Two kinds of pluripotent stem cells can be derived from early mouse embryos (Fig. 1.1). ES cells are derived from the preimplantation ICM (Evans & Kaufman, 1981; Martin, 1981), or from the EPI lineage after EPI–PE segregation (Brook & Gardner, 1997). ES cells can differentiate into all three germ layers *in vitro* and can integrate into the (preimplantation) EPI *in vivo*, contributing to the entire adult organism, including the germ line (Beddington & Robertson, 1989; Bradley, Evans, Kaufman, & Robertson, 1984).

EpiSC are generated from the early postimplantation EPI (Brons et al., 2007; Tesar et al., 2007) or the preimplantation EPI under certain conditions (Najm et al., 2011), and demonstrate pluripotency in *in vitro* assays, similar to ES cells. Although they are unable to integrate into the preimplantation EPI *in vivo*, they contribute to all three germ layers and the germ line of chimeric animals when transplanted into postimplantation embryos (Huang, Osorno,

Tsakiridis, & Wilson, 2012). The differential capacity of ES and EpiSCs to integrate into the mouse embryo *in vivo* suggests that, although both originate from the EPI lineage, they likely represent different developmental stages within this lineage.

While both cell types share some common pluripotency features (*Oct4*, *Sox2*, and *Nanog* expression), they also bear striking differences, such as different requirements for exogenous signaling during derivation and maintenance *in vitro*, distinct colony morphology (ES—dome; EpiSC—flattened), colony formation efficiency from single cells (ES—efficient; EpiSC—inefficient), and epigenetic states (e.g., X inactivation) (Brons et al., 2007; Guo et al., 2009; Hayashi, de Sousa Lopes, Tang, & Surani, 2008; Tesar et al., 2007). This is also reflected in significant differences in global gene expression profiles, where ES cells express higher levels of *Nanog* and *Esrrb*, while EpiSCs express higher levels of *Fgf5*, *Lefty1*, and *Cer1*.

Several studies have shown that ES cells cultured in leukemia inhibitory factor (LIF) and serum conditions show a certain degree of heterogeneity. A subpopulation of ES cells expresses PE markers such as *Gata6* (Singh et al., 2007), *Hex* (Canham, Sharov, Ko, & Brickman, 2010), and *Sox17* (Niakan et al., 2010), while other subpopulations spontaneously express postimplantation EPI/germ layer markers (*Brachyury*, *Fgf5*) (Suzuki et al., 2006; Toyooka, Shimosato, Murakami, Takahashi, & Niwa, 2008), suggesting that these cells are potentially primed for differentiation into an endodermal or a mature EPI-like lineage. *Nanog* was observed to have temporal fluctuations of expression, and it was hypothesized that a low level of this pluripotency marker transiently primes cells for differentiation without committing their fate (Chambers et al., 2007; Kalmar et al., 2009). Therefore, it was suggested that heterogeneous *Nanog* levels are a marker of inherent heterogeneity in ES cell populations.

An attractive hypothesis explaining the fluctuating levels of *Nanog* is based on the observation that *Nanog* is mono-allelicly expressed in ES cells cultured in the presence of LIF (Miyanari & Torres-Padilla, 2012). LIF has been shown to activate opposing signaling pathways that regulate *Nanog* expression through the activity of *Tbx3*, a T-box TF. *Tbx3* can be induced by the PI3K pathway and repressed by the MAPK/Erk pathway, both of which are activated by LIF signaling (see later). Therefore, the *Nanog* level, sensitized by its mono-allelic expression, may fluctuate depending on the relative activity of the two LIF-activated pathways (Nakai-Futatsugi & Niwa, 2013).

A complementary hypothesis is the stochastic autocrine activation of the FGF/MAPK/Erk pathway in ES cell populations, similar to the model described in the preimplantation EPI cell population (Davies et al., 2013; Kunath et al., 2007; Lanner & Rossant, 2010; Yamanaka et al., 2010), which is supported by the bi-allelic expression of *Nanog* when ES cells are cultured in the presence of an FGF inhibitor (Miyanari & Torres-Padilla, 2012). Furthermore, the different levels of Wnt signaling activities among ES cells have been implicated to enhance self-renewal (ten Berge et al., 2011, and reviewed by Atlasi, Looijenga, & Fodde, 2013), acting through a key pluripotency network component, *Esrrb* (Festuccia et al., 2012; Martello et al., 2012). It is therefore conceivable that 2-inhibitor (2i) culture conditions (FGF inhibition and Wnt signaling activation) serve to reduce *Nanog* fluctuations to generate a more pluripotent and homogeneous ES cell population.

However, the extent of *Nanog*'s fluctuations and instability has recently been challenged by two publications, which monitored *Nanog* levels through reporter lines that do not perturb the endogenous expression and function of the gene (Faddah et al., 2013; Filipczyk et al., 2013). While both studies observed fluctuations in *Nanog*, neither study detected the previously reported mono-allelic expression (Miyanari & Torres-Padilla, 2012), nor did either find *Nanog* levels to vary more than levels of other genes expressed in ES cells, such as *Stella* (Hayashi et al., 2008). An explanation of the previously observed instabilities is that they had captured stochastic bursts of *Nanog* transcription, which is consistent with the idea that transcriptional noise in the ES pluripotency network could lead to the observed fluctuations in *Nanog* expression levels (Kalmar et al., 2009). However, it remains unclear if such fluctuations are an *in vitro* feature of ES cell maintenance, or whether they would have functional significance *in vivo*.

The heterogeneous PE marker expression in ES cell populations is reminiscent of the progressive segregation of the EPI and PE in the ICM, where an initial metastable state of heterogeneous gene expression gradually resolves into two committed lineages. Although ES cells share more characteristics with the EPI than the PE lineage, the expression of PE markers in a small population of ES cells, possibly because of inherent culture conditions, may hint at their plasticity. It has yet to be examined whether ICM- or EPI-derived ES or EpiSC lines show different heterogeneities and/or potential to acquire PE-like (or mature EPI-like) fates, but these studies might offer valuable insights into the separation of committed EPI and PE fates *in vivo*.

6. GROWTH FACTOR SIGNALING REQUIREMENTS IN EPI-DERIVED STEM CELLS: *IN VIVO* SIMILARITIES AND DIFFERENCES

Exogenous growth factors play important roles in the maintenance of EPI-derived cells. ES cell maintenance can be achieved by the activation of

LIF (Smith et al., 1988; Williams et al., 1988) (normally produced by feeder-layer cells) and bone morphogenic protein (BMP) signaling (Ying, Nichols, Chambers, & Smith, 2003) (with the ligand provided in serum), and is enhanced by inhibition of FGF signaling and GSK3β activity (Ying et al., 2008). In contrast, EpiSCs require active FGF and Activin/Nodal signaling, with LIF or BMP signaling activity being dispensable (Tesar et al., 2007). Here, we present the major signaling pathways involved in EPI-derived stem cell maintenance and assess their requirements in the pre- and postimplantation embryo (Fig. 1.4).

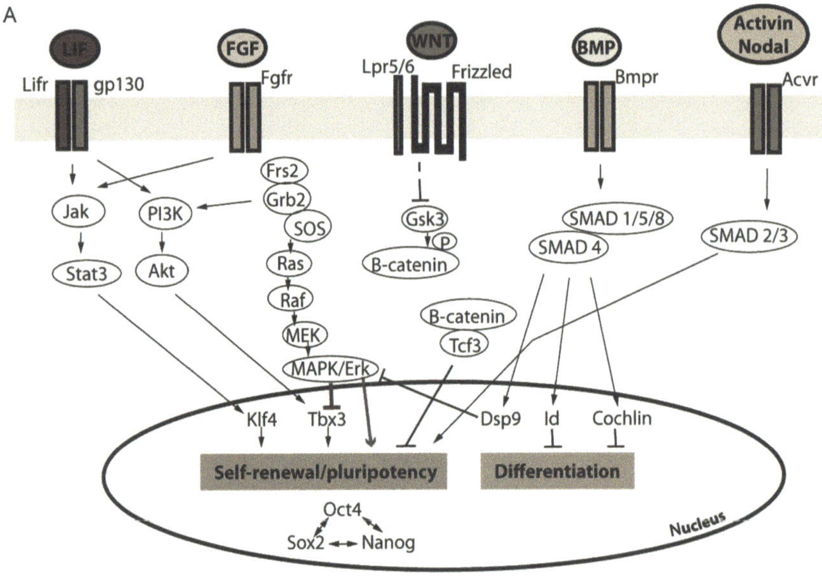

B

	In vitro		*In vivo*	
	ES	**EpiSC**	**Preimplantation EPI**	**Postimplantation EPI**
LIF	✓	X	✓ only in diapause and LIF-independent Stat3 activation	X
BMP	✓	⊘	X	✓/⊘ required for EPI; also required for gastrulaion in combination with other cues
WNT	✓/X with LIF	⊘	X but may enhance proliferation	⊘ induces gastrulation
FGF	⊘	✓	⊘/X inhibitory up to E3.75; not required after	✓
Activin/Nodal	⊘	✓	X	✓

✓ **Required** X **Not required** ⊘ **Inhibitory**

Figure 1.4 Signaling pathways involved in regulating self-renewal and pluripotency in the mouse. (A) Cross talk among intracellular signaling cascades involved in regulating self-renewal and pluripotency. Red line ES specific, blue line EpiSC specific. (B) Signaling pathways utilized in the maintenance of self-renewal and pluripotency *in vitro* (ES and EpiSCs) and *in vivo* (pre- and postimplantation EPI). (See color plate)

6.1. Leukemia inhibitory factor

LIF has been utilized extensively in the derivation and culture of ES cells and constitutes a central signaling pathway in self-renewal and pluripotency. It signals through the LIFR and gp130 heterodimer receptor, which activates several pathways: (a) the JAK/STAT, (b) the PI3K, and (c) the MAPK/Erk pathway (Nakai-Futatsugi & Niwa, 2013). The parallel activation of these pathways has differing effects on ES cell maintenance. Phosphorylation of *Stat3* by JAK enables its translocation into the nucleus to activate downstream targets such as *Klf4*, while the PI3K pathway activates the TF, *Tbx3*. Both *Klf4* and *Tbx3* activate the expression of *Sox2*, *Nanog*, and *Oct4*, which drives the establishment and maintenance of the pluripotency network. Interestingly, the MAPK/Erk pathway negatively affects pluripotency through inhibition of *Tbx3* expression. The activities of these parallel pathways must be precisely regulated to ensure maintenance of pluripotency by expression of *Sox2*, *Nanog*, and *Oct4*, and also to provide a mechanism by which cells can exit from the pluripotency network and engage in differentiation when needed.

The importance of LIF signaling in ES cells has led researchers to investigate its potential role in the maintenance of the EPI lineage *in vivo*. In the preimplantation embryo, LIF and LIFR/gp130 are expressed from the morula stage onward and downregulated when the EPI differentiates shortly after implantation. Intriguingly, LIF is expressed in the TE, while LIFR and gp130 are expressed by the ICM (Nichols et al., 1996), which may result in paracrine signaling between TE and ICM cells. However, ablation or inactivation of LIF signaling does not appear to affect preimplantation development, with no observable defects at these stages in LIF (Stewart et al., 1992), LIFR, and gp130 mutants (Li, Sendtner, & Smith, 1995; Nakashima et al., 1999; Ware et al., 1995; Yoshida et al., 1996). Several models have been proposed to explain these discrepancies. One model hypothesizes the existence of an alternative receptor–ligand *in vivo* that compensates for LIF signaling. This is supported by the inability of *Stat3* mutants to maintain the ICM *in vivo*, and to properly specify EPI and PE lineages (Do et al., 2013).

An alternative model suggests that LIF signaling is crucial only in cells under long-term pluripotency maintenance (e.g., *in vitro*), but dispensable during the rapid and transient pluripotent phase of ICM/EPI cells during preimplantation development. This is supported by the failure of EPI maintenance in gp130 mutant embryos undergoing diapause (delayed implantation of the blastocyst in lactating rodents) (Nichols, Chambers, Taga, & Smith, 2001). In normal diapaused embryos, the EPI enters into a vegetative state and pauses development as it awaits implantation, and therefore requires "long-term maintenance" of the pluripotency network. This *in vivo* prolonged pluripotency phase seems to resemble *in vitro* ES cells, which might explain their LIF signaling-requirement.

6.2. Bone morphogenic protein

In addition to LIF, the presence of BMPs in serum was also identified as critical for self-renewal and pluripotency in ES cell cultures (Ying et al., 2003). BMPs (members of the TGF-β family) bind to their cell surface receptors, which phosphorylate SMAD factors 1/5/8 [family of SMA/MAD homology proteins (related to the sma proteins in *Caenorhabditis elegans* and the mothers against decapentaplegic protein in *Drosophila melanogaster*)]. These receptor-activated SMADs form a complex with *Smad4*, translocate into the nucleus, and induce transcriptional changes. In ES cells, SMAD complexes occupy the promoters of a large number of differentiation inducing genes, and appear to repress their expression (Fei, Xia, et al., 2010). SMAD complexes also activate factors such as *Id1/2* (inhibitor of differentiation) and *Cochlin* (coagulation factor C homolog), both of which inhibit differentiation of ES cells (Tropepe et al., 2001; Ying et al., 2003; Zhang et al., 2013, 2010). BMP signaling also participates in signaling cross talk through *Dusp9* (dual specificity phosphatase 9), which acts on Erk and inhibits its activity, potentially suppressing LIF-activated MAPK signaling (Li & Chen, 2012; Qi et al., 2004).

However, BMP signaling does not strictly impose pluripotency in ES cells and can guide differentiation in the presence of other growth factors (e.g., Activin/Nodal and Wnt). BMP signaling can drive mesendoderm fate and inhibit neuronal fate when used in *in vitro* differentiation (Nostro, Cheng, Keller, & Gadue, 2008; Suzuki et al., 2006). The contrasting roles of BMP signaling appear to be dependent on the presence or absence of the pluripotency factor *Nanog*. In self-renewing conditions (active LIF/JAK/STAT signaling), *Nanog* is strongly expressed, and binds to and inactivates *Smad1*, preventing its differentiation-inducing capacity. In the absence of *Nanog* (under non-self-renewing conditions), BMP can activate unbound *Smad1*, which in turn will induce *Brachyury* expression and promote differentiation toward mesoderm and endoderm (Suzuki et al., 2006).

Another intriguing observation arose from a recent study investigating the balance between pluripotency and differentiation cues. BMP activation of *Id2* was shown to be inhibited during cellular differentiation by the basic helix–loop–helix TF *Tcf15* (Davies et al., 2013). FGF signaling (discussed later) was able to induce a spike in the expression of *Tcf15*, which interacts with and suppresses *Id2* function. This in turn enabled ES cells to transition from the naïve pluripotent state to a primed state. Interestingly, *Tcf15* was shown to have a transient function in counteracting BMP-mediated activity only during priming of ES cells, and not in the later stages of differentiation. This also demonstrates that the diverse functions of BMPs during development might be due to its cross talk with other extrinsic and intrinsic factors.

BMPs play an important role *in vivo* during postimplantation development. BMP is secreted from trophoblast tissues (extraembryonic ectoderm) and signals to the underlying EPI to enhance proliferation (Lawson et al., 1999; Mishina, Suzuki, Ueno, & Behringer, 1995; Winnier, Blessing, Labosky, & Hogan, 1995). Disruption through deletion of *Bmp4*, *Bmpr1a*, or *Smad4* leads to reduced cell proliferation and loss of pluripotency markers in the postimplantation EPI (Di-Gregorio et al., 2007; Mishina et al., 1995; Sirard et al., 2000; Winnier et al., 1995). BMP4 is also required, in combination with other signals, for the formation of the primitive streak during gastrulation (Mishina et al., 1995; Sirard et al., 2000), which mimics EpiSCs, where BMP induces differentiation into mesoderm (Bernardo et al., 2011). In the absence of a primitive streak, EPI cells are unable to differentiate into mesoderm or endoderm and, instead, precociously express neuronal markers (Di-Gregorio et al., 2007). BMP signaling was also shown to be required for the formation of the visceral endoderm (a PE derivative) *in vivo*, which in turn has important roles for the patterning of the EPI (Coucouvanis & Martin, 1995, 1999).

Although there is no current evidence for BMP signaling requirement during preimplantation stages, we cannot exclude potential maternal contributions or functional redundancy between homologs that masks earlier phenotypes.

6.3. Fibroblast growth factor

Inhibition of FGF signaling, in particular the MAPK/Erk pathway, has been shown to contribute to the maintenance of ES pluripotency (Johnson et al., 2007; Kraushaar et al., 2012; Kraushaar, Yamaguchi, & Wang, 2010; Kunath et al., 2007; Stavridis, Lunn, Collins, & Storey, 2007). However, it must act in concert with either LIF signaling or inhibition of GSK3 (also known as 2i conditions) in sustained cultures (Ying et al., 2008). Suppression of FGF signaling in 2i conditions has been used to achieve ground state pluripotency, which facilitates efficient derivation of mouse ES cells, even from non-permissive genetic backgrounds (Hanna et al., 2009; Nichols & Smith, 2009). A hallmark of the 2i-induced naïve state is genome-wide demethylation through a *Nanog*- and *Prdm14*-dependent mechanism (Ficz et al., 2013) and, therefore, it might enhance epigenetic reprogramming during the establishment and maintenance of ES cells. This is consistent with reports that 2i and LIF can drive ES cells toward a "totipotent" state that contributes to embryonic and extraembryonic tissues (Morgani et al., 2013), although further characterizations are required to determine if individual cells are truly totipotent.

As previously discussed, ES cells form a heterogeneous population (a mix of naïve and primed) when grown in LIF and BMP conditions. This is in part governed by FGF signaling (Lanner & Rossant, 2010; Ying et al., 2008), where a fine balance between self-renewing signals and autocrine priming is needed to keep cells in fluctuating, interconvertible states. This intricate interplay between signals serves to maintain both pluripotent and self-renewing capacities of ES cells, and poise them for subsequent differentiation. If the balance is shifted toward active FGF signaling, it will slowly drive cells toward differentiation.

Does FGF-mediated priming direct ES cells toward a specific lineage? Several studies show that FGF signaling promotes an endodermal fate; with spontaneous PE-marker expression detected in a subpopulation of ES cells in LIF and BMP culture conditions (see section 5). Ectopic activation of *Fgfr2* (Santostefano, Hamazaki, Pardo, Kladde, & Terada, 2012) or Ras (Yoshida-Koide et al., 2004), or treatment with a tyrosine phosphatase inhibitor, sodium vanadate (Hamazaki, Kehoe, Nakano, & Terada, 2006) leads to a global upregulation of FGF signaling in ES cells, resulting in downregulation of *Nanog* in a Mek-dependent manner and PE differentiation. The absence of the downstream adapter, *Grb2*, prevents ES cells from generating PE cells during embryoid body (EB) differentiation (Cheng et al., 1998). Intriguingly, this can be rescued by *Gata6* overexpression (Wang et al., 2011). These studies show that the FGFR/Grb2/Ras/MAPK/Erk pathway plays an essential role in PE fate specification *in vitro*, perhaps acting through *Gata6* upregulation, a model that is in concordance with its *in vivo* role during EPI–PE development.

However, FGF signaling does not exclusively drive differentiation toward a PE fate, and it has been proposed that activation of the FGF pathway could drive conversion of ES cells to a state analogous to the postimplantation EPI. In addition to naïve ES cells (marked by high levels of *Nanog*, *Tbx3*, *Klf4*, *Tcl1*), it is possible to isolate a *Rex1*-negative ES population that expresses postimplantation EPI/germ layer markers (e.g., *Brachyury*, *Fgf5*) (Toyooka et al., 2008). These spontaneously occurring populations are dependent on endogenous heparin sulfate proteoglycans (HSPG) (coactivators of FGFRs) during EB differentiation, which induces *Gata6* as well as *Fgf5* and *Brachyury* (Lanner et al., 2010). Addition of FGF4 to *Fgf4*-null ES cells also restores their potential for differentiation into germ layer fates (neuronal and mesodermal) (Kunath et al., 2007), suggesting that FGF4 signaling (perhaps mediated by HSPG) is essential to drive ES cells into a primed state that can generate either PE or embryonic germ layers. This is consistent with the requirement of exogenous FGF in the maintenance of EpiSCs (Brons et al., 2007; Tesar et al., 2007), in both inhibiting differentiation into neuronal lineages and blocking reversion back to ES cells (Greber et al., 2010; Stavridis, Collins, & Storey, 2010).

Taken together, these data suggest that the FGF/MAPK/Erk pathway primes ES cells to enter a transitional stage, from which multiple fates (PE or EPI maturation/germ layer) are possible depending on downstream signaling conditions, corresponding to the *in vivo* differentiation potential of early ICM cells. The decision driving the eventual fate of these primed ES cells might be driven by FGF-regulated downstream factors, such as *Tcf15*, which, when overexpressed, induce mature EPI markers, but do not activate PE markers (Davies et al., 2013).

6.4. Wnt

Wnt signaling has been suggested to play an important role in the maintenance of the pluripotent state in ES cells *in vitro* (reviewed Atlasi et al., 2013), but recent studies on embryos lacking Porcupine argue against any major roles for Wnt signaling in establishing blastocyst cell fate (Biechele, Cockburn, Lanner, Cox, & Rossant, 2013). Porcupine is an

O-acyl transferase, which is required to modify Wnt proteins and permit their passage from the Golgi to the cell membrane for secretion. Blocking Porcupine activity blocks all Wnt secretion and, thus, all downstream Wnt signaling. Maternal and zygotic ablations of Porcupine (and Wnt secretion) show no defects during preimplantation, and while expression of an activated β-catenin (downstream effector of Wnt signaling) marginally increased the number of ICM cells, no effect on cell fate specification was observed. These results suggest that Wnt may regulate self-renewal but does not drive the pluripotent state.

Wnt signaling is detected in the primitive streak that forms from postimplantation EPI cells (Currier et al., 2010; Maretto et al., 2003; Mohamed, Clarke, & Dufort, 2004) and is required for gastrulation (Barrow et al., 2007; Hsieh et al., 2003; Kelly, Pinson, & Skarnes, 2004; Liu et al., 1999). In its absence, EPI cells do not proceed with germ layer differentiation and continue to express pluripotency markers, such as *Oct4* (Barrow et al., 2007; Biechele et al., 2013; Biechele, Cox, & Rossant, 2011; Liu et al., 1999). This is consistent with the role of Wnt signaling in EpiSCs, where its inhibition promotes a more stable pluripotent state, while activation induces mesoderm differentiation (Sumi, Oki, Kitajima, & Meno, 2013).

6.5. Activin and Nodal

Activin and Nodal are members of the TGF-β superfamily that signal through *Smad2* and/or *Smad3*. Like BMP signaling, *Smad2* and *3* also form a complex with *Smad4*, translocate into the nucleus, and induce transcriptional changes. Activin/Nodal signaling has many important roles in the postimplantation embryo. It maintains EPI proliferation and pluripotency, as *Nodal*-null embryos show hypoproliferation and reduction of pluripotency marker expression in the EPI (Mesnard, Guzman-Ayala, & Constam, 2006). It is also involved in embryonic patterning events, in part through the TFs *Eomes* and *Lhx1* (Nowotschin et al., 2013), and later participates with other signals to initiate germ layer specification (Brown, Pietenpol, & Moses, 2007; Shen, 2007).

Nodal's role in postimplantation EPI maintenance is in concordance with Activin/Nodal's important function in maintaining EpiSC pluripotency (Brons et al., 2007; Tesar et al., 2007). It remains unclear how Activin/Nodal directly regulates EpiSC maintenance, but evidence suggests that the pluripotency TFs *Nanog* (Vallier et al., 2009) and *Oct4*

(Lee et al., 2011) are downstream targets. Suppression of this pathway in EpiSCs results in loss of pluripotency and neuroectodermal differentiation (Vallier et al., 2009), in agreement with its role in specification of mesoderm and endoderm germ layers *in vivo* (Conlon et al., 1994; Tremblay, Hoodless, Bikoff, & Robertson, 2000).

In contrast, it is unclear what roles Activin and Nodal signaling play in ES cells. Unlike EpiSCs, ES cells do not appear to require Activin/Nodal signaling for the maintenance of pluripotency when cultured in the presence of LIF, despite their capacity to respond to the ligands (Fei, Xia, et al., 2010; Lee et al., 2011). Inhibition of signaling via chemicals, ectopic expression of endogenous inhibitor, *Smad7* (Ogawa et al., 2007), or knockdown of the cellular transducer, *Smad2* (Fei, Zhu, et al., 2010) does not affect the expression of pluripotency markers, such as *Oct4* or *Nanog*, in ES cells. However, in the absence of LIF, Activin/Nodal appears to enhance proliferation and propagation of the pluripotent cell population (Ogawa et al., 2007) and, eventually, to promote differentiation into mesendodermal lineages (Lee et al., 2011; Pfendler, Catuar, Meneses, & Pedersen, 2005). Intriguingly, it was also reported that phosphorylated *Smad2* binds to the *Oct4* promoter in ES cells grown in chemically defined media (without LIF), and suppression of Activin/Nodal reduces *Oct4* expression and activates certain trophoblast genes (Lee et al., 2011). However, this was not observed in ES cells cultured with LIF (Fei, Xia, et al., 2010). This raises the possibility that Activin/Nodal signaling is important for ES cell maintenance only in the absence of LIF signaling, or that Activin/Nodal drives ES cells to become another self-renewing pluripotent cell type (perhaps similar to EpiSC) under certain culture conditions.

7. CONCLUDING REMARKS

Embryo-derived stem cells have been invaluable in biology, as both an *in vitro* model for understanding the molecular mechanisms of pluripotency and differentiation, and a widely employed tool for performing genetic manipulations and lineage-directed differentiation.

Recent years have seen great advances in our understanding of signaling pathways involved in segregation of pluripotent lineages in the embryo. In ICM–TE specification, it is the different position of cells in the embryo (inside or outside) that translates into differential Hippo signaling activities, which drive separation of the lineages. An intriguing question remains regarding the "signal" that determines if a cell is inside or outside: Are positional differences dictated by exposure to the external environment, or through the "abundance" of cell-to-cell contact?

Another mode of generating differential signaling responses—employed by the EPI–PE specification event—is through initial heterogeneities in gene expression, which lead to small local changes in FGF signaling among inner cells. This subsequently creates a bias toward cell fate, which is eventually determined through amplification of the signaling cascades and activation of downstream TFs. However, the underlying mechanisms that generate the initial heterogeneity and the signaling "threshold" required for PE commitment remain to be elucidated.

In contrast with specification, our understanding of pluripotency maintenance is based masinly on studies of embryo-derived stem cells. ES cells and EpiSCs are considered as proxies for their *in vivo* cell type of origin and enable us to capture very transient *in vivo* phases in culture. However, an important question remains regarding how accurately they recapitulate their *in vivo* counterparts. EpiSCs appear to resemble the early postimplantation EPI in many ways. They share significant overlap in both gene expression and responses to exogenous signaling activities. However, the postimplantation EPI is a dynamic cell population, with subtle differences at various developmental stages. Therefore, EpiSC might represent a specific temporal stage, and not be representative of all postimplantation EPI. Studies addressing these questions will also provide further insights into the molecular and developmental properties of human ES cells, which closely resemble mouse EpiSCs.

In contrast, there is evidence that ES cells do not directly recapitulate the molecular or signaling properties of their tissue of origin. Instead, they make use of some signaling pathways that are not expressed or required *in vivo* in the blastocyst, such as LIF, BMP, and Wnt signaling. However, we wish to argue that these differences are less drastic than previously thought. ES cells, like EPI cells, exhibit similar responses to FGF signaling, which mediates differentiation. Despite showing differential requirement for LIF, recent evidence suggests that downstream signaling is essential in both ES cells *in vitro* and EPI *in vivo*, with the latter activating JAK/STAT in an LIF-independent manner. Furthermore, differences in cell numbers have been observed in both ES cells and *in vivo* EPI in response to activation of the Wnt pathway, suggesting that its signaling might be a regulator of proliferation in the EPI lineage. However, it remains unclear if LIF signaling activity and 2i conditions (FGF suppression and Wnt activation) operate redundantly or synergistically in maintaining pluripotency in ES cells. This uncertainty may contribute to our inability to dissect their requirements in the complex signaling environment of the preimplantation EPI.

It has been demonstrated that heterogeneity is present in ES cells in culture, and may actually represent a dynamic (naïve and primed) snapshot of the *in vivo* EPI lineage. However, the precise molecular characterization of these subpopulations remains elusive, likely because of their low numbers and their sensitivity to culture conditions. It is also increasingly clear that EpiSCs show heterogeneity, not only within their population, but also between EpiSC cultures derived from the same genetic background (Bernemann et al., 2011; Han et al., 2010). This heterogeneity could potentially manifest in differential responses to differentiation signals and developmental biases toward different germ layer lineages. Therefore, a major challenge for the pluripotent stem cell field is to further identify and dissect these subpopulations (in ES cells and EpiSC) to better understand their differences in self-renewal and developmental potential.

In this review, we have treated the mechanisms of regulating pluripotency to be identical in all genetic backgrounds. However, it is known that the genetic background has a major influence on the efficiency of ES cell derivation (Batlle-Morera, Smith, & Nichols, 2008; Brook & Gardner, 1997; Buehr & Smith, 2003). Although variations in efficiency of derivation have been circumvented by the development of the 2i culture conditions, it raises the possibility that intrinsic genetic variation might modulate entry into or stabilization of pluripotency. Human embryos and stem cells are known to differ from mouse in a number of regulative properties (Cockburn & Rossant, 2010; De Los Angeles, Loh, Tesar, & Daley, 2012). This in part could be the result of genetic variation, which is commonplace in humans. Thus, exploring the influence of genetic variation could have major implications for the understanding of human pluripotent cell lines, such as human ES cells and induced pluripotent stem cells.

Although we have elucidated many of the basic mechanisms that underlie the establishment and maintenance of pluripotency, there are still many questions, in both the precise molecular mechanisms governing responses to exogenous signaling, and the influence of different genetic and epigenetic effects on the cellular and embryonic response to the decision between pluripotency and differentiation.

A greater understanding of these processes will be critical to our ability to manipulate and regulate pluripotent cells, and will enable us to fully realize their therapeutic potential in generating cells of interest for cell-based therapies.

ACKNOWLEDGMENTS

We thank Katie Cockburn and Fredrik Lanner for their critical comments on the manuscript. O. H. T. is funded by the Human Frontier Science Program; E. P. is funded by Restracomp (The Hospital for Sick Children, Toronto).

REFERENCES

Aksoy, I., Jauch, R., Chen, J., Dyla, M., Divakar, U., Bogu, G. K., et al. (2013). Oct4 switches partnering from Sox2 to Sox17 to reinterpret the enhancer code and specify endoderm. *The EMBO Journal, 32*(7), 938–953.

Alarcon, V. B. (2010). Cell polarity regulator PARD6B is essential for trophectoderm formation in the preimplantation mouse embryo. *Biology of Reproduction, 83*(3), 347–358.

Arman, E. E., Haffner-Krausz, R. R., Chen, Y. Y., Heath, J. K. J., & Lonai, P. P. (1998). Targeted disruption of fibroblast growth factor (FGF) receptor 2 suggests a role for FGF signaling in pregastrulation mammalian development. *Proceedings of the National Academy of Sciences of the United States of America, 95*(9), 5082–5087.

Arnold, S. J., & Robertson, E. J. (2009). Making a commitment: Cell lineage allocation and axis patterning in the early mouse embryo. *Nature Reviews Molecular Cell Biology, 10*(2), 91–103.

Atlasi, Y., Looijenga, L., & Fodde, R. (2013). Cancer stem cells, pluripotency, and cellular heterogeneity: A WNTer perspective. *Current Topics in Developmental Biology, 107*, 373–404.

Barrow, J. R., Howell, W. D., Rule, M., Hayashi, S., Thomas, K. R., Capecchi, M. R., et al. (2007). Wnt3 signaling in the epiblast is required for proper orientation of the anteroposterior axis. *Developmental Biology, 312*(1), 312–320.

Batlle-Morera, L., Smith, A., & Nichols, J. (2008). Parameters influencing derivation of embryonic stem cells from murine embryos. *Genesis, 46*(12), 758–767.

Beddington, R. S. R., & Robertson, E. J. E. (1989). An assessment of the developmental potential of embryonic stem cells in the midgestation mouse embryo. *Development, 105*(4), 733–737.

Bernardo, A. S., Faial, T., Gardner, L., Niakan, K. K., Ortmann, D., Senner, C. E., et al. (2011). BRACHYURY and CDX2 mediate BMP-induced differentiation of human and mouse pluripotent stem cells into embryonic and extraembryonic lineages. *Cell Stem Cell, 9*(2), 144–155.

Bernemann, C., Greber, B., Ko, K., Sterneckert, J., Han, D. W., Araúzo-Bravo, M. J., et al. (2011). Distinct developmental ground states of epiblast stem cell lines determine different pluripotency features. *Stem Cells, 29*(10), 1496–1503.

Biechele, S., Cockburn, K., Lanner, F., Cox, B. J., & Rossant, J. (2013). Porcn-dependent Wnt signaling is not required prior to mouse gastrulation. *Development, 140*(14), 2961–2971.

Biechele, S., Cox, B. J., & Rossant, J. (2011). Porcupine homolog is required for canonical Wnt signaling and gastrulation in mouse embryos. *Developmental Biology, 355*(2), 275–285.

Bradley, A., Evans, M., Kaufman, M. H., & Robertson, E. (1984). Formation of germ-line chimaeras from embryo-derived teratocarcinoma cell lines. *Nature, 309*(5965), 255–256.

Brons, I. G. M., Smithers, L. E., Trotter, M. W. B., Rugg-Gunn, P., Sun, B., de Sousa, Chuva, et al. (2007). Derivation of pluripotent epiblast stem cells from mammalian embryos. *Nature, 448*(7150), 191–195.

Brook, F. A., & Gardner, R. L. (1997). The origin and efficient derivation of embryonic stem cells in the mouse. *Proceedings of the National Academy of Sciences of the United States of America, 94*(11), 5709–5712.

Brown, K. A., Pietenpol, J. A., & Moses, H. L. (2007). A tale of two proteins: Differential roles and regulation of Smad2 and Smad3 in TGF-β signaling. *Journal of Cellular Biochemistry, 101*(1), 9–33.

Buehr, M., & Smith, A. (2003). Genesis of embryonic stem cells. *Philosophical Transactions of the Royal Society B: Biological Sciences, 358*(1436), 1397–1402.

Cai, K. Q. K., Capo-Chichi, C. D. C., Rula, M. E. M., Yang, D.-H. D., & Xu, X.-X. X. (2008). Dynamic GATA6 expression in primitive endoderm formation and maturation in early mouse embryogenesis. *Developmental Dynamics, 237*(10), 2820–2829.

Canham, M. A., Sharov, A. A., Ko, M. S. H., & Brickman, J. M. (2010). Functional heterogeneity of embryonic stem cells revealed through translational amplification of an early endodermal transcript. *PLoS Biology, 8*(5), http://dx.doi.org/10.1371/journal.pbio.1000379.

Chambers, I., Silva, J., Colby, D., Nichols, J., Nijmeijer, B., Robertson, M., et al. (2007). Nanog safeguards pluripotency and mediates germline development. *Nature, 450*(7173), 1230–1234.

Chazaud, C., Yamanaka, Y., Pawson, T., & Rossant, J. (2006). Early lineage segregation between epiblast and primitive endoderm in mouse blastocysts through the Grb2-MAPK pathway. *Developmental Cell, 10*(5), 615–624.

Cheng, A. M., Saxton, T. M., Sakai, R., Kulkarni, S., Mbamalu, G., Vogel, W., et al. (1998). Mammalian Grb2 regulates multiple steps in embryonic development and malignant transformation. *Cell, 95*(6), 11.

Cockburn, K., Biechele, S., Garner, J., & Rossant, J. (2013). The Hippo pathway member Nf2 is required for inner cell mass specification. *Current Biology, 23*(13), 1195–1201.

Cockburn, K., & Rossant, J. (2010). Making the blastocyst: Lessons from the mouse. *The Journal of Clinical Investigation, 120*(4), 995–1003.

Conlon, F. L., Lyons, K. M., Takaesu, N., Barth, K. S., Kispert, A., Herrmann, B., et al. (1994). A primary requirement for nodal in the formation and maintenance of the primitive streak in the mouse. *Development, 120*(7), 1919–1928.

Coucouvanis, E., & Martin, G. R. (1995). Signals for death and survival: A two-step mechanism for cavitation in the vertebrate embryo. *Cell, 83*(2), 279–287.

Coucouvanis, E., & Martin, G. R. (1999). BMP signaling plays a role in visceral endoderm differentiation and cavitation in the early mouse embryo. *Development, 126*(3), 535–546.

Cross, J. C. (2005). How to make a placenta: Mechanisms of trophoblast cell differentiation in mice—A review. *Placenta, 26*, S3–S9.

Currier, N., Chea, K., Hlavacova, M., Sussman, D. J., Seldin, D. C., & Dominguez, I. (2010). Dynamic expression of a LEF-EGFP Wnt reporter in mouse development and cancer. *Genesis, 48*, 183–194.

Dailey, L., Ambrosetti, D., Mansukhani, A., & Basilico, C. (2005). Mechanisms underlying differential responses to FGF signaling. *Cytokine & Growth Factor Reviews, 16*(2), 233–247.

Davies, O. R., Lin, C.-Y., Radzisheuskaya, A., Zhou, X., Taube, J., Blin, G., et al. (2013). Tcf15 primes pluripotent cells for differentiation. *Cell Reports, 3*(2), 472–484.

De Los Angeles, A., Loh, Y.-H., Tesar, P. J., & Daley, G. Q. (2012). Accessing naïve human pluripotency. *Current Opinion in Genetics & Development, 22*(3), 272–282.

Dietrich, J.-E. J., & Hiiragi, T. T. (2007). Stochastic patterning in the mouse pre-implantation embryo. *Development, 134*(23), 4219–4231.

Di-Gregorio, A., Sancho, M., Stuckey, D. W., Crompton, L. A., Godwin, J., Mishina, Y., et al. (2007). BMP signalling inhibits premature neural differentiation in the mouse embryo. *Development, 134*(18), 3359–3369.

Dillon, N. (2012). Factor mediated gene priming in pluripotent stem cells sets the stage for lineage specification. *BioEssays, 34*(3), 194–204.

Do, D. V., Ueda, J., Messerschmidt, D. M., Lorthongpanich, C., Zhou, Y., Feng, B., et al. (2013). A genetic and developmental pathway from STAT3 to the OCT4-NANOG circuit is essential for maintenance of ICM lineages in vivo. *Genes and Development, 27*(12), 1378–1390.

Ducibella, T., Ukena, T., Karnovsky, M., & Anderson, E. (1977). Changes in cell surface and cortical cytoplasmic organization during early embryogenesis in the preimplantation mouse embryo. *The Journal of Cell Biology, 74*(1), 153–167.

Evans, M. J., & Kaufman, M. H. (1981). Establishment in culture of pluripotential cells from mouse embryos. *Nature, 292*(5819), 154–156.

Faddah, D. A., Wang, H., Cheng, A. W., Katz, Y., Buganim, Y., & Jaenisch, R. (2013). *Cell Stem Cell, 13*(1), 23–29.

Fei, T., Xia, K., Li, Z., Zhou, B., Zhu, S., Chen, H., et al. (2010). Genome-wide mapping of SMAD target genes reveals the role of BMP signaling in embryonic stem cell fate determination. *Genome Research, 20*(1), 36–44.

Fei, T., Zhu, S., Xia, K., Zhang, J., Li, Z., Han, J.-D. J., et al. (2010). Smad2 mediates Activin/Nodal signaling in mesendoderm differentiation of mouse embryonic stem cells. *Cell Research, 20*(12), 1306–1318.

Feldman, B., Poueymirou, W., Papaioannou, V. E., DeChiara, T. M., & Goldfarb, M. (1995). Requirement of FGF-4 for postimplantation mouse development. *Science, 267*(5195), 246–249.

Festuccia, N., Osorno, R., Halbritter, F., Karwacki-Neisius, V., Navarro, P., Colby, D., et al. (2012). Esrrb is a direct nanog target gene that can substitute for nanog function in pluripotent cells. *Cell Stem Cell, 11*(4), 477–490.

Ficz, G., Hore, T. A., Santos, F., Lee, H. J., Dean, W., Arand, J., et al. (2013). *Cell Stem Cell, 13*(1), 1–9.

Filipczyk, A., Gkatzis, K., Fu, J., Hoppe, P. S., Lickert, H., Anastassiadis, K., et al. (2013). Biallelic expression of Nanog protein in mouse embryonic stem cells. *Cell Stem Cell, 13*(1), 12–13.

Frankenberg, S., Gerbe, F., Bessonnard, S., Belville, C., Pouchin, P., Bardot, O., et al. (2011). Primitive endoderm differentiates via a three-step mechanism involving Nanog and RTK signaling. *Developmental Cell, 21*(6), 1005–1013.

Frum, T., Halbisen, M. A., Wang, C., Amiri, H., Robson, P., & Ralston, A. (2013). Oct4 cell-autonomously promotes PE development. *Developmental Cell, 25*, 1–13.

Fujikura, J. J., Yamato, E. E., Yonemura, S. S., Hosoda, K. K., Masui, S. S., Nakao, K. K., et al. (2002). Differentiation of embryonic stem cells is induced by GATA factors. *Genes and Development, 16*(7), 784–789.

Gerbe, F., Cox, B., Rossant, J., & Chazaud, C. (2008). Dynamic expression of Lrp2 pathway members reveals progressive epithelial differentiation of primitive endoderm in mouse blastocyst. *Developmental Biology, 313*(2), 594–602.

Goldin, S. N., & Papaioannou, V. E. (2003). Paracrine action of FGF4 during periimplantation development maintains trophectoderm and primitive endoderm. *Genesis, 36*(1), 40–47.

Greber, B., Wu, G., Bernemann, C., Joo, J. Y., Han, D. W., Ko, K., et al. (2010). Conserved and divergent roles of FGF signaling in mouse epiblast stem cells and human embryonic stem cells. *Cell Stem Cell, 6*(3), 215–226.

Guo, Guoji, Huss, M., Tong, G. Q., Wang, C., Sun, L. L., Clarke, N. D., et al. (2010). Resolution of cell fate decisions revealed by single-cell gene expression analysis from zygote to blastocyst. *Developmental Cell, 18*(4), 675–685.

Guo, Ge, Yang, J., Nichols, J., Hall, J. S., Eyres, I., Mansfield, W., et al. (2009). Klf4 reverts developmentally programmed restriction of ground state pluripotency. *Development, 136*(7), 1063–1069.

Hamazaki, T., Kehoe, S. M., Nakano, T., & Terada, N. (2006). The Grb2/Mek pathway represses Nanog in murine embryonic stem cells. *Molecular and Cellular Biology, 26*(20), 7539–7549.

Han, D. W., Tapia, N., Joo, J. Y., Greber, B., Araúzo-Bravo, M. J., Bernemann, C., et al. (2010). Epiblast stem cell subpopulations represent mouse embryos of distinct pregastrulation stages. *Cell, 143*(4), 617–627.

Handyside, A. H. A. (1980). Distribution of antibody- and lectin-binding sites on dissociated blastomeres from mouse morulae: Evidence for polarization at compaction. *Journal of Embryology and Experimental Morphology, 60*, 99–116.

Hanna, J., Markoulaki, S., Mitalipova, M., Cheng, A. W., Cassady, J. P., Staerk, J., et al. (2009). Metastable pluripotent states in NOD-mouse-derived ESCs. *Cell Stem Cell, 4*(6), 513–524.

Hartshorn, C., Eckert, J. J., Hartung, O., & Wangh, L. J. (2007). Single-cell duplex RT-LATE-PCR reveals Oct4 and Xist RNA gradients in 8-cell embryos. *BMC Biotechnology*, 7(1), 87.

Hay, D. C. D., Sutherland, L. L., Clark, J. J., & Burdon, T. T. (2004). Oct-4 knockdown induces similar patterns of endoderm and trophoblast differentiation markers in human and mouse embryonic stem cells. *Stem Cells, 22*(2), 225–235.

Hayashi, K., de Sousa Lopes, S. M. C., Tang, F., & Surani, M. A. (2008). Dynamic equilibrium and heterogeneity of mouse pluripotent stem cells with distinct functional and epigenetic states. *Cell Stem Cell, 3*(4), 11.

Hirate, Y., Hirahara, S., Inoue, K.-I., Suzuki, A., Alarcon, V. B., Akimoto, K., et al. (2013). Polarity-dependent distribution of angiomotin localizes hippo signaling in preimplantation embryos. *Current Biology, 23*(13), 1181–1194.

Hsieh, J.-C., Lee, L., Zhang, L., Wefer, S., Brown, K., DeRossi, C., et al. (2003). Mesd encodes an LRP5/6 chaperone essential for specification of mouse embryonic polarity. *Cell, 112*(3), 13.

Huang, Y., Osorno, R., Tsakiridis, A., & Wilson, V. (2012). In vivo differentiation potential of epiblast stem cells revealed by chimeric embryo formation. *Cell Reports, 2*(6), 1571–1578.

Jauch, R., Aksoy, I., Hutchins, A. P., Ng, C. K. L., Tian, X. F., Chen, J., et al. (2011). Conversion of Sox17 into a pluripotency reprogramming factor by reengineering its association with Oct4 on DNA. *Stem Cells, 29*(6), 940–951.

Johnson, C. E., Crawford, B. E., Stavridis, M., Ten Dam, G., Wat, A. L., Rushton, G., et al. (2007). Essential alterations of heparan sulfate during the differentiation of embryonic stem cells to Sox1-enhanced green fluorescent protein-expressing neural progenitor cells. *Stem Cells, 25*(8), 1913–1923.

Johnson, M. H., & Ziomek, C. A. (1981). Induction of polarity in mouse 8-cell blastomeres: Specificity, geometry, and stability. *The Journal of Cell Biology, 91*(1), 303–308.

Johnson, M. H., & Ziomek, C. A. (1983). Cell interactions influence the fate of mouse blastomeres undergoing the transition from the 16- to the 32-cell stage. *Developmental Biology, 95*(1), 211–218.

Johnston, R. J., & Desplan, C. (2010). Stochastic mechanisms of cell fate specification that yield random or robust outcomes. *Annual Review of Cell and Developmental Biology, 26*, 689–719.

Kalmar, T., Lim, C., Hayward, P., Muñoz-Descalzo, S., Nichols, J., Garcia-Ojalvo, J., et al. (2009). Regulated fluctuations in nanog expression mediate cell fate decisions in embryonic stem cells. *PLoS Biology*, 7(7).

Kang, M. M., Piliszek, A. A., Artus, J. J., & Hadjantonakis, A.-K. A. (2013). FGF4 is required for lineage restriction and salt-and-pepper distribution of primitive endoderm factors but not their initial expression in the mouse. *Development, 140*(2), 267–279.

Karwacki-Neisius, V., Göke, J., Osorno, R., Halbritter, F., Ng, J. H., Weiße, A. Y., et al. (2013). Reduced Oct4 expression directs a robust pluripotent state with distinct signaling activity and increased enhancer occupancy by Oct4 and Nanog. *Cell Stem Cell, 12*(5), 531–545.

Kelly, S. J. (1977). Studies of the developmental potential of 4- and 8-cell stage mouse blastomeres. *Journal of Experimental Zoology, 200*(3), 365–376.
Kelly, O. G., Pinson, K. I., & Skarnes, W. C. (2004). The Wnt co-receptors Lrp5 and Lrp6 are essential for gastrulation in mice. *Development, 131*(12), 2803–2815.
Koutsourakis, M., Langeveld, A., Patient, R., Beddington, R., & Grosveld, F. (1999). The transcription factor GATA6 is essential for early extraembryonic development. *Development, 126*(9), 723–732.
Kraushaar, D. C., Rai, S., Condac, E., Nairn, A., Zhang, S., Yamaguchi, Y., et al. (2012). Heparan sulfate facilitates FGF and BMP signaling to drive mesoderm differentiation of mouse embryonic stem cells. *The Journal of Biological Chemistry, 287*(27), 22691–22700.
Kraushaar, D. C., Yamaguchi, Y., & Wang, L. (2010). Heparan sulfate is required for embryonic stem cells to exit from self-renewal. *The Journal of Biological Chemistry, 285*(8), 5907–5916.
Kunath, T., Arnaud, D., Uy, G. D., Okamoto, I., Chureau, C., Yamanaka, Y., et al. (2005). Imprinted X-inactivation in extra-embryonic endoderm cell lines from mouse blastocysts. *Development, 132*(7), 1649–1661.
Kunath, T., Saba-El-Leil, M. K., Almousailleakh, M., Wray, J., Meloche, S., & Smith, A. (2007). FGF stimulation of the Erk1/2 signalling cascade triggers transition of pluripotent embryonic stem cells from self-renewal to lineage commitment. *Development, 134*(16), 2895–2902.
Kurimoto, K., Yabuta, Y., Ohinata, Y., Ono, Y., Uno, K. D., Yamada, R. G., et al. (2006). An improved single-cell cDNA amplification method for efficient high-density oligonucleotide microarray analysis. *Nucleic Acids Research, 34*(5), e42.
Lanner, F., Lee, K. L., Sohl, M., Holmborn, K., Yang, H., Wilbertz, J., et al. (2010). Heparan sulfation-dependent fibroblast growth factor signaling maintains embryonic stem cells primed for differentiation in a heterogeneous state. *Stem Cells, 28*(2), 191–200.
Lanner, F., & Rossant, J. (2010). The role of FGF/Erk signaling in pluripotent cells. *Development, 137*(20), 3351–3360.
Lawson, K. A., Dunn, N. R., Roelen, B. A., Zeinstra, L. M., Davis, A. M., Wright, C. V., et al. (1999). Bmp4 is required for the generation of primordial germ cells in the mouse embryo. *Genes and Development, 13*(4), 424–436.
Lawson, K. A., Meneses, J. J., & Pedersen, R. A. (1991). Clonal analysis of epiblast fate during germ layer formation in the mouse embryo. *Development, 113*(3), 891–911.
Lee, K. L., Lim, S. K., Orlov, Y. L., Yit, L. Y., Yang, H., Ang, L. T., et al. (2011). Graded nodal/activin signaling titrates conversion of quantitative phospho-Smad2 levels into qualitative embryonic stem cell fate decisions. *PLoS Genetics*, 7(6), http://dx.doi.org/10.1371/journal.pbio.1000149.
Li, Z., & Chen, Y.-G. (2012). Fine-tune of intrinsic ERK activity by extrinsic BMP signaling in mouse embryonic stem cells. *Protein & Cell, 3*(6), 401–404.
Li, M. M., Sendtner, M. M., & Smith, A. A. (1995). Essential function of LIF receptor in motor neurons. *Nature, 378*(6558), 724–727.
Lian, I., Kim, J., Okazawa, H., Zhao, J., Zhao, B., Yu, J., et al. (2010). The role of YAP transcription coactivator in regulating stem cell self-renewal and differentiation. *Genes and Development, 24*(11), 1106–1118.
Liu, P., Wakamiya, M., Shea, M. J., Albrecht, U., Behringer, R. R., & Bradley, A. (1999). Requirement for Wnt3 in vertebrate axis formation. *Nature Genetics, 22*(4), 361–365.
Lorthongpanich, C., Doris, T. P. Y., Limviphuvadh, V., Knowles, B. B., & Solter, D. (2012). Developmental fate and lineage commitment of singled mouse blastomeres. *Development, 139*(20), 3722–3731.
Lorthongpanich, C., Messerschmidt, D. M., Chan, S. W., Hong, W., Knowles, B. B., & Solter, D. (2013). Temporal reduction of LATS kinases in the early preimplantation embryo prevents ICM lineage differentiation. *Genes and Development, 27*(13), 1441–1446.

Magnúsdóttir, E., Gillich, A., Grabole, N., & Surani, M. A. (2012). Combinatorial control of cell fate and reprogramming in the mammalian germline. *Current Opinion in Genetics & Development*, *22*(5), 466–474.

Maretto, S. S., Cordenonsi, M. M., Dupont, S. S., Braghetta, P. P., Broccoli, V. V., Hassan, A. B. A., et al. (2003). Mapping Wnt/beta-catenin signaling during mouse development and in colorectal tumors. *Proceedings of the National Academy of Sciences of the United States of America*, *100*(6), 3299–3304.

Maro, B., Johnson, M. H., Pickering, S. J., & Louvard, D. (1985). Changes in the distribution of membranous organelles during mouse early development. *Journal of Embryology and Experimental Morphology*, *90*, 287–309.

Martello, G., Sugimoto, T., Diamanti, E., Joshi, A., Hannah, R., Ohtsuka, S., et al. (2012). Esrrb is a pivotal target of the Gsk3/Tcf3 axis regulating embryonic stem cell self-renewal. *Cell Stem Cell*, *11*(4), 491–504.

Martin, G. R. (1981). Isolation of a pluripotent cell line from early mouse embryos cultured in medium conditioned by teratocarcinoma stem cells. *Proceedings of the National Academy of Sciences of the United States of America*, *78*(12), 7634–7638.

Meilhac, S. M., Adams, R. J., Morris, S. A., Danckaert, A., Le Garrec, J.-F., & Zernicka-Goetz, M. (2009). Active cell movements coupled to positional induction are involved in lineage segregation in the mouse blastocyst. *Developmental Biology*, *331*(2), 210–221.

Mesnard, D. D., Guzman-Ayala, M. M., & Constam, D. B. D. (2006). Nodal specifies embryonic visceral endoderm and sustains pluripotent cells in the epiblast before overt axial patterning. *Development*, *133*(13), 2497–2505.

Messerschmidt, D. M., & Kemler, R. (2010). Nanog is required for primitive endoderm formation through a non-cell autonomous mechanism. *Developmental Biology*, *344*(1), 129–137.

Mishina, Y., Suzuki, A., Ueno, N., & Behringer, R. R. (1995). Bmpr encodes a type I bone morphogenetic protein receptor that is essential for gastrulation during mouse embryogenesis. *Genes and Development*, *9*(24), 3027–3037.

Mitsui, K. K., Tokuzawa, Y. Y., Itoh, H. H., Segawa, K. K., Murakami, M. M., Takahashi, K. K., et al. (2003). The homeoprotein Nanog is required for maintenance of pluripotency in mouse epiblast and ES cells. *Cell*, *113*(5), 12.

Miyanari, Y., & Torres-Padilla, M.-E. (2012). Control of ground-state pluripotency by allelic regulation of Nanog. *Nature*, *483*(7390), 470–473.

Mohamed, O. A., Clarke, H. J., & Dufort, D. (2004). Beta-catenin signaling marks the prospective site of primitive streak formation in the mouse embryo. *Developmental Dynamics*, *231*(2), 416–424.

Morgani, S. M., Canham, M. A., Nichols, J., Sharov, A. A., Migueles, R. P., Ko, M. S. H., et al. (2013). Totipotent embryonic stem cells arise in ground-state culture conditions. *Cell Reports*, *3*(6), 1945–1957.

Morris, S. A., Teo, R. T. Y., Li, H., Robson, P., Glover, D. M., & Zernicka-Goetz, M. (2010). Origin and formation of the first two distinct cell types of the inner cell mass in the mouse embryo. *Proceedings of the National Academy of Sciences of the United States of America*, *107*(14), 6364–6369.

Morrisey, E. E., Tang, Z., Sigrist, K., Lu, M. M., Jiang, F., Ip, H. S., et al. (1998). GATA6 regulates HNF4 and is required for differentiation of visceral endoderm in the mouse embryo. *Genes and Development*, *12*(22), 3579–3590.

Najm, F. J., Chenoweth, J. G., Anderson, P. D., Nadeau, J. H., Redline, R. W., McKay, R. D. G., et al. (2011). Isolation of epiblast stem cells from preimplantation mouse embryos. *Cell Stem Cell*, *8*(3), 318–325.

Nakai-Futatsugi, Y. Y., & Niwa, H. H. (2013). Transcription factor network in embryonic stem cells: Heterogeneity under the stringency. *Biological & Pharmaceutical Bulletin*, *36*(2), 166–170.

Nakashima, K. K., Wiese, S. S., Yanagisawa, M. M., Arakawa, H. H., Kimura, N. N., Hisatsune, T. T., et al. (1999). Developmental requirement of gp130 signaling in neuronal survival and astrocyte differentiation. *The Journal of Neuroscience, 19*(13), 5429–5434.

Niakan, K. K., Ji, H., Maehr, R., Vokes, S. A., Rodolfa, K. T., Sherwood, R. I., et al. (2010). Sox17 promotes differentiation in mouse embryonic stem cells by directly regulating extraembryonic gene expression and indirectly antagonizing self-renewal. *Genes and Development, 24*(3), 312–326.

Nichols, J., Chambers, I., Taga, T., & Smith, A. (2001). Physiological rationale for responsiveness of mouse embryonic stem cells to gp130 cytokines. *Development, 128*(12), 2333–2339.

Nichols, J., Davidson, D., Taga, T., Yoshida, K., Chambers, I., & Smith, A. (1996). Complementary tissue-specific expression of LIF and LIF-receptor mRNAs in early mouse embryogenesis. *Mechanisms of Development, 57*(2), 123–131.

Nichols, J., Silva, J., Roode, M., & Smith, A. (2009). Suppression of Erk signalling promotes ground state pluripotency in the mouse embryo. *Development, 136*(19), 3215–3222.

Nichols, J., & Smith, A. (2009). Naive and primed pluripotent states. *Cell Stem Cell, 4*(6), 487–492.

Nishioka, N., Inoue, K.-I., Adachi, K., Kiyonari, H., Ota, M., Ralston, A., et al. (2009). The Hippo signaling pathway components Lats and Yap pattern Tead4 activity to distinguish mouse trophectoderm from inner cell mass. *Developmental Cell, 16*(3), 398–410.

Niswander, L., & Martin, G. R. (1992). Fgf-4 expression during gastrulation, myogenesis, limb and tooth development in the mouse. *Development, 114*(3), 755–768.

Niwa, H., Toyooka, Y., Shimosato, D., Strumpf, D., Takahashi, K., Yagi, R., et al. (2005). Interaction between Oct3/4 and Cdx2 determines trophectoderm differentiation. *Cell, 123*(5), 917–929.

Nostro, M. C., Cheng, X., Keller, G. M., & Gadue, P. (2008). Wnt, activin, and BMP signaling regulate distinct stages in the developmental pathway from embryonic stem cells to blood. *Cell Stem Cell, 2*(1), 60–71.

Nowotschin, S., Costello, I., Piliszek, A., Kwon, G. S., Mao, C.-A., Klein, W. H., et al. (2013). The T-box transcription factor Eomesodermin is essential for AVE induction in the mouse embryo. *Genes and Development, 27*(9), 997–1002.

Nowotschin, S., & Hadjantonakis, A.-K. (2010). Cellular dynamics in the early mouse embryo: From axis formation to gastrulation. *Current Opinion in Genetics & Development, 20*(4), 420–427.

Ogawa, K., Saito, A., Matsui, H., Suzuki, H., Ohtsuka, S., Shimosato, D., et al. (2007). Activin-Nodal signaling is involved in propagation of mouse embryonic stem cells. *Journal of Cell Science, 120*(1), 55–65.

Palmieri, S. L. S., Peter, W. W., Hess, H. H., & Schöler, H. R. H. (1994). Oct-4 transcription factor is differentially expressed in the mouse embryo during establishment of the first two extraembryonic cell lineages involved in implantation. *Developmental Biology, 166*(1), 259–267.

Pfendler, K. C., Catuar, C. S., Meneses, J. J., & Pedersen, R. A. (2005). Overexpression of Nodal promotes differentiation of mouse embryonic stem cells into mesoderm and endoderm at the expense of neuroectoderm formation. *Stem Cells and Development, 14*(2), 162–172.

Pietersen, A. M., & van Lohuizen, M. (2008). Stem cell regulation by polycomb repressors: Postponing commitment. *Current Opinion in Cell Biology, 20*(2), 201–207.

Plusa, B., Frankenberg, S., Chalmers, A., Hadjantonakis, A.-K., Moore, C. A., Papalopulu, N., et al. (2005). Downregulation of Par3 and aPKC function directs cells towards the ICM in the preimplantation mouse embryo. *Journal of Cell Science, 118*(3), 505–515.

Plusa, B., Piliszek, A., Frankenberg, S., Artus, J., & Hadjantonakis, A.-K. (2008). Distinct sequential cell behaviours direct primitive endoderm formation in the mouse blastocyst. *Development, 135*(18), 3081–3091.

Pratt, H. P., Ziomek, C. A., Reeve, W. J., & Johnson, M. H. (1982). Compaction of the mouse embryo: An analysis of its components. *Journal of Embryology and Experimental Morphology, 70,* 113–132.

Qi, X., Li, T.-G., Hao, J., Hu, J., Wang, J., Simmons, H., et al. (2004). BMP4 supports self-renewal of embryonic stem cells by inhibiting mitogen-activated protein kinase pathways. *Proceedings of the National Academy of Sciences of the United States of America, 101*(16), 6027–6032.

Radzisheuskaya, A., Le Bin Chia, G., dos Santos, R. L., Theunissen, T. W., Castro, L. F. C., Nichols, J., et al. (2013). A defined Oct4 level governs cell state transitions of pluripotency entry and differentiation into all embryonic lineages. *Nature Cell Biology, 15*(6), 579–590.

Ralston, A., & Rossant, J. (2008). Cdx2 acts downstream of cell polarization to cell-autonomously promote trophectoderm fate in the early mouse embryo. *Developmental Biology, 313*(2), 614–629.

Rappolee, D. A., Basilico, C., Patel, Y., & Werb, Z. (1994). Expression and function of FGF-4 in peri-implantation development in mouse embryos. *Development, 120*(8), 2259–2269.

Reeve, W. J. W., & Ziomek, C. A. C. (1981). Distribution of microvilli on dissociated blastomeres from mouse embryos: Evidence for surface polarization at compaction. *Journal of Embryology and Experimental Morphology, 62,* 339–350.

Rossant, J., & Cross, J. C. (2001). Placental development: Lessons from mouse mutants. *Nature Reviews Genetics, 2*(7), 538–548.

Rossant, J., & Tam, P. P. L. (2009). Blastocyst lineage formation, early embryonic asymmetries and axis patterning in the mouse. *Development, 136*(5), 701–713.

Santostefano, K. E., Hamazaki, T., Pardo, C. E., Kladde, M. P., & Terada, N. (2012). Fibroblast growth factor receptor 2 homodimerization rapidly reduces transcription of the pluripotency gene Nanog without dissociation of activating transcription factors. *The Journal of Biological Chemistry, 287*(36), 30507–30517.

Schrode, N., Xenopoulos, P., Piliszek, A., Frankenberg, S., Plusa, B., & Hadjantonakis, A.-K. (2013). Anatomy of a blastocyst: Cell behaviors driving cell fate choice and morphogenesis in the early mouse embryo. *Genesis, 51*(4), 219–233.

Shen, M. M. M. (2007). Nodal signaling: Developmental roles and regulation. *Development, 134*(6), 1023–1034.

Silva, J., Nichols, J., Theunissen, T. W., Guo, G., van Oosten, A. L., Barrandon, O., et al. (2009). Nanog is the gateway to the pluripotent ground state. *Cell, 138*(4), 722–737.

Singh, A. M., Hamazaki, T., Hankowski, K. E., & Terada, N. (2007). A heterogeneous expression pattern for Nanog in embryonic stem cells. *Stem Cells, 25*(10), 2534–2542.

Sirard, C., Kim, S., Mirtsos, C., Tadich, P., Hoodless, P. A., Itie, A., et al. (2000). Targeted disruption in murine cells reveals variable requirement for Smad4 in transforming growth factor β-related signaling. *Journal of Biological Chemistry, 275*(3), 2063–2070.

Skamagki, M., Wicher, K. B., Jedrusik, A., Ganguly, S., & Zernicka-Goetz, M. (2013). Asymmetric localization of Cdx2 mRNA during the first cell-fate decision in early mouse development. *Cell Reports, 3,* 1–16.

Smith, A. G., Heath, J. K., Donaldson, D. D., Wong, G. G., Moreau, J., Stahl, M., et al. (1988). Inhibition of pluripotential embryonic stem cell differentiation by purified polypeptides. *Nature, 336*(6200), 688–690.

Stavridis, M. P., Collins, B. J., & Storey, K. G. (2010). Retinoic acid orchestrates fibroblast growth factor signalling to drive embryonic stem cell differentiation. *Development, 137*(6), 881–890.

Stavridis, M. P., Lunn, J. S., Collins, B. J., & Storey, K. G. (2007). A discrete period of FGF-induced Erk1/2 signalling is required for vertebrate neural specification. *Development*, *134*(16), 2889–2894.

Stephenson, R. O., Rossant, J., & Tam, P. P. L. (2012). Intercellular interactions, position, and polarity in establishing blastocyst cell lineages and embryonic axes. *Cold Spring Harbor Perspectives in Biology*, *4*(11), http://dx.doi.org/10.1101/cshperspect.a008235.

Stephenson, R. O. R., Yamanaka, Y. Y., & Rossant, J. J. (2010). Disorganized epithelial polarity and excess trophectoderm cell fate in preimplantation embryos lacking E-cadherin. *Development*, *137*(20), 3383–3391.

Stewart, C. L., Kaspar, P., Brunet, L. J., Bhatt, H., Gadi, I., Köntgen, F., et al. (1992). Blastocyst implantation depends on maternal expression of leukaemia inhibitory factor. *Nature*, *359*(6390), 76–79.

Strumpf, D., Mao, C.-A., Yamanaka, Y., Ralston, A., Chawengsaksophak, K., Beck, F., et al. (2005). Cdx2 is required for correct cell fate specification and differentiation of trophectoderm in the mouse blastocyst. *Development*, *132*(9), 2093–2102.

Sumi, T., Oki, S., Kitajima, K., & Meno, C. (2013). Epiblast ground state is controlled by canonical Wnt/β-catenin signaling in the postimplantation mouse embryo and epiblast stem cells. *PLoS One*, *8*(5).

Suzuki, A., Raya, Á., Kawakami, Y., Morita, M., Matsui, T., Nakashima, K., et al. (2006). Nanog binds to Smad1 and blocks bone morphogenetic protein-induced differentiation of embryonic stem cells. *Proceedings of the National Academy of Sciences of the United States of America*, *103*(27), 10294–10299.

Tabansky, I., Lenarcic, A., Draft, R. W., Loulier, K., Keskin, D. B., Rosains, J., et al. (2013). Developmental bias in cleavage-stage mouse blastomeres. *Current Biology*, *23*(1), 21–31.

Tamm, C., Böwer, N., & Annerén, C. (2011). Regulation of mouse embryonic stem cell self-renewal by a Yes-YAP-TEAD2 signaling pathway downstream of LIF. *Journal of Cell Science*, *124*(7), 1136–1144.

Tanaka, S. (1998). Promotion of trophoblast stem cell proliferation by FGF4. *Science*, *282*(5396), 2072–2075.

Tarkowski, A. K. (1959). Experiments on the development of isolated blastomeres of mouse eggs. *Nature*, *184*, 1286–1287.

Tarkowski, A. K., & Wróblewska, J. (1967). Development of blastomeres of mouse eggs isolated at the 4- and 8-cell stage. *Journal of Embryology and Experimental Morphology*, *18*(1), 155–180.

ten Berge, D., Kurek, D., Blauwkamp, T., Koole, W., Maas, A., Eroglu, E., et al. (2011). Embryonic stem cells require Wnt proteins to prevent differentiation to epiblast stem cells. *Nature Cell Biology*, *13*(9), 1070–1075.

Tesar, P. J., Chenoweth, J. G., Brook, F. A., Davies, T. J., Evans, E. P., Mack, D. L., et al. (2007). New cell lines from mouse epiblast share defining features with human embryonic stem cells. *Nature*, *448*(7150), 196–199.

Torres-Padilla, M.-E., Parfitt, D.-E., Kouzarides, T., & Zernicka-Goetz, M. (2007). Histone arginine methylation regulates pluripotency in the early mouse embryo. *Nature*, *445*(7124), 214–218.

Toyooka, Y., Shimosato, D., Murakami, K., Takahashi, K., & Niwa, H. (2008). Identification and characterization of subpopulations in undifferentiated ES cell culture. *Development*, *135*(5), 909–918.

Tremblay, K. D., Hoodless, P. A., Bikoff, E. K., & Robertson, E. J. (2000). Formation of the definitive endoderm in mouse is a Smad2-dependent process. *Development*, *127*(14), 3079–3090.

Tropepe, V., Hitoshi, S., Sirard, C., Mak, T. W., Rossant, J., & van der Kooy, D. (2001). Direct neural fate specification from embryonic stem cells—A primitive mammalian neural stem cell stage acquired through a default mechanism. *Neuron*, *30*(1), 14.

Vallier, L., Mendjan, S., Brown, S., Chng, Z., Teo, A., Smithers, L. E., et al. (2009). Activin/Nodal signalling maintains pluripotency by controlling Nanog expression. *Development*, *136*(8), 1339–1349.
Wang, Y. Y., Smedberg, J. L. J., Cai, K. Q. K., Capo-Chichi, D. C. D., & Xu, X.-X. X. (2011). Ectopic expression of GATA6 bypasses requirement for Grb2 in primitive endoderm formation. *Developmental Dynamics*, *240*(3), 566–576.
Wang, C. C., & Song, B. B. (1996). Cell-type-specific expression of the platelet-derived growth factor alpha receptor: A role for GATA-binding protein. *Molecular and Cellular Biology*, *16*(2), 712–723.
Ware, C. B., Horowitz, M. C., Renshaw, B. R., Hunt, J. S., Liggitt, D., Koblar, S. A., et al. (1995). Targeted disruption of the low-affinity leukemia inhibitory factor receptor gene causes placental, skeletal, neural and metabolic defects and results in perinatal death. *Development*, *121*(5), 1283–1299.
Williams, R. L., Hilton, D. J., Pease, S., Willson, T. A., Stewart, C. L., Gearing, D. P., et al. (1988). Myeloid leukaemia inhibitory factor maintains the developmental potential of embryonic stem cells. *Nature*, *336*(6200), 684–687.
Winnier, G., Blessing, M., Labosky, P. A., & Hogan, B. L. (1995). Bone morphogenetic protein-4 is required for mesoderm formation and patterning in the mouse. *Genes and Development*, *9*(17), 2105–2116.
Wu, G., Gentile, L., Fuchikami, T., Sutter, J., Psathaki, K., Esteves, T. C., et al. (2010). Initiation of trophectoderm lineage specification in mouse embryos is independent of Cdx2. *Development*, *137*(24), 4159–4169.
Yagi, R., Kohn, M. J., Karavanova, I., Kaneko, K. J., Vullhorst, D., DePamphilis, M. L., et al. (2007). Transcription factor TEAD4 specifies the trophectoderm lineage at the beginning of mammalian development. *Development*, *134*(21), 3827–3836.
Yamanaka, Y., Lanner, F., & Rossant, J. (2010). FGF signal-dependent segregation of primitive endoderm and epiblast in the mouse blastocyst. *Development*, *137*(5), 715–724.
Yamanaka, Y. Y., Ralston, A. A., Stephenson, R. O. R., & Rossant, J. J. (2006). Cell and molecular regulation of the mouse blastocyst. *Developmental Dynamics*, *235*(9), 2301–2314.
Ying, Q.-L., Nichols, J., Chambers, I., & Smith, A. (2003). BMP induction of Id proteins suppresses differentiation and sustains embryonic stem cell self-renewal in collaboration with STAT3. *Cell*, *115*(3), 281–292.
Ying, Q.-L., Wray, J., Nichols, J., Batlle-Morera, L., Doble, B., Woodgett, J., et al. (2008). The ground state of embryonic stem cell self-renewal. *Nature*, *453*(7194), 519–523.
Yoshida, K. K., Taga, T. T., Saito, M. M., Suematsu, S. S., Kumanogoh, A. A., Tanaka, T. T., et al. (1996). Targeted disruption of gp130, a common signal transducer for the interleukin 6 family of cytokines, leads to myocardial and hematological disorders. *Proceedings of the National Academy of Sciences of the United States of America*, *93*(1), 407–411.
Yoshida-Koide, U., Matsuda, T., Saikawa, K., Nakanuma, Y., Yokota, T., Asashima, M., et al. (2004). Involvement of Ras in extraembryonic endoderm differentiation of embryonic stem cells. *Biochemical and Biophysical Research Communications*, *313*(3), 475–481.
Zernicka-Goetz, M., Morris, S. A., & Bruce, A. W. (2009). Making a firm decision: Multifaceted regulation of cell fate in the early mouse embryo. *Nature Publishing Group*, *10*(7), 467–477.
Zhang, J., Fei, T., Li, Z., Zhu, G., Wang, L., & Chen, Y. G. (2013). BMP induces cochlin expression to facilitate self-renewal and suppress neural differentiation of mouse embryonic stem cells. *Journal of Biological Chemistry*, *288*(12), 8053–8060.
Zhang, K., Li, L., Huang, C., Shen, C., Tan, F., Xia, C., et al. (2010). Distinct functions of BMP4 during different stages of mouse ES cell neural commitment. *Development*, *137*(13), 2095–2105.

CHAPTER TWO

Hematopoietic Stem Cell Development: An Epigenetic Journey

Sean M. Cullen[*,†,1], **Allison Mayle**[*,‡,1], **Lara Rossi**[§], **Margaret A. Goodell**[*,†,‡,2]
[*]Stem Cells and Regenerative Medicine Center, Baylor College of Medicine, Houston, Texas, USA
[†]Program in Developmental Biology, Baylor College of Medicine, Houston, Texas, USA
[‡]Department of Molecular and Human Genetics, Baylor College of Medicine, Houston, Texas, USA
[§]Department of Experimental, Diagnostic, and Specialty Medicine (DIMES), Institute of Hematology "L. and A. Seràgnoli", University of Bologna, Bologna, Italy
[1]These authors contributed equally to this work
[2]Corresponding author: e-mail address: goodell@bcm.edu

Contents

Abstract

Hematopoietic development and homeostasis are based on hematopoietic stem cells (HSCs), a pool of ancestor cells characterized by the unique combination of self-renewal and multilineage potential. These two opposing forces are finely orchestrated by several regulatory mechanisms, comprising both extrinsic and intrinsic factors. Over the past decades, several studies have contributed to dissect the key role of niche factors, signaling transduction pathways, and transcription factors in HSC development and maintenance. Accumulating evidence, however, suggests that a higher level of intrinsic regulation exists; epigenetic marks, by controlling chromatin accessibility, directly shape HSC developmental cascades, including their emergence during embryonic development, maintenance of self-renewal, lineage commitment, and aging. In addition,

Current Topics in Developmental Biology, Volume 107
ISSN 0070-2153
http://dx.doi.org/10.1016/B978-0-12-416022-4.00002-0

aberrant epigenetic marks have been found in several hematological malignancies, consistent with clinical findings that mutations targeting epigenetic regulators promote leukemogenesis. In this review, we will focus on both normal and malignant hematopoiesis, covering recent findings that illuminate the epigenetic life of HSCs.

1. INTRODUCTION

Hematopoiesis is the highly dynamic process sustaining the life-long production of blood, one of the most highly regenerative tissues. All hematopoietic lineages—including erythrocytes, platelets, myelocytes, and lymphocytes—derive from a pool of multipotent hematopoietic stem cells (HSCs) residing in the bone marrow (BM). HSCs are characterized by their ability to both self-renew and differentiate: while self-renewal guarantees the life-long maintenance of the stem cell compartment, differentiation involves the sequential steps leading to the production of mature blood cells.

Multilineage hematopoietic stem and progenitor cells (HSPCs) emerge early in ontogeny. After a first wave of primitive blood cells arising from the mesoderm, multilineage HSCs start to emerge from the extraembryonic yolk sac and placenta, followed by the aorta-gonad-mesonephros (AGM) region of the embryo (Moore & Metcalf, 1970; Samokhvalov, Samokhvalova, & Nishikawa, 2007). As gestation progresses, HSCs migrate to the fetal liver, which becomes the major site of definitive hematopoiesis until the latest stages of embryonic development. Shortly before birth, blood cell production emerges in the BM, the final and predominant site of hematopoiesis throughout adulthood (Fig. 2.1A) (Lux et al., 2008). In adult mammals, definitive hematopoiesis is sustained by a pool of long-term HSCs (LT-HSCs), from which short-term HSCs (ST-HSCs) and multipotent progenitors (MPPs) are derived; these stem and progenitor cells present a progressively decreased self-renewal potential, but still hold a multipotential differentiation capacity. Downstream of MPPs are lineage-restricted progenitors, responsible for generating a large pool of terminally differentiated cells eventually released into the peripheral blood (Fig. 2.1B).

Several factors, both extrinsic and intrinsic, regulate the progression of HSCs through the different phases of their development. Cell-extrinsic cues are provided by the stem cell niche and include cytokines, growth factors, chemokines, oxygen tension, and nutrients (Smith & Calvi, 2013; Suda, Takubo, & Semenza, 2011; Wilson & Trumpp, 2006). These signals merge into a network of intrinsic regulators, comprising signaling pathways,

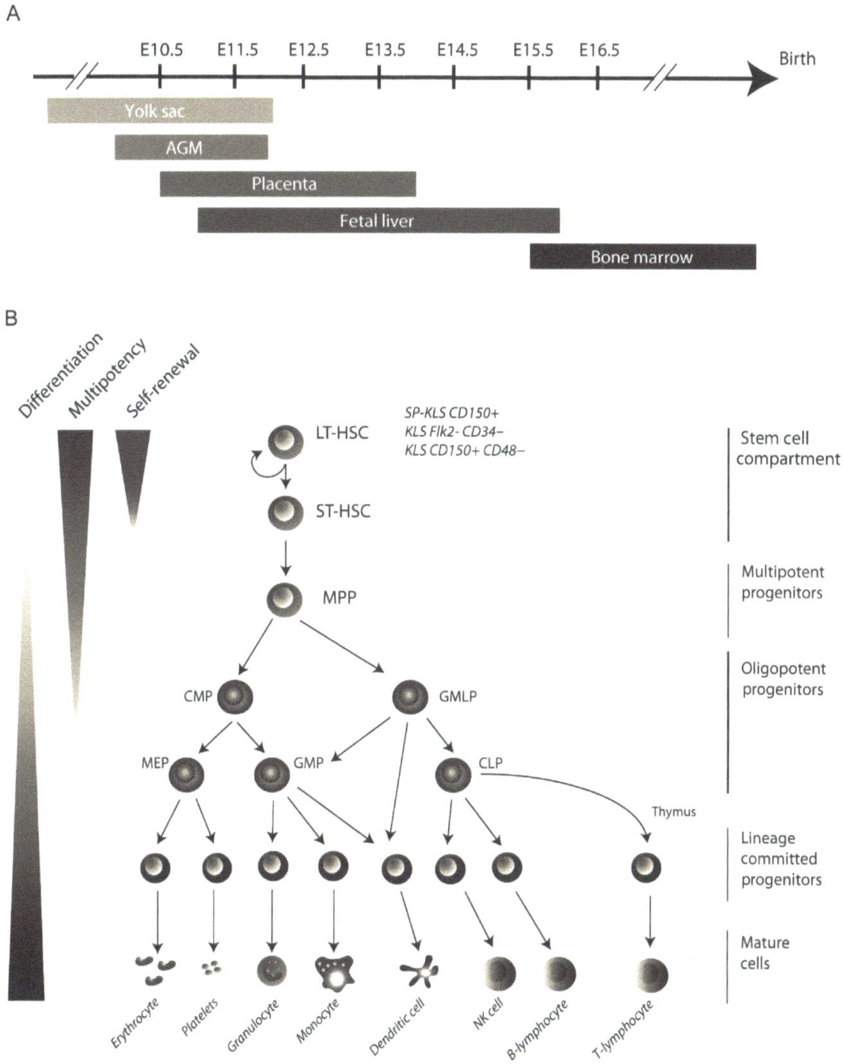

Figure 2.1 (A) Embryonic sites of blood development. The emergence of different sites of hematopoietic activity during development is highly dynamic, involving the coordinated decline of hematopoietic potential in the embryonic site previously occupied. Hematopoietic activity begins in the mesoderm of the extraembryonic yolk sac, followed by the placenta and the aorta-gonad-mesonephros (AGM) region. Hematopoietic cells then migrate to the fetal liver (FL), which becomes the major site of definitive hematopoiesis until the latest stages of embryonic development, when HSCs are recruited to the BM. (B) Overview of the hematopoietic hierarchy. Long-term hematopoietic stem cells (LT-HSCs) produce a continuous supply of cells with progressively decreased multipotency and self-renewal capabilities: short-term HSCs (ST-HSCs), multipotent progenitors (MPPs), and lineage-restricted progenitors. Through rapid proliferation, lineage-restricted progenitors generate mature cells, which are eventually released into the peripheral blood. Phenotypic markers commonly used to identify HSCs are indicated. SP, side population; KSL, c-Kit+Sca-1+Lineage-; CMPs, common myeloid progenitors; CLPs, common lymphoid progenitors; GMLPs, granulocyte–macrophage–lymphocyte progenitors; GMPs, granulocyte–macrophage progenitors; MEPs, megakaryocyte–erythrocyte progenitors; NK, natural killer. (See color plate.)

transcription factors, and epigenetic marks. By controlling chromatin conformations and accessibility, epigenetic mechanisms tune the expression of genes involved in HSC development and help orchestrate the balance between stemness and lineage commitment. Over the past decade, analyses of knockout (KO) mice have contributed to unveiling a number of genes vital for HSC development and function (Rossi et al., 2012), including, but not limited to: DNA-methylating enzymes, Polycomb-Group (PcG) complexes, histone modifiers, and factors involved in microRNA synthesis (summarized in Tables 2.1–2.4). Deletion of these genes in KO mice has been associated with a variety of phenotypes—from hematopoietic failure and repopulation defects to hyperproliferation and leukemia—reinforcing the hypothesis that epigenetic marks concur to mold developmental programs in HSCs. In this review, we will illustrate how epigenetic regulators contribute to the different stages of HSC development—from their embryonic emergence to adult life—focusing on the mechanisms that contribute to HSC self-renewal, lineage commitment, aging, and leukemogenesis.

2. PRINCIPLES OF EPIGENETIC REGULATION

Epigenetic marks include DNA methylation, covalent histone modification, and chromatin remodeling. In addition to these mechanisms, microRNAs and long noncoding RNAs (lncRNAs) have recently emerged as important regulators of transcriptional and epigenetic programs, playing a pivotal role in early development, lineage specification, and differentiation. In this review, we will focus mainly on DNA and chromatin modifications; for a detailed description of microRNA and lncRNA in hematopoietic development, we refer readers to several excellent reviews (Bissels, Bosio, & Wagner, 2012; O'Connell & Baltimore, 2012).

2.1. DNA methylation

DNA methylation plays a key role in gene expression regulation and is associated with a number of phenomena requiring transcriptional silencing, including X-chromosome inactivation and genomic imprinting (Schaefer, Ooi, Bestor, & Bourc'his, 2007). In mammalian cells, DNA methylation involves the addition of a methyl group to the 5′ position (C5) of the cytidine ring, generally within the context of CpG dinucleotides. This modification is catalyzed by DNA methyltransferases (Dnmts), a family of enzymes comprising a maintenance methyltransferase (Dnmt1), which targets hemimethylated DNA sequences and plays a key role in maintaining

Table 2.1 Roles of DNA modifiers in hematopoiesis

Gene	Type of mouse	Pre-Tp defect	Tp defect	Increased engraftment	Development	Self-renewal	Differentiation	Lineage specification	Cells tested	PMID
Dnmt1	cKO		++		+	+	+	+	BM	19801979
Dnmt3a	cKO			+	+	+	+		HSC	22138693
Tet2	KO, cKO	+		+		+			HSC, KSL	21723200
Tet2	KO	+		+					KSL	21873190
Tet2	GT, cKO	+		+					HSC, KSL	21723201
Dnmt3b[a]			nl						HSC	17420264
Tet1[a]		nl								21816367
Tet3[a]										

Summary of hematopoietic phenotypes in mouse models after mutation of genes encoding DNA modifiers. KO, knockout; cKO, conditional KO; GT, gene rap; Tp, transplantation; nl, apparently normal. Tp defect indicates a phenotype observed upon transplantation with ++ signifying a severe defect (reduction in PB output to less than 20% normal). Other columns indicate the processes the gene has been demonstrated to play a role in. BM, whole bone marrow; PMID, PubMed ID number.
[a]Indicates genes that require further investigation into hematopoietic-specific phenotypes.

Table 2.2 Roles of PRC members in hematopoiesis

Gene	Type of mouse	Pre-Tp defect	Tp defect	Increased engraftment	Development	Self-renewal	Differentiation	Lineage specification	Cells tested	PMID
Bmi1	KO					+	+	+	HSC	15589172
Bmi1	KO		++						HSC	12714971
Cbx2	KO					+	+		FL	15589172
Hemp	KO	+	+						FL HSC	21252303
Pcgf2 (*Mel18*)	KO					+			HSC	15183898
Rae28	KO		++						FL	11901201
Cbx2	OE		++			+			5FU BM	23502315
Cbx4	OE		++			+			5FU BM	23502315
Cbx7	OE			+		+			5FU BM	23502315
Cbx8	OE		++			+			5FU BM	23502315
Ezh2	OE			+		+			BM	16293602
Ezh2	het[a]					+			BM	20445021

Eed	het[a]		ND		+	BM	20445021
Suz12	LOF[a]			+	+	BM	20445021
Ezh1	KO	+	++		+	BM	23122289
Asxl1	OE[a]				+	BM	22897849

Summary of hematopoietic phenotypes in mouse models after mutation of genes involved in PRCs. KO, knockout; cKO, conditional KO; OE, over expression; het, heterozygous null; LOF, loss-of-function point mutation; Tp, transplantation. Tp defect indicates a phenotype observed upon transplantation with + signifying a moderate defect (reduction in peripheral blood (PB) output to between 20% and 70% of normal), ++ signifies a severe defect (reduction in PB output to less than 20% normal); ND, Tp not done. Other columns indicate the processes the gene has been demonstrated to play a role in. FL, fetal liver; 5FU BM, whole bone marrow harvested after 5FU treatment (enriched for HSPCs); BM, whole bone marrow; PMID, PubMed ID number.

[a]Indicates the mutation was on an *mpl1* KO background.

Table 2.3 Roles of histone modifiers in hematopoiesis

Gene	Type of mouse	Pre-Tp defect	Tp defect	Increased engraftment	Development	Self-renewal	Differentiation	Lineage specification	Cells tested	PMID
Dot1L	GT	+			+			+	YS	20798234
Dot1L	cKO	+	++			+	+		BM	21398221
Mll1	KO, cKO	+	+		+	+			BM	18371367
Mll1	cKO		++			+			KSL	18371366
Mll1	cKO	+	++		+	+	+		FL	20724987
Myst3 (*Moz*)	KO	+	++						FL HSC	16651658
Myst3	KO	+	++						FL KSL	16702405
Carm1	KO	+	++				+		FL, FBM	23248263
Bap1	cKO		++						Lin- BM	22878500

Summary of hematopoietic phenotypes in mouse models after mutation of genes encoding histone modifiers. KO, knockout; cKO, conditional KO; GT, gene trap; Tp, transplantation. Tp defect indicates a phenotype observed upon transplantation with + signifying a moderate defect (reduction in peripheral blood (PB) output to between 20% and 70% of normal), ++ signifies a severe defect (reduction in PB output to less than 20% normal). Other columns indicate the processes the gene has been demonstrated to play a role in. YS, yolk sac; BM, whole bone marrow; FL, fetal liver; PMID, PubMed ID number.

Table 2.4 Roles of RNA-binding proteins in hematopoiesis

Gene	Type of mouse	Pre-Tp defect	Tp defect	Increased engraftment	Development	Self-renewal	Differentiation	Lineage specification	Cells tested	PMID
Dicer	cKO	+							HSC	20305640
Zfp36	KO	+							HSC	21270394
Zfp36l2	KO		++						FL KSL	19633199
Msi2	GT		+						BM	21613258

Summary of hematopoietic phenotypes in mouse models after mutation of genes encoding RNA-processing proteins. KO, knockout; cKO, conditional KO; GT, gene trap; Tp, transplantation. Tp defect indicates a phenotype observed upon transplantation with + signifying a moderate defect (reduction in peripheral blood (PB) output to between 20% and 70% of normal), ++ signifies a severe defect (reduction in PB output to less than 20% normal). FL, fetal liver; BM, whole bone marrow; PMID, PubMed ID number.

methylation patterns during DNA replication, and two *de novo* methyltransferases (Dnmt3a and Dnmt3b), which react with unmethylated sequences (Okano, Xie, & Li, 1998). In the mammalian genome, CpG dinucleotides are statistically underrepresented when compared to other dinucleotides, but reach almost the expected frequency in some regulatory regions. These CpG-rich genomic segments—referred to as CpG islands (CGIs)—are found in the promoters of about 60% of human genes (Antequera & Bird, 1993). While CpG dinucleotides dispersed in intragenic nonregulatory regions are usually methylated, CGIs are predominantly unmethylated (or hypomethylated), thus facilitating the interaction with transcriptional complexes to promote gene expression. CGI methylation marks are impressed early during development, but are not permanent and their status can be modulated over time by developmental programs (Calvanese et al., 2011). For instance, as discussed in detail in the following section, the unfolding of myeloid differentiation programs requires the expression of myeloid-specific genes and the concomitant silencing of genes associated with other hematopoietic lineages.

The mechanism linking DNA methylation to gene silencing has been the focus of intense research over the past decades. Based on the current findings, methylated CGIs recruit proteins containing specific methyl-CpG-binding domains to promoter regions, which in turn attract multiprotein complexes containing histone deacetylases (HDACs) and histone methyltransferases (HMTs). Together, these enzymes lead to an increased compaction of chromatin filaments, repressing gene expression in that DNA region. In contrast, unmethylated CGIs are characterized by an open chromatin structure, which permits the recruitment of transcription factors and RNA polymerases to target genes (Esteller, 2007). While DNA methylation has been the focus of intense studies, mechanisms allowing DNA demethylation have remained elusive for a long time. Recently, 10–11 translocation (Tet) proteins (Tet-1, -2, and -3) have been identified as 5-methylcytosine (5-mC) hydrolylases (Wu & Zhang, 2011). Through three consecutive reactions of oxidation, Tet proteins convert 5-mCs into 5-hydroxymethylcytosines (5-hmCs), 5-formylcytosine (5fC), and 5-carboxylcytosine (5caC). These modified cytosines represent key intermediates toward DNA demethylation, preventing methyl-DNA-binding proteins from interacting with DNA sequences (Ko et al., 2011; Quivoron et al., 2011).

2.2. Covalent histone modifications

Nucleosomes represent the basic unit of chromatin architecture and allow the organization of DNA strands into higher orders of compaction. Individual core nucleosomes undergo specific posttranslational changes that lead to the acetylation, methylation, phosphorylation, ADP-ribosylation, sumoylation, and ubiquitination of their N-terminal tail. These covalent modifications, collectively referred to as the *histone code*, impact the overall structural organization of DNA fibers and determine, for a specific genomic region, how permissive chromatin will be to binding of the transcriptional machinery (Bannister & Kouzarides, 2011).

Among the most extensively investigated histone modifications are acetylation and methylation. Histone acetylation, occurring by the addition of one acetyl group to a residue of lysine, correlates with an open chromatin structure and the activation of gene expression. Conversely, a decreased level of histone acetylation (hypoacetylation) leads to a more condensed chromatin and transcriptional silencing. The level of histone acetylation is determined by the opposing activity of two classes of enzymes: histone acetyltransferases (HATs), which catalyze the transfer of acetyl groups from acetyl-CoA to lysine residues, and HDACs, which remove acetylation marks from histone tails. Both HATs and HDACs form multiprotein complexes with transcription factors and nuclear receptors, orchestrating the transcriptional fate of DNA sequences (Berger, 2007).

A more complex level of gene regulation is provided by histone methylation, achieved by the attachment of methyl groups to both lysines (which can bear up to three methyl groups) and arginines (mono- or dimethylated). To add complexity to the system, the effects of histone methylation are not unidirectional; while arginine methylation is usually associated with an open chromatin state, lysine methylation can produce opposing effects depending on both the specific residues involved and the degree of methylation. Two families of enzymes are involved in modeling the methylation landscape of histones: HMTs and histone demethylases (HDMs), responsible for appending or detaching methyl groups, respectively. Similar to acetylating/deacetylating enzymes, HMTs and HDMs also form multiprotein complexes, bringing together and orchestrating the activity of different enzymes modulating the histone code (Marmorstein & Trievel, 2009).

Among the epigenetic factors influencing DNA methylation, a key role is played by PcG complexes, which are involved in repressive histone modifications (Sashida & Iwama, 2012). In mammals, PcGs assemble into two major multiprotein chromatin-associated complexes—named Polycomb-Repressive Complex (PRC) 1 and PRC2—acting, at least partly, as interdependent transcriptional repressors. PRC2 activity is mediated by two HMTs (EZH1 and EZH2), responsible for the di-/trimethylation of histone H3 (H3K27) (Cao et al., 2002). PRC2-dependent H3K27 methylation leads to the recruitment of PRC1 (core components: Bmi1, Ring1a/b, Cbx, Mel18, and Mph), which in turn is responsible for histone H2A monoubiquitination, chromatin condensation, and ultimately, gene silencing (Endoh et al., 2012; Li et al., 2006). Over the years, KO studies have greatly contributed to unveil the key role of PcG complexes in hematopoietic development and differentiation, and both PRC1 and PRC2 components may represent targets for leukemogenetic events.

Epigenetic marks—established early during development—undergo significant modulation as developmental programs unfold and actively help shape the transcription profile of differentiating tissues. On the other hand, aberrant methylation patterns and the unbalanced activity of enzymes involved in histone modifications may result in hematological malignancies (Tables 2.1–2.4). The reactivation of genes silenced by development could represent one of the earliest events involved in tumorigenesis, and aberrant epigenetic profiles are now regarded as useful prognostic markers. In the following section, we will discuss how the epigenetic mechanisms described above are involved in HSC development, and how murine models have shed light on how they regulate HSC self-renewal, lineage choice, and aging. Finally, we will briefly review examples of aberrant epigenetic patterns in human hematological malignancies and discuss how altered epigenetic marks may contribute to leukemogenesis.

3. EPIGENETIC REGULATION OF HEMATOPOIESIS

3.1. Hematopoietic development

3.1.1 DNA methylation

The importance of proper DNA methylation throughout mammalian development is made evident by germline deletion of Dnmts: Dnmt1-null mice die at gastrulation (Li, Bestor, & Jaenisch, 1992), Dnmt3b-null mice die at roughly E9.5, and Dnmt3a-null mice die at roughly 3 weeks of age (Okano, Bell, Haber, & Li, 1999). Additional research has shown DNA methylation to be remarkably fluid early in development: present in gametes, nearly completely erased in the early morula, and then largely reestablished in a wave of *de novo* methylation around implantation

(Kafri et al., 1992; Monk, Boubelik, & Lehnert, 1987). From this point forward, DNA methylation changes occur in a tissue-specific fashion, with regions undergoing sequence-specific changes in methylation, as part of the overall developmental program. In the hematopoietic system, for example, these changes involve developmental decisions, impacting processes such as lineage specification and self-renewal maintenance. While DNA methylation enzymes are essential for the function of normal adult HSCs (see below), whether they are required during hematopoietic development is not known.

3.1.2 Histone methylation

3.1.2.1 Polycomb-repressive complexes

Genes encoding members of the Polycomb complexes, which introduce the repressive K3K27me3 mark, have been well studied for their role in hematopoietic development. In embryonic stem cells (ESCs), forced expression of Bmi1, a PRC1 component, has been shown to promote hematopoietic cell development by inducing *Gata2*, an essential component of primitive hematopoiesis. In addition, Bmi1 expression leads to enhanced hematopoietic cell survival, with extensive proliferation of ESC-derived hematopoietic progenitor cells *in vitro* (Ding, Lin, Ensenat-Waser, Rose-John, & Zenke, 2012).

Bmi1 has also been shown in *in vivo* studies to play a role in early hematopoietic function. Fetal liver HSCs isolated from *Bmi1* KO mice and transplanted into irradiated donors can only transiently contribute to hematopoietic output (Park et al., 2003). Rae28, another member of the PRC1 complex, has also been shown to play a role in early hematopoietic development. *Rae28*$^{-/-}$ mice are perinatal lethal and embryos exhibit splenic hypoplasia, with a dramatic reduction in nucleated cell number. Upon further analysis, progressive reduction of hematopoietic progenitors in *Rae28*$^{-/-}$ fetal livers was observed via both colony forming and transplantation assays, indicating a requirement for Rae28 in sustaining early HSC and progenitor populations (Ohta et al., 2002). While the PRC1 component KO murine models are somewhat different phenotypically, it appears that the epigenetic role played by the PRC1 complex in an early hematopoietic environment is critical for maintenance of a self-renewing, normally differentiating HSC population.

3.1.2.2 Other histone modifications

While the Polycomb complexes introduce repressive marks, a number of proteins are involved in generating activating marks to promote gene expression, such as H3K4 and H3K79, and some of these have been shown to have a role in embryonic hematopoiesis. For example, disruptor of telomere silencing 1-like (Dot1l) methylates H3K79, and a *Dot1l*-KO mouse model, is embryonic lethal and displays severe anemia. Further characterization of this mouse, however, showed that there was a selective defect in erythroid, but not myeloid, differentiation. Dot1l was found to be a critical regulator of two important transcription factors, Gata2 (promoter of early hematopoiesis) and PU.1 (promoter of myelopoiesis), and functions as part of an erythroid to myeloid differentiation switch in early hematopoiesis (Feng et al., 2010). In addition, deletion of murine Coactivator-associated arginine methyltransferase 1 (*Carm1*), which facilitates arginine-specific histone methylation, led to decreased numbers of LT-HSCs, ST-HSCs, and lymphoid-biased MPPs in E18.5 embryos. *Carm1*$^{-/-}$ fetal liver cells were also severely impaired in their repopulating ability after transplant, suggesting that Carm1 is important for normal hematopoietic development (Li et al., 2013).

An additional epigenetic regulator involved with histone methylation important in fetal hematopoiesis is Mll1 (mixed-lineage leukemia 1), part of the chromatin-remodeling complex leading to H3K4 methylation. *Mll1* KO mice exhibit a twofold reduction in fetal liver cellularity, with a significant shrinking of the LT-HSC and ST-HSC compartments. The number of quiescent HSCs found in *Mll1* KO fetal livers was also dramatically reduced, indicating a requirement for Mll1 expression to generate sufficient numbers of fetal liver HSCs with normal cell-cycle profiles (McMahon et al., 2007). The authors also looked for effects of a hematopoietic-specific (Vav-Cre) *Mll1* deletion, but, surprisingly, found no deficiencies in postnatal hematopoietic differentiation output and maintenance. In contrast to the results of McMahon et al., however, Gan et al., also using Vav-Cre to conditionally KO *Mll1*, discovered a multilineage deficit in differentiation and decrease in hematopoietic progenitor numbers in young postnatal mice, with death resulting around 3 weeks of age (Gan, Jude, Zaffuto, & Ernst, 2010). Fetal hematopoiesis, in contrast, proceeded normally, but Mll1 expression was found to be required for sustaining postnatal hematopoiesis. Conceivably, differences in the alleles used by each group may explain these discrepancies.

It is also worth noting that Mll1 may interact with an lncRNA in the hematopoietic system. In murine ES cells, a lincRNA called Mistral was shown to recruit Mll1 to sites for transcriptional activation. Ultimately, Mistral recruitment of Mll1 led to chromosome conformational changes culminating in expression of the ESC-repressed *Hoxa6* and *Hoxa7* genes important for germ-layer specification (Bertani, Sauer, Bolotin, & Sauer, 2011). Mistral may be expressed and function in a similar fashion in the hematopoietic system; alternatively, it is possible that a different hematopoietic cell-specific lncRNA or miRNA may serve a similar function in the proper context. Similar interconnections between different classes of epigenetic regulators highlight the sophistication of the overall epigenetic web in various cell processes (Fig. 2.2).

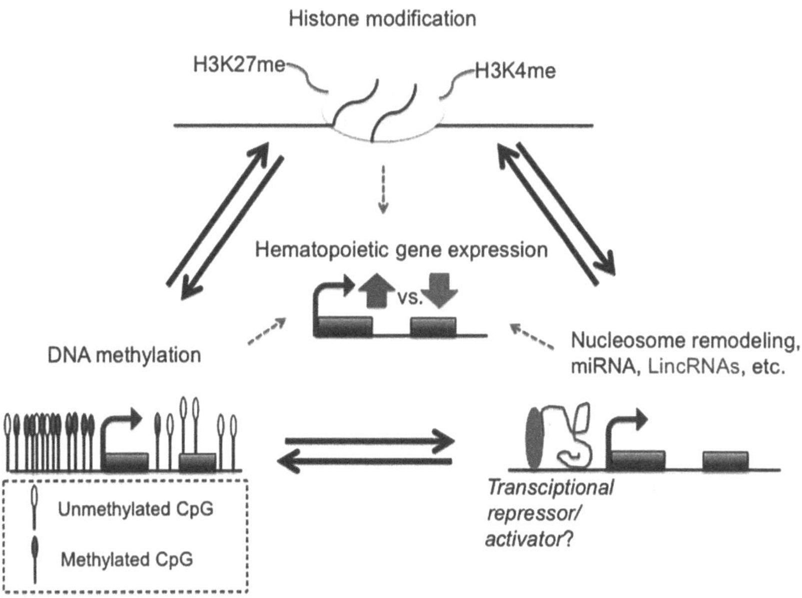

Figure 2.2 Interplay of different classes of epigenetic regulators in hematopoietic stem cells. We propose a model where DNA methylation, histone modification, noncoding RNA, and nucleosome remodeling must all work in concert with one and other in HSCs. Some of these interactions will be of a direct nature, such as histone modifier PRC2, directly interacting with DNA methyltransferases to target specific promoters for DNA methylation. Noncoding RNA molecules, like Mistral in ESC differentiation, could recruit histone modifiers and DNA methyltransferases to sites of activity through direct protein–RNA binding and DNA–RNA interactions. These interconnections will also likely be of a more indirect nature, such as subtle changes in the DNA methylation or histone modification landscape influencing the accessibility and overall effect of other epigenetic regulations. Overall, these interactions will combine to direct changes to hematopoietic gene expression and influence a number of important biological functions in the hematopoietic system, including multipotency maintenance and lineage commitment upon differentiation. (See color plate.)

3.2. Balance between maintenance of stemness in HSCs and lineage commitment

Adult HSCs are characterized by their ability to both self-renew and differentiate. The level of self-renewal is critical to HSCs; too much self-renewal can hinder proper differentiation, while too little self-renewal can lead to a deficiency of HSCs over time. Epigenetic modifiers contribute to balance these opposing forces—self-renewal and differentiation—and unique sets of epigenetic modifications are now emerging as part of the key instructions that direct HSCs toward specific lineage choices.

3.2.1 DNA methylation

Dynamic changes at CpGs have been shown to be prevalent throughout cellular differentiation (Meissner et al., 2008), and both DNA methylation and DNA demethylation influence cellular lineage decisions in differing fashions. For example, DNA methylation has been shown to prevent stem cells from activating differentiation programs prematurely, preserving multipotency. Dnmt1, the methyltransferase required for maintenance methylation, is explicitly required for HSC self-renewal and important for differentiation and niche retention (Bröske et al., 2009; Trowbridge, Snow, Kim, & Orkin, 2009). *Dnmt1* hypomorphic HSCs maintain normal homing capacity, but repopulation potential is severely impaired, revealing that a critical threshold of DNA methylation is required to maintain HSC self-renewal and, concomitantly, to silence specific differentiation programs. *Dnmt1*-null HSCs exhibited an upregulation of genes encoding a number of myeloid-progenitor associated factors including *Gata1*, *Id2*, and *Cepba*, leading to a disproportional increase in myeloid cell output following transplantation. It has been hypothesized that promoters of myeloerythroid regulators are preferentially methylated in HSCs to prevent an overwhelming myeloid output, and permit differentiation of lymphoid tissue appropriately (Bröske et al., 2009).

The *de novo* Dnmts also have an important role in HSC self-renewal and differentiation. Conditional deletion of *Dnmt3a* in the hematopoietic system expanded HSC number upon transplantation while concurrently impairing their differentiation capacity (Challen et al., 2011). It is thought that loss of Dnmt3a leads to an inability of repressing self-renewal genes and fully instigating the differentiation program. Among the genes found to be hypomethylated after *Dnmt3a* KO were *Runx1* and *Vasn*, whose expression is normally restricted to HSCs, and both exhibited incomplete repression in

Dnmt3a-null progeny. However, several enigmatic aspects remain unclear. *Dnmt3a*-null HSCs exhibited both increases and decreases of DNA methylation. The increases could possibly be due to continued expression of Dnmt3b. To date, the independent role of Dnmt3b is unclear.

Data from the *Dnmt3a* KO also showed that Dnmt3a must act very selectively at certain targets, but how those targets are identified is not well understood. Some studies have suggested both direct and indirect mechanisms of action. *Runx1* and *Gata3*, both of which are associated with maintaining HSC stemness (Chen & Zhang, 2001; Okuda, van Deursen, Hiebert, Grosveld, & Downing, 1996), are upregulated in the absence of Dnmt3a and may be bound directly by Dnmt3a. In addition, Dnmt3a activity has been suggested to be modulated by CHD, a single-stranded RNA that is antisense to the E-cadherin promoter. This RNA binds tightly to the catalytic domain of Dnmt3a, inhibiting the activity of Dnmt3a (Holz-Schietinger & Reich, 2012). Dnmt3a, which is itself controlled by other epigenetic mechanisms, works both directly and indirectly to maintain expression of genes important for stem cell function in a stem cell-specific context, and repress them once differentiation begins.

Targeted DNA demethylation and subsequent gene expression changes have also been shown to be involved in HSC differentiation. Genome-wide DNA methylation patterns of hematopoietic cell types from different stages of differentiation have been examined using comprehensive high-throughput array-based relative methylation (CHARM) analysis. From CHARM analysis, various instances of lineage-specific demethylation and concomitant transcriptional upregulation were observed, including *Lck*, responsible for initiating downstream signaling from the T-cell receptor in thymocyte progenitors (DN1-3), *Cxcr2*, encoding a key receptor for neutrophil chemotaxis, and *Gadd45a*, a gene implicated in myeloid development, during the CMP to GMP transition (Ji et al., 2011). Further providing evidence of a role for DNA demethylation in HSC differentiation, mutations in *Tet2*—a member of the TET family of proteins involved in 5-mC conversion to 5-hmC—have also been found to lead to myeloid fate skewing (Moran-Crusio et al., 2011). Stated another way, a mutation in a gene family believed to be part of the pathway of active DNA demethylation that results in a change in HSC differentiation outputs appears to support the hypothesis that proper DNA demethylation, as well methylation, is required for normal HSC differentiation.

Recently, Tet2 has also been implicated in HSC self-renewal (Ito et al., 2010; Koh et al., 2011; Kriaucionis & Heintz, 2009; Tahiliani et al., 2009).

Inactivation of the catalytic activity of Tet2 leads to a phenotypic and functional amplification of HSPCs, and a competitive advantage over wild-type cells (Quivoron et al., 2011). After reduction of Tet2 by shRNA knockdown, or knockout by deletion of a floxed allele, HSPCs retain colony formation ability even after serial plating, while wild-type HSPCs lose the ability after the second round of plating, indicating that cells with reduced Tet2 levels maintain a more primitive "stem-like" phenotype. Of note, *Tet2* KO cells that have been serially replated have a gene expression profile typical of CMPs, along with increased expression of self-renewal factors Meis1 and Evi1, suggesting that they are progenitors that have, through the loss of *Tet2*, gained self-renewal ability. This phenotype is also observed *in vivo*: not only do *Tet2* KO mice at 20 weeks of age have an expanded BM KSL (c-Kit$^+$, Sca1$^+$, Lineage$^-$) compartment (a population highly enriched for murine HSCs), but their BM cells can outcompete the wild-type counterpart after transplantation as well. Taken together, this suggests that Tet2 regulates HSC maintenance, perhaps through known self-renewal factors Meis1 and Evi1 (Moran-Crusio et al., 2011).

Additional studies have shown progressive hypermethylation and transcriptional silencing as HSCs differentiate into MPPs first and then mature differentiated progeny. For example, *Meis1* and *Hoxa9*, which encode transcription factors important for HSCs (Lawrence et al., 1997; Pillay, Forrester, Erickson, Berman, & Waskiewicz, 2010), possess differentially methylated regions (DMRs) that are unmethylated in MPPs, but then become hypermethylated as differentiation occurs (Ji et al., 2011). While the mechanism defining sites for *de novo* DNA methylation is not completely understood, there is some evidence that histone methylation, mediated by PRC2 proteins, exerts some influence on that decision. EZH2, a catalytic component of the PRC2 complex, has been shown to interact with all three DNMTs *in vivo*. In addition, EZH2 was necessary for DNMTs to target PRC-2-specific promoters for DNA methylation, providing a link in overall function of these two epigenetic systems (Viré et al., 2005).

Overall, DNA methylation has been shown to be a quite dynamic process during normal hematopoietic differentiation. A recent study by Bock et al. utilizing reduced representation bisulfite sequencing of 13 highly purified murine hematopoietic progenitor populations and their progeny found a large degree of differential methylation between the different populations. In comparing HSCs to other progenitor populations, 405 regions were found to gain methylation with differentiation, while 2017 regions were found to lose methylation with differentiation. As for the role DNA methylation directly plays in promoting proper HSC differentiation, increasing DNA methylation in progenitor populations was observed at four HSC-specific homeobox genes (*Hoxa9*, *Pbx1*, *Hoxb5*, and *Hoxa5*, some of which are well-known oncogenes), leading to their decreased expression as differentiation progressed. This led the authors to propose that DNA methylation might accumulate at specific genes during differentiation to prevent aberrant activation in downstream populations. The overall picture of DNA methylation changes during HSC differentiation, however, is more complex; as differentiation progresses, a general pattern of decreasing methylation is observed at the lineage-specific regulatory elements. In parallel, a simultaneous increase in methylation at regulatory elements of alternative lineages is registered (Bock et al., 2012). These results highlight the fact that DNA methylation status is both fluid and finely tuned during HSC differentiation, with both DNA methylation and demethylation playing an important role in promoting proper lineage progression.

3.2.2 Chromatin modifications

3.2.2.1 Histone modifiers

Histone modification is another epigenetic mark that, along with DNA methylation, plays an integral role in proper HSC differentiation and self-renewal. Specific chromatin states, for example, facilitate coordinated gene activation and silencing required for proper differentiation into different mature blood cell types. In a process termed "multilineage priming" (Hu et al., 1997), it is hypothesized that, prior to lineage commitment, various chromatin modifications facilitate the preassembly of important transcription factors at promoters of lineage-specific genes in HSCs. According to the model, this priming phase is followed by a coordinated change in activating and repressing histone marks as differentiation occurs (Maes et al., 2008; Orford et al., 2008; Weishaupt, Sigvardsson, & Attema, 2010).

While the low number of HSCs that can be purified has impeded the comprehensive genome-wide description of histone modifications on HSCs, histone modifications have been profiled on more heterogeneous progenitor populations, or at selected sites in more purified HSCs. Most of the features of histone marking that have been identified in ES or other cells have also been found in hematopoietic progenitors. In particular, so-called bivalent genes which exhibit marking with both active (H3K4) and repressive (H3K27) marks have been identified, also at embryonic developmental regulators in adult human hematopoietic progenitors (Weishaupt et al., 2010).

Hematopoietic developmental regulators, such as the gene for Bmp4, were among 994 genes found to possess bivalent promoters. Around 40% of these 994 murine genes were also observed to possess bivalent promoters in human hematopoietic progenitors. The number of promoters with bivalent modifications was greatest in HSCs, and decreased with differentiation, with the largest number of genes losing H3K4me3. Loss of H3K4me3 was often accompanied by an increase in another silencing chromatin mark like H3K9me3, which correlated with decreased gene expression during differentiation. The pattern of changing histone modifications over the course of differentiation was found to be quite complex, as the authors examined a number of additional activating (H3K79me2, H3ac) and silencing (H3K9me3) chromatin marks in genes found to significantly change expression over the course of hematopoietic differentiation. The combinatorial diversity of chromatin modifications available for different lineages indicates a high-order, finely tuned epigenetic toolbox working in chorus throughout hematopoietic differentiation (Weishaupt et al., 2010). Among this diverse pattern of chromatin modifications, another study found developmentally important genes marked with H3K4me2 in hematopoietic progenitors. As lineage commitment progressed, however, increased H3K4me3 was observed at genes expressed in those cells, while silenced genes underwent H3K4 demethylation. For example, *Rag2*, a lymphoid-specific gene, and *Gata1*, an erythroid-specific gene, exhibit H3K4me2 in hematopoietic progenitors; upon differentiation, both genes become enriched for H3K4me3 in the lineages in which their expression is required and are demethylated in others (Orford et al., 2008).

Histone modifiers are also essential for maintaining the gene expression patterns required for HSC self-renewal maintenance. Moz is a histone acetyltransferase in the MYST family. *Moz* KO fetal liver cells have decreased c-Kit levels and cannot reconstitute recipient mice upon transplantation (Katsumoto et al., 2006). Moz-deficient mice have a decreased number of hematopoietic progenitors, but the lineage commitment of cells that are present is unaffected (Thomas et al., 2006).

Like histone acetylation, histone methylation plays an important role in HSC homeostasis. As discussed above, the HMT Mll plays an important role in HSC self-renewal in the fetal liver. In a conditional KO model, mice maintained normal hematopoiesis in the BM, thymus, and spleen, but HSCs were not able to compete against wild-type HSCs upon competitive transplantation (McMahon et al., 2007). A partial tandem duplication knockin *of Mll* led to a decrease in the number of HSPCs due to increased apoptosis, but the cells had a proliferative advantage in both colony forming assays and competitive transplantations, suggesting that ST-HSCs, MPPs, and GMPs gained self-renewal potential in the absence of Mll (Zhang et al., 2012). Further demonstrating the importance of proper histone methylation in HSC self-renewal, a ubiquitous, conditional KO of HMT *Dot1l* in adult mice leads to both pancytopenia and a decrease in functional HSCs, as evidenced by minimal contribution of *Dot1l*-KO whole BM after transplantation (Jo, Granowicz, Maillard, Thomas, & Hess, 2011).

3.2.2.2 Polycomb-repressive complexes

In addition to a general role for histone modifications in hematopoietic differentiation and self-renewal, individual components of chromatin-modifying complexes have been shown to be essential in specific contexts. For example, PRC1 components *Bmi1* and *rae28* both decrease HSC activity when mutated, suggesting a predominant role for PRC1 in promoting HSC/self-renewal activity. Increased expression of *Bmi1* also promotes HSC self-renewal, enhances symmetrical division of HSCs, and increases the probability that a daughter cell will inherit "stemness." Forced expression of *Bmi1* increases *ex vivo* expansion of MPPs, and increases repopulating capacity of HSCs *in vivo*. Conversely, absence of *Bmi1* leads to defects in HSC self-renewal (Iwama et al., 2004). Adult HSCs from *Bmi1* KO mice are decreased in number and can only contribute to transient repopulation after transplantation (Park et al., 2003). In addition, *Bmi1* deletion leads to accelerated lymphoid specification via premature activation of *Ebf1* and *Pax5* (Oguro et al., 2010). Another member of the PRC1 complex, Mel-18, might be expected to have the same effects as Bmi1; indeed MEFs deficient in either *Bmi1* or *Mel-18* have similar upregulation of p16 (a tumor suppressor (TS) that promotes cellular senescence) and induction of quiescence (Jacobs, Kieboom, Marino, DePinho, & van Lohuizen, 1999). When studied in the hematopoietic system, however—unlike the *Bmi1* KO situation, which leads to decreased self-renewal—*Mel-18* KO has been shown to increase Hoxb4, increase HSC quiescence, and promote self-renewal, suggesting that Mel-18 normally functions as a negative regulator of self-renewal (Kajuime, Ninomiya, Ishihara, Kanno, & Kanno, 2004). The different phenotypes observed in the absence of Bmi1 or Mel-18 probably reflect different usage of the proteins in heterogeneous PRC1 complexes.

Cbx family genes are members of the PRC1 complex that provide specificity to the complex. Cbx7 is expressed in HSCs and overexpression of Cbx7 leads to increased self-renewal and induction of leukemia.

Overexpression of Cbx2, 4, or 8, on the other hand, leads to differentiation and exhaustion of the HSC pool. Interestingly, Cbx7 and 8 share many of the same targets, with their activity depending on cell type. Cbx7 preserves self-renewal by repressing progenitor genes in HSCs, while Cbx8, expressed in progenitors, may turn off self-renewal genes to promote differentiation (Klauke et al., 2013). Cbx2 is vital for HSC and progenitor function and regulates expression of p53-controlled cell-cycle regulator p21. Fellow PRC1 complex member Bmi1, however, functions independently of Cbx2, controlling the expression of the *INK4a/ARF* locus, which encodes cellular senescence-related $p16^{INK4a}$ and $p19^{ARF}$ (van den Boom et al., 2013). The Cbx family is an example of how heterogeneous PRC1 complexes function in different cellular contexts to regulate the expression of stem cell or differentiation genes.

Like PRC1, PRC2 is a repressive complex in regard with gene expression, but the two complexes appear to have opposing effects on HSPCs. On an $Mpl1^{-/-}$ background, which is sensitized to HSC changes, heterozygous mutation of PRC2 components *Ezh2*, *Eed*, or *Suz12* lead to an increase in HSPC activity, suggesting that PRC2 normally functions to restrict HSC activity. *Eed* hypomorphic alleles also demonstrated a critical role of PRC2 in restricting proliferation of early lymphoid and myeloid progenitors (Majewski et al., 2010). Overexpression of *Ezh2* prevents exhaustion of HSCs upon serial transplantation, maintaining their stem cell potential (Kamminga et al., 2006). Unlike mutations in *Ezh2*, *Eed*, and *Suz12*, ablation of PRC2 component *Ezh1* induces significant loss of adult HSCs, with concomitant impairment of their self-renewal capacity, due to a potent senescence response (Hidalgo et al., 2012). As observed in PRC1 complexes containing Bmi1 or Mel-18, the range of observed phenotypes from deletion of different PRC2 components highlight a number of possible functions resulting from slight changes to PRC2 complex composition. There is evidence that PRC2 is recruited to specific sites by Asxl1 in hematopoietic cells (Abdel-Wahab et al., 2012), but hematopoietic-specific deletion of *Asxl1* will be required to further understand its effects on hematopoiesis.

ASX, the *Drosophila* homolog of Asxl1, is a part of a PcG complex termed Polycomb-repressive deubiquitinase, along with Calypso, the *Drosophila* homolog of Bap1 (BRCA1-associated protein) (Scheuermann et al., 2010). Conditional loss of *Bap1* in adult mice increased the KSL compartment of both BM and spleen. These cells produced fewer colonies on methylcellulose than wild-type cells, and upon replating, colonies were not well formed, and appeared to be undergoing apoptosis. *Bap1*-deficient cells are unable to reconstitute irradiated recipients, but a homing defect has not been ruled out. Bap1 is important to maintaining normal hematopoiesis, possibly by recruiting other histone-modifying enzymes to control gene expression patterns (Dey et al., 2012).

In summary, DNA methylation and chromatin modifications work together to maintain the transcriptional profile required for maintenance of multipotency and self-renewal of HSCs, and alterations in their reciprocal balance may account for HSC functional impairment. Additionally, upon differentiation, lineage commitment proceeds through a series of cell-type specific alterations that affect both the landscape of DNA methylation and chromatin modifications. These epigenetic changes lead to two interconnected outcomes: (i) repression of stem cell-specific genes, preventing expression in an inappropriate, and potentially harmful, environment; and (ii) expression of genes important in cell fate decisions, including transcription factors and signaling molecules that, in turn, further drive the developmental decisions of differentiation.

3.3. HSC aging

The ability of HSCs to continue production of mature blood cells throughout the life of an organism is an important component in maintaining a homeostatic environment. The process of aging in the hematopoietic system, particularly in HSCs, has been the focus of increased research in recent years. In murine models, HSCs lose functionality with increasing age, especially in regard with repopulating capacity (Kim, Moon, & Spangrude, 2003; Morrison, Wandycz, Akashi, Globerson, & Weissman, 1996; Rossi et al., 2005). Paradoxically, a dramatic increase in the number of HSCs is also observed with aging (Chambers et al., 2007; Morrison et al., 1996). In addition, there is a distinct shift in lineage output toward myeloid progeny with aging at the expense of lymphoid output (Sudo, Ema, Morita, & Nakauchi, 2000), which is readily observed following transplantation of aged HSCs into irradiated recipients, compared to young controls (Guerrettaz, Johnson, & Cambier, 2008; Rossi et al., 2005). Gene expression profiling of aged HSCs, which showed both an increase in expression of genes involved in myeloid specification and a decrease in genes important for lymphoid specification, helps explain the mechanism behind this shift (Rossi et al., 2005). Epigenetic processes, including DNA methylation and chromatin remodeling, have been shown to help regulate these age-dependent functional changes.

In mice, gene expression profiling of aged HSCs showed roughly 1600 genes repressed with age, including a number involved in chromatin remodeling (*Smarca4*, *Smarcb1*), histone deacetylases (*Hdac1*, *5*, and *6*) and *de novo* DNA methyltransferase *Dnmt3b*. This overall dysregulation led to inappropriate transcription from a locus (*IgK*), which required epigenetic modification for expression in aged, but not young, HSCs alone (Chambers et al., 2007).

In humans, purified $CD34^+$ hematopoietic progenitors from both umbilical cord blood and peripheral blood of adult donors showed age-dependent methylation changes, including both hypermethylation and hypomethylation of gene promoters. Hypomethylation of genes expressed during myeloid differentiation were observed, while *de novo* methylation at a number of promoters of Polycomb-Group target genes (PCGTs) occurred (Bocker et al., 2011). These two epigenetic changes are hypothesized to act in concert with each other to both reduce differentiation potential and multipotency in aged HSCs. Additionally, hypermethylation at PCGTs was found to occur in a large cohort of post-menopausal women, and not only in the blood, but across multiple cell types, including ovarian cancer and mesenchymal stem cells. Since a number of these aberrantly methylated genes are already known to be misregulated in cancers, it is hypothesized that the inappropriate silencing of these genes in stem cell populations with increasing age may contribute to carcinogenesis (Teschendorff et al., 2010). In addition, another study found aging-associated differentially methylated regions (aDMRs) in both $CD4^+$ T cells and $CD14^+$ monocytes, suggesting these changes occurred in more immature hematopoietic cell types/stem cells. This hypermethylation was found to be primarily at bivalent promoters described previously, which include important developmental genes known to be hypermethylated in cancer (Rakyan et al., 2010). The interplay between DNA methylation changes and chromatin modifications in the aging hematopoietic system, particularly how it relates to the potential for carcinogenesis, will surely be the subject of further research in the future.

4. ROLE OF EPIGENETIC REGULATORS IN HEMATOPOIETIC MALIGNANCIES

Despite the intense regulatory activity presiding over hematopoietic homeostasis, aberrant clones may arise from HSPCs, giving rise to a pool of malignant cells characterized by unrestrained proliferation and/or abnormal differentiation patterns. Previously viewed as exquisitely genetic diseases, hematopoietic malignancies have now emerged as a deviant developmental process. As such, leukemogenesis can be ascribed to the abnormal activity of the same regulatory mechanisms modulating hematopoiesis under physiologic conditions, including epigenetic regulation. Misregulation in epigenetic patterns and mutations in epigenetic modifiers are frequently observed in different types of human leukemias. Despite the great advantage of studying epigenetic misregulation in a naturally occurring pathological condition, human studies rarely allow a precise dissection of the underlying mechanisms. Over the past decade, *in vivo* KO animal models greatly improved our understanding of the role of epigenetic regulators in hematopoiesis (Tables 2.1–2.4). In this section, we will summarize some of the most recent studies in both human leukemic cells and KO animal models, indicating how aberrant epigenetic patterns and altered expression of epigenetic modifiers may contribute to leukemogenesis.

By sequencing the genome of leukemic cells, several groups recently contributed to link the misregulation of epigenetic marks to the development of hematopoietic malignancies, as recently reviewed by Shih, Abdel-Wahab, Patel, and Levine (2012) and Chung, Schatoff, and Abdel-Wahab (2012). For instance, hypermethylation of the homeobox gene *HOXA4* was reported in both acute myeloid leukemia (AML) (Zangenberg et al., 2009) and chronic myeloid leukemia (CML) patients (Strathdee et al., 2007). In addition, genes controlling cell-cycle dynamics present aberrant epigenetic marks in several types of leukemia: CDKN2B (p15) has been found to be repressed by promoter hypermethylation in patients with AML (Aggerholm, Holm, Guldberg, Olesen, & Hokland, 2006; Deneberg et al., 2010; Tien et al., 2001), pediatric and adult acute lymphoid leukemia (ALL) (Takeuchi et al., 2011; Wong, Ng, Huang, & Lee, 2000), and chronic lymphoid leukemia (CLL), and similar alterations have been reported for CDKN2A (p16) (Batova et al., 1997; Herman et al., 1997; Wong et al., 2000) and CDKN1A (p21). More interestingly, these alterations have been associated with disease progression and poor prognosis; ALL patients with hypermethylated *CDKN1A* (p21), for example, display decreased disease-free survival and this marker is now viewed as a useful tool for tumor classification and patient stratification (Roman-Gomez et al., 2002).

As recently reviewed by Florean et al., the list of TS genes presenting epigenetic misregulation in hematological malignancies is continuing to grow, supporting the hypothesis that TS transcriptional silencing induced by aberrant epigenetic marks may play a role in leukemogenesis. The specific changes leading to altered epigenetic patterns, however, remain largely unknown. Several studies suggest a key role of *de novo* methylation in TS silencing; in AML patients, for instance, *CDKN2B* hypermethylation is frequently associated with higher expression levels of the *de novo* DNA methyltransferase DNMT3B. In these patients, overexpression of DNMT3B (or its catalytically inactive isoform DNMT3B(NC)) correlated with shorter overall survival in AML patients, and inversely correlated with DNMT3A expression level and event-free survival. Based on these findings, DNMT3B could be a statistically significant, independent poor prognostic indicator that may be used to stratify AML patients (Hayette et al., 2012).

More interestingly, recurrent mutations in human *DNMT3A* have been found in a number of patients with AML (Ley et al., 2010), myelodysplastic syndrome (MDS) (Walter et al., 2011), and myeloproliferative neoplasms (MPNs) (Abdel-Wahab, Pardanani, Rampal, et al., 2011). While the most commonly found mutation is a missense mutation at Arginine 882 (R882) in the methyltransferase catalytic domain of DNMT3A, the biology linking *DNMT3A* mutations and leukemogenesis is not completely understood. Some *in vitro* evidence suggests the R882 mutation leads to a reduction of the catalytic activity of DNMT3A (Yamashita et al., 2010), and the prevalence of heterozygous mutations suggests that the mutant protein may act in a dominant-negative fashion. Exon sequencing has led to identification of other genes, such as *NPM1*, *FLT3*, and *IDH1*, that commonly co-occur in AML patients with *DNMT3A* mutations, giving possible mechanistic insight into the function of mutated DNMT3A in leukemogenesis (Patel et al., 2012). A possible mechanism by which these frequently reported *DNMT3A* mutations contribute to the progression of oncogenesis is the elimination of tetramerization and disruption of processivity of the Dnmt activity (Holz-Schietinger, Matje, & Reich, 2012). Nonetheless, recent studies suggest that DNMT3A mutations are not sufficient *per se* to drive leukemogenesis; mice with hematopoietic-specific deletion of *Dnmt3a* do not develop leukemia (Challen et al., 2011), possibly because a second mutation is required for transformation.

More recently, Roller et al. summarized the literature regarding *DNMT3A* mutations in hematological malignancies, and also produced data from 194 CN-AML (cytogenetically normal acute myeloid leukemia) patients, 115 MDS patients, 103 CMML (chronic myelomonocytic leukemia) patients, and 99 T-ALL (T-cell acute lympoblastic leukemia) patients. Importantly, they noted that 21.5% of the mutations they identified were in exons 8–14 of *DNMT3A*, while other studies have primarily focused on exons 15–23. These findings not only emphasize the need to expand the scope of sequencing to identify all relevant mutations but also suggest that mutations in *DNMT3A* may be underreported and their role in the development of hematological malignancies underestimated (Roller et al., 2013).

Despite the accumulating evidence, the mechanisms underlying the link between *DNMT3A* mutations and hematopoietic malignancies remain controversial. Conflicting results as to whether mutations in DNMT3A, and other epigenetic modifiers, could lead to genetic instability have been reported. In one study of 34 adult AML patients, mutations in epigenetic modifiers did not change in any sample from diagnose to relapse, suggesting a clonal role for these genes. At relapse, however, patients with mutations in epigenetic modifiers (especially DNMT3A) presented a high frequency of acquired *FLT3*-ITD mutations. This finding suggests that epigenetic mutations may cause genetic instability, fostering the onset of *FLT3*-ITD mutations that, in turn, can lead to resistance to therapy and relapse (Wakita et al., 2012). In contrast, in another study of paired samples (37 cases of secondary AML that had progressed from MDS or MPN), *FLT3*-ITD and *DNMT3A* mutations occurred independently of one another (Zebisch, Hoefler, Quehenberger, Wölfler, & Sill, 2013). These conflicting results could be reflective of the difference between *de novo* and secondary leukemogenesis, but more studies will be required to confirm any connection between mutations in epigenetic modifies and genetic instability, and their potential role in leukemogenesis.

The discovery of the TET family of DNA-modifying enzymes has now opened a new front of investigation on epigenetic mechanisms associated with leukemia. Mutations in *TET2* have been identified in about 50% of human CMML (Kosmider et al., 2009). More interestingly, *TET2* mutations were identified in the HSPC compartment (identified as CD34+ cells) and some patients presented more than one mutation (Brecqueville et al., 2012; Delhommeau et al., 2009; Smith et al., 2010). In addition, mutated forms of TET2 associate with reduced survival in CMML patients, suggesting that these mutations might represent useful prognostic markers (Kosmider et al., 2009). The role of TET2 in myeloid leukemias has been confirmed by KO studies, although with important differences among the different models. In two studies, conditional *Tet2* loss leads to development of CMML (Moran-Crusio et al., 2011; Quivoron et al., 2011), whereas a different *Tet2* KO model develops a hematological disorder resembling MDS (Li et al., 2011), suggesting that *Tet2* mutation predispose mice to the development of leukemia. Remarkably, a significant latency seems to be required before overt leukemia develops, a feature that may be shared by several mutations targeting epigenetic modifiers (e.g., *Dnmt3a*). Together, these findings suggest that aberrant DNA methylation and hydroxymethylation patterns may foster the emergence of hematopoietic clones that, in the long term, may become the target of secondary events and give rise to leukemia. The specific mechanisms by which altered DNA methylation (and hydroxymethylation) promotes tumorigenesis remain largely unknown, but changes in gene expression and expression of noncoding RNAs are both likely to contribute to transformation.

In addition to aberrant methylation and hydroxymethylation patterns, mutations in histone-modifying enzymes have also been associated with the development of hematopoietic neoplasms. Among the most studied chromatin modifiers are PcG proteins: while PRC1 family member mutations have not been described in hematological malignancies, mutations in PRC2 complex family members have frequently been reported in patients with various types of leukemia. The most interesting findings concern EZH2, the enzymatic component of PRC2, acting as a H3K27 methyltransferase. Activating mutations in EZH2 have been identified in diffuse large B-cell lymphomas of germinal-center origin (Morin et al., 2010), while loss-of-function mutations have been identified in patients with chronic myeloid neoplasms, including MDS and CMML (Jankowska et al., 2011).

Interesting insights have been recently provided by a comprehensive study of mutations in 19 genes in pediatric AML patients, highlighting some key differences between adult and pediatric AML. Based on their findings, Liang et al. suggested that the development of pediatric AML requires fewer mutations than adult AML. Indeed, mutations affecting epigenetic modifiers (called Type IV mutations) were extremely rare in pediatric AML; within a large cohort of children with *de novo* AML, less than 2% of the patients were found to carry identified mutations in *TET2*, *DNMT3A*, *ASXL1*, *IDH1*, or *IDH2*, and no overlap was observed between the five genes (Liang et al., 2013). The apparently lower contribution of epigenetic mutations to pediatric malignancies seems to confirm that mutations in epigenetic modifiers are early mutations that require the accumulation of additional mutations before transformation occurs. Overall, these findings support the emerging idea that even within a certain group of leukemias (e.g., AML), there are actually a wide variety of distinct diseases, responsive to different therapeutic approaches. The various aberrant epigenetic patterns reported in hematological patients reflect the underlying complexity of hematological neoplasms. While the list of misregulated epigenetic marks in leukemia patients continues to grow, an even greater effort will need to be devoted to identifying how these alterations contribute to the global epigenetic landscape at leukemia onset, during progression, and while assessing response to therapeutic protocols. In addition, murine models should continue to provide insights into the development and progression of hematological diseases. For instance, *ASXL1* mutations have been identified in hematopoietic malignancies including MDS, MPN, and AML (Abdel-Wahab, Pardanani, Patel, et al., 2011; Bejar et al., 2011; Gelsi-Boyer et al., 2009), and are associated with adverse survival in MDS and AML (Bejar et al., 2011; Metzeler et al., 2011; Pratcorona et al., 2012; Thol et al., 2011), but many questions remain about the function of this mutation in leukemogenesis. The specific mechanism leading to myeloid transformation has not been elucidated, but ASXL1 appears to act as a TS. In addition, knockdown of *Asxl1* accelerated *NRas*G12D-mediated myeloproliferative disease in a transplanted mouse model (Abdel-Wahab et al., 2012), while mice expressing a null allele of *Asxl1* has defects in lymphopoiesis, but does not develop overt leukemia. A hematopoietic-specific KO and generation of mice expressing mutant forms of Asxl1 similar to the mutations seen in patients will shed light on whether *ASXL1* mutations in patients act as dominant negatives, or whether additional mutations are required for transformation (Fisher et al., 2010).

5. CONCLUSIONS

Over the past decade, the understanding of the molecular mechanisms presiding over epigenetic regulation has greatly improved. Not only have epigenetic marks emerged as pivotal regulators of the different stages of hematopoietic development but they also appear to be key players in leukemogenesis. In the future, the ability to manipulate the epigenetic pattern of genes involved in hematological diseases is expected to provide a huge array of new therapeutic approaches for leukemia. Along this line, HDAC inhibitors have already been approved for cutaneous T-cell lymphoma (Mann, Johnson, Cohen, Justice, & Pazdur, 2007; Piekarz, 2001), hypomethylating agents like 5-azacytidine and decitabine being used in the treatment of MDS and AML (Kantarjian et al., 2012; Wijermans et al., 2000), and more clinical trials are expected to be implemented in the coming years.

Despite the increasing understanding of how the single epigenetic marks are regulated, several key aspects are still waiting to be elucidated. There is little understanding of how different epigenetic marks cooperate in defined genetic regions and their relative importance in affecting gene expression is still under debate. Adding complexity to the picture, the family of factors contributing to sculpt the epigenetic landscape is continuing to grow. Recently, several lncRNAs have been shown to participate in epigenetic regulation by binding chromatin-modifying proteins and recruiting their activity to specific genomic sites. The significant abundance and variety of lncRNAs in the cells, combined with their high specificity, makes lncRNAs an ideal target for therapies focused on epigenetic regulation in leukemic cells.

ACKNOWLEDGMENTS

The authors would like to thank members of the Goodell lab for helpful discussions. The authors are supported by grants DK092883, 5T32HL092332, 1RC2AG036562-01, the Samuel Waxman Foundation, and the Cancer Prevention and Research Institute of Texas (CPRIT RP110028). L. R. was supported by the Italian Leukemia and Lymphoma Association, section of Bologna (BolognaAIL).

REFERENCES

Abdel-Wahab, O., Adli, M., LaFave, L. M., Gao, J., Hricik, T., Shih, A. H., et al. (2012). ASXL1 mutations promote myeloid transformation through loss of PRC2-mediated gene repression. *Cancer Cell*, *22*, 180–193.

Abdel-Wahab, O., Pardanani, A., Patel, J., Wadleigh, M., Lasho, T., Heguy, A., et al. (2011). Concomitant analysis of EZH2 and ASXL1 mutations in myelofibrosis, chronic myelomonocytic leukemia and blast-phase myeloproliferative neoplasms. *Leukemia*, *25*, 1200–1202.

Abdel-Wahab, O., Pardanani, A., Rampal, R., Lasho, T. L., Levine, R. L., & Tefferi, A. (2011). DNMT3A mutational analysis in primary myelofibrosis, chronic myelomonocytic leukemia and advanced phases of myeloproliferative neoplasms. *Leukemia*, *25*, 1219–1220.

Aggerholm, A., Holm, M. S., Guldberg, P., Olesen, L. H., & Hokland, P. (2006). Promoter hypermethylation of p15INK4B, HIC1, CDH1, and ER is frequent in myelodysplastic syndrome and predicts poor prognosis in early-stage patients. *European Journal of Haematology*, *76*, 23–32.

Antequera, F., & Bird, A. (1993). Number of CpG islands and genes in human and mouse. *Proceedings of the National Academy of Sciences of the United States of America*, *90*, 11995–11999.

Bannister, A. J., & Kouzarides, T. (2011). Regulation of chromatin by histone modifications. *Cell Research*, *21*, 381–395.

Batova, A., Diccianni, M. B., Yu, J. C., Nobori, T., Link, M. P., Pullen, J., et al. (1997). Frequent and selective methylation of p15 and deletion of both p15 and p16 in T-cell acute lymphoblastic leukemia. *Cancer Research*, *57*, 832–836.

Bejar, R., Stevenson, K., Abdel-Wahab, O., Galili, N., Nilsson, B., Garcia-Manero, G., et al. (2011). Clinical effect of point mutations in myelodysplastic syndromes. *New England Journal of Medicine*, *364*, 2496–2506.

Berger, S. L. (2007). The complex language of chromatin regulation during transcription. *Nature*, *447*, 407–412.

Bertani, S., Sauer, S., Bolotin, E., & Sauer, F. (2011). The noncoding RNA mistral activates Hoxa6 and Hoxa7 expression and stem cell differentiation by recruiting MLL1 to chromatin. *Molecular Cell*, *43*, 1040–1046.

Bissels, U., Bosio, A., & Wagner, W. (2012). MicroRNAs are shaping the hematopoietic landscape. *Haematologica*, *97*, 160–167.

Bock, C., Beerman, I., Lien, W.-H., Smith, Z. D., Gu, H., Boyle, P., et al. (2012). DNA methylation dynamics during in vivo differentiation of blood and skin stem cells. *Molecular Cell*, *47*, 633–647.

Bocker, M. T., Hellwig, I., Breiling, A., Eckstein, V., Ho, A. D., & Lyko, F. (2011). Genome-wide promoter DNA methylation dynamics of human hematopoietic progenitor cells during differentiation and aging. *Blood*, *117*, 182–189.

Brecqueville, M., Rey, J., Bertucci, F., Coppin, E., Finetti, P., Carbuccia, N., et al. (2012). Mutation analysis of ASXL1, CBL, DNMT3A, IDH1, IDH2, JAK2, MPL, NF1, SF3B1, SUZ12, and TET2 in myeloproliferative neoplasms. *Genes, Chromosomes and Cancer*, *51*, 743–755.

Bröske, A.-M., Vockentanz, L., Kharazi, S., Huska, M. R., Mancini, E., Scheller, M., et al. (2009). DNA methylation protects hematopoietic stem cell multipotency from myeloerythroid restriction. *Nature Genetics*, *41*, 1207–1215.

Calvanese, V., Fernandez, A. F., Urdinguio, R. G., Suarez-Alvarez, B., Mangas, C., Perez-Garcia, V., et al. (2011). A promoter DNA demethylation landscape of human hematopoietic differentiation. *Nucleic Acids Research*, *40*, 116–131.

Cao, R., Wang, L., Wang, H., Xia, L., Erdjument-Bromage, H., Tempst, P., et al. (2002). Role of histone H3 lysine 27 methylation in Polycomb-group silencing. *Science*, *298*, 1039–1043.

Challen, G. A., Sun, D., Jeong, M., Luo, M., Jelinek, J., Berg, J. S., et al. (2011). Dnmt3a is essential for hematopoietic stem cell differentiation. *Nature Genetics*, *44*, 23–31.

Chambers, S. M., Shaw, C. A., Gatza, C., Fisk, C. J., Donehower, L. A., & Goodell, M. A. (2007). Aging hematopoietic stem cells decline in function and exhibit epigenetic dysregulation. *PLoS Biology*, *5*, 1750–1762.

Chen, D., & Zhang, G. (2001). Enforced expression of the GATA-3 transcription factor affects cell fate decisions in hematopoiesis. *Experimental Hematology*, *29*, 971–980.

Chung, Y. R., Schatoff, E., & Abdel-Wahab, O. (2012). Epigenetic alterations in hematopoietic malignancies. *International Journal of Hematology*, *96*, 413–427.

Delhommeau, F., Dupont, S., Della Valle, V., James, C., Trannoy, S., Massé, A., et al. (2009). Mutation in TET2 in myeloid cancers. *New England Journal of Medicine*, *360*, 2289–2301.

Deneberg, S., Grovdal, M., Karimi, M., Jansson, M., Nahi, H., Corbacioglu, A., et al. (2010). Gene-specific and global methylation patterns predict outcome in patients with acute myeloid leukemia. *Leukemia*, *24*, 932–941.

Dey, A., Seshasayee, D., Noubade, R., French, D. M., Liu, J., Chaurushiya, M. S., et al. (2012). Loss of the tumor suppressor BAP1 causes myeloid transformation. *Science*, *337*, 1541–1546.

Ding, X., Lin, Q., Ensenat-Waser, R., Rose-John, S., & Zenke, M. (2012). Polycomb group protein Bmi1 promotes hematopoietic cell development from embryonic stem cells. *Stem Cells and Development*, *21*, 121–132.

Endoh, M., Endo, T. A., Endoh, T., Isono, K., Sharif, J., Ohara, O., et al. (2012). Histone H2A mono-ubiquitination is a crucial step to mediate PRC1-dependent repression of developmental genes to maintain ES cell identity. *PLoS Genetics*, *8*, 1–10.

Esteller, M. (2007). Epigenetic gene silencing in cancer: The DNA hypermethylome. *Human Molecular Genetics*, *16*, 50–59, Spec. No. 1.

Feng, Y., Yang, Y., Ortega, M. M., Copeland, J. N., Zhang, M., Jacob, J. B., et al. (2010). Early mammalian erythropoiesis requires the Dot1L methyltransferase. *Blood*, *116*, 4483–4491.

Fisher, C. L., Pineault, N., Brookes, C., Helgason, C. D., Ohta, H., Bodner, C., et al. (2010). Loss-of-function additional sex combs like 1 mutations disrupt hematopoiesis but do not cause severe myelodysplasia or leukemia. *Blood, 115*, 38–46.

Gan, T., Jude, C. D., Zaffuto, K., & Ernst, P. (2010). Developmentally induced Mll1 loss reveals defects in postnatal haematopoiesis. *Leukemia, 24*, 1732–1741.

Gelsi-Boyer, V., Trouplin, V., Adélaïde, J., Bonansea, J., Cervera, N., Carbuccia, N., et al. (2009). Mutations of polycomb-associated gene ASXL1in myelodysplastic syndromes and chronic myelomonocytic leukaemia. *British Journal of Haematology, 145*, 788–800.

Guerrettaz, L. M., Johnson, S. A., & Cambier, J. C. (2008). Acquired hematopoietic stem cell defects determine B-cell repertoire changes associated with aging. *Proceedings of the National Academy of Sciences of the United States of America, 105*, 11898–11902.

Hayette, S., Thomas, X., Jallades, L., Chabane, K., Charlot, C., Tigaud, I., et al. (2012). High DNA methyltransferase DNMT3B levels: A poor prognostic marker in acute myeloid leukemia. *PLoS One, 7*, 1–10.

Herman, J. G., Civin, C. I., Issa, J. P., Collector, M. I., Sharkis, S. J., & Baylin, S. B. (1997). Distinct patterns of inactivation of p15INK4B and p16INK4A characterize the major types of hematological malignancies. *Cancer Research, 57*, 837–841.

Hidalgo, I., Herrera-Merchan, A., Ligos, J. M., Carramolino, L., Nuñez, J., Martinez, F., et al. (2012). Ezh1 is required for hematopoietic stem cell maintenance and prevents senescence-like cell cycle arrest. *Stem Cell, 11*, 649–662.

Holz-Schietinger, C., Matje, D. M., & Reich, N. O. (2012). Mutations in DNA methyltransferase (DNMT3A) observed in acute myeloid leukemia patients disrupt processive methylation. *Journal of Biological Chemistry, 287*, 30941–30951.

Holz-Schietinger, C., & Reich, N. O. (2012). RNA modulation of the human DNA methyltransferase 3A. *Nucleic Acids Research, 40*, 8550–8557.

Hu, M., Krause, D., Greaves, M., Sharkis, S., Dexter, M., Heyworth, C., et al. (1997). Multilineage gene expression precedes commitment in the hemopoietic system. *Genes and Development, 11*, 774–785.

Ito, S., D'Alessio, A. C., Taranova, O. V., Hong, K., Sowers, L. C., & Zhang, Y. (2010). Role of Tet proteins in 5mC to 5hmC conversion, ES-cell self-renewal and inner cell mass specification. *Nature, 466*, 1129–1133.

Iwama, A., Oguro, H., Negishi, M., Kato, Y., Morita, Y., Tsukui, H., et al. (2004). Enhanced self-renewal of hematopoietic stem cells mediated by the polycomb gene product Bmi-1. *Immunity, 21*, 843–851.

Jacobs, J. J., Kieboom, K., Marino, S., DePinho, R. A., & van Lohuizen, M. (1999). The oncogene and Polycomb-group gene bmi-1 regulates cell proliferation and senescence through the ink4a locus. *Nature, 397*, 164–168.

Jankowska, A. M., Makishima, H., Tiu, R. V., Szpurka, H., Huang, Y., Traina, F., et al. (2011). Mutational spectrum analysis of chronic myelomonocytic leukemia includes genes associated with epigenetic regulation: UTX, EZH2, and DNMT3A. *Blood, 118*, 3932–3941.

Ji, H., Ehrlich, L. I. R., Seita, J., Murakami, P., Doi, A., Lindau, P., et al. (2011). Comprehensive methylome map of lineage commitment from haematopoietic progenitors. *Nature, 467*, 338–342.

Jo, S. Y., Granowicz, E. M., Maillard, I., Thomas, D., & Hess, J. L. (2011). Requirement for Dot1l in murine postnatal hematopoiesis and leukemogenesis by MLL translocation. *Blood, 117*, 4759–4768.

Kafri, T., Ariel, M., Brandeis, M., Shemer, R., Urven, L., McCarrey, J., et al. (1992). Developmental pattern of gene-specific DNA methylation in the mouse embryo and germ line. *Genes and Development, 6*, 705–714.

Kajuime, T., Ninomiya, Y., Ishihara, H., Kanno, R., & Kanno, M. (2004). Polycomb group gene mel-18 modulates the self-renewal activity and cell cycle status of hematopoietic stem cells. *Experimental Hematology, 32*, 571–578.

Kamminga, L. M., Bystrykh, L. V., de Boer, A., Houwer, S., Douma, J., Weersing, E., et al. (2006). The Polycomb group gene Ezh2 prevents hematopoietic stem cell exhaustion. *Blood*, *107*, 2170–2179.

Kantarjian, H. M., Thomas, X. G., Dmoszynska, A., Wierzbowska, A., Mazur, G., Mayer, J., et al. (2012). Multicenter, randomized, open-label, phase III trial of decitabine versus patient choice, with physician advice, of either supportive care or low-dose cytarabine for the treatment of older patients with newly diagnosed acute myeloid leukemia. *Journal of Clinical Oncology*, *30*, 2670–2677.

Katsumoto, T., Aikawa, Y., Iwama, A., Ueda, S., Ichikawa, H., & Ochiya, T. (2006). MOZ is essential for maintenance of hematopoietic stem cells. *Genes and Development*, *20*, 1321–1330.

Kim, M., Moon, H. B., & Spangrude, G. J. (2003). Major age-related changes of mouse hematopoietic stem/progenitor cells. *Annals of the New York Academy of Sciences*, *996*, 195–208.

Klauke, K., Radulović, V., Broekhuis, M., Weersing, E., Zwart, E., Olthof, S., et al. (2013). Polycomb Cbx family members mediate the balance between haematopoietic stem cell self-renewal and differentiation. *Nature Cell Biology*, *15*, 353–362.

Ko, M., Bandukwala, H. S., An, J., Lamperti, E. D., Thompson, E. C., Hastie, R., et al. (2011). Ten-Eleven-Translocation 2 (TET2) negatively regulates homeostasis and differentiation of hematopoietic stem cells in mice. *Proceedings of the National Academy of Sciences of the United States of America*, *108*, 14566–14571.

Koh, K. P., Yabuuchi, A., Rao, S., Huang, Y., Cunniff, K., Nardone, J., et al. (2011). Tet1 and Tet2 regulate 5-hydroxymethylcytosine production and cell lineage specification in mouse embryonic stem cells. *Stem Cell*, *8*, 200–213.

Kosmider, O., Gelsi-Boyer, V., Ciudad, M., Racoeur, C., Jooste, V., Vey, N., et al. (2009). TET2 gene mutation is a frequent and adverse event in chronic myelomonocytic leukemia. *Haematologica*, *94*, 1676–1681.

Kriaucionis, S., & Heintz, N. (2009). The nuclear DNA base 5-hydroxymethylcytosine is present in Purkinje neurons and the brain. *Science*, *324*, 929–930.

Lawrence, H. J., Helgason, C. D., Sauvageau, G., Fong, S., Izon, D. J., Humphries, R. K., et al. (1997). Mice bearing a targeted interruption of the homeobox gene HOXA9 have defects in myeloid, erythroid, and lymphoid hematopoiesis. *Blood*, *89*, 1922–1930.

Ley, T. J., Ding, L., Walter, M. J., McLellan, M. D., Lamprecht, T., Larson, D. E., et al. (2010). DNMT3A mutations in acute myeloid leukemia. *New England Journal of Medicine*, *363*, 2424–2433.

Li, E., Bestor, T. H., & Jaenisch, R. (1992). Targeted mutation of the DNA methyltransferase gene results in embryonic lethality. *Cell*, *69*, 915–926.

Li, Z., Cai, X., Cai, C.-L., Wang, J., Zhang, W., Petersen, B. E., et al. (2011). Deletion of Tet2 in mice leads to dysregulated hematopoietic stem cells and subsequent development of myeloid malignancies. *Blood*, *118*, 4509–4518.

Li, Z., Cao, R., Wang, M., Myers, M. P., Zhang, Y., & Xu, R. M. (2006). Structure of a Bmi-1-Ring1B polycomb group ubiquitin ligase complex. *The Journal of Biological Chemistry*, *281*, 20643–20649.

Li, J., Zhao, Z., Carter, C., Ehrlich, L. I. R., Bedford, M. T., & Richie, E. R. (2013). Coactivator-associated arginine methyltransferase 1 regulates fetal hematopoiesis and thymocyte development. *The Journal of Immunology*, *190*, 597–604.

Liang, D. C., Liu, H. C., Yang, C. P., Jaing, T. H., Hung, I. J., Yeh, T. C., et al. (2013). Cooperating gene mutations in childhood acute myeloid leukemia with special reference on mutations of ASXL1, TET2, IDH1, IDH2, and DNMT3A. *Blood*, *121*, 2988–2995.

Lux, C. T., Yoshimoto, M., McGrath, K., Conway, S. J., Palis, J., & Yoder, M. C. (2008). All primitive and definitive hematopoietic progenitor cells emerging before E10 in the mouse embryo are products of the yolk sac. *Blood*, *111*, 3435–3438.

Maes, J., Maleszewska, M., Guillemin, C., Pflumio, F., Six, E., Andre-Schmutz, I., et al. (2008). Lymphoid-affiliated genes are associated with active histone modifications in human hematopoietic stem cells. *Blood*, *112*, 2722–2729.

Majewski, I. J., Ritchie, M. E., Phipson, B., Corbin, J., Pakusch, M., Ebert, A., et al. (2010). Opposing roles of polycomb repressive complexes in hematopoietic stem and progenitor cells. *Blood*, *116*, 731–739.

Mann, B. S., Johnson, J. R., Cohen, M. H., Justice, R., & Pazdur, R. (2007). FDA approval summary: Vorinostat for treatment of advanced primary cutaneous T-cell lymphoma. *The Oncologist*, *12*, 1247–1252.

Marmorstein, R., & Trievel, R. C. (2009). Histone modifying enzymes: Structures, mechanisms, and specificities. *Biochimica et Biophysica Acta*, *1789*, 58–68.

McMahon, K. A., Hiew, S. Y. L., Hadjur, S., Veiga-Fernandes, H., Menzel, U., Price, A. J., et al. (2007). Mll has a critical role in fetal and adult hematopoietic stem cell self-renewal. *Cell Stem Cell*, *1*, 338–345.

Meissner, A., Mikkelsen, T. S., Gu, H., Wernig, M., Hanna, J., Sivachenko, A., et al. (2008). Genome-scale DNA methylation maps of pluripotent and differentiated cells. *Nature*, *454*, 766–770.

Metzeler, K. H., Becker, H., Maharry, K., Radmacher, M. D., Kohlschmidt, J., Mrozek, K., et al. (2011). ASXL1 mutations identify a high-risk subgroup of older patients with primary cytogenetically normal AML within the ELN favorable genetic category. *Blood*, *118*, 6920–6929.

Monk, M., Boubelik, M., & Lehnert, S. (1987). Temporal and regional changes in DNA methylation in the embryonic, extraembryonic and germ cell lineages during mouse embryo development. *Development*, *99*, 371–382.

Moore, M. A., & Metcalf, D. (1970). Ontogeny of the haemopoietic system: Yolk sac origin of in vivo and in vitro colony forming cells in the developing mouse embryo. *British Journal of Haematology*, *18*, 279–296.

Moran-Crusio, K., Reavie, L., Shih, A., Abdel-Wahab, O., Ndiaye-Lobry, D., Lobry, C., et al. (2011). Tet2 loss leads to increased hematopoietic stem cell self-renewal and myeloid transformation. *Cancer Cell*, *20*, 11–24.

Morin, R. D., Johnson, N. A., Severson, T. M., Mungall, A. J., An, J., Goya, R., et al. (2010). Somatic mutations altering EZH2 (Tyr641) in follicular and diffuse large B-cell lymphomas of germinal-center origin. *Nature Genetics*, *42*, 181–185.

Morrison, S. J., Wandycz, A. M., Akashi, K., Globerson, A., & Weissman, I. L. (1996). The aging of hematopoietic stem cells. *Nature Medicine*, *2*, 1011–1016.

O'Connell, R. M., & Baltimore, D. (2012). MicroRNAs and hematopoietic cell development. *Current Topics in Developmental Biology*, *99*, 145–174.

Oguro, H., Yuan, J., Ichikawa, H., Ikawa, T., Yamazaki, S., Kawamoto, H., et al. (2010). Poised lineage specification in multipotential hematopoietic stem and progenitor cells by the polycomb protein Bmi1. *Stem Cell*, *6*, 279–286.

Ohta, H., Sawada, A., Kim, J. Y., Tokimasa, S., Nishiguchi, S., Humphries, R. K., et al. (2002). Polycomb group gene rae28 is required for sustaining activity of hematopoietic stem cells. *The Journal of Experimental Medicine*, *195*, 759–770.

Okano, M., Bell, D. W., Haber, D. A., & Li, E. (1999). DNA methyltransferases Dnmt3a and Dnmt3b are essential for de novo methylation and mammalian development. *Cell*, *99*, 247–257.

Okano, M., Xie, S., & Li, E. (1998). Cloning and characterization of a family of novel mammalian DNA (cytosine-5) methyltransferases. *Nature Genetics*, *19*, 219–220.

Okuda, T., van Deursen, J., Hiebert, S. W., Grosveld, G., & Downing, J. R. (1996). AML1, the target of multiple chromosomal translocations in human leukemia, is essential for normal fetal liver hematopoiesis. *Cell*, *84*, 321–330.

Orford, K., Kharchenko, P., Lai, W., Dao, M. C., Worhunsky, D. J., Ferro, A., et al. (2008). Differential H3K4 methylation identifies developmentally poised hematopoietic genes. *Developmental Cell, 14*, 798–809.

Park, I.-K., Qian, D., Kiel, M., Becker, M. W., Pihalja, M., Weissman, I. L., et al. (2003). Bmi-1 is required for maintenance of adult self-renewing haematopoietic stem cells. *Nature, 423*, 302–305.

Patel, J. P., Gönen, M., Figueroa, Maria E., Fernandez, H., Sun, Z., Racevskis, J., et al. (2012). Prognostic relevance of integrated genetic profiling in acute myeloid leukemia. *New England Journal of Medicine, 366*, 1079–1089.

Piekarz, R. L. (2001). Inhibitor of histone deacetylation, depsipeptide (FR901228), in the treatment of peripheral and cutaneous T-cell lymphoma: A case report. *Blood, 98*, 2865–2868.

Pillay, L. M., Forrester, A. M., Erickson, T., Berman, J. N., & Waskiewicz, A. J. (2010). The Hox cofactors Meis1 and Pbx act upstream of gata1 to regulate primitive hematopoiesis. *Developmental Biology, 340*, 306–317.

Pratcorona, M., Abbas, S., Sanders, M. A., Koenders, J. E., Kavelaars, F. G., Erpelinck-Verschueren, C. A. J., et al. (2012). Acquired mutations in ASXL1 in acute myeloid leukemia: Prevalence and prognostic value. *Haematologica, 97*, 388–392.

Quivoron, C., Couronne, L., Della Valle, V., Lopez, C. K., Plo, I., Wagner-Ballon, O., et al. (2011). TET2 inactivation results in pleiotropic hematopoietic abnormalities in mouse and is a recurrent event during human lymphomagenesis. *Cancer Cell, 20*, 25–38.

Rakyan, V. K., Down, T. A., Maslau, S., Andrew, T., Yang, T. P., Beyan, H., et al. (2010). Human aging-associated DNA hypermethylation occurs preferentially at bivalent chromatin domains. *Genome Research, 20*, 434–439.

Roller, A., Grossmann, V., Bacher, U., Poetzinger, F., Weissmann, S., Nadarajah, N., et al. (2013). Landmark analysis of DNMT3A mutations in hematological malignancies. *Leukemia, 27*, 1573–1578.

Roman-Gomez, J., Castillejo, J., Jimenez, A., Gonzalez, M., Moreno, F., Rodriguez, M., et al. (2002). 5′ CpG island hypermethylation is associated with transcriptional silencing of the p21CIP1/WAF1/SDI1 gene and confers poor prognosis in acute lymphoblastic leukemia. *Blood, 99*, 2291–2296.

Rossi, D. J., Bryder, D., Zahn, J. M., Ahlenius, H., Sonu, R., Wagers, A. J., et al. (2005). Cell intrinsic alterations underlie hematopoietic stem cell aging. *Proceedings of the National Academy of Sciences of the United States of America, 102*, 9194–9199.

Rossi, L., Lin, K. K., Boles, N. C., Yang, L., King, K. Y., Jeong, M., et al. (2012). Less is more: Unveiling the functional core of hematopoietic stem cells through knockout mice. *Cell Stem Cell, 11*, 302–317.

Samokhvalov, I. M., Samokhvalova, N. I., & Nishikawa, S. (2007). Cell tracing shows the contribution of the yolk sac to adult haematopoiesis. *Nature, 446*, 1056–1061.

Sashida, G., & Iwama, A. (2012). Epigenetic regulation of hematopoiesis. *International Journal of Hematology, 96*, 405–412.

Schaefer, C. B., Ooi, S. K., Bestor, T. H., & Bourc'his, D. (2007). Epigenetic decisions in mammalian germ cells. *Science, 316*, 398–399.

Scheuermann, J. C., de Ayala Alonso, A. G., Oktaba, K., Ly-Hartig, N., McGinty, R. K., Fraterman, S., et al. (2010). Histone H2A deubiquitinase activity of the Polycomb repressive complex PR-DUB. *Nature, 465*, 243–247.

Shih, A. H., Abdel-Wahab, O., Patel, J. P., & Levine, Ross L. (2012). The role of mutations in epigenetic regulators in myeloid malignancies. *Nature Reviews Cancer, 12*, 599–612.

Smith, J. N. P., & Calvi, L. M. (2013). Concise review: Current concepts in bone marrow microenvironmental regulation of hematopoietic stem and progenitor cells. *Stem Cells, 31*, 1044–1050.

Smith, A. E., Mohamedali, A. M., Kulasekararaj, A., Lim, Z., Gaken, J., Lea, N. C., et al. (2010). Next-generation sequencing of the TET2 gene in 355 MDS and CMML patients reveals low-abundance mutant clones with early origins, but indicates no definite prognostic value. *Blood, 116*, 3923–3932.

Strathdee, G., Holyoake, T. L., Sim, A., Parker, A., Oscier, D. G., Melo, J. V., et al. (2007). Inactivation of HOXA genes by hypermethylation in myeloid and lymphoid malignancy is frequent and associated with poor prognosis. *Clinical Cancer Research, 13*, 5048–5055.

Suda, T., Takubo, K., & Semenza, G. L. (2011). Metabolic regulation of hematopoietic stem cells in the hypoxic niche. *Stem Cell, 9*, 298–310.

Sudo, K., Ema, H., Morita, Y., & Nakauchi, H. (2000). Age-associated characteristics of murine hematopoietic stem cells. *Journal of Experimental Medicine, 192*, 1273–1280.

Tahiliani, M., Koh, K. P., Shen, Y., Pastor, W. A., Bandukwala, H., Brudno, Y., et al. (2009). Conversion of 5-methylcytosine to 5-hydroxymethylcytosine in mammalian DNA by MLL partner TET1. *Science, 324*, 930–935.

Takeuchi, S., Matsushita, M., Zimmermann, M., Ikezoe, T., Komatsu, N., Seriu, T., et al. (2011). Clinical significance of aberrant DNA methylation in childhood acute lymphoblastic leukemia. *Leukemia Research, 35*, 1345–1349.

Teschendorff, A. E., Menon, U., Gentry-Maharaj, A., Ramus, S. J., Weisenberger, D. J., Shen, H., et al. (2010). Age-dependent DNA methylation of genes that are suppressed in stem cells is a hallmark of cancer. *Genome Research, 20*, 440–446.

Thol, F., Friesen, I., Damm, F., Yun, H., Weissinger, E. M., Krauter, J., et al. (2011). Prognostic significance of ASXL1 mutations in patients with myelodysplastic syndromes. *Journal of Clinical Oncology, 29*, 2499–2506.

Thomas, T., Corcoran, L. M., Gugasyan, R., Dixon, M. P., Brodnicki, T., Nutt, S. L., et al. (2006). Monocytic leukemia zinc finger protein is essential for the development of long-term reconstituting hematopoietic stem cells. *Genes and Development, 20*, 1175–1186.

Tien, H. F., Tang, J. H., Tsay, W., Liu, M. C., Lee, F. Y., Wang, C. H., et al. (2001). Methylation of the p15(INK4B) gene in myelodysplastic syndrome: It can be detected early at diagnosis or during disease progression and is highly associated with leukaemic transformation. *British Journal of Haematology, 112*, 148–154.

Trowbridge, J. J., Snow, J. W., Kim, J., & Orkin, S. H. (2009). DNA methyltransferase 1 is essential for and uniquely regulates hematopoietic stem and progenitor cells. *Cell Stem Cell, 5*, 442–449.

van den Boom, V., Rozenveld-Geugien, M., Bonardi, F., Malanga, D., van Gosliga, D., Heijink, A. M., et al. (2013). Nonredundant and locus-specific gene repression functions of PRC1 paralog family members in human hematopoietic stem/progenitor cells. *Blood, 121*, 2452–2461.

Viré, E., Brenner, C., Deplus, R., Blanchon, L., Fraga, M., Didelot, C., et al. (2005). The Polycomb group protein EZH2 directly controls DNA methylation. *Nature Cell Biology, 439*, 871–874.

Wakita, S., Yamaguchi, H., Omori, I., Terada, K., Ueda, T., Manabe, E., et al. (2012). Mutations of the epigenetics-modifying gene (DNMT3a, TET2, IDH1/2) at diagnosis may induce FLT3-ITD at relapse in de novo acute myeloid leukemia. *Leukemia, 27*, 1044–1052.

Walter, M. J., Ding, L., Shen, D., Shao, J., Grillot, M., Mclellan, M., et al. (2011). Recurrent DNMT3A mutations in patients with myelodysplastic syndromes. *Leukemia, 25*, 1153–1158.

Weishaupt, H., Sigvardsson, M., & Attema, J. L. (2010). Epigenetic chromatin states uniquely define the developmental plasticity of murine hematopoietic stem cells. *Blood, 115*, 247–256.

Wijermans, P., Lübbert, M., Verhoef, G., Bosly, A., Ravoet, C., Andre, M., et al. (2000). Low-dose 5-aza-2'-deoxycytidine, a DNA hypomethylating agent, for the treatment of

high-risk myelodysplastic syndrome: A multicenter phase II study in elderly patients. *Journal of Clinical Oncology*, *18*, 956–962.
Wilson, A., & Trumpp, A. (2006). Bone-marrow haematopoietic-stem-cell niches. *Nature Reviews Immunology*, *6*, 93–106.
Wong, I. H., Ng, M. H., Huang, D. P., & Lee, J. C. (2000). Aberrant p15 promoter methylation in adult and childhood acute leukemias of nearly all morphologic subtypes: Potential prognostic implications. *Blood*, *95*, 1942–1949.
Wu, H., & Zhang, Y. (2011). Mechanisms and functions of Tet protein-mediated 5-methylcytosine oxidation. *Genes and Development*, *25*, 2436–2452.
Yamashita, Y., Yuan, J., Suetake, I., Suzuki, H., Ishikawa, Y., Choi, Y. L., et al. (2010). Array-based genomic resequencing of human leukemia. *Oncogene*, *29*, 3723–3731.
Zangenberg, M., Grubach, L., Aggerholm, A., Silkjaer, T., Juhl-Christensen, C., Nyvold, C. G., et al. (2009). The combined expression of HOXA4 and MEIS1 is an independent prognostic factor in patients with AML. *European Journal of Haematology*, *83*, 439–448.
Zebisch, A., Hoefler, G., Quehenberger, F., Wölfler, A., & Sill, H. (2013). Mutant DNMT3A in acute myeloid leukemia: Guilty of inducing genetic instability? *Leukemia*, *27*, 1777–1778.
Zhang, Y., Yan, X., Sashida, G., Zhao, X., Rao, Y., Goyama, S., et al. (2012). Stress hematopoiesis reveals abnormal control of self-renewal, lineage bias, and myeloid differentiation in Mll partial tandem duplication (Mll-PTD) hematopoietic stem/progenitor cells. *Blood*, *120*, 1118–1129.

CHAPTER THREE

Intestinal Stem Cells and Their Defining Niche

David Wei-Min Tan[*], **Nick Barker**[*,†,1]
[*]Institute of Medical Biology, Singapore
[†]MRC Centre for Regenerative Medicine, The University of Edinburgh, Edinburgh, United Kingdom
[1]Corresponding author: e-mail address: nicholas.barker@imb.a-star.edu.sg

Contents

Abstract

The intestinal epithelium is a classic example of a rapidly self-renewing tissue fueled by dedicated resident stem cells. These stem cells reside at the crypt base, generating committed progeny that mature into the various functional epithelial lineages while following a rapid migratory path toward the villi. Two models of intestinal stem cell location were proposed half a century ago and data have been presented in support of both models, dividing the scientific community. Molecular markers have been identified and validated using new techniques such as *in vivo* lineage tracing and

Current Topics in Developmental Biology, Volume 107
ISSN 0070-2153
http://dx.doi.org/10.1016/B978-0-12-416022-4.00003-2

ex vivo organoid culture. The intestinal stem cell niche comprises both epithelial cells, in particular the Paneth cell, and the stromal compartment, where cell-associated ligands and soluble factors regulate stem cell behavior. This review highlights the recent advances in identifying and characterizing the intestinal stem cells and their defining niche.

1. THE MAMMALIAN SMALL INTESTINE

The mammalian small intestine functions to complete digestion, absorb nutrients from food passing through from the stomach, and also forms a protective barrier against pathogens in the intestinal lumen. These functions are performed by a simple columnar epithelium lining the small intestine that is ordered into invaginations, known as crypts of Lieberkühn, interspersed with finger-like protrusions, called villi, which vastly expand the absorptive surface area of the gut (Marshman, Booth, & Potten, 2002). The intestinal epithelium displays the highest turnover of all solid organs, undergoing complete renewal every 4–5 days (Gregorieff et al., 2005). Importantly, the intestinal epithelium is able to maintain a constant and effective barrier function during this high turnover. Rapid self-renewal is likely to be a mechanism by which the epithelium deals with the harsh intestinal environment (Barker, van Oudenaarden, & Clevers, 2012). However, misregulation of this finely tuned system of epithelial regeneration can lead to cancerous growth (Barker et al., 2009). The stereotypical organization and high rate of self-renewal of the intestinal epithelium has made it an attractive system to study stem cell biology.

In the last decade, the development of transgenic mouse models has provided powerful genetic tools with which to probe the intestinal stem cell system, leading to important insight into intrinsic and extrinsic factors that govern stem cell self-renewal and the dynamic homeostatic maintenance of the epithelium (Barker et al., 2012). These tools include transgenic mouse strains that incorporate reporter constructs such as a β-galactosidase or a fluorescent reporter gene downstream of a cell-specific promoter, allowing for direct visualization of these cells within the tissue context. Inducible forms of transgene reporters have been developed that facilitate the tracking of cells of interest and their progeny: a technique known as lineage tracing. Through these technological advances, new candidate molecular markers of stem cells have been identified, some of which have been extensively validated by different means.

The availability of validated stem cell markers has allowed researchers to accurately define the physiological location of the stem cell populations and to determine how they interact with their local microenvironment (the niche). Isolation and purification of stem cells by flow cytometric sorting using cell surface markers has also led to the development of near-physiological *in vitro* culture systems that sustain the development and long-term maintenance of epithelial structures displaying a remarkably similar architecture to that of the original organ. These major discoveries not only enhance our understanding of the basic biology that underpins stem cell identity and behavior but also have important implications for clinical therapy to treat human diseases.

In this review, we summarize recent findings on the identity and location of intestinal stem cells and provide a critical appraisal of how different components of the surrounding niche contribute to their self-renewal and maintenance. While a number of putative stem cell markers have been identified recently, controversy still exists over the specificity of some of these markers. We aim to offer a balanced and critical overview of the advances in the field during the last few years.

1.1. The intestinal epithelium

The intestinal epithelium is a classic example of a self-renewing tissue fueled by dedicated stem cells (Barker et al., 2008). Proliferation within the crypt supplies new cells to drive the self-renewal process. Each crypt comprises approximately 250 cells and an equivalent number of cells are generated each day (Barker et al., 2012). Stem cells considered to reside within the crypt base generate actively proliferating progenitors termed transit-amplifying cells, which rapidly move upward toward the crypt–villus border (Vries, Huch, & Clevers, 2010). As the cells migrate, they differentiate into cell types of the absorptive lineage (enterocytes and goblet cells) or the secretory lineage (enteroendocrine cells, tuft cells, and Paneth cells; Fig. 3.1). At the villus tip, cells die and are shed into the intestinal lumen. Paneth cells are the only differentiated intestinal cell type that do not move upward out of the crypt but instead migrate downward to reside at the bottom of the crypt for up to 6–8 weeks (van der Flier & Clevers, 2009).

2. INTESTINAL STEM CELLS

2.1. The long search for the elusive intestinal stem cells

Almost 50 years ago, two models of intestinal stem cell (ISC) location were proposed: the classical +4 model and the stem cell zone model. Since then,

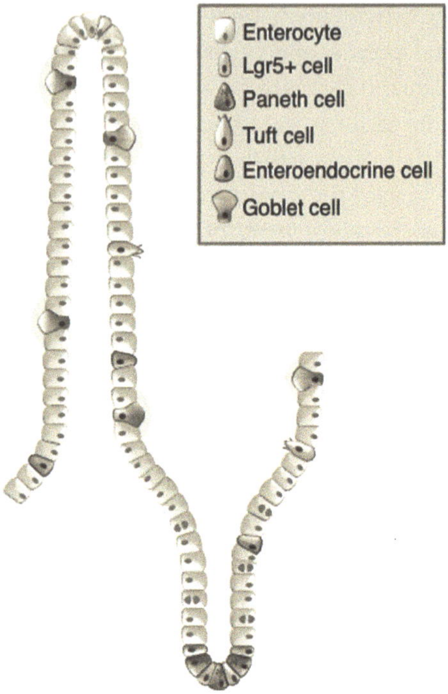

Figure 3.1 Cartoon representation of the intestinal epithelium showing the crypt where stem cells reside, and the villi along which differentiated epithelial cells migrate upward toward the tip. (See color plate.)

data have been presented in support of both models and until today, scientists are somewhat divided over the true identity of the stem cells.

The classical model (+4 model) of intestinal crypt stem cells was first proposed after rudimentary cell tracking experiments suggested a cell of origin situated directly above the Paneth cells in the "+4 position," as determined by counting Paneth cell nuclei from the crypt bottom (Cairnie, Lamerton, & Steel, 1965). These putative +4 stem cells were found to retain a DNA label, and to be actively cycling and sensitive to radiation damage (Potten, 1977; Potten, Kovacs, & Hamilton, 1974), characteristics that prompted the proposal of the +4 DNA label-retaining cell (LRC) as the ISC. Radiosensitivity was considered a beneficial trait of long-lived stem cells to avoid accumulation of genetic damage. Long-term DNA-label retention, for example with BrdU, has been a commonly used surrogate marker of stem cell identity based on the assumption that while their transit-amplifying daughters actively proliferate and dilute out an incorporated DNA label, stem cells are generally quiescent (Kiel et al., 2007).

The ability of actively cycling +4 stem cells to retain DNA labels has proven difficult to reconcile experimentally. One model proposes that the +4 cells achieve this via selective segregation of template DNA strands during mitosis, ensuring the retention of BrdU-labeled template strands within the stem cell (Potten, Owen, & Booth, 2002). In support of this hypothesis, there has been evidence reported for asymmetric retention of label-retaining DNA in ISCs by perpendicular orientation of mitotic spindles to the apical cell surface (Quyn et al., 2010). Conversely, it has also been reported that ISCs segregate their chromosomes randomly during mitosis (Escobar et al., 2011; Schepers, Vries, van den Born, van de Wetering, & Clevers, 2011), refuting the selective-segregation model of cycling stem cells.

In the stem cell zone model, slender cells wedged between Paneth cells at the crypt base were proposed to represent the intestinal crypt stem cell compartment (Cheng & Leblond, 1974). When treated with [^{3}H]-thymidine, crypt-base columnar (CBC) cells that survived were found to phagocytose damaged cells nearby. Radioactively labeled differentiated cells of the various epithelial lineages were subsequently observed, leading to the conclusion that they had originated from CBC cells. By chemical mutagenesis of the *Dlb-1* locus to randomly mark intestinal cells, long-lived marked clones comprising cells of all epithelial lineages were observed to invariably include a marked CBC cell, implicating the CBC as being the cell-of-origin of the various lineages (Bjerknes & Cheng, 1999, 2002). Following these early studies, in 2007, the first specific marker for CBC cells was identified that facilitated in-depth lineage tracing analysis of this candidate stem cell population, described later in the text.

2.2. Assaying for adult stem cells

Historically, putative stem cells have been identified using presumptive characteristics such as DNA-label retention, proliferation status, location, and radiosensitivity. When taken out of the body, stem cells grown in suitable culture conditions are capable of forming large proliferative clones that typically can be maintained for multiple passages. Such a feature of stem cells has been used to assay for "stemness," compared to their transit-amplifying progeny that can only generate small, differentiated clones that eventually cease growing. More recently, a robust culture system has been developed that supports the long-term growth of near-physiological intestinal epithelia from single stem cells. This *ex vivo* system provides an excellent way to evaluate the stem cell identity of candidate epithelial populations extracted from living tissue (Sato et al., 2009).

Despite the undoubted value of these surrogate markers and assays, the definitive identification, isolation, and characterization of endogenous stem cells rely on the availability of robust cell surface markers. These markers allow us to visualize stem cell populations in their native environment as a prerequisite to studying their contributions to both tissue homeostasis and disease, as well as deciphering their interactions with the local microenvironment. Similarly, markers of tumor resident stem cell populations (cancer stem cells) are crucial for identifying and validating these important therapeutic targets. Such markers will also facilitate the successful clinical exploitation of the potential of normal stem cells in regenerative medicine and the successful elimination of cancer stem cell populations as more effective cancer therapies.

In the last six years, multiple markers of putative CBC or +4 ISCs have been identified. Although many of these markers have not been well validated, a few have been shown to selectively mark cells exhibiting endogenous stem cell functions *in vivo* using the technique of *in vivo* lineage tracing (Fig. 3.2). This technique facilitates the introduction of a heritable genetic mark into a candidate stem cell via Cre–Lox recombination in living tissues.

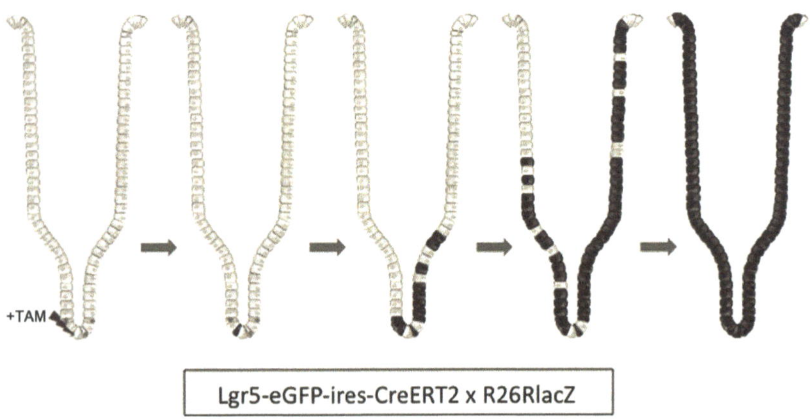

Figure 3.2 How lineage tracing works. By low-frequency genetic marking of cells in a given system, single marked cells transmit the mark to their progeny. From a marked stem cell, either by virtue of its proliferation or by that of its progeny, a labeled clone develops. Intestinal stem cells in which the lacZ transgene has been activated generate progeny that express the β-galactosidase gene. These cells extend from the crypt base to the villus tip and can be visualized as ribbons of blue cells after treatment with X-gal. (See color plate.)

The subsequent appearance and maintenance of labeled progeny of all cell lineages within the associated epithelium identify the original cell as being a self-renewing, multipotent stem cell (Barker & Clevers, 2010). The markers discussed in the next section have been reported to mark ISCs in the +4 position as well as at the crypt base. However, controversy surrounds several of the markers with regard to the specificity of their expression, an important issue that has yet to be resolved.

2.3. CBC markers

Lgr5, the first specific marker for CBC cells (Barker et al., 2007), was initially identified as a member of the Wnt target gene program that is constitutively active in human colon cancer cell lines (van de Wetering et al., 2002). The same target gene program was later found to be active within healthy crypts of the small intestine (Munoz et al., 2012; van der Flier et al., 2007). Lgr5 emerged as a candidate stem cell marker when it was found to be selectively expressed on a small population of slender, morphologically immature cells interspersed with the Paneth cells at the very crypt base. These Lgr5-expressing cells were identified as being the CBC cells, a lineage that had previously been championed as a candidate stem cell population. *Lgr5* was later shown to encode a seven-transmembrane protein that functions as a facultative component of the Wnt receptor complex and a receptor for R-spondins, a small family of Wnt pathway agonists (Carmon, Gong, Lin, Thomas, & Liu, 2011; de Lau et al., 2011; Glinka et al., 2011; van der Flier et al., 2007).

To validate the stem cell status of the *Lgr5*-expressing CBC cells, a transgenic mouse model was generated with an eGFP reporter gene and a gene encoding a tamoxifen-regulated Cre protein inserted downstream of the *Lgr5* promoter (Lgr5-EGFP-ires-CreERT2). This mouse line facilitated both visualization of endogenous Lgr5+ cells in living tissue and direct evaluation of Lgr5+ stem cell identity via *in vivo* lineage tracing. Each crypt was found to contain approximately 10–15 Lgr5+ cells, 10% of which occupied the +4 position. Lgr5+ cells were slender, with scant cytoplasm, and invariably in contact with adjacent Paneth cells, and they were found to cycle once a day on average (Barker et al., 2007). By crossing the Lgr5-EGFP-ires-CreERT2 mouse model to a Cre-activatable R26R-lacZ reporter mouse, *in vivo* lineage tracing was initiated by application of a low dose of tamoxifen to achieve stochastic activation of the *lacZ* reporter gene in isolated Lgr5+ cells. Over time, clonal ribbons of lacZ-positive cells were seen extending from the crypt base to the villus tip, which contained all cell lineages and persisted throughout life, providing evidence of multipotency and longevity of Lgr5+ cells. In the colon, Lgr5+ cells at the crypt base were also subsequently identified as adult stem cells (Barker et al., 2008).

By microarray profiling of the murine Lgr5+ stem cell gene expression signature (Munoz et al., 2012; van der Flier et al., 2007), additional markers of the stem cell population were discovered, including *Ascl2*, *Olfm4*, *Rnf43*, *Znrf3*, and *Smoc2*. Achaete-scute-like 2 (Ascl2) is a basic helix–loop–helix transcription factor whose expression is Wnt-dependent (Sansom et al., 2004; van der Flier et al., 2007). *Ascl2* was identified as a master regulator of the CBC stem cell when conditional *Ascl2* ablation was found to cause rapid Lgr5 stem cell death and transgenic overexpression of *Ascl2* led to dramatic expansion of the stem cell compartment (van der Flier et al., 2009). Olfactomedin-4 (*Olfm4*) is a Wnt-independent gene that encodes a secreted extracellular matrix glycoprotein involved in mediating cell adhesion (Liu, Chen, Zhu, & Rodgers, 2006; Zhang et al., 2002). Olfm4 is enriched in human colon crypts (Kosinski et al., 2007), and increased expression has been reported in the epithelium of inflamed colonic mucosa caused by inflammatory bowel disease (Shinozaki et al., 2001). Rnf43 and Znrf3 are both transmembrane E3 ligases expressed on Lgr5+ stem cells that promote the internalization of Frizzled ligands from the cell surface, thus fine-tuning Wnt signaling strengths to control the stem cell zone (Hao et al., 2012; Koo et al., 2012). According to these studies, deletion of *Rnf43* and *Znrf3* resulted in the rapid formation of adenomas harboring expanded stem cell and Paneth cell compartments. Another member of the Lgr5+ stem cell signature (a BMP-signaling inhibitor) was independently shown to mark endogenous stem cells via *in vivo* lineage tracing, thus validating the accuracy of the expression profiling approach (Munoz et al., 2012). *Troy*, a member of the tumor necrosis factor receptor superfamily, was also shown to selectively mark the Lgr5+ cells, where it functions as a negative modulator of Wnt signaling (Fafilek et al., 2013).

Although the discovery of stem cell markers for the mouse intestine is undoubtedly a major breakthrough, it remains to be established whether these markers can be applied to humans. A formal demonstration of stemness for the *Lgr5*-expressing cells of the human intestine requires anti-Lgr5 antibodies for their isolation. However, it has proven difficult to generate such antibodies capable of recognizing endogenous Lgr5 protein, in part because of the low expression levels and limited number of protein domains available for antibody binding at the cell surface. Although there are multiple commercial antibodies available against human LGR5, none of them are suitable for flow sorting of cells expressing endogenous levels of LGR5. In the absence of a direct flow sorting strategy, another Wnt target gene, the receptor tyrosine kinase *EPHB2*, has been used to enrich for human ISCs by virtue of its gradient of expression from the crypt base toward the differentiated cell compartment (Jung et al., 2011; Merlos-Suarez et al., 2011). Cells expressing high levels of *EPHB2* were shown to have organoid-forming capacity *ex vivo*, implicating them as human ISCs. Importantly, these cells expressed the highest levels of LGR5, indicating that LGR5 is also likely to selectively mark ISCs in human intestine. The ISC gene expression signature identified by this strategy overlapped with the Lgr5 stem cell signature and was found to be expressed by a stem-like cell population in human colorectal tumors (Merlos-Suarez et al., 2011). Recently, three new antibodies have been developed that were used to show that LGR5 is a functional cancer stem cell marker in human colorectal tumors (Kemper et al., 2012). Efforts are underway to try to directly isolate *Lgr5*-expressing cells from human tissues using these antibodies to confirm their stem cell status in *ex vivo* assays and to evaluate their regenerative medicine potential for treating human disorders.

Musashi-1 (Msi1) is an RNA-binding protein proposed to mark CBC cells (Kayahara et al., 2003; Potten et al., 2003) that was confirmed by mass spectrometry-based proteomic profiling to be specifically expressed by Lgr5+ cells (Munoz et al., 2012). In adult APCmin mice exhibiting elevated Wnt signaling levels, *in situ* hybridization showed that Msi1 expression broadened to encompass the proliferative compartment, with a decreasing gradient of signal along the base of the villus (van der Flier et al., 2007). A recently developed Msi1–eGFP mouse model has been used to isolate two major cell populations that reportedly correspond to the Lgr5- and Bmi1/mTert-expressing cells (Cambuli, Rezza, Nadjar, & Plateroti, 2013). While the data provide circumstantial evidence that Msi1 is expressed both by CBC cells and cells at the +4 position, further investigation is needed to clarify the specificity of Msi1 as a CBC marker.

2.4. +4 Markers

The first +4 stem cell marker validated by *in vivo* lineage tracing was *Bmi1* (Sangiorgi & Capecchi, 2008), which encodes a component of a Polycomb transcriptional repressor complex. mRNA *in situ* hybridization showed that Bmi1 marked cells located at the +4 position in the proximal small intestine.

Lineage tracing using a Bmi1-CreERT2 model documented the endogenous stem cell identity of these +4 Bmi1+ cells during tissue homeostasis. Importantly, ablation of the Bmi1+ population with targeted diphtheria toxin resulted in rapid crypt death, indicating loss of the stem cell compartment. In another study, the Bmi1+ population was characterized as being highly radiation-resistant and relatively quiescent, in contrast to the Lgr5+ stem cell population (Yan et al., 2012). Interestingly, radiation-induced injury-activated Bmi1+ cells to proliferate and effect rapid crypt regeneration *in vivo*. In culture, single Bmi1+ cells generated long-lived self-renewing organoids that included Lgr5+ cells. Yan et al. concluded that while Lgr5+ cells maintain the intestinal epithelium under homeostasis, Bmi1+ cells remained quiescent in reserve until called upon to drive injury-induced regeneration.

High telomerase expression is believed to be a stem cell trait that protects the cells from entering senescence. Using an mTert-GFP mouse model, Breault observed rare GFP+ cells within the intestinal crypt that were BrdU-label retaining (Breault et al., 2008). Further studies revealed that the mTert+ cell population was radiation-resistant, slowly cycling, and distinct from the Lgr5+ population, yet capable of generating all differentiated intestinal cell types, including Lgr5+ stem cells (Montgomery et al., 2011). Another marker of +4 cells is Hopx, which reportedly colocalizes with Bmi1- and mTert-expressing cells and identifies a slow-cycling and LRC population distinct from the proliferating Lgr5+ cell population (Takeda et al., 2011). Hopx is an atypical homeobox protein that marks +4 cells in a Hopx-lacZ transgenic knock-in mouse model. In lineage tracing experiments, Hopx+ cells displayed a bidirectional lineage relationship with CBCs where either cell type could give rise to the other (Takeda et al., 2011).

2.5. Stem cell markers: A specificity debate

The discovery of markers for the long-elusive ISCs has undoubtedly driven rapid advances in our understanding of adult stem cell biology. However, it has also led to a great deal of controversy regarding the accuracy of the models and methods used to identify and validate these markers (summarized in Fig. 3.3). One example of this is the *Bmi1* +4 stem cell marker. Despite the initial report of Bmi1 being restricted to +4 cells and the functional validation of this marker via *in vivo* lineage tracing, other studies have documented endogenous expression throughout the entire crypt, including the Lgr5+ CBC stem cell compartment (Montgomery et al., 2011; Munoz

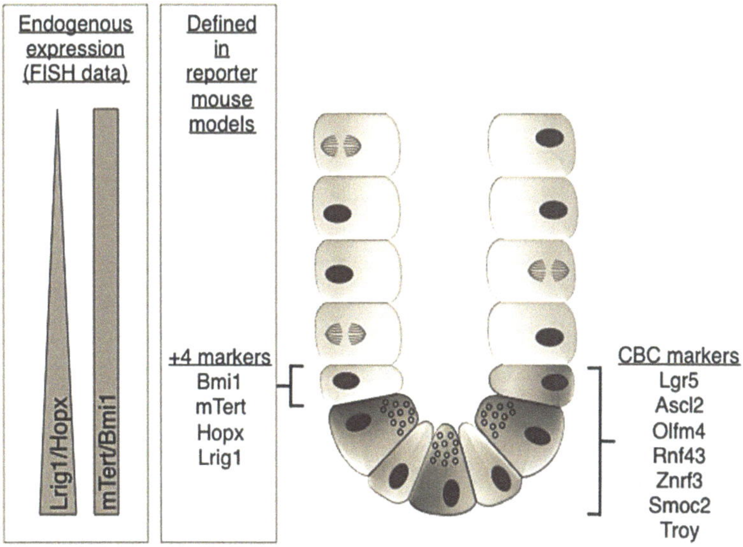

Figure 3.3 Cartoon of an intestinal crypt outlining expression of putative and validated stem cell markers, as well as conflicting data showing broad expression of several markers. CBC markers have been validated and shown to be specifically expressed while data have been reported that show a broader expression of some putative +4 markers. Markers such as Bmi1 and mTert have been observed to be uniformly expressed within the crypt while genes such as Lrig1 and Hopx have been reported to be enriched at the crypt base with a gradient of expression as cells extend toward the crypt/villus border. (See color plate.)

et al., 2012; Powell et al., 2012; van der Flier et al., 2009). By single mRNA molecule FISH, Itzkovitz et al. saw that *Bmi1* mRNA was expressed in all proliferative crypt cells (Itzkovitz et al., 2011), an expression pattern that was mirrored by antibody detection of Bmi1 protein (Munoz et al., 2012; Takeda et al., 2011). Independent analysis of Bmi1-driven lineage tracing using the same Bmi1-CreERT2 model appeared to corroborate this widespread expression, when tracing was found to originate at random throughout the crypt, including the Lgr5+ stem cell compartment (Munoz et al., 2012). This overlapping expression with Lgr5 readily explains the appearance of long-lived tracing units in the original study of Sangiorgi and Capecchi. However, the new data would appear to disqualify *Bmi1* as a bona fide marker of +4 stem cells in mice. Similar to *Bmi1*, *mTert* has been reported to be broadly expressed within the crypt (Itzkovitz et al., 2011; Schepers et al., 2011), rather than marking rare slow-cycling cells, thus raising doubts about the accuracy of the mTert-GFP transgenic mouse model. Also in contradiction to the mouse reporter studies, endogenous *Hopx* expression has been documented to exist as a broad gradient within the crypt, with the highest levels in Lgr5 + cells (Munoz et al., 2012; Powell et al., 2012). Although it is difficult to definitively reconcile the contradictory findings, much of the confusion may stem from an overreliance on the use of targeted knock-in mice to document gene expression *in vivo*. Disruption of key regulatory elements such as enhancers, silencers, miRNA binding sites, etc., is an inherent risk during gene targeting, potentially skewing expression of the targeted reporter genes. This highlights the importance of accurately documenting endogenous expression profiles (via FISH or similar methods) of candidate marker genes to confirm the accuracy of reporter mouse models.

Conflicting results have also been recently reported for other proposed markers of ISCs. Lrig1 is a negative-feedback regulator of the ErbB family that has been shown to control stem cell proliferation in both human and murine epidermis (Jensen et al., 2009; Jensen & Watt, 2006). According to the first study, Lrig1 expression was observed to form a gradient within the intestinal crypt, with the highest levels in Lgr5+ stem cells that were continuously proliferating, and functioned to regulate the amplitude of growth factor signaling in controlling the size of the ISC niche (Wong et al., 2012). In contrast, a second independent study using a different mouse model (Powell et al., 2012) showed that Lrig1 + marked stem cells that were transcriptionally distinct from the Lgr5+ population and capable of long-term lineage tracing. Interestingly, the two markers were rarely coexpressed in this mouse model and Lrig1 + cells were mostly quiescent until irradiation injury stimulated them to proliferate. Again, the discrepancy between reports could be because of the mouse models used as well as the different antibodies against Lrg1, which remains to be clarified.

In a second example, single radiation-sensitive Dclk1 + cells were found scattered in intestinal crypts at or near the +4 position, prompting speculation that they might be +4 stem cells (May et al., 2008). Double cortin-like kinase 1 (*Dclk1*) was first identified as a putative ISC marker from microarray profiling of gastric epithelial progenitors and whole stomach libraries (Giannakis et al., 2006). However, a separate study subsequently showed that *Dclk1* marked cells also express typical tuft cell differentiation markers, indicating that *Dclk1* was not a stem cell marker (Gerbe, Brulin, Makrini, Legraverend, & Jay, 2009).

Evidently, the hunt for markers of ISCs is a challenging process with many potential pitfalls (for a more comprehensive review, refer to Barker et al., 2012). Candidate markers must be validated using several approaches including lineage tracing and endogenous expression profiling to provide compelling evidence of their specificity. Until the controversies are clarified, the debate will continue over the +4 model versus the stem cell zone model.

3. CHARACTERIZING STEM CELL BEHAVIOR DURING HOMEOSTASIS AND DISEASE

Mammalian stem cell populations are thought to be maintained in a largely quiescent state, possibly to preserve their long-term proliferation potential and to guard against DNA replication errors during cell division (Fuchs, 2009). For example, in the hematopoietic stem cell (HSC) system, dormant LRCs exhibit long-term regenerative potential whereas more actively dividing HSCs demonstrate only shorter term repopulation potential (Wilson et al., 2008). In the hair follicle, which undergoes cyclical episodes of regeneration (anagen), degeneration (catagen) and dormancy (telogen), slow-cycling LRCs reside in the "bulge" region and are able to generate all lineages of the hair follicle and sebaceous gland (Tumbar et al., 2004).

However, an obligate requirement for stem cells to be quiescent has been challenged by the discovery of actively proliferating Lgr5+ stem cells in a range of tissues, including the intestine, stomach and hair follicle (Barker et al., 2010, 2007; Jaks et al., 2008). High-resolution clonal fate mapping experiments have been performed to determine how these actively cycling stem cells achieve an optimal balance between tissue homeostasis and stem cell population maintenance. In the interfollicular epidermis (IFE) and the esophageal epithelium, an actively cycling single progenitor cell adopts random fate choices allowing for homeostatic maintenance on a tissue-wide level (Clayton et al., 2007; Doupe et al., 2012). In the intestine, clonal fate mapping experiments have shown that Lgr5+ stem cells typically undergo symmetrical division to generate either committed transit-amplifying (TA) progeny or additional Lgr5+ stem cells. Although this mechanism dictates the regular extinction of individual Lgr5+ stem cells, a balanced supply of TA progeny and new stem cells is achieved at the population level (Lopez-Garcia, Klein, Simons, & Winton, 2010; Snippert et al., 2010). Clones from a single stem cell follow a "neutral drift" process whereby clone sizes randomly increase and decrease, and clones can be extinguished by chance. At the population level, the average size of the clones, in general, increases and eventually, the entire crypt base stem cell population derives from a single clone. By a process of neutral competition, ISCs can continuously turn over while keeping the total stem cell number constant.

Stem cell identity is likely to be defined by a combination of cell intrinsic and environmental (niche) factors. It is becoming apparent that crypt populations can exhibit a striking degree of plasticity in response to tissue wounding, facilitating rapid epithelial regeneration following catastrophic loss of the regular stem cell pool. This was first proposed in the 1970s, when mathematical models predicted that the earliest TA cell progeny could drop back into the stem cell niche and acquire stem cell identity following ablation of the regular stem cells through acute injury (Cheng & Leblond, 1974; Potten, 1977). Phenotypic conversion of TA cells to stem cells is likely driven by local niche signals including Wnt and Notch. Functional evidence of this was provided when Lgr5-TA cells that normally do not form organoids *in vitro* were converted into Lgr5+ stem cells capable of efficient organoid formation following exposure to Wnt3A (Sato, van Es, et al., 2011).

Loss of the Lgr5+ stem cell compartment appears to be compensated by other cell populations in the crypt, as evidenced by ablation experiments using a human diphtheria toxin receptor (*DTR*) knocked into the Lgr5 locus (Tian et al., 2011). The absence of Lgr5+ cells did not perturb epithelial homeostasis, and Bmi1-expressing cells could give rise to new Lgr5+ cells as well as directly generate all intestinal lineages. In another example, lineage tracing studies of early daughters of Lgr5+ CBCs expressing the marker *Dll1* in and around position +5 showed these cells to be short-lived progenitors that generated small, short-lived clones of secretory cells during homeostasis. However, upon radiation-induced loss of Lgr5+ stem cells, Dll1+ progenitors could revert to Lgr5+ stem cells, with subsequent stem cell lineage tracing events (van Es et al., 2012).

The existence of plasticity within the crypt suggests that the niche plays an important role in providing a suitable microenvironment for progenitors to revert to stem cells. Following tissue damage where the stem cell pool is lost, the remaining stem cell niche provides the requisite signals to stimulate reversion of progenitors to repopulate the stem cell compartment. Evidently, the niche is a key component in regulating stem cell behavior not only in homeostasis but also during tissue regeneration following injury. It is therefore vital to understand the stem cell niche in order to safely harness adult stem cells for therapeutic applications in regenerative medicine.

4. RECONCILING THE OPPOSING ISC MODELS

As discussed earlier, the intestinal crypts are thought to harbor independent pools of cycling stem cells marked by Lgr5 (the CBC cells) and relatively quiescent stem cells marked by *Bmi1*, *mTert*, and *Lrig1* (+4 cells).

In an attempt to directly isolate the quiescent +4 cells from the small intestine, independent groups employed an H2B-YFP pulse-chase model to mark LRCs in the crypts (Buczacki et al., 2013; Roth et al., 2012). Expression profiling of the isolated LRC population revealed a robust expression of many +4 markers, including Lrig1, mTert, and Hopx, together with a set of Paneth cell markers. In addition, Buczacki et al. documented coexpression of *Lgr5*, leading to speculation that some Lgr5+ cells at the crypt base may represent Paneth cell precursors. However, this finding was not confirmed in the study of Roth et al.

Using an elegant lineage tracing approach, Buczacki et al. selectively marked the LRCs and tracked their contribution to epithelial renewal during homeostasis and following damage. As expected, the marked LRCs remained quiescent during homeostasis, expressing markers of early Paneth cells. However, following epithelial injury, the cells began to proliferate and contributed progeny to all epithelial cell lineages, consistent with them having adopted a true stem cell identity. Collectively, these observations supported a model in which LRCs function as Lgr5+ stem cell-derived Paneth cell precursors during regular epithelial homeostasis, but the same LRC population demonstrates plasticity following epithelial injury, reverting to a true stem cell state to effect tissue regeneration (Fig. 3.4).

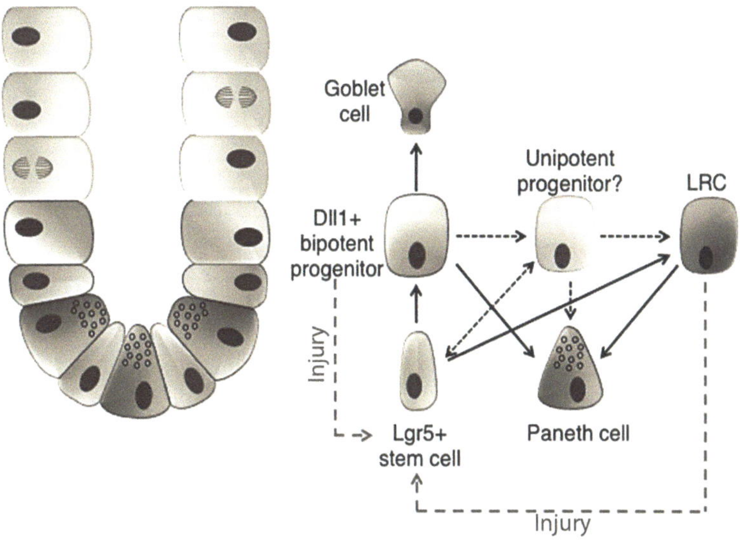

Figure 3.4 Cartoon of the intestinal crypt base showing position and relationships between Lgr5+ CBC cells, Dll1+ progenitor cells, label-retaining cells, and Paneth cells. (See color plate.)

These data point to a model where actively cycling Lgr5+ stem cells maintain homeostatic renewal of the intestinal epithelium while the quiescent LRC population can act as a "reserve" stem cell pool to be activated upon tissue damage. Importantly, this model provides insight into the bidirectional lineage relationship between Lgr5+ cells and +4 cells previously observed by others (Montgomery et al., 2011; Takeda et al., 2011; Yan et al., 2012). A picture is emerging where different cell types are, in fact, capable of reversion to a stem cell phenotype when given the right stimulus.

5. THE STEM CELL NICHE

5.1. Concept of a niche

The term "stem cell niche" was first coined by Ray Schofield in 1978 to describe the association of a stem cell with other cells, which determine its behavior (Schofield, 1978). The postulation of a microenvironment necessary for maintenance of stem cells has gained traction over the years with generation of data that indicate that a stem cell niche may comprise cells or cells together with extracellular matrix, and act as a source of growth factors that regulate self-renewal and proliferation (Lander et al., 2012). It is believed that because of their key role in maintaining and replenishing a tissue, stem cells are necessarily long-lived and capable of life-long self-renewal with minimal DNA damage. For this reason, stem cells might rely on a niche to support a particular metabolic state and to provide feedback control to allow for adequate proliferation during homeostatic maintenance, while remaining poised for rapid response to injury (Lander et al., 2012). Stem cell niche components could derive from stem cell progeny or from nonstem-cell-derived lineages. Within the niche, the communication between stem cells and the niche components is bidirectional and influences both compartments.

The first example of a stem cell niche was demonstrated in *Caenorhabditis elegans*, where a single mesenchymal "distal tip cell" (DTC) provides stem cells with the essential microenvironment for generation of adult germline tissue (Kimble & White, 1981). Notch signaling from the DTC to germline stem cells maintains their stemness, resulting in a gradient of maturation as stem cells get further from the DTC (Byrd & Kimble, 2009). Wnt signaling controls properties of the niche, and modulation of this signaling pathway results in the formation of ectopic niches (Lam, Chesney, & Kimble, 2006). In the hematopoietic system, stem cells reside in the bone marrow, where several cell types act together to produce niche signaling molecules that control HSC function and maintenance (reviewed by Trumpp, Essers, & Wilson, 2010). These cell types include osteoblastic cells, macrophages, *CXCL12*-expressing CAR cells, and nestin+mesenchymal stem cells. In the skin, also an epithelium, distinct pools of stem cells are believed to maintain the different compartments (Watt, Lo Celso, & Silva-Vargas, 2006). In the hair follicle stem cell niche, known as the "bulge," progeny of hair follicle stem cells transmit BMP and FGF signals to maintain quiescence in stem cells (Hsu, Pasolli, & Fuchs, 2011), while cross talk between the dermal papilla and the bulge controls the activating signals required to switch stem cells from a quiescent to an active state (Driskell, Giangreco, Jensen, Mulder, & Watt, 2009). In the human IFE, stem cells reside in clusters at the tops of rete ridges in the basal layer (Legg, Jensen, Broad, Leigh, & Watt, 2003), and single-cell profiling of primary human keratinocytes have indicated that heterogeneity results in differences in cellular interaction with their environment (Tan et al., 2013).

5.2. ISC niche

In the small intestine, stem cells reside at the base of crypts, adjacent to Paneth cells, and are surrounded by a stromal microenvironment. The stroma consists of multiple components and cell types, including extracellular matrix, fibroblasts, myofibroblasts, smooth muscle cells, neural cells, endothelial cells, lymphocytes, and macrophages. The ISC niche is likely to comprise several different cell types, each of which contributes both cell-associated ligands and soluble cytokines, chemokines, and growth factors that regulate stem cell behavior. In this section, we discuss two main aspects of the ISC niche: the stromal microenvironment and the Paneth cell.

5.3. Stromal microenvironment

The lamina propria, located beneath the intestinal crypts, contains cells such as subepithelial myofibroblasts and stromal cells, which have been shown to signal to the ISC compartment during intestinal morphogenesis and to regulate proliferation and differentiation (Fig. 3.5; McLin, Henning, & Jamrich, 2009; Powell, Pinchuk, Saada, Chen, & Mifflin, 2011). Intestinal subepithelial myofibroblasts exhibit qualities of both smooth muscle cells and fibroblasts (Pinchuk, Mifflin, Saada, & Powell, 2010; Samuel et al., 2009). Such myofibroblasts, from both mouse and human, have been reported to support the growth, differentiation, and expansion of human intestinal epithelium, although the factors that facilitate this support are yet to be

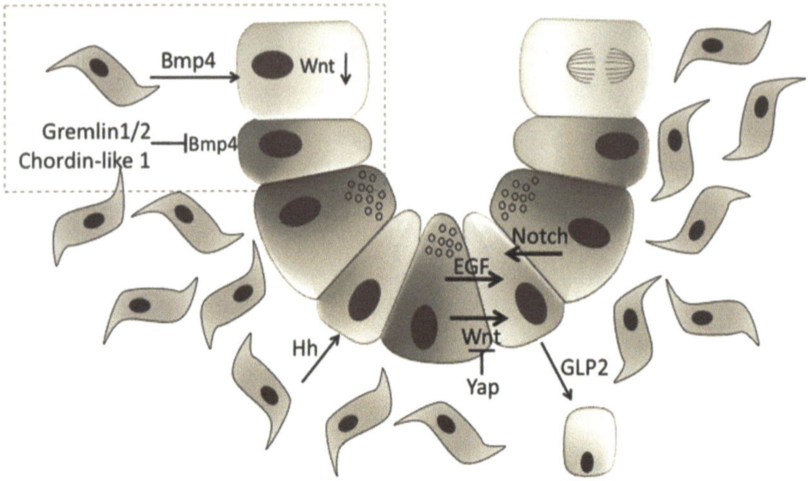

Figure 3.5 Signaling within the stem cell niche. Key niche factors produced by both the epithelial and the stromal compartments mediate cross talk that controls proliferation and differentiation. Dashed box encloses signaling components involved in producing a Wnt gradient along the crypt–villus axis to ensure stem cell maintenance and proliferation in the crypt base while promoting differentiation in cells that migrate upwards. (See color plate.)

exactly determined (Lahar et al., 2011). Lamina propria stromal cells produce Bmp4, a signaling protein of the TGFβ superfamily that antagonizes Wnt signaling, inhibiting stem cell expansion and promoting epithelial differentiation (Miyazono, Kamiya, & Morikawa, 2010). While Bmp signaling is important for stimulating differentiation of progenitor cells as they migrate up the crypt–villus axis, stem cells in the crypt need to avoid this stimulus. To counter Bmp signaling in the crypt, myofibroblasts and muscularis mucosae cells surrounding the crypt base secrete gremlin1, gremlin2, and chordin-like 1, which are potent Bmp inhibitors (Kosinski et al., 2007).

Hedgehog (Hh) signaling is another important component of the epithelial–mesenchymal cross talk. Hh signaling in the stroma reduces epithelial proliferation and increases the myofibroblast and smooth muscle cell compartment (van Dop et al., 2009). In contrast, attenuation of Hh signaling expands the epithelial stem and progenitor cell compartments while depleting myofibroblasts and smooth muscle cells (Kosinski et al., 2010; Zacharias et al., 2010). The nervous system has also been shown to play a key role in a signaling feedback loop regulating epithelial growth. Glucagon-like peptide 2 (GLP-2) produced by enteroendocrine cells in the intestinal epithelium signals to underlying enteric neurons that express the GLP-2 receptor, which then stimulate the proliferation of enterocyte progenitors (Bjerknes & Cheng, 2001).

5.4. The Paneth cell

Because of the apposition of stem cells with Paneth cells within the epithelial layer, Paneth cells are believed to be a key component of the ISC niche. Recently published data in support and against this hypothesis are discussed later.

Paneth cells were first described more than a century ago as small-intestinal columnar epithelial cells of the secretory lineage located at the base of crypts, with cytoplasm filled with large granules (Paneth, 1887; Schwalbe, 1872). They contain extensive endoplasmic reticulum and Golgi networks that produce antimicrobial peptides and enzymes such as lysozyme (Deckx, Vantrappen, & Parein, 1967) and cryptdins to protect the epithelium from the potentially lethal microbial load of the gut lumen (Ouellette, Miller, Henschen, & Selsted, 1992). These proteins are stored in large granules within the cytoplasm and are eventually secreted into the crypt lumen. Paneth cells have previously been found to have some measure of radioresistance, allowing them to survive sublethal exposure to radiation (Brennan, Carr, Seed, & McCullough, 1998). Lgr5+ CBC cells are interdigitated between Paneth cells in tight association, which has led to the hypothesis that Paneth cells are the ISC niche. Paneth cells produce growth factors, such as EGF, Notch, and Wnt ligands, which promote stem cell proliferation and maintenance (Sato, van Es, et al., 2011), evidence in support of the hypothesis that Paneth cells provide essential niche signals to Lgr5+ stem cells.

5.4.1 Paneth cell function

Evidence that Paneth cells are essential to Lgr5+ stem cells has come largely from *in vitro* investigations. Single flow-sorted Lgr5^{hi} stem cells are capable of forming "mini-guts," or organoid structures with a single-layer epithelium surrounding a hollow lumen (Sato et al., 2009). Lgr5+ stem cells in isolation typically formed short-lived, cystic clusters, with only 7% of cells forming long-lived GFP+ organoids. However, when stem cells were combined with Paneth cells, identified by the expression of the marker CD24, the rate of organoid formation increased by 11-fold to 77% (Sato, van Es, et al., 2011). Stem cell–Paneth cell doublets also demonstrated greater plating efficiencies of about 25% compared to single stem cells ($<1\%$), an effect that could be mimicked by the addition of Wnt3A during the first 3 days in culture (Sato, van Es, et al., 2011).

In contrast to the small intestine, the colon is largely devoid of Paneth cells, implying that other cells must be supplying niche factors to sustain the Lgr5+ stem cells. Both CD24+ cells and c-Kit+ colonic goblet cells, located adjacent to the Lgr5+ stem cells at the crypt base, have been proposed as essential niche components in the colon (Rothenberg et al., 2012; Sato, van Es, et al., 2011). Notably, the c-Kit+ cells were found to be regulated by Notch signaling and could enhance the ability of colonic Lgr5+ cells to generate organoids *ex vivo*. Collectively, these data indicate that c-Kit+ cells perform a similar niche function to support colonic stem cell function *in vivo* (Rothenberg et al., 2012).

Genetic ablation experiments indicate that Paneth cells may not be essential niche components *in vivo*, because of the existence of partially redundant sources of key niche signals such as Wnt (see next section). However, Yilmaz et al. have recently proposed a key role for Paneth cells in sensing calorie restriction and relaying this information to the Lgr5+ stem cell pool to effect changes in population sizes (Yilmaz et al., 2012). An evolutionary advantage of such a system could be for adaptation to fluctuations in food availability. Calorie restriction reduced mTORC1 signaling in Paneth cells but not in stem cells, an effect that could be mimicked by rapamycin. Increased expression of bone stromal antigen (*Bst1*) in Paneth cells revealed a role for this ectoenzyme in mediating the effects of calorie restriction on the stem cell compartment.

In addition to its role as an ISC niche component, the Paneth cell is reportedly capable of dedifferentiation into Bmi1+ stem cells to effect tissue regeneration following injury-induced ablation of the Lgr5+ cycling stem cells (Roth et al., 2012). In summary, the Paneth cell not only plays an important part in intestinal function by producing antimicrobial peptides and calorie restriction sensing, but also provides key niche signals to the Lgr5+ population during epithelial turnover and in response to changes in the local microenvironment.

5.5. Ablating Paneth cells *in vivo*

To conclusively determine whether the Paneth cell is an essential niche component, efforts have been focused on ablating these cells *in vivo* to evaluate the effects on the ISC pool. This has typically been achieved via transgenic expression of the DTR in Paneth cells or by knocking out key transcription factors that control Paneth cell lineage commitment. Controversy about the dispensability of Paneth cells has arisen because of the difficulty in achieving complete Paneth cell ablation.

Garabedian et al. expressed the DTR specifically in Paneth cells under the control of the cryptdin2 promoter (CR2-tox176). Upon applying diphtheria toxin, Paneth cell numbers dropped by 82% and there was an increase in cells that appeared to be undifferentiated CBC cells, based on electron microscopy analyses (Garabedian, Roberts, McNevin, & Gordon, 1997). The authors concluded that maintenance of stem cells in the crypt might not be defined by instructive interactions involving the Paneth cell. However, when this model was revisited by Sato et al., they confirmed the ablation of the majority of Paneth cells, but found that CBC numbers were instead reduced (Sato, van Es, et al., 2011). Importantly, more than 90% of crypts analyzed still contained at least one Paneth cell, suggesting that because of the incomplete ablation of Paneth cells, the ISC compartment could still be maintained, albeit at a reduced level.

A knockout mouse model deficient for the transcriptional repressor Gfi1 was described to lack Paneth cells (Shroyer, Wallis, Venken, Bellen, & Zoghbi, 2005). However, this was questioned in two subsequent reports, when intestinal crypts from adult *Gfi1−/−* mice were found to contain Paneth cells, albeit at a reduced frequency (Bjerknes & Cheng, 2010; Sato, van Es, et al., 2011). Sato et al. observed that this partial reduction in Paneth cell numbers was mirrored by a decrease in the size of the Lgr5+ stem cell pool. Surviving stem cells were invariably adjacent to the remaining Paneth cells. The vast majority of crypts analyzed were found to harbor at least one Paneth cell, similar to the CR2-tox176 mouse model. Taken together, it appears that it was incomplete ablation of Paneth cells that previously led to erroneous conclusions that the Paneth cell was dispensable for ISC maintenance.

In support of the hypothesis that a decrease in Paneth cell numbers adversely affects ISC populations, mice deficient for the hematopoietic growth factor receptor Csf1r had far fewer mature Paneth cells and a markedly reduced stem cell proliferation, which could be rescued by locally restricted transgenic expression of *Csf1* (Huynh et al., 2009). An inducible model of Csf1 receptor deletion in adult mice showed a similar phenotype, as well as reduced organoid formation *in vitro* (Akcora et al., 2013). In another study, deletion of *Sox9* by an inducible Ah-Cre led to virtually complete Paneth cell ablation after 8 weeks (Sato, van Es, et al., 2011), whereupon stem cell numbers were reduced and any left were found to be crowded around the remaining Paneth cells.

Recently, two studies have reportedly achieved complete Paneth cell ablation by inducibly knocking out *Math1* (*Atoh1*; Durand et al., 2012; Kim, Escudero, & Shivdasani, 2012), a transcription factor necessary for intestinal secretory cell differentiation (Shroyer et al., 2007). In the first study, Kim et al. observed that in the sustained absence of Paneth cells, Lgr5+ stem cells were found to function as normal (Kim et al., 2012). Durand et al. confirmed these findings and reported that while Math1-deficient crypt cells tolerated *in vivo* Paneth cell loss and maintained active β-catenin signaling, they required exogenous Wnt when grown in organoid culture (Durand et al., 2012). Interestingly, Farin et al. found that conditional deletion of *Wnt3* in adult mice had no effect on stem cell function, although it was essential in organoid cultures (Farin, Van Es, & Clevers, 2012). Together, it appears that while the Paneth cell may well provide many niche factors *in vivo*, there may be redundancy with the stromal microenvironment. This would explain why ISCs can tolerate Paneth cell loss *in vivo*, but *in vitro* where no stroma is present, Paneth cells are essential.

5.6. Signaling within the niche

The canonical Wnt signaling pathway is crucial to epithelial proliferation, and key Wnt signaling factors including Wnts, Frizzleds, and low-density lipoprotein receptor-related proteins are expressed by epithelial cells in the intestine (Gregorieff et al., 2005). Paneth cells support Lgr5+ stem cells in the crypt by providing Wnt, Notch, and EGF signals (Fig. 3.5; Sato, van Es, et al., 2011). By expressing the Notch ligands Dll1 and Dll4 on the cell surface, Paneth cells trigger Notch signaling in adjacent Lgr5+ cells to ensure maintenance of their stem cell identity and prevent terminal differentiation toward the secretory lineage (Pellegrinet et al., 2011). As transit-amplifying cells are produced and move upward along the crypt–villus axis, they lose contact with Notch ligands. Stochastic loss of Notch receptor expression results in upregulation of Dll1, and this cell can then stimulate Notch signaling in surrounding transit-amplifying cells. By lateral inhibition, the Dll1-expressing cell is fated toward the secretory lineage, while its neighboring cells that have active Notch signaling develop along the enterocyte lineage (van Es et al., 2012).

Expression of the yes-associated protein (YAP) induces growth in many tissues; however, Villin-driven YAP expression causes degeneration of the small intestine and colon associated with the loss of proliferating crypts as well as mislocalization and eventual disappearance of Paneth cells (Barry et al., 2013). The crypt-loss phenotype was found to be associated with repression of CD44, a Wnt target gene, loss of nuclear β-catenin, and down-regulation of the stem cell marker genes *Lgr5*, *Olfm4*, *Ascl2*, and *Smoc2*. Conversely, upon genetic ablation of *Yap* followed by irradiation-induced injury, crypt hyperplasia and overgrowth was observed, along with upregulation of Wnt target genes and increased numbers of Paneth cells. Yap was found to modulate Wnt signaling independently of the canonical Axin–APC–GSK3β complex, where Yap restricts nuclear translocation of DVL2 and DVL3 to block their ability to induce Wnt signaling.

The crypt is influenced by an array of external signals that regulate homeostasis of the ISC population. Interleukin-22 was recently shown to maintain epithelial integrity and protect stem cells against damage, with an important role in graft-versus-host disease (Hanash et al., 2012). IL-22 is produced by RORϒt+ innate lymphoid cells that reside in cryptopatches situated in the small intestine lamina propria underlying the crypts (Spits & Di Santo, 2011) and signals to both Paneth cells and stem cells that express the IL-22 receptor. Innate lymphoid cell-produced IL-22 regulates the expression of the antimicrobial lectin Reg3γ by Paneth cells (Hanash et al., 2012).

R-spondin 1, a Wnt pathway agonist, has been shown to amplify the local response to short-range Wnt signaling from Paneth cells, a mechanism to control Wnt signaling within the crypt (Carmon et al., 2011; de Lau et al., 2011). This way, only the Lgr5+ stem cells that are neighbors of Paneth cells would receive strong Wnt signals, which are subsequently amplified by R-spondin 1 (Sato, van Es, et al., 2011). Exactly which Wnt ligands regulate stem cells *in vivo* is not yet fully understood. While Paneth cell-secreted Wnt3 is an essential niche signal in organoid cultures, Wnt3 was found to be dispensable *in vivo*, suggesting that alternative nonepithelial Wnt ligands compensate as a redundant source of Wnt (Farin et al., 2012). *Wnt11* was not expressed by Lgr5+ stem cells and *Wnt3* expression was decreased in these cells upon Paneth cell depletion (Kim et al., 2012).

6. STUDYING ISCS *EX VIVO*

Efforts to grow intestinal crypts *ex vivo* by either culturing pieces of intestinal tissue or growing crypts from single stem cells facilitate the detailed study of specific niche components as a prerequisite to understanding their *in vivo* function. Apart from the "mini-gut" organoid culture system described earlier (Sato et al., 2009), other systems include one that incorporates an air–liquid interface together with underlying stromal elements that allowed for prolonged intestinal epithelial expansion as sphere-like organoids that could proliferate and differentiate into the various lineages. Growing intestinal epithelium *in vitro* also has important regenerative medicine applications, an example of which involves implanting organoid units on biodegradable scaffolds to create whole tissue-engineered small intestines (Sala et al., 2011).

A particularly striking example of the clinical potential of these epithelial culture systems was provided when colonic organoids grown *in vitro* from Lgr5+ stem cells were shown to help repair damaged colonic epithelium following administration via a simple enema (Yui et al., 2012). After 4 weeks, the donor-derived cells had formed a single-layered epithelium that was not only functionally and histologically normal, but also persisted long term. The culture system has also been translated for use with human tissue. In one study, small intestine and colonic tissue could be cultured in a similar fashion to mouse tissue with the addition of nicotinamide, an Alk inhibitor and a p38 inhibitor, which also allowed growth of human colorectal cancer cells and human metaplastic epithelia from regions of Barrett's esophagus (Sato, Stange, et al., 2011). In another study, human colonic stem cells purified according to their cell surface levels of EPHB2 could be expanded and maintained *in vitro* in an undifferentiated state (Jung et al., 2011).

7. CONCLUSION

Our current understanding of the intestinal crypt is that rapidly proliferating stem cells reside at the crypt base flanked by Paneth cells. These cells express markers such as *Lgr5*, *Ascl2*, *Olfm4*, *Rnf43*, *Znrf3*, *Smoc2*, and *Troy*. More quiescent stem cells are generally detected at the +4 position and these act as a reserve population that is activated to proliferate in response to injury. While the Paneth cells work in conjunction with the surrounding mesenchyme to provide key niche signals to maintain the CBC stem cell compartment, it is yet to be determined whether there is a distinct niche for the +4 reserve stem cell population. The CBC is long-lived and multipotent, and single CBC cells can be stimulated to proliferate and differentiate in culture to generate organoids for a variety of purposes including *in vitro* assays and transplantation for medical therapy. Most intriguingly, the Lgr5+ stem cell identity appears to be a state that is largely controlled by external factors such as Wnt signaling than a hard-wired genetic program. A re-evaluation of the concept of a "stem cell" may be in order, particularly in view of efforts to harvest the regenerative capability of "stem cells." If indeed modulation of the niche microenvironment could push progenitor cells to function as stem cells, then far more cells harbor regenerative medicine potential in the intestine than previously thought.

ACKNOWLEDGMENTS

We thank all the members of the Barker lab, and the Agency for Science, Technology and Research (A*STAR) for funding.

REFERENCES

Akcora, D., Huynh, D., Lightowler, S., Germann, M., Robine, S., de May, J. R., et al. (2013). The CSF-1 receptor fashions the intestinal stem cell niche. *Stem Cell Research, 10*(2), 203–212.

Barker, N., & Clevers, H. (2010). Lineage tracing in the intestinal epithelium. *Current Protocols in Stem Cell Biology, Chapter 5*, Unit 5A.4.

Barker, N., Huch, M., Kujala, P., van de Wetering, M., Snippert, H. J., van Es, J. H., et al. (2010). Lgr5(+ve) stem cells drive self-renewal in the stomach and build long-lived gastric units in vitro. *Cell Stem Cell, 6*(1), 25–36.

Barker, N., Ridgway, R. A., van Es, J. H., van de Wetering, M., Begthel, H., van den Born, M., et al. (2009). Crypt stem cells as the cells-of-origin of intestinal cancer. *Nature, 457*(7229), 608–611.

Barker, N., van Es, J. H., Jaks, V., Kasper, M., Snippert, H., Toftgard, R., et al. (2008). Very long-term self-renewal of small intestine, colon, and hair follicles from cycling Lgr5 +ve stem cells. *Cold Spring Harbor Symposia on Quantitative Biology, 73*, 351–356.

Barker, N., van Es, J. H., Kuipers, J., Kujala, P., van den Born, M., Cozijnsen, M., et al. (2007). Identification of stem cells in small intestine and colon by marker gene Lgr5. *Nature, 449*(7165), 1003–1007.

Barker, N., van Oudenaarden, A., & Clevers, H. (2012). Identifying the stem cell of the intestinal crypt: Strategies and pitfalls. *Cell Stem Cell, 11*(4), 452–460.

Barry, E. R., Morikawa, T., Butler, B. L., Shrestha, K., de la Rosa, R., Yan, K. S., et al. (2013). Restriction of intestinal stem cell expansion and the regenerative response by YAP. *Nature, 493*, 106–110.

Bjerknes, M., & Cheng, H. (1999). Clonal analysis of mouse intestinal epithelial progenitors. *Gastroenterology, 116*(1), 7–14.

Bjerknes, M., & Cheng, H. (2001). Modulation of specific intestinal epithelial progenitors by enteric neurons. *Proceedings of the National Academy of Sciences of the United States of America, 98*(22), 12497–12502.

Bjerknes, M., & Cheng, H. (2002). Multipotential stem cells in adult mouse gastric epithelium. *American Journal of Physiology Gastrointestinal and Liver Physiology, 283*(3), G767–G777.

Bjerknes, M., & Cheng, H. (2010). Cell lineage metastability in Gfi1-deficient mouse intestinal epithelium. *Developmental Biology, 345*(1), 49–63.

Breault, D. T., Min, I. M., Carlone, D. L., Farilla, L. G., Ambruzs, D. M., Henderson, D. E., et al. (2008). Generation of mTert-GFP mice as a model to identify and study tissue progenitor cells. *Proceedings of the National Academy of Sciences of the United States of America, 105*(30), 10420–10425.

Brennan, P. C., Carr, K. E., Seed, T., & McCullough, J. S. (1998). Acute and protracted radiation effects on small intestinal morphological parameters. *International Journal of Radiation Biology, 73*(6), 691–698.

Buczacki, S. J., Zecchini, H. I., Nicholson, A. M., Russell, R., Vermeulen, L., Kemp, R., et al. (2013). Intestinal label-retaining cells are secretory precursors expressing Lgr5. *Nature, 495*(7439), 65–69.

Byrd, D. T., & Kimble, J. (2009). Scratching the niche that controls Caenorhabditis elegans germline stem cells. *Seminars in Cell & Developmental Biology, 20*(9), 1107–1113.

Cairnie, A. B., Lamerton, L. F., & Steel, G. G. (1965). Cell proliferation studies in the intestinal epithelium of the rat. I. Determination of the kinetic parameters. *Experimental Cell Research, 39*(2), 528–538.

Cambuli, F. M., Rezza, A., Nadjar, J., & Plateroti, M. (2013). Musashi1-Egfp mice, a new tool for differential isolation of the intestinal stem cell populations. *Stem Cells, 31*(10), 2273–2278.

Carmon, K. S., Gong, X., Lin, Q., Thomas, A., & Liu, Q. (2011). R-spondins function as ligands of the orphan receptors LGR4 and LGR5 to regulate Wnt/beta-catenin signaling. *Proceedings of the National Academy of Sciences of the United States of America, 108*(28), 11452–11457.

Cheng, H., & Leblond, C. P. (1974). Origin, differentiation and renewal of the four main epithelial cell types in the mouse small intestine. V. Unitarian Theory of the origin of the four epithelial cell types. *The American Journal of Anatomy, 141*(4), 537–561.

Clayton, E., Doupe, D. P., Klein, A. M., Winton, D. J., Simons, B. D., & Jones, P. H. (2007). A single type of progenitor cell maintains normal epidermis. *Nature, 446*(7132), 185–189.

Deckx, R. J., Vantrappen, G. R., & Parein, M. M. (1967). Localization of lysozyme activity in a Paneth cell granule fraction. *Biochimica et Biophysica Acta, 139*(1), 204–207.

de Lau, W., Barker, N., Low, T. Y., Koo, B. K., Li, V. S., Teunissen, H., et al. (2011). Lgr5 homologues associate with Wnt receptors and mediate R-spondin signalling. *Nature, 476*(7360), 293–297.

Doupe, D. P., Alcolea, M. P., Roshan, A., Zhang, G., Klein, A. M., Simons, B. D., et al. (2012). A single progenitor population switches behavior to maintain and repair esophageal epithelium. *Science, 337*(6098), 1091–1093.

Driskell, R. R., Giangreco, A., Jensen, K. B., Mulder, K. W., & Watt, F. M. (2009). Sox2-positive dermal papilla cells specify hair follicle type in mammalian epidermis. *Development, 136*(16), 2815–2823.

Durand, A., Donahue, B., Peignon, G., Letourneur, F., Cagnard, N., Slomianny, C., et al. (2012). Functional intestinal stem cells after Paneth cell ablation induced by the loss of transcription factor Math1 (Atoh1). *Proceedings of the National Academy of Sciences of the United States of America, 109*(23), 8965–8970.

Escobar, M., Nicolas, P., Sangar, F., Laurent-Chabalier, S., Clair, P., Joubert, D., et al. (2011). Intestinal epithelial stem cells do not protect their genome by asymmetric chromosome segregation. *Nature Communications, 2*, 258.

Fafilek, B., Krausova, M., Vojtechova, M., Pospichalova, V., Tumova, L., Sloncova, E., et al. (2013). Troy, a tumor necrosis factor receptor family member, interacts with Lgr5 to inhibit Wnt signaling in intestinal stem cells. *Gastroenterology, 144*, 381–391.

Farin, H. F., Van Es, J. H., & Clevers, H. (2012). Redundant sources of wnt regulate intestinal stem cells and promote formation of Paneth cells. *Gastroenterology, 143*(6), 1518–1529, e7.

Fuchs, E. (2009). The tortoise and the hair: Slow-cycling cells in the stem cell race. *Cell, 137*(5), 811–819.

Garabedian, E. M., Roberts, L. J., McNevin, M. S., & Gordon, J. I. (1997). Examining the role of Paneth cells in the small intestine by lineage ablation in transgenic mice. *The Journal of Biological Chemistry, 272*(38), 23729–23740.

Gerbe, F., Brulin, B., Makrini, L., Legraverend, C., & Jay, P. (2009). DCAMKL-1 expression identifies Tuft cells rather than stem cells in the adult mouse intestinal epithelium. *Gastroenterology, 137*(6), 2179–2180, author reply 2180–2181.

Giannakis, M., Stappenbeck, T. S., Mills, J. C., Leip, D. G., Lovett, M., Clifton, S. W., et al. (2006). Molecular properties of adult mouse gastric and intestinal epithelial progenitors in their niches. *The Journal of Biological Chemistry, 281*(16), 11292–11300.

Glinka, A., Dolde, C., Kirsch, N., Huang, Y. L., Kazanskaya, O., Ingelfinger, D., et al. (2011). LGR4 and LGR5 are R-spondin receptors mediating Wnt/beta-catenin and Wnt/PCP signalling. *EMBO Reports, 12*(10), 1055–1061.

Gregorieff, A., Pinto, D., Begthel, H., Destree, O., Kielman, M., & Clevers, H. (2005). Expression pattern of Wnt signaling components in the adult intestine. *Gastroenterology, 129*(2), 626–638.

Hanash, A. M., Dudakov, J. A., Hua, G., O'Connor, M. H., Young, L. F., Singer, N. V., et al. (2012). Interleukin-22 protects intestinal stem cells from immune-mediated tissue damage and regulates sensitivity to graft versus host disease. *Immunity, 37*(2), 339–350.

Hao, H. X., Xie, Y., Zhang, Y., Charlat, O., Oster, E., Avello, M., et al. (2012). ZNRF3 promotes Wnt receptor turnover in an R-spondin-sensitive manner. *Nature, 485*(7397), 195–200.

Hsu, Y. C., Pasolli, H. A., & Fuchs, E. (2011). Dynamics between stem cells, niche, and progeny in the hair follicle. *Cell, 144*(1), 92–105.

Huynh, D., Dai, X. M., Nandi, S., Lightowler, S., Trivett, M., Chan, C. K., et al. (2009). Colony stimulating factor-1 dependence of Paneth cell development in the mouse small intestine. *Gastroenterology, 137*(1), 136–144, e1–3.

Itzkovitz, S., Lyubimova, A., Blat, I. C., Maynard, M., van Es, J., Lees, J., et al. (2011). Single-molecule transcript counting of stem-cell markers in the mouse intestine. *Nature Cell Biology, 14*(1), 106–114.

Jaks, V., Barker, N., Kasper, M., van Es, J. H., Snippert, H. J., Clevers, H., et al. (2008). Lgr5 marks cycling, yet long-lived, hair follicle stem cells. *Nature Genetics, 40*(11), 1291–1299.

Jensen, K. B., Collins, C. A., Nascimento, E. N., Tan, D. W., Frye, M., Itami, S., et al. (2009). Lrig1 expression defines a distinct multipotent stem cell population in mammalian epidermis. *Cell Stem Cell, 4*(5), 427–439.

Jensen, K. B., & Watt, F. M. (2006). Single-cell expression profiling of human epidermal stem and transit-amplifying cells: Lrig1 is a regulator of stem cell quiescence. *Proceedings of the National Academy of Sciences of the United States of America, 103*(32), 11958–11963.

Jung, P., Sato, T., Merlos-Suarez, A., Barriga, F. M., Iglesias, M., Rossell, D., et al. (2011). Isolation and in vitro expansion of human colonic stem cells. *Nature Medicine, 17*(10), 1225–1227.

Kayahara, T., Sawada, M., Takaishi, S., Fukui, H., Seno, H., Fukuzawa, H., et al. (2003). Candidate markers for stem and early progenitor cells, Musashi-1 and Hes1, are expressed in crypt base columnar cells of mouse small intestine. *FEBS Letters, 535*(1–3), 131–135.

Kemper, K., Prasetyanti, P. R., De Lau, W., Rodermond, H., Clevers, H., & Medema, J. P. (2012). Monoclonal antibodies against Lgr5 identify human colorectal cancer stem cells. *Stem Cells, 30*(11), 2378–2386.

Kiel, M. J., He, S., Ashkenazi, R., Gentry, S. N., Teta, M., Kushner, J. A., et al. (2007). Haematopoietic stem cells do not asymmetrically segregate chromosomes or retain BrdU. *Nature, 449*(7159), 238–242.

Kim, T. H., Escudero, S., & Shivdasani, R. A. (2012). Intact function of Lgr5 receptor-expressing intestinal stem cells in the absence of Paneth cells. *Proceedings of the National Academy of Sciences of the United States of America, 109*(10), 3932–3937.

Kimble, J. E., & White, J. G. (1981). On the control of germ cell development in Caenorhabditis elegans. *Developmental Biology, 81*(2), 208–219.

Koo, B. K., Spit, M., Jordens, I., Low, T. Y., Stange, D. E., van de Wetering, M., et al. (2012). Tumour suppressor RNF43 is a stem-cell E3 ligase that induces endocytosis of Wnt receptors. *Nature, 488*(7413), 665–669.

Kosinski, C., Li, V. S., Chan, A. S., Zhang, J., Ho, C., Tsui, W. Y., et al. (2007). Gene expression patterns of human colon tops and basal crypts and BMP antagonists as intestinal stem cell niche factors. *Proceedings of the National Academy of Sciences of the United States of America*, *104*(39), 15418–15423.

Kosinski, C., Stange, D. E., Xu, C., Chan, A. S., Ho, C., Yuen, S. T., et al. (2010). Indian hedgehog regulates intestinal stem cell fate through epithelial-mesenchymal interactions during development. *Gastroenterology*, *139*(3), 893–903.

Lahar, N., Lei, N. Y., Wang, J., Jabaji, Z., Tung, S. C., Joshi, V., et al. (2011). Intestinal subepithelial myofibroblasts support in vitro and in vivo growth of human small intestinal epithelium. *PLoS One*, *6*(11), e26898.

Lam, N., Chesney, M. A., & Kimble, J. (2006). Wnt signaling and CEH-22/tinman/Nkx2.5 specify a stem cell niche in C. elegans. *Current Biology*, *16*(3), 287–295.

Lander, A. D., Kimble, J., Clevers, H., Fuchs, E., Montarras, D., Buckingham, M., et al. (2012). What does the concept of the stem cell niche really mean today? *BMC Biology*, *10*, 19.

Legg, J., Jensen, U. B., Broad, S., Leigh, I., & Watt, F. M. (2003). Role of melanoma chondroitin sulphate proteoglycan in patterning stem cells in human interfollicular epidermis. *Development*, *130*(24), 6049–6063.

Liu, W., Chen, L., Zhu, J., & Rodgers, G. P. (2006). The glycoprotein hGC-1 binds to cadherin and lectins. *Experimental Cell Research*, *312*(10), 1785–1797.

Lopez-Garcia, C., Klein, A. M., Simons, B. D., & Winton, D. J. (2010). Intestinal stem cell replacement follows a pattern of neutral drift. *Science*, *330*(6005), 822–825.

Marshman, E., Booth, C., & Potten, C. S. (2002). The intestinal epithelial stem cell. *Bioessays*, *24*(1), 91–98.

May, R., Riehl, T. E., Hunt, C., Sureban, S. M., Anant, S., & Houchen, C. W. (2008). Identification of a novel putative gastrointestinal stem cell and adenoma stem cell marker, doublecortin and CaM kinase-like-1, following radiation injury and in adenomatous polyposis coli/multiple intestinal neoplasia mice. *Stem Cells*, *26*(3), 630–637.

McLin, V. A., Henning, S. J., & Jamrich, M. (2009). The role of the visceral mesoderm in the development of the gastrointestinal tract. *Gastroenterology*, *136*(7), 2074–2091.

Merlos-Suarez, A., Barriga, F. M., Jung, P., Iglesias, M., Cespedes, M. V., Rossell, D., et al. (2011). The intestinal stem cell signature identifies colorectal cancer stem cells and predicts disease relapse. *Cell Stem Cell*, *8*(5), 511–524.

Miyazono, K., Kamiya, Y., & Morikawa, M. (2010). Bone morphogenetic protein receptors and signal transduction. *Journal of Biochemistry*, *147*(1), 35–51.

Montgomery, R. K., Carlone, D. L., Richmond, C. A., Farilla, L., Kranendonk, M. E., Henderson, D. E., et al. (2011). Mouse telomerase reverse transcriptase (mTert) expression marks slowly cycling intestinal stem cells. *Proceedings of the National Academy of Sciences of the United States of America*, *108*(1), 179–184.

Munoz, J., Stange, D. E., Schepers, A. G., van de Wetering, M., Koo, B. K., Itzkovitz, S., et al. (2012). The Lgr5 intestinal stem cell signature: Robust expression of proposed quiescent '+4' cell markers. *The EMBO Journal*, *31*(14), 3079–3091.

Ouellette, A. J., Miller, S. I., Henschen, A. H., & Selsted, M. E. (1992). Purification and primary structure of murine cryptdin-1, a Paneth cell defensin. *FEBS Letters*, *304*(2–3), 146–148.

Paneth, J. (1887). Ueber die secernirenden Zellen des Dünndarm-Epithels. *Archiv für Mikroskopische Anatomie*, *31*(1), 113–191.

Pellegrinet, L., Rodilla, V., Liu, Z., Chen, S., Koch, U., Espinosa, L., et al. (2011). Dll1- and dll4-mediated notch signaling are required for homeostasis of intestinal stem cells. *Gastroenterology*, *140*(4), 1230–1240, e1–7.

Pinchuk, I. V., Mifflin, R. C., Saada, J. I., & Powell, D. W. (2010). Intestinal mesenchymal cells. *Current Gastroenterology Reports*, *12*(5), 310–318.

Potten, C. S. (1977). Extreme sensitivity of some intestinal crypt cells to X and gamma irradiation. *Nature, 269*(5628), 518–521.

Potten, C. S., Booth, C., Tudor, G. L., Booth, D., Brady, G., Hurley, P., et al. (2003). Identification of a putative intestinal stem cell and early lineage marker; musashi-1. *Differentiation, 71*(1), 28–41.

Potten, C. S., Kovacs, L., & Hamilton, E. (1974). Continuous labelling studies on mouse skin and intestine. *Cell and Tissue Kinetics*, 7(3), 271–283.

Potten, C. S., Owen, G., & Booth, D. (2002). Intestinal stem cells protect their genome by selective segregation of template DNA strands. *Journal of Cell Science, 115*(Pt 11), 2381–2388.

Powell, D. W., Pinchuk, I. V., Saada, J. I., Chen, X., & Mifflin, R. C. (2011). Mesenchymal cells of the intestinal lamina propria. *Annual Review of Physiology, 73*, 213–237.

Powell, A. E., Wang, Y., Li, Y., Poulin, E. J., Means, A. L., Washington, M. K., et al. (2012). The pan-ErbB negative regulator Lrig1 is an intestinal stem cell marker that functions as a tumor suppressor. *Cell, 149*(1), 146–158.

Quyn, A. J., Appleton, P. L., Carey, F. A., Steele, R. J., Barker, N., Clevers, H., et al. (2010). Spindle orientation bias in gut epithelial stem cell compartments is lost in precancerous tissue. *Cell Stem Cell, 6*(2), 175–181.

Roth, S., Franken, P., Sacchetti, A., Kremer, A., Anderson, K., Sansom, O., et al. (2012). Paneth cells in intestinal homeostasis and tissue injury. *PLoS One*, 7(6), e38965.

Rothenberg, M. E., Nusse, Y., Kalisky, T., Lee, J. J., Dalerba, P., Scheeren, F., et al. (2012). Identification of a cKit(+) colonic crypt base secretory cell that supports Lgr5(+) stem cells in mice. *Gastroenterology, 142*(5), 1195–1205, e6.

Sala, F. G., Matthews, J. A., Speer, A. L., Torashima, Y., Barthel, E. R., & Grikscheit, T. C. (2011). A multicellular approach forms a significant amount of tissue-engineered small intestine in the mouse. *Tissue Engineering Part A, 17*(13–14), 1841–1850.

Samuel, S., Walsh, R., Webb, J., Robins, A., Potten, C., & Mahida, Y. R. (2009). Characterization of putative stem cells in isolated human colonic crypt epithelial cells and their interactions with myofibroblasts. *American Journal of Physiology Cell Physiology, 296*(2), C296–C305.

Sangiorgi, E., & Capecchi, M. R. (2008). Bmi1 is expressed in vivo in intestinal stem cells. *Nature Genetics, 40*(7), 915–920.

Sansom, O. J., Reed, K. R., Hayes, A. J., Ireland, H., Brinkmann, H., Newton, I. P., et al. (2004). Loss of Apc in vivo immediately perturbs Wnt signaling, differentiation, and migration. *Genes & Development, 18*(12), 1385–1390.

Sato, T., Stange, D. E., Ferrante, M., Vries, R. G., Van Es, J. H., Van den Brink, S., et al. (2011). Long-term expansion of epithelial organoids from human colon, adenoma, adenocarcinoma, and Barrett's epithelium. *Gastroenterology, 141*(5), 1762–1772.

Sato, T., van Es, J. H., Snippert, H. J., Stange, D. E., Vries, R. G., van den Born, M., et al. (2011). Paneth cells constitute the niche for Lgr5 stem cells in intestinal crypts. *Nature, 469*(7330), 415–418.

Sato, T., Vries, R. G., Snippert, H. J., van de Wetering, M., Barker, N., Stange, D. E., et al. (2009). Single Lgr5 stem cells build crypt-villus structures in vitro without a mesenchymal niche. *Nature, 459*(7244), 262–265.

Schepers, A. G., Vries, R., van den Born, M., van de Wetering, M., & Clevers, H. (2011). Lgr5 intestinal stem cells have high telomerase activity and randomly segregate their chromosomes. *The EMBO Journal, 30*(6), 1104–1109.

Schofield, R. (1978). The relationship between the spleen colony-forming cell and the haemopoietic stem cell. *Blood Cells, 4*(1–2), 7–25.

Schwalbe, G. (1872). Beiträge zur Kenntniss der Drüsen in den Darmwandungen, ins besondere der Brunnerschen Drüsen. *Archiv für Mikroskopische Anatomie, 8*, 92–140.

Shinozaki, S., Nakamura, T., Iimura, M., Kato, Y., Iizuka, B., Kobayashi, M., et al. (2001). Upregulation of Reg 1alpha and GW112 in the epithelium of inflamed colonic mucosa. *Gut, 48*(5), 623–629.
Shroyer, N. F., Helmrath, M. A., Wang, V. Y., Antalffy, B., Henning, S. J., & Zoghbi, H. Y. (2007). Intestine-specific ablation of mouse atonal homolog 1 (Math1) reveals a role in cellular homeostasis. *Gastroenterology, 132*(7), 2478–2488.
Shroyer, N. F., Wallis, D., Venken, K. J., Bellen, H. J., & Zoghbi, H. Y. (2005). Gfi1 functions downstream of Math1 to control intestinal secretory cell subtype allocation and differentiation. *Genes & Development, 19*(20), 2412–2417.
Snippert, H. J., van der Flier, L. G., Sato, T., van Es, J. H., van den Born, M., Kroon-Veenboer, C., et al. (2010). Intestinal crypt homeostasis results from neutral competition between symmetrically dividing Lgr5 stem cells. *Cell, 143*(1), 134–144.
Spits, H., & Di Santo, J. P. (2011). The expanding family of innate lymphoid cells: Regulators and effectors of immunity and tissue remodeling. *Nature Immunology, 12*(1), 21–27.
Takeda, N., Jain, R., LeBoeuf, M. R., Wang, Q., Lu, M. M., & Epstein, J. A. (2011). Interconversion between intestinal stem cell populations in distinct niches. *Science, 334*(6061), 1420–1424.
Tan, D. W., Jensen, K. B., Trotter, M. W., Connelly, J. T., Broad, S., & Watt, F. M. (2013). Single-cell gene expression profiling reveals functional heterogeneity of undifferentiated human epidermal cells. *Development, 140*(7), 1433–1444.
Tian, H., Biehs, B., Warming, S., Leong, K. G., Rangell, L., Klein, O. D., et al. (2011). A reserve stem cell population in small intestine renders Lgr5-positive cells dispensable. *Nature, 478*(7368), 255–259.
Trumpp, A., Essers, M., & Wilson, A. (2010). Awakening dormant haematopoietic stem cells. *Nature Reviews Immunology, 10*(3), 201–209.
Tumbar, T., Guasch, G., Greco, V., Blanpain, C., Lowry, W. E., Rendl, M., et al. (2004). Defining the epithelial stem cell niche in skin. *Science, 303*(5656), 359–363.
van der Flier, L. G., & Clevers, H. (2009). Stem cells, self-renewal, and differentiation in the intestinal epithelium. *Annual Review of Physiology, 71*, 241–260.
van der Flier, L. G., Sabates-Bellver, J., Oving, I., Haegebarth, A., De Palo, M., Anti, M., et al. (2007). The Intestinal Wnt/TCF Signature. *Gastroenterology, 132*(2), 628–632.
van der Flier, L. G., van Gijn, M. E., Hatzis, P., Kujala, P., Haegebarth, A., Stange, D. E., et al. (2009). Transcription factor achaete scute-like 2 controls intestinal stem cell fate. *Cell, 136*(5), 903–912.
van de Wetering, M., Sancho, E., Verweij, C., de Lau, W., Oving, I., Hurlstone, A., et al. (2002). The beta-catenin/TCF-4 complex imposes a crypt progenitor phenotype on colorectal cancer cells. *Cell, 111*(2), 241–250.
van Dop, W. A., Uhmann, A., Wijgerde, M., Sleddens-Linkels, E., Heijmans, J., Offerhaus, G. J., et al. (2009). Depletion of the colonic epithelial precursor cell compartment upon conditional activation of the hedgehog pathway. *Gastroenterology, 136*(7), 2195–2203, e1–7.
van Es, J. H., Sato, T., van de Wetering, M., Lyubimova, A., Nee, A. N., Gregorieff, A., et al. (2012). Dll1 + secretory progenitor cells revert to stem cells upon crypt damage. *Nature Cell Biology, 14*(10), 1099–1104.
Vries, R. G., Huch, M., & Clevers, H. (2010). Stem cells and cancer of the stomach and intestine. *Molecular Oncology, 4*(5), 373–384.
Watt, F. M., Lo Celso, C., & Silva-Vargas, V. (2006). Epidermal stem cells: An update. *Current Opinion in Genetics & Development, 16*(5), 518–524.
Wilson, A., Laurenti, E., Oser, G., van der Wath, R. C., Blanco-Bose, W., Jaworski, M., et al. (2008). Hematopoietic stem cells reversibly switch from dormancy to self-renewal during homeostasis and repair. *Cell, 135*(6), 1118–1129.

Wong, V. W., Stange, D. E., Page, M. E., Buczacki, S., Wabik, A., Itami, S., et al. (2012). Lrig1 controls intestinal stem-cell homeostasis by negative regulation of ErbB signalling. *Nature Cell Biology*, *14*(4), 401–408.

Yan, K. S., Chia, L. A., Li, X., Ootani, A., Su, J., Lee, J. Y., et al. (2012). The intestinal stem cell markers Bmi1 and Lgr5 identify two functionally distinct populations. *Proceedings of the National Academy of Sciences of the United States of America*, *109*(2), 466–471.

Yilmaz, O. H., Katajisto, P., Lamming, D. W., Gultekin, Y., Bauer-Rowe, K. E., Sengupta, S., et al. (2012). mTORC1 in the Paneth cell niche couples intestinal stem-cell function to calorie intake. *Nature*, *486*(7404), 490–495.

Yui, S., Nakamura, T., Sato, T., Nemoto, Y., Mizutani, T., Zheng, X., et al. (2012). Functional engraftment of colon epithelium expanded in vitro from a single adult Lgr5(+) stem cell. *Nature Medicine*, *18*(4), 618–623.

Zacharias, W. J., Li, X., Madison, B. B., Kretovich, K., Kao, J. Y., Merchant, J. L., et al. (2010). Hedgehog is an anti-inflammatory epithelial signal for the intestinal lamina propria. *Gastroenterology*, *138*(7), 2368–2377, e1–4.

Zhang, J., Liu, W. L., Tang, D. C., Chen, L., Wang, M., Pack, S. D., et al. (2002). Identification and characterization of a novel member of olfactomedin-related protein family, hGC-1, expressed during myeloid lineage development. *Gene*, *283*(1–2), 83–93.

CHAPTER FOUR

Epithelial Stem Cells in Adult Skin

Ana Mafalda Baptista Tadeu, Valerie Horsley[1]
Department of Molecular, Cell and Developmental Biology, Yale University, New Haven, Connecticut, USA
[1]Corresponding author: e-mail address: valerie.horsley@yale.edu

Contents

Abstract

The skin is the first line of defense against dehydration and external environmental aggressions. It constantly renews itself throughout adult life mainly due to the activity of tissue-specific stem cells. In this review, we discuss fundamental characteristics of different stem cell populations within the skin and how they are able to contribute to normal skin homeostasis. We also examine the most recent results regarding the cell-intrinsic and -extrinsic components of the stem cell niche within the adult skin epithelium. Finally, we address the recent efforts to understand how abnormal regulation of stem cell activity contributes to the initiation and progression of skin-associated cancers.

1. INTRODUCTION

The skin serves as a highly dynamic and adaptable outer coating for the bodies of many animal species, protecting against the external environment and providing tactile function for touch sensation. These roles are engendered by multiple types of differentiated cells within the interfollicular epidermis (IFE) and by the formation of epidermal appendages such as hair

Current Topics in Developmental Biology, Volume 107
ISSN 0070-2153
http://dx.doi.org/10.1016/B978-0-12-416022-4.00004-4

follicles (HFs), sebaceous glands (SGs), and sweat glands. In order for organisms to navigate a continuously changing external environment, specialized cell types in the skin continually regenerate through the action of several distinct epithelial stem cell (SC) populations that self-renew and generate cells with unipotent and multipotent differentiation potential (Lee & Tumbar, 2012; Sennett & Rendl, 2012).

The IFE and its appendages interact with the dermis, which is rich with connective tissue and a multitude of cells that confer structure and function to the epithelial cells (Fig. 4.1). Beneath the basement membrane, three main cell layers exist to support the epithelium. The uppermost papillary dermis contains fine matrix fibers, while a second layer of the reticular dermis is composed of large fibers of matrix molecules (Dick, 1947). A thick layer of dermal adipocytes resides below the reticular dermis (Chase, Montagna, & Malone, 1953). These layers are permeated with additional cell types including inflammatory cells, neurons, blood vessels, and muscle cells. The function of the dermal cell types in controlling epithelial SCs in the skin is just starting to emerge.

This review will describe the organization and cellular hierarchy of epithelial SCs in the skin. We will highlight the cellular and molecular mechanisms that regulate epithelial SC populations within the outermost IFE and its appendages, HFs, SGs, and sweat glands with an emphasis on recent work in the area. Finally, we will also highlight recent work that sheds light into mechanisms of SC deregulation and their contribution to epidermal cancer formation and progression.

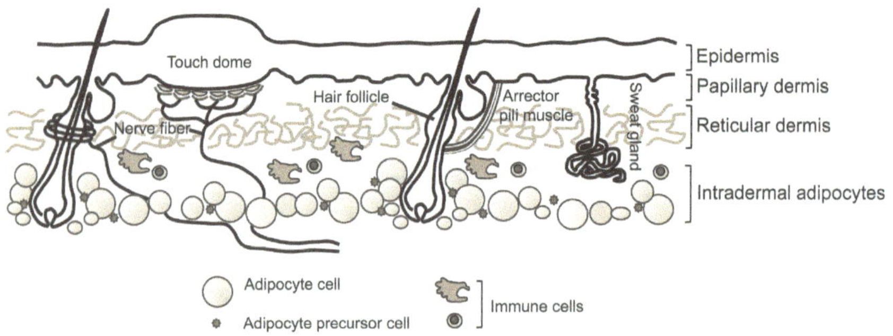

Figure 4.1 Schematic cross-section representation of mammalian skin. The skin is composed of a multitude of cell types and skin appendages that need to interact efficiently and accurately to ensure normal tissue homeostasis. (See color plate.)

2. STEM CELLS IN THE INTERFOLLICULAR EPIDERMIS

The outermost layer of mammalian skin is comprised of a multilayered or stratified epidermis of the IFE that is anchored to the underlying papillary dermis via integrin-mediated adhesion to a basement membrane (reviewed in Blanpain & Fuchs, 2006). The epidermal cells that adhere to the basement membrane are proliferative keratinocytes of the basal layer. Epidermal keratinocytes are formed during embryonic development from the surface ectoderm and generate differentiated suprabasal cells through asymmetric cell divisions (Lechler & Fuchs, 2005). Cells in the outermost epidermal layer (stratum corneum) tightly adhere to one other and form a protein–lipid matrix that ultimately creates the skin's essential barrier (reviewed in Sandilands, Sutherland, Irvine, & McLean, 2009). The cells of the stratum corneum are constantly shed and thus, proliferative basal cells fuel the continual reformation of these dedicated cells of the IFE.

Classic experiments analyzing IFE homeostasis via morphology and proliferation proposed the existence of an epidermal proliferative unit (EPU) in which a central slow-cycling basal cell generates a defined number of rapidly dividing progenitor cells that differentiate into a restricted number of "units" (Loeffler, Potten, & Wichmann, 1987; Mackenzie, 1969, 1970; Potten, 1981; Potten, Wichmann, Loeffler, Dobek, & Major, 1982). More recently, extensive and quantitative analyses of basal cell progeny using genetic lineage tracing was performed in several mouse models (Clayton et al., 2007; Doupé, Klein, Simons, & Jones, 2010; Mascré et al., 2012). The groundbreaking initial studies used mouse models expressing tamoxifen-regulated cre recombinase driven by an inducible CYP1A1 promoter (AhcreERT), crossed to a YFP reporter strain (Clayton et al., 2007; Doupé et al., 2010). Low-dose tamoxifen administration allowed single-cell labeling within the tail and ear IFE and the ability to follow clone generation long term. Interestingly, the average size of persisting clones increased linearly with time, which is contrary to the previously proposed restricted size of the EPU. Furthermore, mathematic analysis of the clone generation in these studies suggested that basal cells could generate proliferative or differentiated progeny stochastically. However, whether these experiments labeled the most primitive SC within the IFE was unclear.

More recently, comparing lineage tracing in the IFE of mouse models expressing either an inducible CreER driven by the keratin 14 (K14) promoter or a fragment of the Involucrin (Inv) promoter reveal a hieracherical and heterogeneous nature of progenitor cells in the IFE (Mascré et al., 2012). In the InvCreER mouse model, persistent labeled clones followed the same cell-fate dynamics and linear growth patterns as the clones generated in the AhCre model (Clayton et al., 2007; Doupé et al., 2010). By contrast, the persistent clones generated in the K14CreER mouse model displayed stochastic fate decisions but were restricted in their growth potential, consistent with an EPU-type model and supporting the existence of a slow-cycling SC within the IFE. Molecular characterization of basal cells labeled in the K14CreER and InvCreER mouse models further supported the labeling of two distinct cell types within the IFE. Together, these studies reveal a hierarchy of cells within the IFE: a slow-cycling SC (marked by K14CreER) that gives rise to more rapidly cycling committed progenitors (marked by AhCreER or InvCreER) that subsequently undergo terminal differentiation (Fig. 4.2). These studies provide significant knowledge regarding the molecular characteristics of the slow-cycling IFE progenitors that will allow the future identification of novel markers as well as the further analysis of the molecular regulation of these cells during skin homeostasis.

Heterogeneity also exists within the slow-cycling SCs of the IFE. Within the majority of the murine epidermis, orthokeratotic differentiation generates cells that are spinous, granular, and have lost nuclei in the stratum corneum (Didierjean, Wrench, & Saurat, 1983; Schweizer & Marks, 1977). In the tail, however, IFE regions between the organized arrays of HFs postnatally develop a parakeratotic program postnatally, resulting in IFE regions lacking the granular layer and retention of nuclei within the stratum corneum (Didierjean et al., 1983; Schweizer & Marks, 1977). Careful genetic lineage analysis of the clonogenic behavior of K14CreER marked cells in these different regions revealed that IFE SCs are restricted to a particular compartment and display distinct proliferative behavior, suggesting that heterogeneous, unipotent SCs exist within the IFE (Gomez, Chua, Miremadi, Quist, & Headon, 2013).

Within the adult IFE, a separate epithelial SC niche exists for Merkel cells, specialized sensory cells that allow mammals to respond to mechanical stimuli during touch sensations (Fig. 4.2). Mature Merkel cells reside in touch domes, which are clusters of cells that are innervated by afferent somatosensory nerve fibers at the dermal–epidermal border in specialized skin regions (Merkel, 1875). These unique cells express both cytokeratins and neuroendocrine proteins and thus their developmental origin was unclear until several recent studies demonstrated that these cells derive from K14 expressing cells of the developing epidermis (Morrison, Miesegaes,

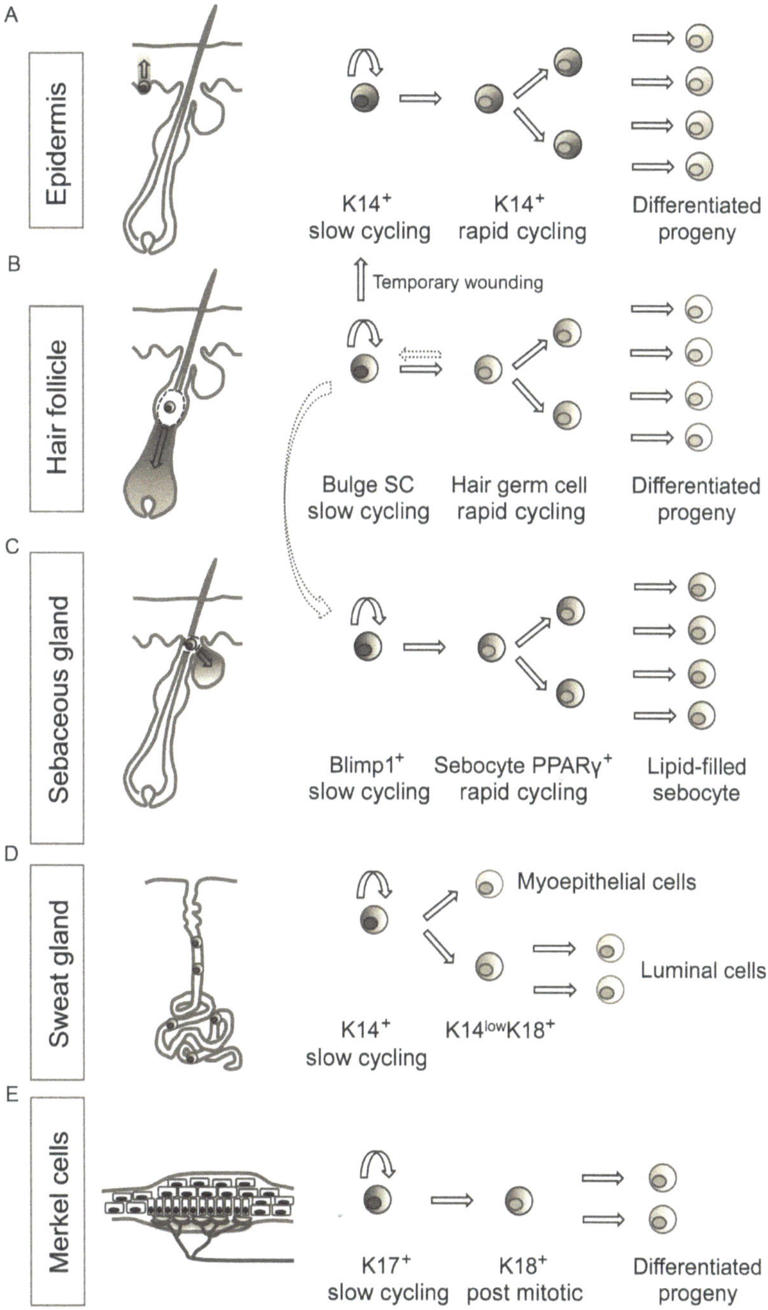

Figure 4.2 Different stem cell populations regulate normal epidermal homeostasis. Schematic representation of the different stem cell population found within different

(Continued)

Lumpkin, & Maricich, 2009; Van Keymeulen et al., 2009). Once mature touch domes are formed in the epidermis, $K17^+$ keratinocytes within the epidermal touch dome maintain mature Merkel cells during homeostasis, turning over every 2 months in adult skin (Doucet, Woo, Ruiz, & Owens, 2013). Ablation of $K17^+$ cells results in Merkel cell loss and the inability of sensory afferents to innervate the skin, indicating a functional role for Merkel cell progenitor cells in maintaining the neuronal niche in the skin (Doucet et al., 2013). How the surrounding keratinocytes and interacting neurons control Merkel cell progenitors will be an interesting area for future investigation.

3. STEM CELLS IN THE PILOSEBACEOUS UNIT

During epidermal development, basal progenitor cells are specified to appendage cell fates such as the pilosebaceous unit containing the HF and the SG (Millar, 2002). The SG is a continually regenerative gland that produces sebum, specialized lipids that are released into the hair canal onto the skin's surface through lysis of differentiated sebocytes (Niemann & Horsley, 2012). The HF is maintained through a three-stage regenerative process of the hair cycle, which begins with HF growth (anagen), where the lower portion of the HF grows into the dermis and produces differentiated lineages that allow hair production (Lee & Tumbar, 2012; Sennett & Rendl, 2012). Eventually, hair growth ends and a destructive phase (catagen) starts, where the lower portion of the HF dies and regresses, leaving a permanent region, called the bulge. The final stage involves a resting phase (telogen) before the next regenerative hair cycle starts when bulge SCs are activated to start growth of a new follicle.

Figure 4.2—Cont'd skin compartments. (A) Different lineage tracing experiment have shown that the IFE is maintained by the presence of $K14^+$ progenitor basal cells that are able to generate the differentiated lineages that comprise the squamous epithelium of the skin. (B and C) Within the pilosebaceous unit, a slow cycling SC residing in the bulge is able to give rise to all the different HF lineages and regenerate a new hair follicle during anagen. They can also contribute to the IFE after wounding and to the SG population that is usually maintained by a resident pool of $Blimp1^+$ slow-cycling residing progenitors. (D) In the sweat gland, a $K14^+$ progenitor was shown to be able to regenerate a full functioning gland capable of sebum production. Finally, within the touch dome structure (E), $K17^+$ slow-cycling cells have been shown to give rise to $K18^+$ rapid-cycling progenitor that can further generate differentiated progeny. (See color plate.)

The initial identification of the SC properties of bulge cells took advantage of their slow-cycling nature, which allowed retention of nucleotide analogs such as BrdU and tritiated thymidine during pulse-chase experiments (Bickenbach, 1981; Cotsarelis, Sun, & Lavker, 1990). Innovative genetic mouse models were also developed that allowed fluorescent labeling of histone H2B in epithelial cells and after a lengthy chase period, bulge cells were the primary cell-type labeled within the skin (Tumbar et al., 2004). These experiments demonstrated that the cell cycle of bulge cells was slower than the other epithelial cells of the skin.

Genetic lineage tracing experiments using multiple mouse models with cell-type-specific expression of cre recombinase have demonstrated that several populations of cells within the pilosebaceous unit have the capacity to generate lineages of the HF and SG (Fig. 4.2) (Jaks et al., 2008; Jensen et al., 2009; Morris et al., 2004; Petersson et al., 2011; Snippert et al., 2010), as well as the epidermis after wounding (Ito et al., 2005; Levy, Lindon, Zheng, Harfe, & Morgan, 2007; Snippert et al., 2010). The isolation and transplantation of putative follicular SCs based on microdissection or based on fluorescence-activated cell sorting for bulge markers such CD34 and α6-integrin also demonstrated the multipotency of SCs within the HF bulge (Blanpain, Lowry, Geoghegan, Polak, & Fuchs, 2004; Claudinot, Nicolas, Oshima, Rochat, & Barrandon, 2005). In sum, these studies have demonstrated that bulge cells are multipotent, contributing temporarily to the epidermis during wounding and primarily to the homeostasis of HF and SG.

The activation of HF SCs during hair cycling involves several cellular processes including proliferation and migration of bulge cells and hair germ cells (Greco, Chen, Rendl, Schober, & Pasolli, 2009; Rompolas et al., 2012; Zhang, Cheong, Ciapurin, McDermitt, & Tumbar, 2009). The activated SCs generate proliferative progeny and differentiate to form the inner root sheath and the HF shaft (Ito et al., 2005; Oshima, Rochat, Kedzia, Kobayashi, & Barrandon, 2001; Tumbar et al., 2004).

Similar to the heterogeneity in IFE SCs (Gomez et al., 2013), the permanent portion of the HF contains heterogeneous group of cells with differing capacities for contributing to homeostasis of epithelial lineages of the skin (Goldstein & Horsley, 2012). At the junction between the IFE and HF, cells expressing SC antigen-1 and/or leucine-rich repeats and immunoglobulin-like domain 1 (Lrig1) exist and are able to generate IFE and SG cells but are limited in their contribution to the HF lineage (Jensen et al., 2009, 2008). Below the junctional zone and at the base of the SG, several slow-cycling cells exist such as MTS24 expressing cells (Nijhof et al., 2006) and also leucine-rich repeat containing G protein-coupled receptor 6 or B-lymphocyte-induced maturation protein 1 (Blimp1) positive cells that can generate SG lineages (Horsley et al., 2006; Snippert et al., 2010). The cells that primarily generate HF lineages are located at the base of the isthmus and throughout the $CD34^+$, $K15^+$ region of the bulge (Blanpain et al., 2004; Brownell, Guevara, Bai, Loomis, & Joyner, 2011; Morris et al., 2004; Snippert et al., 2010; Tumbar et al., 2004). At the base of this compartment and within the hair germ, leucine-rich repeat containing G protein-coupled receptor 5 (Lgr5) positive cells can give rise to different HF lineages (Jaks et al., 2008). These populations are not restricted but dynamically interact: $K15^+$ bulge cells can migrate to the SG where they locally self-renew and continue to contribute to the normal gland homeostasis (Petersson et al., 2011). Additionally, $Lgr5^+$descendants of bulge cells can repopulate the bulge at the end of hair HF regression (Hsu, Pasolli, & Fuchs, 2011).

4. STEM CELLS IN THE SWEAT GLAND

Eccrine sweat glands are an additional epidermal appendage that generate sweat for mammalian thermoregulation. Sweat fluids are generated by specialized epithelial cells in a coiled structure and are secreted into luminal structures into a duct that opens directly onto the skin surface (Lobitz & Dobson, 1961). Similar to the mammary gland, the secretory coil of the sweat gland contains an outer basal layer of myoepithelial cells that express K5, K14, and smooth muscle actin. An inner suprabasal layer of luminal cells is positive for K8, K18, and K19 (Langbein et al., 2005; Moll & Moll, 1992; Schön, Benwood, O'Connell-Willstaedt, & Rheinwald, 1999). In contrast to the dramatic cellular changes during puberty and pregnancy in the mammary gland (Richert, Schwertfeger, Ryder, & Anderson, 2000), the sweat gland presents little sign of continual renewal (Lu et al., 2012).

Initial experiments to address the existence of sweat gland progenitors identified cell divisions in basal cells of the gland's duct region (Lobitz, Holyoke, & Montagna, 1954). Other studies analyzed the regenerative potential of sweat glands *in vitro* by showing proliferation of dissociated gland cells in culture and *in vivo* by demonstrating the contribution of sweat gland for epidermal reconstitution after superficial skin injuries (Biedermann et al., 2010; Miller, Burke, Rader, Coulombe, & Lavker, 1998). Recent work has identified $K14^+$ progenitors that can differentiate into myoepithelial cells and can also stratify to form the suprabasal layer of $K14^{low}/K18^{+}$ cell that give rise to luminal cells (Lu et al., 2012) (Fig. 4.2). Furthermore, lineage tracing and EdU pulse-chase experiments were able to show that both sweat gland progenitor cell types participate in repair of the epidermis after wounding but the sweat gland itself remains relatively quiescent throughout the wound repair process (Lu et al., 2012). The plasticity of sweat gland progenitor cells was illustrated by transplantation studies in which progenitor cells were injected into cleared mammary fat pads and were able to reconstitute an entire sweat gland. By contrast, luminal cells were unable to contribute to *de novo* formation of sweat glands (Lu et al., 2012). Further studies will be necessary to define the molecular mechanisms that regulate sweat glands homeostasis and their contribution to epidermal healing.

5. COMPONENTS OF ADULT STEM CELL NICHES IN THE SKIN

The identification, isolation, and characterization of different populations of SCs within the adult skin epithelium identified layers of molecular mechanisms that act intrinsically to regulate SC behavior. Recent reviews have discussed several studies that identify molecular mechanisms that regulate SC behavior in the skin (Arwert, Hoste, & Watt, 2012; Blanpain & Fuchs, 2006; Hsu et al., 2011; Lee & Dai, 2013; Lee & Tumbar, 2012). Several signaling pathways are involved in regulating SC behavior in the skin including Wnt, BMP, FGF, and PDGF (Festa et al., 2011; Greco et al., 2009; Kandyba et al., 2013; Kobielak, Stokes, de la Cruz, Polak, & Fuchset, 2007; Myung, Takeo, Ito, & Atit, 2013; Oshimori & Fuchs, 2012; Plikus et al., 2008). In addition, the involvement of dermal cell types has recently come to light (Goldstein & Horsley, 2012). Here, we will highlight recent developments regarding the intrinsic and extrinsic regulation of epithelial SCs.

5.1. Intrinsic regulation of stem cell function

Cellular metabolism and environmental assaults can result in DNA damage, especially in the skin, which directly receives UV irradiation from the sun and environmental mutagens that can induce genomic instability. Interestingly, HFSCs are more resistant to radiation-induced damage compared to other epithelial skin cells (Sotiropoulou et al., 2010). To attain this resistance, HFSCs express high levels of the antiapoptotic protein B-cell lymphoma 2 (Bcl2) and transiently express p53 to promote survival. In addition, breast cancer 1 (Brca1) is essential for DNA damage repair (Gudmundsdottir & Ashworth, 2006; Moynahan & Jasin, 2010) and epidermal deletion of Brca1 leads to defects in HF formation as well as induction of caspase-dependent apoptosis that leads to hyperproliferation and subsequent exhaustion of adult SCs (Sotiropoulou et al., 2013). The differential regulation of DNA damage in SCs also exists within other tissues (Mandal, Blanpain, & Rossi, 2011) and may be similar to the mechanisms that act in SCs in the IFE and other epidermal appendages.

Several transcriptional regulators of SC function in the HF have been identified and are shared among SCs of other tissues including transcription factor 3 and 4 (TCF3/4), nuclear factor of activated T-cells 1 (NFATc1) and sex determining region Y-box 9 (Sox9) (Blanpain & Fuchs, 2006; Nguyen et al., 2009; Nguyen, Rendl, & Fuchs, 2006; Nowak, Polak, Pasolli, & Fuchs, 2008). In addition, Lgr5 (Barker et al., 2007) and the atypical HOP homeobox protein Hopx are expressed by an intestinal SC epithelial pool at the base of the crypt (Takeda, Jain, LeBoeuf, Wang, & Lu, 2011). In the HF, Hopx is expressed within bulge cells and can contribute to all HF lineages upon HF growth as well as to IFE cells upon wounding (Takeda, Jain, LeBoeuf, & Padmanabhan, 2013). Lower bulge cells expressing the SC marker Lgr5 also express Hopx, are able to escape apoptosis during the HF death phase and contribute long-term to bulge cell maintenance (Takeda et al., 2013).

The transcription factor LIM homeobox protein 2 (Lhx2) is another homeobox protein that has been implicated in regulating morphogenesis and patterning of ectodermal derivatives and in SC maintenance and quiescence within the HF SC niche (Mardaryev et al., 2011; Rhee, Polak, & Fuchs, 2006; Törnqvist, Sandberg, Hägglund, & Carlsson, 2010). Lhx2 is expressed in the bulge and secondary hair germ where it co-localizes with the SC markers Sox9, Tcf4, and Lgr5. In response to skin injury, $Lhx2^{+}$ cells within the bulge and secondary hair germ proliferate and contribute to skin re-epithelialization via positive regulation of *Sox9* and *Tcf4* while inhibiting HF cycling through negatively regulating *Lgr5* (Mardaryev et al., 2011). These and many other studies have provided novel insights of how Wnt and BMP signaling pathways and transcriptional regulation networks modulate activity of epithelial SCs during normal homeostasis and in response to injury (Blanpain & Fuchs, 2006; Lee & Tumbar, 2012; Sennett & Rendl, 2012).

Epithelial SCs are also regulated post-transcriptionally and translationally in part by microRNAs (miRNAs), which are small noncoding RNAs that alter RNA translation or stability to control gene expression. Complete ablation of miRNA production by deletion of the upstream processing enzyme Dicer in mice results in perinatal lethality and severe HF defects (Andl et al., 2006; Yi et al., 2006). Among these defects are undeveloped and misaligned HFs, increased apoptosis and lack of $K15^+$ and $CD34^+$ cells within the bulge compartment suggesting that miRNAs, in general, are important for HF SC maintenance (Andl et al., 2006).

Several miRNAs are spatiotemporally regulated within the IFE and the HFSCs. MiR203 was shown to be preferentially enriched in the IFE versus the HF (Andl et al., 2006; Yi, Poy, Stoffel, & Fuchs, 2008) and is sufficient to promote IFE differentiation and suppress self-renewal in the IFE by controlling the expression of p63 (Andl et al., 2006; Yi et al., 2008).Additionally, miR203 is transcriptionally activated during asymmetric cell division in the developing epidermis, localizing to the differentiated daughter cell, where it promotes cell cycle exit and abolishes self-renewal in a process involving co-suppression of p63, S-phase kinase-associated protein 2 (Skp2), and musashi RNA-binding protein 2 (Msi2) (Jackson et al., 2013).

An additional miRNA, miR125b is sufficient to alter IFE homeostasis and abrogate hair specification (Zhang, Stokes, Polak, & Fuchs, 2011). MiR31 can also alter HFSC activity by targeting fibroblast growth factor 10 (Fgf10), distal-less homeobox 3 (Dlx3), several keratin genes and also components of the Wnt and BMP signaling pathways (Mardaryev et al., 2010). The differential regulation of several miRNAs in the epithelium of the skin suggests that roles for additional miRNAs will be defined as this burgeoning field continues to expand.

Another level of regulation of skin SCs occurs through modification of histones and DNA to epigenetically regulate transcription (Calo & Wysocka, 2013). Several epigenetic factors play a role in epidermal differentiation (Mulder et al., 2012). Histone acetylation and methylation through histone deacetylase and methyltransferase activity, respectively, regulate IFE development (Driskell et al., 2012; LeBoeuf et al., 2010) and homeostasis (Driskell et al., 2012). Maintenance of repressive histone modifications via the polycomb repressor complex, enhancer of zeste homolog 1 (Ezh1) and Ezh2 are essential for IFE differentiation and for HF morphogenesis and maintenance (Bardot et al., 2013; Ezhkova et al., 2011). Merkel cells also require Ezh2 proteins for their maintenance through the regulation of the transcription factor Sox2 (Bardot et al., 2013). Histone methylation controlled by the demethylase Jumonji domain containing 3 (JmjD3) is essential for IFE differentiation (Sen, Webster, Barragan, Chang, & Khavari,

2008), while the demethylase Jumonji/jmjc domain-containing protein 2 (Jarid2) is required to maintain IFE basal progenitors (Mejetta et al., 2011). In addition, the DNA methyltransferase 1 (DNMT1) and the ubiquitin like, containing PHD and RING finger domain-1 (UHRF1) are expressed in basal cells and are downregulated once cells enter the differentiation program suggesting that they are also involved in regulating stemness. Deletion ofDNMT1 in human skin regeneration assays induced premature differentiation of progenitors and progressive tissue loss further demonstrating its importance for self-renewal (Sen, Reuter, Webster, Zhu, & Khavari, 2010).

Additional control of SC function occurs through the regulation of gene expression by altering nucleosome positioning through the action of chromatin remodeling complexes such as the SWI/SNF complex (Kidder, Palmer, & Knott, 2009). By rearranging nucleosome positions within the chromatin, these complexes regulate RNA polymerase II occupancy and thus transcriptional initiation in an ATP-dependent manner (Liu, Balliano, & Hayes, 2011). At the crux of these complexes, brahma-related gene 1 (Brg1) acts as a catalytic subunit and regulates SC proliferation and differentiation. In the HF, it was recently shown that Brg1 is dynamically activated after SC activation in the skin. Deletion of Brg1 with the bulge-specific NFATc1-Cre induced precocious HF regression, loss of HFSCs, and progressive hair loss (Xiong et al., 2013). Molecularly, Brg1 and Shh act in a molecular loop, where Brg1 regulates Shh expression and Shh activates Brg1 expression within the follicle (Xiong et al., 2013). Whether Brg1 regulates additional genes to control HFSC function will be an interesting area of future investigation.

5.2. Cell extrinsic regulation of SC function

The intrinsic regulation of epithelial SCs in the skin is influenced by multiple cell types, including follicle-associated melanocytes and other cells within the dermis (Goldstein & Horsley, 2012). Of primary importance, the dermal papillae (DP) is a mesenchymal cell population that abuts the HF, induces its morphogenesis and remains associated with the follicle throughout its life cycle (Driskell, Clavel, Rendl, & Watt, 2011). The association of the DP with the follicle is essential for the activation of HF growth (Chi, Wu, & Morgan, 2013; Rompolas et al., 2012) and the size of the DP can define the size and shape of the hair follicle (Chi et al., 2013). Several signaling ligands are expressed by the DP (Rendl, Lewis, & Fuchs, 2005) and the identification of DP specific Cre lines (Enshell-Seijffers, Lindon, Kashiwagi, & Morgan, 2010; Grisanti et al., 2013) will allow the identification of the molecular mechanisms by which these cells control HFSC activity.

Below the DP, a depot of dermal adipocytes displays dynamic changes in size during the hair cycle. Both adipocyte hypertrophy and adipogenesis contribute to the growth of the adipocyte depot in the skin. The production of immature adipocyte precursor cells during adipogenesis is both necessary and sufficient to drive hair cycling (Festa et al., 2011). Adipogenesis also occurs after acute wounding and inhibition of adipogenesis can alter fibroblast function in the skin, leading to wound closure failure (Schmidt & Horsley, 2013). The molecular mechanisms by which adipocyte lineage cells function in the skin will reveal novel components of the skin SC niche.

Permeating through the dermal cell layers, somatosensory nerve fibers innervate the touch dome in the IFE and surround the HF in a piloneural collar (Lumpkin, Marshall, & Nelson, 2010). The HF cells of each follicle type may provide unknown cues for the distinct neural innervation. Interestingly, each hair follicle type is innervated by distinct mechanoreceptors which converge within the dorsal horn of the spinal column to process touch sensations (Li et al., 2011). Afferents from the dorsal root ganglion produce glutamate that is essential for the proper development of the piloneural mechanoreceptors. The neural innervation also provides signals for the HF, such as Shh and can promote the contribution of HF cells to wound healing and SG homeostasis (Brownell et al., 2011).

Genome-wide association studies have implicated inflammatory cells in the regulation of hair loss (Petukhova et al., 2010), suggesting that these cells may be instrumental in SC function in the skin. Supporting a role for immune cells in hair cycling, mice lacking γδT cells through deletion of TCRδ have defects in hair cycling (Kloepper, Kawai, Bertolini, Kanekura, & Paus, 2013). Inflammatory cells also play many roles in wound healing (Eming, Krieg, & Davidson, 2007), and γδT cells can induce HF neogenesis from healed epithelial cells after wounding via the production of Fgf9 (Gay et al., 2013). Functional roles of other immune cells may reveal their functions in the control of additional epithelial SCs in the skin.

6. STEM CELLS IN EPITHELIAL SKIN CANCERS

Several types of epithelial cancers form in the skin. Papillomas, basal cell carcinomas (BCCs), and squamous cell carcinomas (SCCs) are found in the IFE, while pilomatriomas, trichofolliculomas, and SG carcinomas are associated with the pilosebaceous unit. Resident epidermal SCs have been proposed to initiate epithelial tumorigenesis. Furthermore, maintenance of tumors following initiation is thought to be driven by tumor cells with SC characteristics such as self-renewal, and slow-cycling, properties (Al-Hajj & Clarke, 2004).

BCCs comprise 80% of epithelial skin cancers and are often associated with activating mutations in the Shh pathway via inactivation of the repressive receptor Patched (Ptch), expression/activation of the transcriptional mediators GLI family zinc finger (Gli) proteins, or activation of the Shh signal transducer, smoothened (Smo) (Athar, Tang, Lee, Kopelovich, & Kim, 2006; Rogers et al., 2010). The morphology of BCCs is variable, forming various subtypes including superficial and nodular tumors, suggesting that different cellular origins may exist.

Based on the similarities of keratin expression between BCCs and HF progenitor cells (Asada, 1993) and the inability of BCCs to develop in irradiated mice with only one allele of Ptch1 during HF SC quiescence (Mancuso et al., 2006), it was initially hypothesized that HF progenitor cells were the cells of origin for BCCs. However, genetic lineage-tracing experiments in mice expressing an active Gli2 demonstrated that BCCs can arise from several epithelial cell types including the bulge, SGs, and IFE (Grachtchouk et al., 2011). Similarly, $Lgr5^{+}$ bulge progeny required wounding to generate BCC-like lesions upon genetic Gli1 activation (Kasper et al., 2011). Expression of a constitutively active form of Smo in a cell-type specific manner was shown to generate BCCs preferentially from cells in the IFE and infundibulum rather than from bulge cells (Youssef et al., 2010). Taken together, these data suggest that BCC may have multiple origins and that wounding promotes BCC expansion.

Abnormal Wnt signaling has been associated with the development of epidermal tumors. In fact expression of a constitutively active form of β-catenin promotes the development of pilomatricomas and trichofolliculomas and mutations of lymphoid enhancer-binding factor 1 (lef1) that normally inactivates β-catenin is associated with human SG tumors (Chan, Gat, McNiff, & Fuchs, 1999; Gat, DasGupta, Degenstein, & Fuchs, 1998; Takeda et al., 2006). Inducible activation of β-catenin under the K14 promoter lead to the formation of lesions similar to pilomatricomas that regress when β-catenin is no longer activated (Lo Celso, Prowse, & Watt, 2004). Furthermore, expression of ΔNLef1 in epidermis induced SG tumors (Niemann, Owens, Hülsken,

Birchmeier, & Watt, 2002; Niemann, Owens, Schettina, & Watt, 2007) and ablation of β-catenin also under the K14 promoter caused a regression of chemically induced papillomas (Malanchi et al., 2008). Different SC populations react differently to Wnt signaling-associated tumor formation stimuli. For example, sustained β-catenin activity under the K15 promoter leads to an increase in proliferation, expression of Wnt target genes within the bulge compartment, and inability to form pilomatricomas even when a wounding stimulus is enforced, while sustained β-catenin activity under a truncated K5 promoter (expressed in SG and HF bulb) causes conversion of SG into HF structures that develop and resemble benign tumors (Baker, Verstuyf, Jensen, & Watt, 2010).

SCCs encompass 20% of epithelia-derived tumors in the skin and have metastatic potential following papilloma formation (Alam & Ratner, 2001). Induction of SCCs in mice can occur via UV irradiation, chemical carcinogenesis protocols using 7,12-dimethylbenz(a)anthracene (DMBA) to induce Ras mutations followed by promotion with 12-*O*-tetradecanoylphorbol-13-acetate or genetic expression of oncogenes within the epithelium. Clone analysis of human skin with p53 mutations suggested that SCCs derived from the dermal–epidermal junction and from HFs (Jonason et al., 1996). Similar results were obtained when a mutated activated form of Ras was expressed in the proliferative cells of the IFE and HF (Brown, Strathdee, Bryson, Lambie, & Balmain, 1998).

To further elucidate the cellular origin of Ras induced SCCs, cell-type-specific expression of mutant Kras was induced in mouse models controlled by bulge-specific promoters, K15 or K19 (Lapouge et al., 2011; White et al., 2011). Activation of Kras in the bulge led to papilloma formation, whereas tumorigenesis was not induced when Kras mutations were targeted to the transiently amplifying cells of the HF. Interestingly, activation of Kras in InvCre-expressing cells also generated papillomas (Lapouge et al., 2011). Since this promoter can drive expression in IFE progenitor cells (Mascré et al., 2012), IFE cells may also contribute to SCC formation. The contribution of bulge cells to papilloma formation was further confirmed by the presence of progeny from $K15^+$ bulge cells in murine papillomas after induction of Ras mutations with DMBA (Kangsamaksin, Park, Trempus, & Morris, 2007).

To determine if cells with SC potential exist within SCCs, transplantation assays of purified cells into immunocompromised mice have revealed that CD34 or α6β1 integrin expression can enrich for tumor-initiating cells in this assay (Malanchi et al., 2008; Schober & Fuchs, 2011). Importantly, ablation of TGFβ signaling enhanced the proliferation of integrin^{hi}, $CD34^{+}$ cells, and their tumorigenic potential (Schober & Fuchs, 2011). By applying quantitative clonal analysis to SCC tumor growth *in vivo*, Blanpain and colleagues found a minority of cells within papillomas with dramatic growth potential (Driessens, Beck, Caauwe, Simons, & Blanpain, 2012). By contrast, metastatic SCCs displayed clones of cells with increased replicative and abrogated differentiation potential. Interestingly, a hierarchical organization of tumor cells also exists in benign intestinal adenomas and metastatic brain tumors such as glioblastomas (Chen et al., 2012; Schepers et al., 2012). In glioblastomas, when the highly proliferative progeny of the cells are ablated with chemotherapeutic drugs, the other cells with SC properties can repopulate the tumors. When these cells with more primitive potential were selectively ablated with genetic tools in the presence of chemotherapeutic drugs, tumor growth was significantly hampered. Together, these studies provide strong evidence for the ability of cancer cells to acquire SC properties to fuel tumorigenesis and suggest that targeting both primitive and more developed cells within tumors will be important for effective cancer therapies.

7. CONCLUDING REMARKS

Several advances have been made toward understanding the cellular and molecular mechanisms that control epidermal SC quiescence and differentiation in multiple lineages in the skin. Important progress achieved in recent years to develop mouse models to target and disrupt distinct SC pools within the epidermis have provided precious information that begins to unravel not only the complexity of the different epidermal SC niches but also the interactions between these niches. Future in-depth studies looking at the different SC niche signals and how they are affected during disease will definitely contribute to the better understanding of epidermal SC biology and the consequent application toward treatment of SC-related pathologies such as cancer.

ACKNOWLEDGMENTS

We thank the Horsley lab members for critical reading of the chapter and valuable discussions. A. M. B. T. was a Fundação para a Ciência e Tecnologia postdoctoral fellow. V. H. is a Pew Scholar in Biomedical Research and is funded by the NIH (AR060295) and the state of CT (12-SCB-YALE-01 and 12-SCA-YALE-09).

REFERENCES

Alam, M., & Ratner, D. (2001). Cutaneous squamous-cell carcinoma. *The New England Journal of Medicine, 344*, 975–983.

Al-Hajj, M., & Clarke, M. F. (2004). Self-renewal and solid tumor stem cells. *Oncogene, 23*, 7274–7282.

Andl, T., Murchison, E. P., Liu, F., Zhang, Y., Yunta-Gonzalez, M., Tobias, J. W., et al. (2006). The miRNA-processing enzyme dicer is essential for the morphogenesis and maintenance of hair follicles. *Current Biology, 16*, 1041–1049.

Arwert, E. N., Hoste, E., & Watt, F. M. (2012). Epithelial stem cells, wound healing and cancer. *Nature Reviews Cancer, 12*, 170–180.

Asada, M., Schaart, F. M., de Almeida, H. L., Jr, Korge, B., Kurokawa, I., Asada, Y., et al. (1993). Solid basal cell epithelioma (BCE) possibly originates from the outer root sheath of the hair follicle. *Acta Dermato Venereologica, 73*(4), 286–292.

Athar, M., Tang, X., Lee, J. L., Kopelovich, L., & Kim, A. L. (2006). Hedgehog signalling in skin development and cancer. *Experimental Dermatology, 15*, 667–677.

Baker, C. M., Verstuyf, A., Jensen, K. B., & Watt, F. M. (2010). Differential sensitivity of epidermal cell subpopulations to beta-catenin-induced ectopic hair follicle formation. *Developmental Biology, 343*, 40–50.

Bardot, E. S., Valdes, V. J., Zhang, J., Perdigoto, C. N., Nicolis, S., Hearn, S. A., et al. (2013). Polycomb subunits Ezh1 and Ezh2 regulate the Merkel cell differentiation program in skin stem cells. *EMBO Journal.*

Barker, N., van Es, J. H., Kuipers, J., Kujala, P., van den Born, M., Cozijnsen, M., et al. (2007). Identification of stem cells in small intestine and colon by marker gene Lgr5. *Nature, 449*, 1003–1007.

Bickenbach, J. R. (1981). Identification and behavior of label-retaining cells in oral mucosa and skin. *Journal of Dental Research, 60*, 1611–1620, Spec. No. C.

Biedermann, T., Pontiggia, L., Böttcher-Haberzeth, S., Tharakan, S., Braziulis, E., Schiestl, C., et al. (2010). Human eccrine sweat gland cells can reconstitute a stratified epidermis. *The Journal of Investigative Dermatology, 130*, 1996–2009.

Blanpain, C., & Fuchs, E. (2006). Epidermal stem cells of the skin. *Annual Review of Cell and Developmental Biology, 22*, 339–373.

Blanpain, C., Lowry, W. E., Geoghegan, A., Polak, L., & Fuchs, E. (2004). Self-renewal, multipotency, and the existence of two cell populations within an epithelial stem cell niche. *Cell, 118*, 635–648.

Brown, K., Strathdee, D., Bryson, S., Lambie, W., & Balmain, A. (1998). The malignant capacity of skin tumours induced by expression of a mutant H-ras transgene depends on the cell type targeted. *Current Biology, 8*, 516–524.

Brownell, I., Guevara, E., Bai, C. B., Loomis, C. A., & Joyner, A. L. (2011). Nerve-derived sonic hedgehog defines a niche for hair follicle stem cells capable of becoming epidermal stem cells. *Cell Stem Cell, 8*, 552–565.

Calo, E., & Wysocka, J. (2013). Modification of enhancer chromatin: What, how, and why? *Mol. Cell, 49*, 825–837.

Chan, E. F., Gat, U., McNiff, J. M., & Fuchs, E. (1999). A common human skin tumour is caused by activating mutations in beta-catenin. *Nature Genetics, 21*, 410–413.

Chase, H. B., Montagna, W., & Malone, J. D. (1953). Changes in the skin in relation to the hair growth cycle. *The Anatomical Record, 116*, 75–81.

Chen, J., Li, Y., Yu, T.-S., McKay, R. M., Burns, D. K., Kernie, S. G., et al. (2012). A restricted cell population propagates glioblastoma growth after chemotherapy. *Nature, 488*, 522–526.

Chi, W., Wu, E., & Morgan, B. A. (2013). Dermal papilla cell number specifies hair size, shape and cycling and its reduction causes follicular decline. *Development, 140*, 1676–1683.

Claudinot, S., Nicolas, M., Oshima, H., Rochat, A., & Barrandon, Y. (2005). Long-term renewal of hair follicles from clonogenic multipotent stem cells. *Proceedings of the National Academy of Sciences of the United States of America*, *102*, 14677–14682.
Clayton, E., Doupé, D. P., Klein, A. M., Winton, D. J., Simons, B. D., & Jones, P. H. (2007). A single type of progenitor cell maintains normal epidermis. *Nature*, *446*, 185–189.
Cotsarelis, G., Sun, T. T., & Lavker, R. M. (1990). Label-retaining cells reside in the bulge area of pilosebaceous unit: Implications for follicular stem cells, hair cycle, and skin carcinogenesis. *Cell*, *61*, 1329–1337.
Dick, J. C. (1947). Observations on the elastic tissue of the skin with a note on the reticular layer at the junction of the dermis and epidermis. *Journal of Anatomy*, *81*, 201–211.
Didierjean, L., Wrench, R., & Saurat, J. H. (1983). Expression of cytoplasmic antigens linked to orthokeratosis during the development of parakeratosis in newborn mouse tail epidermis. *Differentiation*, *23*, 250–255.
Doucet, Y. S., Woo, S.-H., Ruiz, M. E., & Owens, D. M. (2013). The touch dome defines an epidermal niche specialized for mechanosensory signaling. *Cell Reports*, *3*, 1759–1765.
Doupé, D. P., Klein, A. M., Simons, B. D., & Jones, P. H. (2010). The ordered architecture of murine ear epidermis is maintained by progenitor cells with random fate. *Developmental Cell*, *18*, 317–323.
Driessens, G., Beck, B., Caauwe, A., Simons, B. D., & Blanpain, C. (2012). Defining the mode of tumour growth by clonal analysis. *Nature*, *488*, 527–530.
Driskell, R. R., Clavel, C., Rendl, M., & Watt, F. M. (2011). Hair follicle dermal papilla cells at a glance. *Journal of Cell Science*, *124*, 1179–1182.
Driskell, I., Oda, H., Blanco, S., Nascimento, E., Humphreys, P., & Frye, M. (2012). The histone methyltransferase Setd8 acts in concert with c-Myc and is required to maintain skin. *The EMBO Journal*, *31*, 616–629.
Eming, S. A., Krieg, T., & Davidson, J. M. (2007). Inflammation in wound repair: Molecular and cellular mechanisms. *The Journal of Investigative Dermatology*, *127*, 514–525.
Enshell-Seijffers, D., Lindon, C., Kashiwagi, M., & Morgan, B. A. (2010). beta-catenin activity in the dermal papilla regulates morphogenesis and regeneration of hair. *Developmental Cell*, *18*, 633–642.
Ezhkova, E., Lien, W.-H., Stokes, N., Pasolli, H. A., Silva, J. M., & Fuchs, E. (2011). EZH1 and EZH2 cogovern histone H3K27 trimethylation and are essential for hair follicle homeostasis and wound repair. *Genes and Development*, *25*, 485–498.
Festa, E., Fretz, J., Berry, R., Schmidt, B., Rodeheffer, M., Horowitz, M., et al. (2011). Adipocyte lineage cells contribute to the skin stem cell niche to drive hair cycling. *Cell*, *146*, 761–771.
Gat, U., DasGupta, R., Degenstein, L., & Fuchs, E. (1998). De Novo hair follicle morphogenesis and hair tumors in mice expressing a truncated beta-catenin in skin. *Cell*, *95*, 605–614.
Gay, D., Kwon, O., Zhang, Z., Spata, M., Plikus, M. V., Holler, P. D., et al. (2013). Fgf9 from dermal γδ T cells induces hair follicle neogenesis after wounding. *Nature Medicine*, *19*, 916–923.
Goldstein, J., & Horsley, V. (2012). Home sweet home: Skin stem cell niches. *Cellular and Molecular Life Sciences*, *69*, 2573–2582.
Gomez, C., Chua, W., Miremadi, A., Quist, S., & Headon, D. J. (2013). The interfollicular epidermis of adult mouse tail comprises two distinct cell lineages that are differentially regulated by Wnt, Edaradd, and Lrig1. *Stem Cell Reports*, *1*, 19–27.
Grachtchouk, M., Pero, J., Yang, S. H., Ermilov, A. N., Michael, L. E., Wang, A., et al. (2011). Basal cell carcinomas in mice arise from hair follicle stem cells and multiple epithelial progenitor populations. *The Journal of Clinical Investigation*, *121*, 1768–1781.
Greco, V., Chen, T., Rendl, M., Schober, M., & Pasolli, H. A. (2009). A two-step mechanism for stem cell activation during hair regeneration. *Cell Stem Cell*, *4*, 155–169.

Grisanti, L., Clavel, C., Cai, X., Rezza, A., Tsai, S.-Y., Sennett, R., et al. (2013). Tbx18 targets dermal condensates for labeling, isolation, and gene ablation during embryonic hair follicle formation. *The Journal of Investigative Dermatology, 133*, 344–353.

Gudmundsdottir, K., & Ashworth, A. (2006). The roles of BRCA1 and BRCA2 and associated proteins in the maintenance of genomic stability. *Oncogene, 25*, 5864–5874.

Horsley, V., O'Carroll, D., Tooze, R., Ohinata, Y., Saitou, M., Obukhanych, T., et al. (2006). Blimp1 defines a progenitor population that governs cellular input to the sebaceous gland. *Cell, 126*, 597–609.

Hsu, Y.-C., Pasolli, H. A., & Fuchs, E. (2011). Dynamics between stem cells, niche, and progeny in the hair follicle. *Cell, 144*, 92–105.

Ito, M., Liu, Y., Yang, Z., Nguyen, J., Liang, F., Morris, R. J., et al. (2005). Stem cells in the hair follicle bulge contribute to wound repair but not to homeostasis of the epidermis. *Nature Medicine, 11*, 1351–1354.

Jackson, S. J., Zhang, Z., Feng, D., Flagg, M., O'Loughlin, E., Wang, D., et al. (2013). Rapid and widespread suppression of self-renewal by microRNA-203 during epidermal differentiation. *Development, 140*, 1882–1891.

Jaks, V., Barker, N., Kasper, M., van Es, J. H., Snippert, H. J., Clevers, H., et al. (2008). Lgr5 marks cycling, yet long-lived, hair follicle stem cells. *Nature Genetics, 40*, 1291–1299.

Jensen, K. B., Collins, C. A., Nascimento, E., Tan, D. W., Frye, M., Itami, S., et al. (2009). Lrig1 expression defines a distinct multipotent stem cell population in mammalian epidermis. *Cell Stem Cell, 4*, 427–439.

Jensen, U. B., Yan, X., Triel, C., Woo, S.-H., Christensen, R., & Owens, D. M. (2008). A distinct population of clonogenic and multipotent murine follicular keratinocytes residing in the upper isthmus. *Journal of Cell Science, 121*, 609–617.

Jonason, A. S., Kunala, S., Price, G. J., Restifo, R. J., Spinelli, H. M., Persing, J. A., et al. (1996). Frequent clones of p53-mutated keratinocytes in normal human skin. *Proceedings of the National Academy of Sciences of the United States of America, 93*, 14025–14029.

Kandyba, E., Leung, Y., Chen, Y.-B., Widelitz, R., Chuong, C.-M., & Kobielak, K. (2013). Competitive balance of intrabulge BMP/Wnt signaling reveals a robust gene network ruling stem cell homeostasis and cyclic activation. *Proceedings of the National Academy of Sciences of the United States of America, 110*, 1351–1356.

Kangsamaksin, T., Park, H. J., Trempus, C. S., & Morris, R. J. (2007). A perspective on murine keratinocyte stem cells as targets of chemically induced skin cancer. *Molecular Carcinogenesis, 46*, 579–584.

Kasper, M., Jaks, V., Are, A., Bergström, Å., Schwäger, A., Svärd, J., et al. (2011). Wounding enhances epidermal tumorigenesis by recruiting hair follicle keratinocytes. *Proceedings of the National Academy of Sciences of the United States of America, 108*, 4099–4104.

Kidder, B. L., Palmer, S., & Knott, J. G. (2009). SWI/SNF-Brg1 regulates self-renewal and occupies core pluripotency-related genes in embryonic stem cells. *Stem Cells, 27*, 317–328.

Kloepper, J. E., Kawai, K., Bertolini, M., Kanekura, T., & Paus, R. (2013). Loss of $\gamma\delta$ T cells results in hair cycling defects. *The Journal of Investigative Dermatology, 133*, 1666–1669.

Kobielak, K., Stokes, N., de la Cruz, J., Polak, L., & Fuchs, E. (2007). Loss of a quiescent niche but not follicle stem cells in the absence of bone morphogenetic protein signaling. *Proceedings of the National Academy of Sciences of the United States of America, 104*, 10063–10068.

Langbein, L., Rogers, M. A., Praetzel, S., Cribier, B., Peltre, B., Gassler, N., et al. (2005). Characterization of a novel human type II epithelial keratin K1b, specifically expressed in eccrine sweat glands. *The Journal of Investigative Dermatology, 125*, 428–444.

Lapouge, G., Youssef, K. K., Vokaer, B., Achouri, Y., Michaux, C., Sotiropoulou, P. A., et al. (2011). Identifying the cellular origin of squamous skin tumors. *Proceedings of the National Academy of Sciences of the United States of America, 108*, 7431–7436.

LeBoeuf, M., Terrell, A., Trivedi, S., Sinha, S., Epstein, J. A., Olson, E. N., et al. (2010). Hdac1 and Hdac2 act redundantly to control p63 and p53 functions in epidermal progenitor cells. *Developmental Cell*, *19*, 807–818.

Lechler, T., & Fuchs, E. (2005). Asymmetric cell divisions promote stratification and differentiation of mammalian skin. *Nature*, *437*, 275–280.

Lee, B., & Dai, X. (2013). Transcriptional control of epidermal stem cells. *Advances in Experimental Medicine and Biology*, *786*, 157–173.

Lee, J., & Tumbar, T. (2012). Hairy tale of signaling in hair follicle development and cycling. *Seminars in Cell and Developmental Biology*, *23*, 906–916.

Levy, V., Lindon, C., Zheng, Y., Harfe, B. D., & Morgan, B. A. (2007). Epidermal stem cells arise from the hair follicle after wounding. *The FASEB Journal*, *21*, 1358–1366.

Li, L., Rutlin, M., Abraira, V. E., Cassidy, C., Kus, L., Gong, S., et al. (2011). The functional organization of cutaneous low-threshold mechanosensory neurons. *Cell*, *147*, 1615–1627.

Liu, N., Balliano, A., & Hayes, J. J. (2011). Mechanism(s) of SWI/SNF-induced nucleosome mobilization. *Chembiochem*, *12*, 196–204.

Lo Celso, C., Prowse, C., & Watt, F. M. (2004). Transient activation of beta-catenin signalling in adult mouse epidermis is sufficient to induce new hair follicles but continuous activation is required to maintain hair follicle tumours. *Development*, *131*, 1787–1799.

Lobitz, W. C., & Dobson, R. L. (1961). Dermatology: The eccrine sweat glands. *Annual Review of Medicine*, *12*, 289–298.

Lobitz, W. C., Holyoke, J. B., & Montagna, W. (1954). Responses of the human eccrine sweat duct to controlled injury: Growth center of the epidermal sweat duct unit. *The Journal of Investigative Dermatology*, *23*, 329–344.

Loeffler, M., Potten, C. S., & Wichmann, H. E. (1987). Epidermal cell proliferation. II. A comprehensive mathematical model of cell proliferation and migration in the basal layer predicts some unusual properties of epidermal stem cells. *Virchows Archiv B Cell Pathology*, *53*, 286–300.

Lu, C. P., Polak, L., Rocha, A. S., Pasolli, H. A., Chen, S.-C., Sharma, N., et al. (2012). Identification of stem cell populations in sweat glands and ducts reveals roles in homeostasis and wound repair. *Cell*, *150*, 136–150.

Lumpkin, E. A., Marshall, K. L., & Nelson, A. M. (2010). The cell biology of touch. *The Journal of Cell Biology*, *191*, 237–248.

Mackenzie, J. C. (1969). Ordered structure of the stratum corneum of mammalian skin. *Nature*, *222*, 881–882.

Mackenzie, I. C. (1970). Relationship between mitosis and the ordered structure of the stratum corneum in mouse epidermis. *Nature*, *226*, 653–655.

Malanchi, I., Peinado, H., Kassen, D., Hussenet, T., Metzger, D., Chambon, P., et al. (2008). Cutaneous cancer stem cell maintenance is dependent on beta-catenin signalling. *Nature*, *452*, 650–653.

Mandal, P. K., Blanpain, C., & Rossi, D. J. (2011). DNA damage response in adult stem cells: Pathways and consequences. *Nature Reviews Molecular Cell Biology*, *12*, 198–202.

Mardaryev, A. N., Ahmed, M. I., Vlahov, N. V., Fessing, M. Y., Gill, J. H., Sharov, A. A., et al. (2010). Micro-RNA-31 controls hair cycle-associated changes in gene expression programs of the skin and hair follicle. *The FASEB Journal*, *24*, 3869–3881.

Mardaryev, A. N., Meier, N., Poterlowicz, K., Sharov, A. A., Sharova, T. Y., Ahmed, M. I., et al. (2011). Lhx2 differentially regulates Sox9, Tcf4 and Lgr5 in hair follicle stem cells to promote epidermal regeneration after injury. *Development*, *138*, 4843–4852.

Mancuso, M., Leonardi, S., Tanori, M., Pasquali, E., Pierdomenico, M., Rebessi, S., et al. (2006). Hair Cycle–Dependent Basal Cell Carcinoma Tumorigenesis in Ptc1neo67/+ Mice Exposed to Radiation. *Cancer Research*, *66*(13), 6606–6614.

Mascré, G., Dekoninck, S., Drogat, B., Youssef, K. K., Broheé, S., Sotiropoulou, P. A., et al. (2012). Distinct contribution of stem and progenitor cells to epidermal maintenance. *Nature, 489,* 257–262.

Mejetta, S., Morey, L., Pascual, G., Kuebler, B., Mysliwiec, M. R., Lee, Y., et al. (2011). Jarid2 regulates mouse epidermal stem cell activation and differentiation. *The EMBO Journal, 30,* 3635–3646.

Merkel, F. (1875). Tastzellen und Tastkörperchen bei den Haustieren und beim Menschen. *Archiv für mikroskopische Anatomie und Entwicklungsmechanik, 11,* 636–652.

Millar, S. E. (2002). Molecular mechanisms regulating hair follicle development. *The Journal of Investigative Dermatology, 118,* 216–225.

Miller, S. J., Burke, E. M., Rader, M. D., Coulombe, P. A., & Lavker, R. M. (1998). Re-epithelialization of porcine skin by the sweat apparatus. *The Journal of Investigative Dermatology, 110,* 13–19.

Moll, I., & Moll, R. (1992). Changes of expression of intermediate filament proteins during ontogenesis of eccrine sweat glands. *The Journal of Investigative Dermatology, 98,* 777–785.

Morris, R. J., Liu, Y., Marles, L., Yang, Z., Trempus, C., Li, S., et al. (2004). Capturing and profiling adult hair follicle stem cells. *Nature Biotechnology, 22,* 411–417.

Morrison, K. M., Miesegaes, G. R., Lumpkin, E. A., & Maricich, S. M. (2009). Mammalian Merkel cells are descended from the epidermal lineage. *Developmental Biology, 336,* 76–83.

Moynahan, M. E., & Jasin, M. (2010). Mitotic homologous recombination maintains genomic stability and suppresses tumorigenesis. *Nature Reviews Molecular Cell Biology, 11,* 196–207.

Mulder, K. W., Wang, X., Escriu, C., Ito, Y., Schwarz, R. F., Gillis, J., et al. (2012). Diverse epigenetic strategies interact to control epidermal differentiation. *Nature Cell Biology, 14,* 753–763.

Myung, P. S., Takeo, M., Ito, M., & Atit, R. P. (2013). Epithelial Wnt ligand secretion is required for adult hair follicle growth and regeneration. *The Journal of Investigative Dermatology, 133,* 31–41.

Nguyen, H., Merrill, B. J., Polak, L., Nikolova, M., Rendl, M., Shaver, T. M., et al. (2009). Tcf3 and Tcf4 are essential for long-term homeostasis of skin epithelia. *Nature Genetics, 41,* 1068–1075.

Nguyen, H., Rendl, M., & Fuchs, E. (2006). Tcf3 governs stem cell features and represses cell fate determination in skin. *Cell, 127,* 171–183.

Niemann, C., & Horsley, V. (2012). Development and homeostasis of the sebaceous gland. *Seminars in Cell and Developmental Biology, 23,* 928–936.

Niemann, C., Owens, D. M., Hülsken, J., Birchmeier, W., & Watt, F. M. (2002). Expression of DeltaNLef1 in mouse epidermis results in differentiation of hair follicles into squamous epidermal cysts and formation of skin tumours. *Development, 129,* 95–109.

Niemann, C., Owens, D. M., Schettina, P., & Watt, F. M. (2007). Dual role of inactivating Lef1 mutations in epidermis: Tumor promotion and specification of tumor type. *Cancer Research, 67,* 2916–2921.

Nijhof, J. G. W., Braun, K. M., Giangreco, A., van Pelt, C., Kawamoto, H., Boyd, R. L., et al. (2006). The cell-surface marker MTS24 identifies a novel population of follicular keratinocytes with characteristics of progenitor cells. *Development, 133,* 3027–3037.

Nowak, J. A., Polak, L., Pasolli, H. A., & Fuchs, E. (2008). Hair follicle stem cells are specified and function in early skin morphogenesis. *Cell Stem Cell, 3,* 33–43.

Oshima, H., Rochat, A., Kedzia, C., Kobayashi, K., & Barrandon, Y. (2001). Morphogenesis and renewal of hair follicles from adult multipotent stem cells. *Cell, 104,* 233–245.

Oshimori, N., & Fuchs, E. (2012). Paracrine TGF-β signaling counterbalances BMP-mediated repression in hair follicle stem cell activation. *Cell Stem Cell, 10,* 63–75.

Petersson, M., Brylka, H., Kraus, A., John, S., Rappl, G., Schettina, P., et al. (2011). TCF/Lef1 activity controls establishment of diverse stem and progenitor cell compartments in mouse epidermis. *The EMBO Journal, 30*, 3004–3018.

Petukhova, L., Duvic, M., Hordinsky, M., Norris, D., Price, V., Shimomura, Y., et al. (2010). Genome-wide association study in alopecia areata implicates both innate and adaptive immunity. *Nature, 466*, 113–117.

Plikus, M. V., Mayer, J. A., la Cruz, de, D., Baker, R. E., Maini, P. K., Maxson, R., et al. (2008). Cyclic dermal BMP signalling regulates stem cell activation during hair regeneration. *Nature, 451*, 340–344.

Potten, C. S. (1981). Cell replacement in epidermis (keratopoiesis) via discrete units of proliferation. *International Review of Cytology, 69*, 271–318.

Potten, C. S., Wichmann, H. E., Loeffler, M., Dobek, K., & Major, D. (1982). Evidence for discrete cell kinetic subpopulations in mouse epidermis based on mathematical analysis. *Cell and Tissue Kinetics, 15*, 305–329.

Rendl, M., Lewis, L., & Fuchs, E. (2005). Molecular dissection of mesenchymal-epithelial interactions in the hair follicle. *PLoS Biology, 3*, e331.

Rhee, H., Polak, L., & Fuchs, E. (2006). Lhx2 maintains stem cell character in hair follicles. *Science, 312*, 1946–1949.

Richert, M. M., Schwertfeger, K. L., Ryder, J. W., & Anderson, S. M. (2000). An atlas of mouse mammary gland development. *Journal of Mammary Gland Biology and Neoplasia, 5*, 227–241.

Rogers, H. W., Weinstock, M. A., Harris, A. R., Hinckley, M. R., Feldman, S. R., Fleischer, A. B., et al. (2010). Incidence estimate of nonmelanoma skin cancer in the United States, 2006. *Archives of Dermatology, 146*, 283–287.

Rompolas, P., Deschene, E. R., Zito, G., Gonzalez, D. G., Saotome, I., Haberman, A. M., et al. (2012). Live imaging of stem cell and progeny behaviour in physiological hair-follicle regeneration. *Nature, 487*, 496–499.

Sandilands, A., Sutherland, C., Irvine, A. D., & McLean, W. H. I. (2009). Filaggrin in the frontline: Role in skin barrier function and disease. *Journal of Cell Science, 122*, 1285–1294.

Schepers, A. G., Snippert, H. J., Stange, D. E., van den Born, M., van Es, J. H., van de Wetering, M., et al. (2012). Lineage tracing reveals Lgr5+ stem cell activity in mouse intestinal adenomas. *Science, 337*, 730–735.

Schmidt, B. A., & Horsley, V. (2013). Intradermal adipocytes mediate fibroblast recruitment during skin wound healing. *Development, 140*, 1517–1527.

Schober, M., & Fuchs, E. (2011). Tumor-initiating stem cells of squamous cell carcinomas and their control by TGF-β and integrin/focal adhesion kinase (FAK) signaling. *Proceedings of the National Academy of Sciences of the United States of America, 108*, 10544–10549.

Schön, M., Benwood, J., O'Connell-Willstaedt, T., & Rheinwald, J. G. (1999). Human sweat gland myoepithelial cells express a unique set of cytokeratins and reveal the potential for alternative epithelial and mesenchymal differentiation states in culture. *Journal of Cell Science, 112*(Pt. 12), 1925–1936.

Schweizer, J., & Marks, F. (1977). A developmental study of the distribution and frequency of Langerhans cells in relation to formation of patterning in mouse tail epidermis. *The Journal of Investigative Dermatology, 69*, 198–204.

Sen, G. L., Reuter, J. A., Webster, D. E., Zhu, L., & Khavari, P. A. (2010). DNMT1 maintains progenitor function in self-renewing somatic tissue. *Nature, 463*, 563–567.

Sen, G. L., Webster, D. E., Barragan, D. I., Chang, H. Y., & Khavari, P. A. (2008). Control of differentiation in a self-renewing mammalian tissue by the histone demethylase JMJD3. *Genes and Development, 22*, 1865–1870.

Sennett, R., & Rendl, M. (2012). Mesenchymal-epithelial interactions during hair follicle morphogenesis and cycling. *Seminars in Cell and Developmental Biology, 23*, 917–927.

Snippert, H. J., Haegebarth, A., Kasper, M., Jaks, V., van Es, J. H., Barker, N., et al. (2010). Lgr6 marks stem cells in the hair follicle that generate all cell lineages of the skin. *Science*, *327*, 1385–1389.

Sotiropoulou, P. A., Candi, A., Mascré, G., De Clercq, S., Youssef, K. K., Lapouge, G., et al. (2010). Bcl-2 and accelerated DNA repair mediates resistance of hair follicle bulge stem cells to DNA-damage-induced cell death. *Nature Cell Biology*, *12*, 572–582.

Sotiropoulou, P. A., Karambelas, A. E., Debaugnies, M., Candi, A., Bouwman, P., Moers, V., et al. (2013). BRCA1 deficiency in skin epidermis leads to selective loss of hair follicle stem cells and their progeny. *Genes and Development*, *27*, 39–51.

Takeda, N., Jain, R., LeBoeuf, M. R., & Padmanabhan, A. (2013). Hopx expression defines a subset of multipotent hair follicle stem cells and a progenitor population primed to give rise to K6+ niche cells. *Development*, *140*, 1655–1664.

Takeda, N., Jain, R., LeBoeuf, M. R., Wang, Q., & Lu, M. M. (2011). Interconversion between intestinal stem cell populations in distinct niches. *Science*, *334*, 1420–1424.

Takeda, H., Lyle, S., Lazar, A. J. F., Zouboulis, C. C., Smyth, I., & Watt, F. M. (2006). Human sebaceous tumors harbor inactivating mutations in LEF1. *Nature Medicine*, *12*, 395–397.

Törnqvist, G., Sandberg, A., Hägglund, A. C., & Carlsson, L. (2010). Cyclic expression of lhx2 regulates hair formation. *PLoS Genetics*, *6*(4): e1000904.

Tumbar, T., Guasch, G., Greco, V., Blanpain, C., Lowry, W. E., Rendl, M., et al. (2004). Defining the epithelial stem cell niche in skin. *Science*, *303*, 359–363.

Van Keymeulen, A., Mascré, G., Youseff, K. K., Harel, I., Michaux, C., De Geest, N., et al. (2009). Epidermal progenitors give rise to Merkel cells during embryonic development and adult homeostasis. *The Journal of Cell Biology*, *187*, 91–100.

White, A. C., Tran, K., Khuu, J., Dang, C., Cui, Y., Binder, S. W., et al. (2011). Defining the origins of Ras/p53-mediated squamous cell carcinoma. *Proceedings of the National Academy of Sciences of the United States of America*, *108*, 7425–7430.

Xiong, Y., Li, W., Shang, C., Chen, R. M., Han, P., Yang, J., et al. (2013). Brg1 governs a positive feedback circuit in the hair follicle for tissue regeneration and repair. *Developmental Cell*, *25*, 169–181.

Yi, R., O'Carroll, D., Pasolli, H. A., Zhang, Z., Dietrich, F. S., Tarakhovsky, A., et al. (2006). Morphogenesis in skin is governed by discrete sets of differentially expressed microRNAs. *Nature Genetics*, *38*, 356–362.

Yi, R., Poy, M. N., Stoffel, M., & Fuchs, E. (2008). A skin microRNA promotes differentiation by repressing 'stemness'. *Nature*, *452*, 225–229.

Youssef, K. K., Van Keymeulen, A., Lapouge, G., Beck, B., Michaux, C., Achouri, Y., et al. (2010). Identification of the cell lineage at the origin of basal cell carcinoma. *Nature Cell Biology*, *12*, 299–305.

Zhang, Y. V., Cheong, J., Ciapurin, N., McDermitt, D. J., & Tumbar, T. (2009). Distinct self-renewal and differentiation phases in the niche of infrequently dividing hair follicle stem cells. *Cell Stem Cell*, *5*, 267–278.

Zhang, L., Stokes, N., Polak, L., & Fuchs, E. (2011). Specific microRNAs are preferentially expressed by skin stem cells to balance self-renewal and early lineage commitment. *Cell Stem Cell*, *8*, 294–308.

CHAPTER FIVE

The Mammary Stem Cell Hierarchy

Naiyang Fu[*,†], **Geoffrey J. Lindeman**[*,‡,§], **Jane E. Visvader**[*,†,1]
[*]Stem Cells and Cancer Division, The Walter and Eliza Hall Institute of Medical Research, Parkville, Victoria, Australia
[†]Department of Medical Biology, University of Melbourne, Parkville, Victoria, Australia
[‡]Department of Medical Oncology, Royal Melbourne Hospital, Parkville, Victoria, Australia
[§]Department of Medicine, University of Melbourne, Parkville, Victoria, Australia
[1]Corresponding author: e-mail address: visvader@wehi.edu.au

Contents

Abstract

The mammary epithelium undergoes enormous morphogenetic changes during the lifespan of a mammal. The recent elucidation of an epithelial differentiation hierarchy in the mouse mammary gland through classical transplantation and clonogenic assays has pointed to the existence of multipotent mammary stem cells (MaSCs) and at least two distinct luminal progenitor types. Moreover, an analogous functional hierarchy has been defined in human breast tissue. The existence of slow cycling stem cells, both long- and short-term repopulating cells, and a unique fetal MaSC population, imply a complex stem cell compartment within the mammary gland. The recent discovery of unipotent stem-like cells from lineage tracing studies has added a further layer of complexity to the emerging differentiation hierarchy. Although the precise relationships between stem and progenitor cells have yet to be resolved, the epithelial hierarchy has

Current Topics in Developmental Biology, Volume 107
ISSN 0070-2153
http://dx.doi.org/10.1016/B978-0-12-416022-4.00005-6

provided an important framework for elucidating the roles of molecular regulators of mammary gland ontogeny and understanding potential cells of origin in breast cancer.

1. INTRODUCTION

Development of the mammary ductal tree primarily occurs during puberty under the governance of ovarian hormones and cytokines (reviewed in Macias & Hinck, 2012). At the onset of puberty, the mammary epithelial ducts branch and elongate to yield an extensive network that fills the mammary fat pad. This branching process largely occurs via the highly proliferative terminal end buds that are situated at the ends of branches. With each estrus cycle, the mammary epithelium is subjected to cycles of proliferation and differentiation, accompanied by the waxing and waning of alveolar buds. In pregnancy, the mammary epithelium undergoes dramatic expansion with the formation of alveolar structures that undergo progressive differentiation in late pregnancy to enable milk production during lactation. At the cessation of lactation, the mammary gland involutes through apoptosis and proteinase-driven tissue remodeling, resulting in a gland that resembles the prepregnant state. The mammary epithelium is organized into two primary lineages, the luminal and myoepithelial. The luminal lineage can be subdivided into the ductal and alveolar cell sublineages that line the lumina and form the milk-producing alveolar units, respectively. Myoepithelial cells are located in a basal position adjacent to the basement membrane and have contractile properties. Each cell lineage is characterized by distinct markers: luminal cells express steroid hormone receptors and "luminal" cytokeratin proteins (CK8, CK18, and CK19), among other markers, while myoepithelial cells express smooth muscle actin and "basal" cytokeratins (CK5 and CK14).

The dynamic nature of the mammary epithelium and its profound regenerative capacity evident upon successive rounds of reproduction suggest the presence of long-lived and renewable mammary stem cells (MaSCs). The earliest evidence to support the existence of MaSCs came from the pioneering studies of De Ome, Faulkin, Bern, and Blair (1959) who developed the fat pad transplantation assay. This technique involves the implantation of mammary epithelial cells or tissue fragments into the "cleared" mammary fat pads of recipient mice and assessment of repopulating potential several weeks later in either virgin or pregnant hosts. Importantly, although the fat pad is de-epithelialized, the stromal environment remains intact to support growth of the transplanted cells. Early studies used either mammary explants or cell suspensions to demonstrate reconstitution of the ductal tree (Daniel, De Ome, Young, Blair, & Faulkin, 1968; Hoshino & Gardner, 1967; Smith, 1996). Moreover, these could be serially transplanted for several generations, thus indicating the presence of stem-like cells. Retroviral marking of donor mammary tissue fragments suggested that the entire mammary gland could originate from a single stem cell (Kordon & Smith, 1998). In 2006, the mouse MaSC was prospectively identified using the "gold standard" transplantation assay, and shown to reconstitute a fully functional mammary gland capable of milk production (Shackleton et al., 2006; Stingl et al., 2006). More recent studies have turned to *in vivo* lineage tracing, which have revealed an unexpected degree of complexity within the hierarchy (van Amerongen, Bowman, & Nusse, 2012; Van Keymeulen et al., 2011). These studies have suggested that unipotent cells are the main drivers of homeostasis in the postnatal mammary gland. In this review, we will focus on the evolving hierarchy in the normal mammary gland and efforts to identify important molecular regulators of different steps along the hierarchy. Finally, implications of the MaSC hierarchy for understanding breast cancer heterogeneity and potential "cells of origin" will be discussed.

2. IDENTIFICATION OF ADULT MOUSE MaSCs

The primary strategy used to delineate cell types that can reconstitute mammary tissue has been adapted from the hematopoietic system. It relies on the isolation of cellular subpopulations through the use of specific combinations of antibodies against cell surface proteins for fluorescence-activated cell sorting of single cell suspensions, followed by direct testing of their repopulating capacity by transplantation into an orthotopic site (the fat pad). Using this approach, MaSCs have been found to be highly enriched in the $CD29^{hi}CD49f^{hi}CD24^{+}EpCAM^{+}Sca\text{-}1^{-}$ fraction (Shackleton et al., 2006; Sleeman, Kendrick, Ashworth, Isacke, & Smalley, 2006; Stingl et al., 2006; Table 5.1). These cells can generate all mature cell types (myoepithelial, ductal, and alveolar) found in the mammary gland and, moreover, the primary outgrowths contain daughter cells with the same *in vivo* repopulating capacity as the original stem cell transplanted. Therefore, these cells fulfill the two hallmarks of stem cells, multilineage differentiation and self-renewing capability *in vivo*.

Table 5.1 Current markers for the isolation of distinct epithelial cell subpopulations from adult mouse and human mammary tissue, as determined by transplantation and clonogenic assays

Population definition	Phenotype[a]	References
Mouse		
MaSC/basal	$CD29^{hi}/CD49f^{hi}CD24^{+}$	Shackleton et al. (2006)
		Stingl et al. (2006)
	$CD24^{+}Sca\text{-}1^{-}$	Sleeman et al. (2006)
	$CD49f^{hi}EpCAM^{med}$	Shehata et al. (2012)
Slow cycling	$CD24^{+}CD29^{hi}CD1d^{+}$	Dos Santos et al. (2013)
Luminal progenitors		
$ER^{+/-}PR^{+/-}$	$CD29^{lo}CD24^{+}CD61^{+}/c\text{-}Kit^{+}$	Asselin-Labat et al. (2007, 2011)
$ER^{+}PR^{+}$	$CD24^{hi}Sca1^{+}CD133^{+}c\text{-}Kit^{+}$	Sleeman et al. (2007)
		Regan et al. (2012)
	$EpCAM^{hi}CD49f^{+}Sca\text{-}1^{+}CD49b^{+}$	Shehata et al. (2012)
	$CD29^{lo}CD24^{+}Sca\text{-}1^{+}CD49b^{+}$	Pal et al. (2013)
$ER^{-}PR^{-}$	$CD24^{+/hi}Sca1^{-}CD133^{-}c\text{-}Kit^{+/-}$	Regan et al. (2012)
	$CD29^{lo}CD24^{+}CD14^{+}c\text{-}Kit^{lo}$	Asselin-Labat et al. (2011)
	$EpCAM^{hi}CD49f^{+}Sca\text{-}1^{-}CD49b^{+}$	Shehata et al. (2012)
	$CD29^{lo}CD24^{+}Sca\text{-}1^{-}CD49b^{+}$	Pal et al. (2013)

Mature luminal	$CD29^{lo}CD24^{+}CD14^{-}c\text{-}Kit^{-}Sca\text{-}1^{+}$	Asselin-Labat et al. (2011) Regan et al. (2012)
	$EpCAM^{hi}CD49f^{+}Sca\text{-}1^{+}CD49b^{-}$	Shehata et al. (2012)
Human		
MaSC/basal	$CD49f^{hi}EpCAM^{-/lo}$	Eirew et al. (2008)
	$Muc1^{+}CD133^{+}ALDH^{+}$	Lim et al. (2009)
	$CD24^{hi}CD49f^{hi}DNER^{hi}$	Pece et al. (2010)
Luminal progenitor	$CD49f^{+}EpCAM^{+}$	Eirew et al. (2008, 2012)
		Villadsen et al. (2007)
		Lim et al. (2009)
		Shehata et al. (2012)
Mature luminal	$CD49f^{-}EpCAM^{+}$	Eirew et al. (2008)
		Lim et al. (2009)

[a]This is not intended as a complete list of all markers used to date, but rather those that have been widely used and validated in transplantation and colony forming cell (CFC) assays.

Furthermore, a single microscopically visualized cell could be demonstrated to reconstitute a complete mammary gland.

The frequency of MaSCs in the $CD29^{hi}CD49f^{hi}CD24^{+}$ population is estimated to be around 5%, with the remainder of the population containing mature myoepithelial cells and in all likelihood other basal cell intermediates. The basal population also harbors a distinct stem cell pool that is induced in pregnancy (see below), suggesting that a hierarchy of stem cells may exist in the mammary gland. The top 5% of most brightly stained $CD49f^{+}$ cells appears to be further enriched for MaSCs but this remains a somewhat arbitrary strategy for MaSC purification (Stingl et al., 2006). Moreover, EpCAM may be preferable to CD24 for the purification of MaSCs, particularly when used in combination with CD49f (Shehata et al., 2012). With continuing improvements to the tissue dissociation and sorting protocols to enhance cell viability, the frequency of MaSCs in the basal population may increase further. The embedding of cells within Matrigel for injection into the fat pad enhances the frequency of repopulation (Moraes et al., 2007; Vaillant, Lindeman, & Visvader, 2011) and is now commonly used for transplantation. Nevertheless, the basal population remains heterogeneous since it has proven difficult to segregate myoepithelial from stem cells. This likely reflects their close gene expression profiles (Stingl et al., 2006). Indeed, the basal population is the most highly conserved across human and mouse at the gene expression level (Lim et al., 2010).

The quest for further purification of MaSCs has continued using transgenic reporter mice. Given the importance of Wnt signaling to stem cell function in multiple organs, mice carrying the Wnt target gene *Axin2* linked to a LacZ reporter were used to address the role of Wnt-responsive cells in the mammary gland (Zeng & Nusse, 2010). Axin2-$LacZ^{+}$ cells were found to reside exclusively in the MaSC-enriched basal population, of which they comprised about 5%. However, the observation that Axin2-$LacZ^{+}$ cells exhibit only twofold higher *in vivo* repopulating activity than LacZ-negative cells indicates that MaSCs are not restricted to the Wnt-responsive subset. *Lgr5* is another Wnt/β-catenin target gene that has previously been identified as a marker of cycling stem cells in organs such as the intestine. Interestingly, evaluation of the Lgr5CreER-GFP reporter activity in the mammary gland revealed that $Lgr5^{+}$ cells are restricted to a rare subpopulation within the basal fraction and that they have markedly enriched repopulating activity compared to $Lgr5^{-}$ basal cells (<100 cells), which failed to reconstitute a mammary gland (Plaks et al., 2013). Indeed, a single $Lgr5^{+}$ cell could efficiently produce a mammary outgrowth and specific ablation of these cells demonstrated their essential role in postnatal mammary gland development. Using the same reporter mice, however, another group observed repopulating capacity for both $Lgr5^+$ and $Lgr5^-$ cells, with a slightly higher frequency for $Lgr5^+$ cells (de Visser et al., 2012). It remains unclear whether Lgr5 is a true target of the Wnt/β-catenin signaling pathway in the mammary gland and whether Axin2 and Lgr5 mark the same subset of basal cells. Lgr5 antibodies suitable for cell sorting and prospective isolation may be needed to further address whether MaSCs are restricted to $Lgr5^+$ cells.

Interestingly, the MaSC-enriched population in both human and mouse expresses a number of core epithelial–mesenchymal transition (EMT) genes including *Snai2/Slug*, *Twist*, and *Vimentin*. Indeed, a link between the EMT and acquisition of stem cell properties has been described (Guo et al., 2012; Mani et al., 2008). Using a reporter transgenic line in which the endogenous promoter of Slug drives expression of the *YFP* gene, sorted Slug-YFP^+ cells were shown to have 24-fold higher repopulating capacity than those not expressing Slug (Guo et al., 2012). Nonetheless, as more than 80% of cells in the basal population express Slug-YFP, this gene presumably marks myoepithelial/basal cells in addition to MaSCs.

Other attempts to visualize stem cells *in vivo* have focused on the *s-SHIP* gene, initially identified in embryonic and hematopoietic stem cells. The s-SHIP promoter-driven reporter was shown to label cap cells of the pubertal mammary gland but not mature ductal cells, and basal cells of the alveoli during pregnancy (Bai & Rohrschneider, 2010). The s-SHIP GFP^+ cells exhibited a basal cell phenotype, were actively dividing and exhibited potent regenerative capacity when sorted from either the cap layer or alveoli. These data suggest that the *s-SHIP* gene marks active MaSCs that are recruited during puberty and pregnancy to drive expansion. Quiescent MaSCs may also reside in the mammary gland based on the discovery of slowly cycling cells using either PKH26 labeling of mammosphere cultures (Cicalese et al., 2009) or an inducible histone 2b promoter linked to a GFP reporter (K5tTA-H2b-GFP) (Dos Santos et al., 2013). About 0.2% of cells in the basal population were found to be slowly cycling *in vivo*, consistent with the enrichment of long-lived label-retaining cells in the MaSC population (Shackleton et al., 2006). RNA-seq analysis of the quiescent H2b-GFP^{hi} subpopulation led to the identification of novel cell surface antigens, including CD1d. Interestingly, the combination of CD1d with CD24 and CD29 improved the frequency of repopulation such that 1 in 8 cells could reconstitute a ductal outgrowth, although a comparison between $CD1d^+$ versus

$CD1d^-$ cells in the transplantation assay was not described. The mouse MaSC-enriched population also has lower concentrations of reactive oxygen species, consistent with the presence of a quiescent MaSC pool (Diehn et al., 2009). Earlier studies by Stingl et al. (2006) revealed that all cells within the sorted basal subpopulation were cycling. It is possible that dissociation of the tissue into single cells prior to sorting leads to the activation of quiescent stem cells.

3. THE ISOLATION OF DISTINCT MAMMARY LUMINAL PROGENITOR CELLS

In contrast to stem cells that exhibit extensive self-renewal *in vivo*, progenitor cells are not capable oflong-term tissue reconstitution and are usually defined by their clonogenic activity *in vitro*. In the context of mammary progenitor cells, significantly more advances have been made in dissecting those along the luminal versus the myoepithelial lineage. At least two discrete luminal progenitor subtypes have been identified in the mouse mammary gland using various cell surface markers. CD61/β3-integrin was initially used to subdivide the total luminal population into committed progenitor cells ($CD29^{lo}CD24^+CD61^+$) that lack regenerative capacity *in vivo* and mature luminal cells ($CD29^{lo}CD24^+CD61^-$) (Asselin-Labat et al., 2007). It should be noted that while CD61 clearly delineates progenitor cells in the mammary glands from mice on a FVB/N background, there appear to be strain differences as CD61 does not identify luminal progenitors in C57BL/6 mice (M-L Asselin-Labat and N Fu, unpublished data). In some cases, a small progenitor population can also be detected within the $CD61^-$ population. On the basis of combined staining for CD14 and c-kit, a signature gene of luminal progenitor cells (Lim et al., 2010), it seems likely that the $CD29^{lo}CD24^+CD14^+$ c-Kit^+ and $CD29^{lo}CD24^+CD14^+$c-$Kit^{-/lo}$ subsets correspond to ductal and alveolar-restricted progenitors, respectively. The latter is primed for milk protein synthesis and may fluctuate in size in virgin animals according to the estrus cycle and the presence of alveolar buds (Asselin-Labat et al., 2011). Luminal progenitor cells in the adult mammary gland have also been identified as $CD133^+$Sca-1^- (Sleeman et al., 2007).

More recently, it has been possible to segregate hormone receptor-positive cells, which comprise a very small fraction, from hormone receptor-negative luminal progenitor cells. Sca-1 in combination with c-kit expression identifies ER-positive luminal progenitors ($CD24^{hi}$Sca-1^+ c-Kit^+) (Regan et al., 2012), while staining of the luminal population for

CD49b and Sca-1 expression resolves two distinct progenitor subsets: ER-positive ($EpCAM^{hi}CD49f^{+}Sca\text{-}1^{+}CD49b^{+}$) and ER-negative ($EpCAM^{hi}CD49f^{+}Sca\text{-}1^{-}CD49b^{+}$) (Shehata et al., 2012). Both of these subsets also demonstrate ALDH enzyme activity (Shehata et al., 2012). The subsets defined by Sca-1 and CD49b expression represent the most refined luminal progenitor subsets isolated to date. The ER-negative luminal progenitor cell appears to represent the alveolar progenitor based on molecular profiling (Shehata et al., 2012) and probably corresponds to the alveolar-biased cell described above ($CD29^{lo}CD24^{+}CD14^{+}c\text{-}kit^{-/lo}$). Table 5.1 presents a list of currently used markers for the delineation of luminal progenitor as well as other populations. Curiously, the luminal progenitor populations exhibit plasticity and can generate progeny representing all epithelial types when injected with Matrigel, albeit at an exceedingly low frequency (Shehata et al., 2012). Limited outgrowths were also observed infrequently on transplantation of large numbers of luminal cells (Sleeman et al., 2006). However, it remains unclear whether this potential is utilized in a physiological context.

4. *IN VIVO* LINEAGE TRACING IN THE MAMMARY GLAND

Lineage tracing is being increasingly applied to stem cell biology to monitor the fate of specific cells within a tissue or organ. An integral part of this strategy involves expression of a recombinase enzyme in a cell- or tissue-specific manner to activate expression of a reporter gene, ideally in a spatially and temporally controlled fashion. All daughter cells of the labeled parental cell thus carry the mark permanently. In early experiments using a WAP-Cre/Rosa26-LacZ model, Boulanger, Wagner, and Smith (2005) identified a fraction of cells termed parity-identified cells (PI-MECs) that persisted following remodeling of the gland after involution and could repopulate a gland following the transplantation of explants or cultured cells. It is not yet clear whether the PI-MECs reflect stem or progenitor cells.

Recent lineage tracing studies have indicated that unipotent rather than bipotent stem cells play a major role in maintaining mammary epithelium in the adult gland. These findings have added a new layer of complexity to the evolving cellular hierarchy and further suggested that cell-fate switches occur at different points of the morphogenetic cycle. In the first study by Van Keymeulen et al. (2011), inducible cre reporter strains were used to track the fate of mammary stem/progenitor cells during embryogenesis, adulthood, and pregnancy. While embryonic K14-expressing progenitors were found to give rise to all lineages in the adult gland, the luminal ($K8^+/K18^+$) and basal ($K5^+/K14^+$) lineages appeared to be uncoupled in the adult gland and in pregnancy. That is, unipotent luminal ($K8^+$) and basal ($K14^+$) stem cells rather than bipotent MaSCs were found to contribute to maintenance and expansion of the epithelia in the adult gland. The second lineage tracing study using an Axin2-CreER line to track the fate of Wnt-responsive cells has produced an even more complicated picture (van Amerongen et al., 2012). Their findings indicate the existence of both unipotent and bipotent stem/progenitor cells in the adult gland but only luminal-restricted stem cells in the embryo. Following a single labeling pulse in prepubescent mice, the $Axin2^+$ cells appeared to switch fate and commit solely to the basal lineage. Although both luminal and myoepithelial cells labeled in puberty, the data suggest that they derive from separate stem or progenitor cells based on their different locations within ducts. When the $Axin2^+$ cells were pulsed postpuberty, these cells only gave rise to basal cells. In pregnancy, however, $Axin2^+$ cells contributed to both luminal and myoepithelial cells constituting the alveolar structures, suggesting that myoepithelial-restricted stem cells in the virgin gland can be switched to bipotent stem cells during pregnancy. Cell-fate switching from the luminal to basal lineage during the first 12 days of postnatal development was also recently reported in the case of Lgr5-expressing cells, although very small numbers of cells were labeled (de Visser et al., 2012).

How do the cell types identified through lineage tracing relate to those prospectively isolated from the mouse mammary gland? Two models are presented to accommodate the data that have emerged from transplantation assays and *in vivo* lineage tracking studies in the mammary gland (Fig. 5.1). Transplantation studies clearly identify a multipotent stem cell that can give rise to all mature epithelial cells (myoepithelial, ductal, and alveolar luminal) via a series of intermediates, such as the luminal progenitor subsets that have also been prospectively identified. In contrast, lineage tracing studies have suggested that bipotent or multipotent stem cells may not be present or active in the resting mammary gland. Instead, unipotent cells are proposed to maintain the mammary epithelium in the adult. In pregnancy, however, Axin2-labeled cells were found to contribute to both lineages. These cells may correspond to the expanded MaSC subset identified by transplantation assays at mid-pregnancy. Nevertheless, lineage tracing studies have raised questions about the existence of a multipotent stem cell in the resting or nonpregnant adult gland.

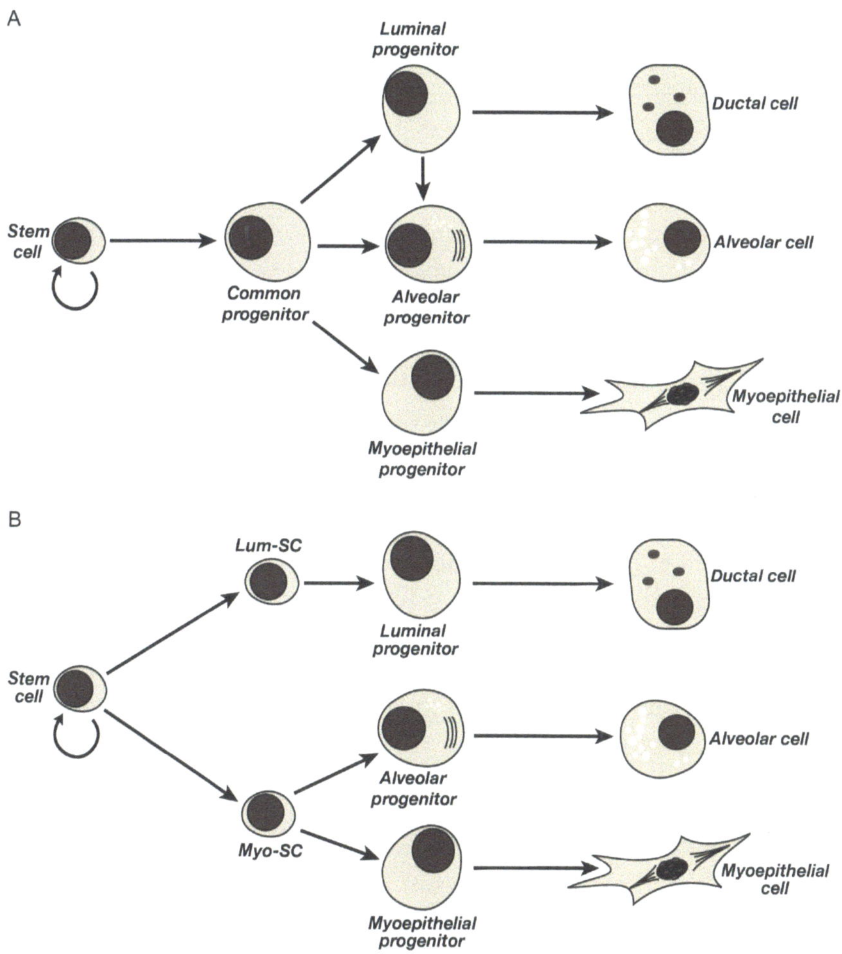

Figure 5.1 Schematic diagram showing two potential models of the epithelial hierarchy within the mammary gland. (A) Multipotent stem cells give rise to committed progenitor cells for the myoepithelial and luminal (ductal and alveolar) epithelial lineages but the precise number of progenitor cells is yet to be determined. Luminal progenitors are restricted to either a ductal or alveolar cell fate, and the alveolar-restricted progenitor is thought to be hormone receptor-negative for ER and PR. A bipotent cellular intermediate presumably lies upstream of the lineage-restricted progenitor cells. The prospective isolation of MaSCs, committed luminal progenitors, and mature luminal cells from mouse and human mammary tissue provides strong support for this hierarchical organization. (B) Recent lineage tracing studies suggest that bipotent MaSCs only exist during embryogenesis and that there is a switch to unipotent stem cell populations at birth. Two types of unipotent cells (lum-SC and myo-SC) restricted to either a luminal or myoepithelial fate are postulated to play a major role in the expansion and maintenance of mammary glands in the postnatal gland. (See color plate.)

How do we reconcile the complexities that have emerged from transplantation and cell lineage tracing studies? It is plausible that multipotent stem cells are more difficult to label than progenitors, perhaps reflecting features of quiescent cells and resulting in inefficient uptake of doxycycline or tamoxifen. On the other hand, MaSCs may be more sensitive to tamoxifen, similar to Bmi1-expressing stem cells in the intestinal crypt that underwent apoptosis in response to tamoxifen, and resulted in unintentional bias during lineage tracing (Zhu, Huang, Kek, & Bulavin, 2013). Tamoxifen doses will be even more relevant in the context of the mammary gland since ER plays a crucial role in mammary development. Another important aspect is that lineage tracing is reliant on the use of highly specific promoters that accurately mimic expression of the endogenous gene. In cases where the percentage of cells labeled for tracking has been very inefficient, such as the low degree of labeling of luminal cells observed in the K18CreER-mediated reporter model (Van Keymeulen et al., 2011), then questions are raised about the fate of the majority of cells within the lineage and the fidelity of the promoter. A further perplexing issue is that uni-lineage-restricted cells isolated from prepubescent glands could assume bipotency upon transplantation (van Amerongen et al., 2012). Thus, transplantation appears to have unlocked a multipotent regenerative potential that is not used physiologically, at least under homeostatic conditions.

5. EVIDENCE FOR THE EXISTENCE OF FETAL MaSCs

The mammary epithelium is first established at E12.5 as a result of inductive interactions between the epithelium and underlying mesenchyme. Moreover, epithelium isolated from mammary primordia at E13.0 can generate a fully functional gland (Sakakura, Sakagami, & Nishizuka, 1979), suggesting the presence of primordial MaSCs. Spike et al. (2012) recently showed that there are one or fewer stem cells per mammary rudiment at E13.5, and that this number increases dramatically (at least 200-fold) between E13.5 and E18.5. Using the same combination of cell surface markers previously identified for adult MaSCs (CD24 and CD49f), a population of cells was isolated from E18.5 mammary rudiments and shown to have reconstituting capacity in cleared fat pads at a frequency of 1 in 60 cells. Notably, the epithelial cells in the fetal rudiments did not comprise distinct luminal and basal populations as evident in the adult gland, but rather a single population with very high CD24 and CD49f expression, of which 30% expressed both K14 and K8. It is plausible that K14/K8 double-positive cells represent fetal MaSCs. Fetal MaSCs also displayed a gene expression program that differs from their adult counterpart. Despite these differences between the adult and fetal populations, fetal MaSCs could generate morphologically normal mammary glands and self-renew in serial transplantation assays. It remains to be determined how long these cells persist in the postnatal animal and what regulates the switch from a fetal to adult MaSC.

6. ELUCIDATION OF A FUNCTIONALLY ANALOGOUS EPITHELIAL HIERARCHY IN HUMAN BREAST

Human breast tissue has recently been established to have a similar hierarchical organization to the mouse mammary gland. Extensive *in vitro* clonogenic assays have revealed discrete progenitors in human breast tissue: myoepithelial-restricted, luminal-restricted, and uncommitted bipotent progenitors (Dontu et al., 2003; Gudjonsson et al., 2002; Stingl, Eaves, Zandieh, & Emerman, 2001; Villadsen et al., 2007). Limiting dilution analyses of sorted human breast subpopulations using EpCAM and CD49f antibodies after depletion of heterotypic (blood and endothelial) cells has demonstrated that only the $CD49f^{hi}EpCAM^{lo}$ subset has regenerative capacity in collagen gels placed under the kidney capsule (Eirew et al., 2008). Similarly, only the $CD49f^{hi}EpCAM^{-/lo}$ subset could yield mammary outgrowths when transplanted into "humanized" mammary fat pads (Lim et al., 2009); extensive lobular structures reminiscent of those in normal breast tissue were evident as well as branching ducts.

Characterization of sorted subpopulations using *in vitro* colony formation assays has defined a luminal progenitor ($CD49f^{+}EpCAM^{+}$) population comprising both ER-positive and ER-negative cells, which appear to be analogous to the murine $CD29^{lo}CD24^{+}CD61^{+}$ population. More recent analyses of mammary epithelial subsets have demonstrated that ALDH activity (primarily the ALDH1A3 isoform) is a feature of committed luminal progenitor cells in human breast tissue, whereas the stem cell population shows exceedingly low ALDH activity (Eirew et al., 2008; Shehata et al., 2012). Members of this enzyme family have been implicated as markers of primitive cells in blood and other tissues. Of note, the $ALDH^{+}$ subpopulation in human breast tissue appears to be functionally analogous to the ER-negative subset defined in the mouse mammary gland (Shehata et al., 2012). In human breast, a third subset of luminal progenitor cells characterized by lack of ERBB3 expression and ALDH activity was described but only in 25% of patient samples for reasons that are not understood at present (Shehata et al., 2012).

Interestingly, there appears to be an age-dependent increase in c-KIT$^+$ epithelial cells cultured from human breast tissue, suggesting an accumulation of c-KIT$^+$ luminal progenitor cells (Garbe et al., 2012).

7. CONTROL OF MaSCs BY STEROID HORMONES

The ovarian steroid hormones estrogen and progesterone play essential roles in mammary gland development and their cognate receptors (ERα and PR) are important prognostic markers in breast cancer. Hormone deprivation elicited through ovariectomy has a profound effect on the size and colony-forming activity of the luminal population; however, the ER$^+$ progenitor subfraction appears to be the least sensitive to hormone withdrawal (Shehata et al., 2012). Notably, the mouse and human MaSC-enriched subsets lack expression of estrogen receptor (ER), progesterone receptor (PR) and ErbB2 (Asselin-Labat et al., 2006; Lim et al., 2009), reminiscent of the triple negative subtype of breast cancer. Despite their negative hormone receptor status, these cells are highly responsive to steroid hormones (Asselin-Labat et al., 2010; Joshi et al., 2010). Studies in the mouse have shown that MaSC function is sensitive to either steroid hormone deprivation or stimulation. Progesterone promotes transient expansion of the MaSC population during the diestrus phase of each cycle, while exogenous estrogen and progesterone also increase the repopulating activity of the MaSC pool. Conversely, ovariectomy or treatment with the aromatase inhibitor letrozole (an inhibitor of estrogen biosynthesis) severely compromised MaSC activity, thus underscoring the importance of estrogen for normal stem cell function. During pregnancy, a dramatic increase in both the size and repopulating potential of the MaSC subset was evident, suggesting that this expanded pool may underlie the short-term increase in breast cancer associated with pregnancy.

A distinct type of MaSC appears to be expanded during pregnancy, based on its repopulating potential and gene expression signature (Asselin-Labat et al., 2010). Pregnancy-induced MaSCs show lower self-renewal on secondary transplantation compared to MaSCs derived from virgin females, suggesting that they have intermediate properties akin to the short-term repopulating hemopoietic stem cell. Intriguingly, transcriptome comparisons have revealed that the expression of alveolar-restricted genes is dramatically induced in the MaSC/basal population during pregnancy, suggesting that the stem cells are primed for alveolar differentiation. The genes that form part of this "lineage priming" signature include *Elf5*, *c-Kit*, *WAP*, *β-casein*, and *α-casein*, accompanied by a marked decrease in *Lgr5* levels (Asselin-Labat et al., 2010). Notably, the expression of all these genes could be correlated with H3K27 trimethylation marks (Pal et al., 2013). While pregnancy is associated with a transient increase in the risk of breast cancer (perhaps reflecting expansion of a mutant clone that arose during puberty), an early pregnancy is protective against this disease. Recent findings have shown that the MaSC/basal population was markedly diminished in parous mice, both in terms of the transplantability and colony-forming capacity (Meier-Abt et al., 2013). Thus, the MaSC population in parous mice seems to exist in an altered state that likely contributes to the observed decrease in breast cancer risk (see below).

8. MOLECULAR REGULATORS OF THE MAMMARY EPITHELIAL HIERARCHY

Many genes have been shown to exert important effects on mammary gland development using either loss-of-function or gain-of-function studies. This section focuses on transcription factors and developmental genes, whose function has been examined using sorted mammary epithelial subpopulations to define where they act along the differentiation hierarchy.

8.1. Regulators of MaSC function

The Wnt/β-catenin pathway governs the self-renewal of stem cells in diverse organs including the mammary gland. Purified Wnt3A protein clonally expands MaSCs for multiple generations and thereby enhances their ability to generate functional glands in transplantation assays (Zeng & Nusse, 2010). The Wnt co-receptor Lrp5 also plays a crucial role in maintaining stem cell activity in the mammary gland (Badders et al., 2009). In accordance with the Wnt pathway regulating MaSC self-renewal, premalignant mammary tissue from MMTV-wnt-1 mice contains higher MaSC numbers (Shackleton et al., 2006). There is also evidence that the Wnt pathway has a role in regulating the human MaSC subset isolated from reduction mammoplasties. In contrast to mouse MaSCs, the expansion of human MaSCs *in vitro* requires collaboration between paracrine signals from the Wnt and TGF-β pathways apparently to induce an EMT program (Scheel et al., 2011). Interestingly, the histone methylation reader Pygo-2, a coactivator of the Wnt pathway, was recently shown to be important for suppressing luminal differentiation of the MaSC-enriched population by coordinating Wnt and Notch pathway activity (Gu, Watanabe, Sun, Fallahi, & Dai, 2013).

Both Slug and Sox9 have emerged as key determinants of the MaSC state. These genes are highly expressed in the MaSC/basal population and knockdown of either Slug or Sox9 in primary mammary epithelial cells reduces the activity of MaSCs (Guo et al., 2012). Remarkably, transient coexpression of Slug and Sox9 was sufficient to convert differentiated luminal cells into MaSCs that exhibited long-term reconstituting ability *in vivo*, while overexpression of Slug alone in murine luminal progenitors could promote a basal cell-fate. In cultured human breast epithelial cells, ectopic expression of other EMT transcription factors, Twist or Snail, could generate cells with stem cell-like properties (Mani et al., 2008). Physiologically, *Slug*-deficiency leads to a developmental delay in puberty, and *Slug*$^{-/-}$ epithelial cells isolated from mouse mammary glands lose their potential to generate secondary mammospheres. Consistent with an important role for Slug in maintaining the MaSC/basal cell population, the expression of luminal genes including *Gata-3* and *ER* was induced in *Slug*$^{-/-}$ mammary epithelial cells (Nassour et al., 2012).

Other potential positive regulators of MaSCs include the polycomb group proteins Bmi1 (Pietersen et al., 2008) and Ezh2 (Pal et al., 2013), the transcription factor Stat3 (Staniszewska et al., 2012), and the tumor-suppressor BRCA1 (Smart et al., 2011). In each case, targeted deletion of the gene resulted in a reduced repopulating frequency *in vivo*. It should be noted, however, that it is difficult to distinguish between a defect in stem cells and a downstream daughter cell in the transplantation assay, as both would influence the formation of ductal outgrowths in the fat pad assay. Indeed, targeting of *Ezh2* in the mouse mammary gland profoundly impairs the clonogenic activity of all basal and luminal progenitor populations and thus the 14-fold decrease in repopulating ability *in vivo* may reflect impaired stem or progenitor cells (Pal et al., 2013). Negative regulators of MaSCs include Cbf-1/Rbp-Jκ, a canonical effector in the Notch pathway, whereby knockdown resulted in increased *in vivo* repopulating activity (Bouras et al., 2008), and p53 whose loss led to the expansion of stem cells both *in vitro* and *in vivo* (Cicalese et al., 2009). Furthermore, analysis of the partitioning of the cell-fate determinant Numb revealed that cultured PKH^{hi} cells within the MaSC-enriched population predominantly divide through asymmetric division and that the absence of p53 promotes this form of renewal.

8.2. Regulators of the luminal lineages

Several transcriptional regulators have been established to be important for the ductal and/or alveolar luminal lineages. The ETS transcription factor Elf5 is a major regulator of alveolar morphogenesis and differentiation although it has no impact on ductal morphogenesis during mouse mammary development. The accumulation of $CD61^{+}$ luminal progenitors in *Elf5*-null glands during pregnancy is indicative of a block in their differentiation to mature milk-producing alveolar cells (Chakrabarti, Hwang, et al., 2012; Chakrabarti, Wei, et al., 2012; Oakes, Rogers, Naylor, & Ormandy, 2012, 2008). Similarly, Gata-3 is essential for the differentiation of luminal progenitor cells into alveolar cells during pregnancy (Asselin-Labat et al., 2007; Kouros-Mehr, Slorach, Sternlicht, & Werb, 2006), but also regulates the differentiation of progenitors into mature ductal cells and the activity of MaSCs or their immediate descendants (Asselin-Labat et al., 2011). Furthermore, Gata-3 behaves as a master regulator of the luminal lineage since ectopic expression of this gene in the MaSC subset induced milk protein gene synthesis in the absence of a lactogenic stimulus (Asselin-Labat et al., 2007). Despite some similarities between the phenotypes of mammary glands deficient in *Elf5*, *Gata-3*, or *Stat5a*, the latter have a reduced pool of luminal progenitor cells, suggesting that loss of Stat5a might prevent the generation of alveolar progenitor cells from stem cells (Yamaji et al., 2009). Another Stat family member, Stat3, also participates in controlling luminal progenitor function (Staniszewska et al., 2012). The tumor-suppressor *Brca1* is abundantly expressed in the luminal lineage where it plays a key role. *Brca1*-deficient mouse mammary glands accumulate luminal progenitors that exhibit defective differentiation, elevated expression of c-Kit, and factor-independent growth properties *in vitro* (Lim et al., 2009; Regan et al., 2012; Smart et al., 2011).

Elf5 is not only a key specifier of alveolar fate but appears to be a regulator of basal to luminal cell-fate decisions (Chakrabarti, Hwang, et al., 2012; Chakrabarti, Wei, et al., 2012). Loss of Elf5 leads to an EMT-like phenotype during pregnancy and lactation, with the upregulation of EMT regulators such as Twist1/2, Zeb1/2, and Slug. Indeed, Elf5 was shown to directly repress the transcription of *Slug*, suggesting that it represses a basal lineage-determining program. Moreover, loss of *Elf5* in the mammary epithelium triggered a modest increase in repopulating cells, suggesting that Elf5 is important for preventing the dedifferentiation of luminal progenitors to stem-like cells. The Notch signaling pathway was also activated upon loss of *Elf5* (day 1 of lactation), consistent with the increase in committed luminal progenitor cells at this stage.

8.3. Paracrine signaling amongst different epithelial subsets

Estrogen and progesterone mediate their effects on the mammary epithelium through paracrine signaling, in which ER/PR-expressing cells act as "transmitter" cells by instructing neighboring ER/PR$^-$ cells to proliferate. Progesterone, a key pregnancy hormone, induces adjacent PR$^+$ cells to release paracrine factors such as RANKL and Wnt4, both of which are direct targets of PR (Fernandez-Valdivia & Lydon, 2012). RANKL is important for alveologenesis (Fata et al., 2000) and is induced in the luminal population in pregnancy, where it plays an important role in controlling the activity of MaSCs *in vivo* and *in vitro* (Asselin-Labat et al., 2010; Fig 5.2). In mouse models of tumorigenesis, the synthetic progesterone derivative medroxyprogesterone acetate was found to increase the MaSC-enriched population prior to carcinogen exposure and, moreover, this expansion was suppressed by blockade of RANKL signaling (Gonzalez-Suarez et al., 2010; Schramek et al., 2010). Conditional inactivation of the RANK receptor in the mammary epithelium indicated that these effects were intrinsic to epithelial cells. Interestingly, the induction of *Elf5* that occurs in luminal progenitors in response to progesterone was recently demonstrated to be mediated via RANKL (Lee et al., 2013).

The mammary epigenome is highly sensitive to steroid hormones at specific developmental stages (Pal et al., 2013). Intriguingly, the H3K27me3 (and not H3K4me3) epigenome of luminal but not MaSC/basal cells was shown to undergo profound changes in the mammary gland during pregnancy. Progesterone is likely to be a key player in mediating expansion of the alveolar luminal population through modulation of the epigenome. Indeed, progesterone was identified as an important *in vivo* regulator of the H3K27 methyltransferase Ezh2, thus linking hormonal cues to changes in chromatin structure. In addition to Rank modulating MaSC activity, it may also transmit hormonal cues to luminal progenitors. In support of this, Rank was highly expressed in the PR-negative but not PR-positive luminal progenitors, and a striking increase in *Ezh2* was seen in PR-negative luminal progenitor cells in response to *in vivo* treatment with progesterone (Pal et al., 2013).

In parous mice, altered paracrine signaling has been implicated in changing MaSC function. The MaSC-enriched subset in parous mice was recently shown to have an altered Wnt: Notch pathway ratio, with decreased expression of Wnt target genes and increased expression of Notch pathway members (Meier-Abt et al., 2013). Decreased Wnt4 levels in luminal cells was

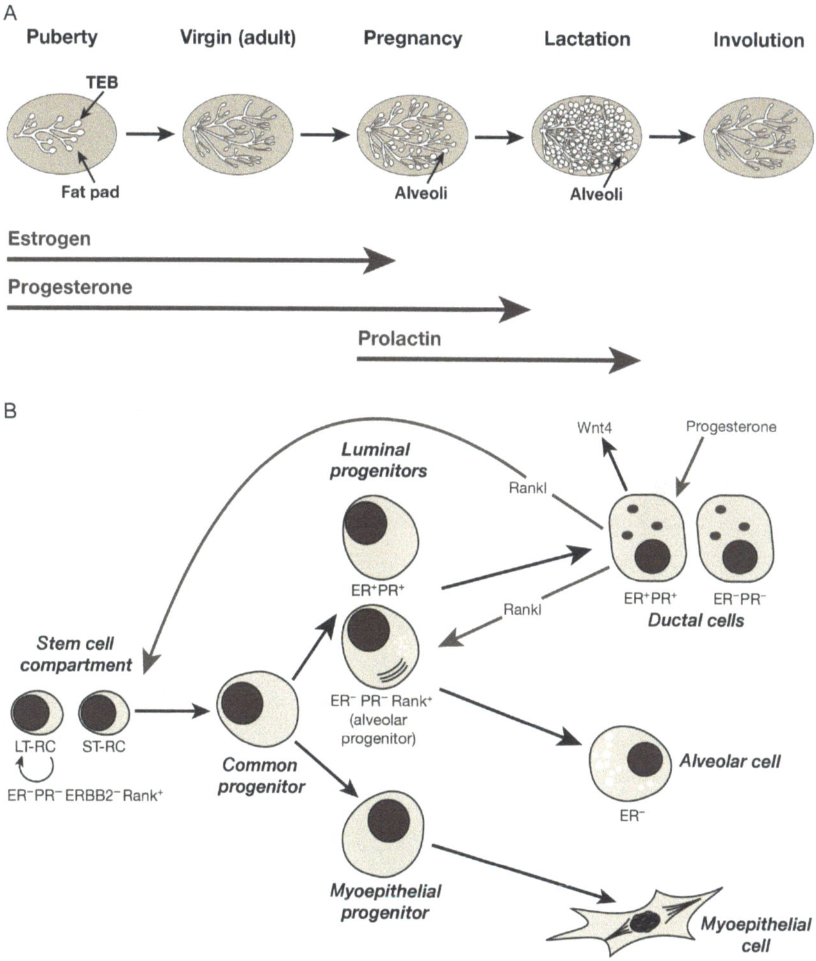

Figure 5.2 Hormonal regulation of the epithelial hierarchy. (A) Schematic diagram depicting the primary developmental stages where estrogen and progesterone (steroid hormones), and prolactin (a peptide hormone) act. Estrogen and progesterone are essential for puberty and estrus cycling, while progesterone is important throughout pregnancy. Prolactin plays key roles during pregnancy and lactation. (B) RANKL, an important paracrine mediator induced by progesterone, influences the activity of both mammary stem and progenitor cells. Short-term repopulating cells (ST-RC) are induced in mid-pregnancy. Although PR has many targets including RANKL and Wnt4, RANKL seems to be a likely candidate for transmitting signals from mature luminal cells to MaSCs in response to elevated progesterone levels during pregnancy and diestrus. These short-term MaSCs likely contribute to massive expansion of the gland during pregnancy. Progesterone is also likely to regulate the expansion of alveolar progenitor cells through paracrine signaling from RANKL-producing mature luminal cells to RANK-positive and PR-negative progenitors. (See color plate.)

suggested to attenuate the self-renewing activity of MaSCs by altering their transcriptome, since this population was enriched in differentiation genes but depleted of proliferation-associated genes. The increased expression of Notch target genes in the MaSC/basal population of parous mice is compatible with the Notch pathway repressing MaSC function and instructing luminal lineage commitment (Bouras et al., 2008). Overall, these data suggest that pregnancy, perhaps via epigenetic mechanisms, changes the MaSC pool to a more differentiated or less active status, thereby providing a long-term decrease in the risk of developing breast cancer.

8.4. Noncoding RNAs in the epithelial hierarchy

There is increasing evidence that noncoding RNAs, such as miRNAs and long noncoding RNAs, control specific mammary epithelial subtypes. Using the mouse epithelial cell line Comma-Dβ, which contains a heterogeneous progenitor cell population characterized by Sca-1 expression, Ibarra, Erlich, Muthuswamy, Sachidanandam, and Hannon (2007) provided the first evidence for involvement of miRNAs in the mammary gland. miRNA-205 and miRNA-22 were highly expressed in the Sca-1^{+} progenitor subset, whereas let-7 and miR-93 were the most abundant miRNAs in more differentiated cells. Accordingly, exogenous expression of let-7 in the Comma-Dβ cell line led to depletion of the progenitor population. In primary mouse mammary epithelial cells, miR-205 was highly expressed in the MaSC-enriched population and *PTEN*, *Zeb1*, and *Zeb2* were revealed as potential targets (Greene, Gunaratne, Hammond, & Rosen, 2010). However, the physiological role of miR-205 in the MaSC hierarchy has not been formally addressed. In a human context, miR-93 levels were significantly higher in luminal progenitor cells than the MaSC-enriched population and overexpression of this miR biased these cells toward a luminal fate (Liu et al., 2012). Expression analysis of miRNAs in both normal mouse and human mammary tissue has revealed three conserved clusters of miRNAs, miR-200C-141, miR-200b-200a-429, and miR-183-96-182, that appear to be downregulated in MaSCs and putative breast cancer stem cells. In transplantation assays, mouse mammary cells overexpressing miR-200c often yielded small disorganized structures that may result from reduced levels of Bmi1, identified to be a target of miR-200c (Shimono et al., 2009).

The long noncoding RNA *mPINC* appears to play a critical role in luminal progenitors in which it is exclusively expressed and shown to be specifically upregulated in the luminal population during pregnancy. The finding that mPINC forms a complex with polycomb repressive complex 2 suggests that mPINC may regulate or block luminal differentiation through modulating the epigenome (Shore et al., 2012), although this is yet to be proven *in vivo*.

9. IMPLICATIONS OF THE DIFFERENTIATION HIERARCHY FOR BREAST CANCER

Breast cancer is very heterogeneous disease at both the histological and molecular levels, with five definitive molecular subtypes defined thus far: luminal A, luminal B, HER2-positive, claudin-low, and basal-like (Prat & Perou, 2011). Targeted therapies that rely on the ER and ErbB2 receptors have been very effective in the treatment of luminal and HER2-positive breast cancers. However, basal-like breast cancers and the rare subclass of claudin-low tumors (Herschkowitz et al., 2007) are predominately negative for ER, PR, and ErbB2 receptors (referred to as a triple negative phenotype) and are associated with poor prognosis. Evidence is accumulating that cancers of distinct subtypes within an organ may derive from different "cells of origin" (Visvader, 2011). The characterization of target cell populations may enable earlier detection of malignancies and the development of preventative therapies for individuals at high risk of developing cancer.

The epithelial gene expression signatures established for the different mammary epithelial populations have been instrumental in understanding potential relationships between normal epithelial cells and cells of origin for breast cancer subtypes. Stem cells have been speculated as the favored candidates for targets of transformation during breast tumorigenesis because of their inherent capacity for self-renewal and longevity, which would allow the sequential accumulation of mutations. Molecular profiling studies have revealed that MaSC/basal signature is most closely aligned with the claudin-low subtype (Herschkowitz et al., 2007; Lim et al., 2009; Prat et al., 2010), suggesting that this is the likely target population for claudin-low cancer. Surprisingly, the expression profiles of basal-like cancers were most concordant with the expression signature of the luminal progenitor population and not that of the MaSC-enriched subset, as had been widely presumed. While the luminal cancer subgroups were closest to mature ductal cells, it has not been possible to correlate the HER2-positive subtype with a specific subpopulation as yet. Genetic alterations also contribute to tumor heterogeneity.

For example, loss of the tumor-suppressor retinoblastoma Rb is frequently associated with luminal B and triple negative tumors. Both of these tumor types could be recapitulated by inactivation of Rb1 in mammary epithelial progenitor cells using mouse models, emphasizing that genetic lesions also influence tumor histopathology (Jiang et al., 2010).

Proof of these gene expression correlations across normal and cancerous cells ultimately demands *in vivo* biological studies. Functional evidence supporting MaSCs as the cells of origin for claudin-low tumors was recently provided by genetically transforming cell populations enriched for basal and luminal cells and investigating their tumor-initiating capacity (Keller et al., 2012). Transformation of basal epithelial cells resulted in the development of rare metaplastic tumors reminiscent of the claudin-low subtype, whereas transformation of luminal cells resulted in tumors that were either ER-positive or ER-negative. As the starting populations were not pure, it remains to be determined whether MaSCs in the basal population are the actual targets of transformation for claudin-low tumors.

Different approaches have been used to address "cells of origin" for basal-like breast cancers arising in *BRCA1* mutation carriers and each has underscored the importance of luminal progenitor cells. Cell sorting of fresh preneoplastic tissue from individuals harboring mutations in the *BRCA1* tumor-suppressor gene initially showed that these luminal progenitor cells were expanded and had altered growth properties (Lim et al., 2009). Importantly, significant similarities were evident between the gene expression profiles of normal breast luminal progenitors, preneoplastic tissue from *BRCA1*-mutation carriers, and basal-like breast cancers. In more recent studies using mouse models of mammary tumorigenesis, inactivation of *Brca1* (and *p53*) in either luminal or basal cells of the mouse mammary gland demonstrated that only the luminal cell population initiated basal-like cancers reminiscent of those arising in *BRCA1*-mutation carriers (Molyneux et al., 2010). Furthermore, Proia et al. used human tissue from *BRCA1* mutation carriers to show that luminal epithelial cells from *BRCA1* heterozygous but not wild-type individuals produced tumors with basal characteristics. They concluded that the basal-like phenotype of *BRCA1*-associated tumors results from an aberrant luminal progenitor population that has a propensity to differentiate toward a basal state and may in part be mediated through upregulation of Slug at the protein level (Proia et al., 2011). Perturbed differentiation was also observed for human breast epithelial cells (Liu et al., 2008) and in mice heterozygous for *Brca1* where luminal to basal-like tumor transformation was apparent (Bai, Smith, Chan, & Pei, 2012).

10. CONCLUDING REMARKS

A clearer understanding of the mammary epithelial hierarchy will rely on the continued prospective isolation of refined subsets together with cell-fate mapping studies that utilize highly specific and well-characterized reporter systems. This in turn depends on the identification of more precise sublineage markers to facilitate the construction of improved reported strains for lineage tracing studies. Ultimately, these combined approaches should help resolve the controversy between the transplantation and lineage tracking data and yield a more precise picture of the linear relationships between normal mammary epithelial cells. The isolation of functionally distinct epithelial subsets from murine and human mammary tissue at greater purity also has the potential to reveal clinically relevant biomarkers for diagnosis and therapy, where the hemopoietic system has served as a paradigm. More global strategies such as functional screens in mammary epithelial cells using genome-wide shRNA libraries will likely be required to define important regulators of the hierarchy and may potentially lead to new therapeutic targets in breast cancer.

ACKNOWLEDGMENTS

We are grateful to P. Maltezos for figure preparation. This work was supported by the National Health and Medical Research Council, Australia; NHMRC IRIISS; the Victorian State Government through Operational Infrastructure Support; the Australian Cancer Research Foundation; and The National Breast Cancer Foundation (NBCF). N. F. was supported by a NBCF/Cure Cancer Postdoctoral Fellowship, G. J. L. by a NHMRC Research Fellowship, and J. E. V. by a NHMRC Australia Fellowship.

REFERENCES

Asselin-Labat, M. L., Shackleton, M., Stingl, J., Vaillant, F., Forrest, N. C., Eaves, C. J., et al. (2006). Steroid hormone receptor status of mouse mammary stem cells. *Journal of the National Cancer Institute, 98*, 1011–1014.

Asselin-Labat, M. L., Sutherland, K. D., Barker, H., Thomas, R., Shackleton, M., Forrest, N. C., et al. (2007). Gata-3 is an essential regulator of mammary-gland morphogenesis and luminal-cell differentiation. *Nature Cell Biology, 9*, 201–209.

Asselin-Labat, M. L., Sutherland, K. D., Vaillant, F., Gyorki, D. E., Wu, D., Holroyd, S., et al. (2011). Gata-3 negatively regulates the tumor-initiating capacity of mammary luminal progenitor cells and targets the putative tumor suppressor caspase-14. *Molecular and Cellular Biology, 31*, 4609–4622.

Asselin-Labat, M. L., Vaillant, F., Sheridan, J. M., Pal, B., Wu, D., Simpson, E. R., et al. (2010). Control of mammary stem cell function by steroid hormone signalling. *Nature, 465*, 798–802.

Badders, N. M., Goel, S., Clark, R. J., Klos, K. S., Kim, S., Bafico, A., et al. (2009). The Wnt receptor, Lrp5, is expressed by mouse mammary stem cells and is required to maintain the basal lineage. *PLoS One, 4*, e6594.

Bai, L., & Rohrschneider, L. R. (2010). s-SHIP promoter expression marks activated stem cells in developing mouse mammary tissue. *Genes & Development, 24*, 1882–1892.

Bai, F., Smith, M. D., Chan, H. L., & Pei, X. H. (2012). Germline mutation of Brca1 alters the fate of mammary luminal cells and causes luminal-to-basal mammary tumor transformation. *Oncogene, 32*, 2715–2725.

Boulanger, C. A., Wagner, K. U., & Smith, G. H. (2005). Parity-induced mouse mammary epithelial cells are pluripotent, self-renewing and sensitive to TGF-beta1 expression. *Oncogene, 24*, 552–560.

Bouras, T., Pal, B., Vaillant, F., Harburg, G., Asselin-Labat, M. L., Oakes, S. R., et al. (2008). Notch signaling regulates mammary stem cell function and luminal cell-fate commitment. *Cell Stem Cell, 3*, 429–441.

Chakrabarti, R., Hwang, J., Andres Blanco, M., Wei, Y., Lukacisin, M., Romano, R. A., et al. (2012). Elf5 inhibits the epithelial-mesenchymal transition in mammary gland development and breast cancer metastasis by transcriptionally repressing Snail2. *Nature Cell Biology, 14*, 1212–1222.

Chakrabarti, R., Wei, Y., Romano, R. A., DeCoste, C., Kang, Y., & Sinha, S. (2012). Elf5 regulates mammary gland stem/progenitor cell fate by influencing notch signaling. *Stem Cells, 30*, 1496–1508.

Cicalese, A., Bonizzi, G., Pasi, C. E., Faretta, M., Ronzoni, S., Giulini, B., et al. (2009). The tumor suppressor p53 regulates polarity of self-renewing divisions in mammary stem cells. *Cell, 138*, 1083–1095.

Daniel, C. W., De Ome, K. B., Young, J. T., Blair, P. B., & Faulkin, L. J., Jr. (1968). The in vivo life span of normal and preneoplastic mouse mammary glands: A serial transplantation study. *Proceedings of the National Academy of Sciences of the United States of America, 61*, 53–60.

De Ome, K. B., Faulkin, L. J., Jr., Bern, H. A., & Blair, P. B. (1959). Development of mammary tumors from hyperplastic alveolar nodules transplanted into gland-free mammary fat pads of female C3H mice. *Cancer Research, 19*, 515–520.

de Visser, K. E., Ciampricotti, M., Michalak, E. M., Tan, D. W., Speksnijder, E. N., Hau, C. S., et al. (2012). Developmental stage-specific contribution of LGR5(+) cells to basal and luminal epithelial lineages in the postnatal mammary gland. *The Journal of Pathology, 228*, 300–309.

Diehn, M., Cho, R. W., Lobo, N. A., Kalisky, T., Dorie, M. J., Kulp, A. N., et al. (2009). Association of reactive oxygen species levels and radioresistance in cancer stem cells. *Nature, 458*, 780–783.

Dontu, G., Abdallah, W. M., Foley, J. M., Jackson, K. W., Clarke, M. F., Kawamura, M. J., et al. (2003). In vitro propagation and transcriptional profiling of human mammary stem/progenitor cells. *Genes & Development, 17*, 1253–1270.

Dos Santos, C. O., Rebbeck, C., Rozhkova, E., Valentine, A., Samuels, A., Kadiri, L. R., et al. (2013). Molecular hierarchy of mammary differentiation yields refined markers of mammary stem cells. *Proceedings of the National Academy of Sciences of the United States of America, 110*, 7123–7130.

Eirew, P., Stingl, J., Raouf, A., Turashvili, G., Aparicio, S., Emerman, J. T., et al. (2008). A method for quantifying normal human mammary epithelial stem cells with in vivo regenerative ability. *Nature Medicine, 14*, 1384–1389.

Fata, J. E., Kong, Y. Y., Li, J., Sasaki, T., Irie-Sasaki, J., Moorehead, R. A., et al. (2000). The osteoclast differentiation factor osteoprotegerin-ligand is essential for mammary gland development. *Cell, 103*, 41–50.

Fernandez-Valdivia, R., & Lydon, J. P. (2012). From the ranks of mammary progesterone mediators, RANKL takes the spotlight. *Molecular and Cellular Endocrinology, 357*, 91–100.
Garbe, J. C., Pepin, F., Pelissier, F. A., Sputova, K., Fridriksdottir, A. J., Guo, D. E., et al. (2012). Accumulation of multipotent progenitors with a basal differentiation bias during aging of human mammary epithelia. *Cancer Research, 72*, 3687–3701.
Gonzalez-Suarez, E., Jacob, A. P., Jones, J., Miller, R., Roudier-Meyer, M. P., Erwert, R., et al. (2010). RANK ligand mediates progestin-induced mammary epithelial proliferation and carcinogenesis. *Nature, 468*, 103–107.
Greene, S. B., Gunaratne, P. H., Hammond, S. M., & Rosen, J. M. (2010). A putative role for microRNA-205 in mammary epithelial cell progenitors. *Journal of Cell Science, 123*, 606–618.
Gu, B., Watanabe, K., Sun, P., Fallahi, M., & Dai, X. (2013). Chromatin effector Pygo2 mediates Wnt-Notch crosstalk to suppress luminal/alveolar potential of mammary stem and basal cells. *Cell Stem Cell, 13*, 48–61.
Gudjonsson, T., Villadsen, R., Nielsen, H. L., Ronnov-Jessen, L., Bissell, M. J., & Petersen, O. W. (2002). Isolation, immortalization, and characterization of a human breast epithelial cell line with stem cell properties. *Genes & Development, 16*, 693–706.
Guo, W., Keckesova, Z., Donaher, J. L., Shibue, T., Tischler, V., Reinhardt, F., et al. (2012). Slug and Sox9 cooperatively determine the mammary stem cell state. *Cell, 148*, 1015–1028.
Herschkowitz, J. I., Simin, K., Weigman, V. J., Mikaelian, I., Usary, J., Hu, Z., et al. (2007). Identification of conserved gene expression features between murine mammary carcinoma models and human breast tumors. *Genome Biology, 8*, R76.
Hoshino, K., & Gardner, W. U. (1967). Transplantability and life span of mammary gland during serial transplantation in mice. *Nature, 213*, 193–194.
Ibarra, I., Erlich, Y., Muthuswamy, S. K., Sachidanandam, R., & Hannon, G. J. (2007). A role for microRNAs in maintenance of mouse mammary epithelial progenitor cells. *Genes & Development, 21*, 3238–3243.
Jiang, Z., Deng, T., Jones, R., Li, H., Herschkowitz, J. I., Liu, J. C., et al. (2010). Rb deletion in mouse mammary progenitors induces luminal-B or basal-like/EMT tumor subtypes depending on p53 status. *The Journal of Clinical Investigation, 120*, 3296–3309.
Joshi, P. A., Jackson, H. W., Beristain, A. G., Di Grappa, M. A., Mote, P. A., Clarke, C. L., et al. (2010). Progesterone induces adult mammary stem cell expansion. *Nature, 465*, 803–807.
Keller, P. J., Arendt, L. M., Skibinski, A., Logvinenko, T., Klebba, I., Dong, S., et al. (2012). Defining the cellular precursors to human breast cancer. *Proceedings of the National Academy of Sciences of the United States of America, 109*, 2772–2777.
Kordon, E. C., & Smith, G. H. (1998). An entire functional mammary gland may comprise the progeny from a single cell. *Development, 125*, 1921–1930.
Kouros-Mehr, H., Slorach, E. M., Sternlicht, M. D., & Werb, Z. (2006). GATA-3 maintains the differentiation of the luminal cell fate in the mammary gland. *Cell, 127*, 1041–1055.
Lee, H. J., Gallego-Ortega, D., Ledger, A., Schramek, D., Joshi, P., Szwarc, M. M., et al. (2013). Progesterone drives mammary secretory differentiation via RankL-mediated induction of Elf5 in luminal progenitor cells. *Development, 140*, 1397–1401.
Lim, E., Vaillant, F., Wu, D., Forrest, N. C., Pal, B., Hart, A. H., et al. (2009). Aberrant luminal progenitors as the candidate target population for basal tumor development in BRCA1 mutation carriers. *Nature Medicine, 15*, 907–913.
Lim, E., Wu, D., Pal, B., Bouras, T., Asselin-Labat, M. L., Vaillant, F., et al. (2010). Transcriptome analyses of mouse and human mammary cell subpopulations reveal multiple conserved genes and pathways. *Breast Cancer Research, 12*, R21.
Liu, S., Ginestier, C., Charafe-Jauffret, E., Foco, H., Kleer, C. G., Merajver, S. D., et al. (2008). BRCA1 regulates human mammary stem/progenitor cell fate. *Proceedings of the National Academy of Sciences of the United States of America, 105*, 1680–1685.

Liu, S., Patel, S. H., Ginestier, C., Ibarra, I., Martin-Trevino, R., Bai, S., et al. (2012). MicroRNA93 regulates proliferation and differentiation of normal and malignant breast stem cells. *PLoS Genetics*, *8*, e1002751.

Macias, H., & Hinck, L. (2012). Mammary gland development. *Wiley Interdisciplinary Reviews Developmental Biology*, *1*, 533–557.

Mani, S. A., Guo, W., Liao, M. J., Eaton, E. N., Ayyanan, A., Zhou, A. Y., et al. (2008). The epithelial-mesenchymal transition generates cells with properties of stem cells. *Cell*, *133*, 704–715.

Meier-Abt, F., Milani, E., Roloff, T., Brinkhaus, H., Duss, S., Meyer, D. S., et al. (2013). Parity induces differentiation and reduces Wnt/Notch signaling ratio and proliferation potential of basal stem/progenitor cells isolated from mouse mammary epithelium. *Breast Cancer Research*, *15*, R36.

Molyneux, G., Geyer, F. C., Magnay, F. A., McCarthy, A., Kendrick, H., Natrajan, R., et al. (2010). BRCA1 basal-like breast cancers originate from luminal epithelial progenitors and not from basal stem cells. *Cell Stem Cell*, *7*, 403–417.

Moraes, R. C., Zhang, X., Harrington, N., Fung, J. Y., Wu, M. F., Hilsenbeck, S. G., et al. (2007). Constitutive activation of smoothened (SMO) in mammary glands of transgenic mice leads to increased proliferation, altered differentiation and ductal dysplasia. *Development*, *134*, 1231–1242.

Nassour, M., Idoux-Gillet, Y., Selmi, A., Come, C., Faraldo, M. L., Deugnier, M. A., et al. (2012). Slug controls stem/progenitor cell growth dynamics during mammary gland morphogenesis. *PLoS One*, *7*, e53498.

Oakes, S. R., Rogers, R. L., Naylor, M. J., & Ormandy, C. J. (2008). Prolactin regulation of mammary gland development. *Journal of Mammary Gland Biology and Neoplasia*, *13*, 13–28.

Pal, B., Bouras, T., Shi, W., Vaillant, F., Sheridan, J. M., Fu, N., et al. (2013). Global changes in the mammary epigenome are induced by hormonal cues and coordinated by Ezh2. *Cell Reports*, *3*, 411–426.

Pece, S., Tosoni, D., Confalonieri, S., Mazzarol, G., Vecchi, M., Ronzoni, S., et al. (2010). Biological and molecular heterogeneity of breast cancers correlates with their cancer stem cell content. *Cell*, *140*, 62–73.

Pietersen, A. M., Evers, B., Prasad, A. A., Tanger, E., Cornelissen-Steijger, P., Jonkers, J., et al. (2008). Bmi1 regulates stem cells and proliferation and differentiation of committed cells in mammary epithelium. *Current Biology*, *18*, 1094–1099.

Plaks, V., Brenot, A., Lawson, D. A., Linnemann, J. R., Van Kappel, E. C., Wong, K. C., et al. (2013). Lgr5-expressing cells are sufficient and necessary for postnatal mammary gland organogenesis. *Cell Reports*, *3*, 70–78.

Prat, A., Parker, J. S., Karginova, O., Fan, C., Livasy, C., Herschkowitz, J. I., et al. (2010). Phenotypic and molecular characterization of the claudin-low intrinsic subtype of breast cancer. *Breast Cancer Research*, *12*, R68.

Prat, A., & Perou, C. M. (2011). Deconstructing the molecular portraits of breast cancer. *Molecular Oncology*, *5*, 5–23.

Proia, T. A., Keller, P. J., Gupta, P. B., Klebba, I., Jones, A. D., Sedic, M., et al. (2011). Genetic predisposition directs breast cancer phenotype by dictating progenitor cell fate. *Cell Stem Cell*, *8*, 149–163.

Regan, J. L., Kendrick, H., Magnay, F. A., Vafaizadeh, V., Groner, B., & Smalley, M. J. (2012). c-Kit is required for growth and survival of the cells of origin of Brca1-mutation-associated breast cancer. *Oncogene*, *31*, 869–883.

Sakakura, T., Sakagami, Y., & Nishizuka, Y. (1979). Persistence of responsiveness of adult mouse mammary gland to induction by embryonic mesenchyme. *Developmental Biology*, *72*, 201–210.

Scheel, C., Eaton, E. N., Li, S. H., Chaffer, C. L., Reinhardt, F., Kah, K. J., et al. (2011). Paracrine and autocrine signals induce and maintain mesenchymal and stem cell states in the breast. *Cell, 145*, 926–940.

Schramek, D., Leibbrandt, A., Sigl, V., Kenner, L., Pospisilik, J. A., Lee, H. J., et al. (2010). Osteoclast differentiation factor RANKL controls development of progestin-driven mammary cancer. *Nature, 468*, 98–102.

Shackleton, M., Vaillant, F., Simpson, K. J., Stingl, J., Smyth, G. K., Asselin-Labat, M. L., et al. (2006). Generation of a functional mammary gland from a single stem cell. *Nature, 439*, 84–88.

Shehata, M., Teschendorff, A., Sharp, G., Novcic, N., Russell, A., Avril, S., et al. (2012). Phenotypic and functional characterization of the luminal cell hierarchy of the mammary gland. *Breast Cancer Research, 14*, R134.

Shimono, Y., Zabala, M., Cho, R. W., Lobo, N., Dalerba, P., Qian, D., et al. (2009). Down-regulation of miRNA-200c links breast cancer stem cells with normal stem cells. *Cell, 138*, 592–603.

Shore, A. N., Kabotyanski, E. B., Roarty, K., Smith, M. A., Zhang, Y., Creighton, C. J., et al. (2012). Pregnancy-induced noncoding RNA (PINC) associates with polycomb repressive complex 2 and regulates mammary epithelial differentiation. *PLoS Genetics, 8*, e1002840.

Sleeman, K. E., Kendrick, H., Ashworth, A., Isacke, C. M., & Smalley, M. J. (2006). CD24 staining of mouse mammary gland cells defines luminal epithelial, myoepithelial/basal and non-epithelial cells. *Breast Cancer Research, 8*, R7.

Sleeman, K. E., Kendrick, H., Robertson, D., Isacke, C. M., Ashworth, A., & Smalley, M. J. (2007). Dissociation of estrogen receptor expression and in vivo stem cell activity in the mammary gland. *The Journal of Cell Biology, 176*, 19–26.

Smart, C. E., Wronski, A., French, J. D., Edwards, S. L., Asselin-Labat, M. L., Waddell, N., et al. (2011). Analysis of Brca1-deficient mouse mammary glands reveals reciprocal regulation of Brca1 and c-kit. *Oncogene, 30*, 1597–1607.

Smith, G. H. (1996). Experimental mammary epithelial morphogenesis in an in vivo model: Evidence for distinct cellular progenitors of the ductal and lobular phenotype. *Breast Cancer Research and Treatment, 39*, 21–31.

Spike, B. T., Engle, D. D., Lin, J. C., Cheung, S. K., La, J., & Wahl, G. M. (2012). A mammary stem cell population identified and characterized in late embryogenesis reveals similarities to human breast cancer. *Cell Stem Cell, 10*, 183–197.

Staniszewska, A. D., Pensa, S., Caffarel, M. M., Anderson, L. H., Poli, V., & Watson, C. J. (2012). Stat3 is required to maintain the full differentiation potential of mammary stem cells and the proliferative potential of mammary luminal progenitors. *PLoS One, 7*, e52608.

Stingl, J., Eaves, C. J., Zandieh, I., & Emerman, J. T. (2001). Characterization of bipotent mammary epithelial progenitor cells in normal adult human breast tissue. *Breast Cancer Research and Treatment, 67*, 93–109.

Stingl, J., Eirew, P., Ricketson, I., Shackleton, M., Vaillant, F., Choi, D., et al. (2006). Purification and unique properties of mammary epithelial stem cells. *Nature, 439*, 993–997.

Vaillant, F., Lindeman, G. J., & Visvader, J. E. (2011). Jekyll or Hyde: Does Matrigel provide a more or less physiological environment in mammary repopulating assays? *Breast Cancer Research, 13*, 108.

van Amerongen, R., Bowman, A. N., & Nusse, R. (2012). Developmental stage and time dictate the fate of Wnt/beta-catenin-responsive stem cells in the mammary gland. *Cell Stem Cell, 11*, 387–400.

Van Keymeulen, A., Rocha, A. S., Ousset, M., Beck, B., Bouvencourt, G., Rock, J., et al. (2011). Distinct stem cells contribute to mammary gland development and maintenance. *Nature, 479*, 189–193.

Villadsen, R., Fridriksdottir, A. J., Ronnov-Jessen, L., Gudjonsson, T., Rank, F., LaBarge, M. A., et al. (2007). Evidence for a stem cell hierarchy in the adult human breast. *The Journal of Cell Biology*, *177*, 87–101.

Visvader, J. E. (2011). Cells of origin in cancer. *Nature*, *469*, 314–322.

Yamaji, D., Na, R., Feuermann, Y., Pechhold, S., Chen, W., Robinson, G. W., et al. (2009). Development of mammary luminal progenitor cells is controlled by the transcription factor STAT5A. *Genes & Development*, *23*, 2382–2387.

Zeng, Y. A., & Nusse, R. (2010). Wnt proteins are self-renewal factors for mammary stem cells and promote their long-term expansion in culture. *Cell Stem Cell*, *6*, 568–577.

Zhu, Y., Huang, Y. F., Kek, C., & Bulavin, D. V. (2013). Apoptosis differently affects lineage tracing of Lgr5 and Bmi1 intestinal stem cell populations. *Cell Stem Cell*, *12*, 298–303.

CHAPTER SIX

Satellite Cells: The Architects of Skeletal Muscle

Natasha C. Chang[*,†], **Michael A. Rudnicki**[*,†,1]
[*]The Sprott Centre for Stem Cell Research, Ottawa Hospital Research Institute, Ottawa, Ontario, Canada
[†]Department of Cellular and Molecular Medicine, Faculty of Medicine, University of Ottawa, Ottawa, Ontario, Canada
[1]Corresponding author: e-mail address: mrudnicki@ohri.ca

Contents

Abstract

The outstanding regenerative capacity of skeletal muscle is attributed to the resident muscle stem cell termed satellite cell. Satellite cells are essential for skeletal muscle regeneration as they ultimately provide the myogenic precursors that rebuild damaged muscle tissue. Satellite cells characteristically are a heterogeneous population of stem cells and committed progenitor cells. Delineation of cellular hierarchy and understanding how lineage fate choices are determined within the satellite cell population will be invaluable for the advancement of muscle regenerative therapies.

1. INTRODUCTION

Satellite cells were first described by Alexander Mauro as mononucleated cells "wedged" between the basement membrane and plasma membrane of the muscle fiber as determined by electron microscopy of muscle fibers isolated from frog (Mauro, 1961). While initially defined by this distinct anatomical location, our current understanding of the muscle stem cell has expanded considerably with the discovery of unique cell surface

Current Topics in Developmental Biology, Volume 107
ISSN 0070-2153
http://dx.doi.org/10.1016/B978-0-12-416022-4.00006-8

markers and transcription factors as well as more sophisticated methods of satellite cell isolation and visualization. At present, it is evident that satellite cells are responsible for muscle growth as well as for the repair of muscle in response to stresses such as exercise, injury, or disease.

Satellite cell progression through the myogenic program is tightly regulated by a hierarchy of transcription factors. Under normal physiological conditions, adult satellite cells remain quiescent and reside dormant within their sublaminal niche adjacent to the myofiber (Schultz, Gibson, & Champion, 1978). All adult quiescent satellite cells express the paired-homeobox transcription factor *Pax7* (Seale et al., 2000); *Pax3*, a paralogue of *Pax7*, is also expressed in a subset of satellite cells of certain muscles (Day, Shefer, Richardson, Enikolopov, & Yablonka-Reuveni, 2007; Relaix et al., 2006). While Pax3 plays a critical role during embryonic myogenesis, most satellite cells, however, downregulate *Pax3* before birth (Horst et al., 2006; Kassar-Duchossoy et al., 2005). In contrast, Pax7 appears to be dispensable for embryonic myogenesis and is uniquely required in satellite cells after birth as *Pax7*-null mice are viable but lack functional satellite cells (Kuang, Charge, Seale, Huh, & Rudnicki, 2006; Seale et al., 2000). Thus, in skeletal muscle, Pax7 and Pax3 play overlapping yet nonredundant roles with respect to myogenesis. Recent genome-wide binding site comparative analyses between Pax7 and Pax3 confirm that Pax7 plays a dominant role in regulating transcription in adult skeletal muscle cells. Pax7 and Pax3 display differential binding affinities for paired versus homeobox motifs, suggesting that differences in DNA binding and chromatin status may contribute to the distinct developmental and functional differences between these similar transcription factors (Soleimani et al., 2012).

The overall myogenic differentiation pathway includes the activation of quiescent satellite cells, commitment to differentiation and proliferation, fusion to form myotubes, and ultimately maturation into myofibers (Fig. 6.1). Upon stimulation from exercise or following muscle injury, satellite cells become activated and enter mitosis, undergoing cellular division and giving rise to myogenic progenitor cells, a highly proliferative transient amplifying cell population also known as myoblasts. Pax7 and Pax3 both target genes that promote proliferation and commitment to the myogenic lineage, while repressing genes that induce terminal myogenic differentiation (Soleimani et al., 2012). Downstream of the Pax transcription factors, the myogenic regulatory factors (MRFs) are responsible for both the specification of satellite cells to the myogenic lineage as well as for the regulation

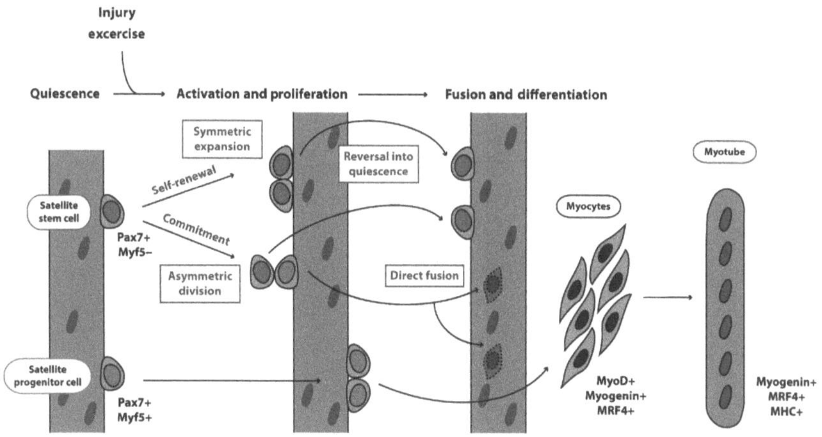

Figure 6.1 Schematic illustrating the activation and differentiation of satellite cells. Satellite cells are a heterogeneous population of satellite stem cells (Pax7+/Myf5−) and satellite progenitor cells (Pax7+/Myf5+). Upon muscle injury or exercise, satellite cells become activated and exit quiescence. Satellite stem cells (depicted as cells with blue nuclei) can self-renew and undergo either planar symmetric cell division or asymmetric cell division along the apical–basal axis. Expression of *Myf5* indicates myogenic commitment (depicted as cells with green nuclei), and these progenitor cells can contribute to muscle regeneration either by direct fusion into the regenerating myofiber or through the formation of new myotubes, which are characterized by the upregulation of genes that encode for sarcomeric proteins such as myosin heavy chain (MHC). (See color plate.)

of myogenic differentiation. The MRFs belong to a family of basic helix–loop–helix transcription factors, which include myogenic factor 5 (Myf5), myoblast determination protein (MyoD), myogenin, and muscle-specific regulatory factor 4 (Mrf4, also known as Myf6). *Myf5* and/or *MyoD* expression are rapidly upregulated upon satellite cell activation (Cooper et al., 1999; Cornelison & Wold, 1997). Pax7 and Pax3 regulate and induce the expression of both *Myf5* and *MyoD* by directly binding to distal enhancer elements and the proximal promoter of these MRF genes, respectively (Bajard et al., 2006; Hu, Geles, Paik, DePinho, & Tjian, 2008; McKinnell et al., 2008). Following sufficient generation of myogenic progenitor cells, *Pax7* is downregulated before terminal differentiation (Olguin & Olwin, 2004). *Myogenin*, a direct target of MyoD, together with *Mrf4*, initiates terminal differentiation, transition from myoblasts to elongated myocytes, fusion into myotubes, and maturation into myofibers, the multinucleated cells of skeletal muscle (reviewed in Rudnicki, Le Grand, McKinnell, & Kuang, 2008).

Importantly, a subpopulation of activated satellite cells can return to quiescence after division, in order to preserve the satellite cell pool for future regeneration. A seminal study by Collins and colleagues ascertained that satellite cells are able to maintain their own population through self-renewal. Transplantation of single myofibers into radiation-ablated muscles of regeneration-deficient dystrophic mice demonstrated that transplantation of as few as seven satellite cells was able to generate hundreds of new myofibers with associated satellite cells and repopulate the host muscle with new satellite cells. Moreover, the newly generated satellite cells are functionally competent and were able to sustain muscle regeneration through subsequent rounds of injury (Collins et al., 2005). Thus, satellite cells are muscle resident stem cells that are capable of self-renewal and generating myogenic precursors that differentiate to regenerate damaged muscle.

A series of recent studies using different *Pax7* transgenic alleles to genetically ablate satellite cells in mice have demonstrated that satellite cells are indispensible for muscle regeneration (Lepper, Partridge, & Fan, 2011; Murphy, Lawson, Mathew, Hutcheson, & Kardon, 2011; Sambasivan et al., 2011). In the absence of satellite cells, injured skeletal muscle failed to regenerate and the damaged muscle tissue was replaced with inflammatory and adipogenic cells (Sambasivan et al., 2011). Importantly, nonsatellite cell types were unable to form new muscle, suggesting that satellite cells are the primary contributors to skeletal muscle regeneration. These findings are consistent with the phenotype of *Pax7*-null mice, which exhibit significant depletion of the satellite cell pool as well as abrogated muscle repair in response to injury (Kuang et al., 2006; Oustanina, Hause, & Braun, 2004; Seale et al., 2000).

Accumulating evidence demonstrates that satellite cells are not a uniform population, but rather represent a heterogeneous population of cells that differ in terms of gene expression signature, self-renewal capability, myogenic commitment, and transplantation potential. Differences between satellite cells with different embryonic origins and from different muscle types as well as intrinsic differences between satellite cells from the same myofiber suggest many potential sources of heterogeneity. The concept of satellite cell heterogeneity is a subject that has been under intense investigation for the past few decades; however, the muscle field has yet to consolidate the findings from multiple groups and generate a unifying model to describe satellite cell hierarchy. Here we review the biology of satellite cells from embryonic development to adult homeostasis with a focus on the reported sources of cellular heterogeneity.

2. EMBRYONIC ORIGIN OF SATELLITE CELLS

Muscle development and maturation is carried out through several distinct and successive phases. In the initial stage of embryonic myogenesis (E10.5–E12.5 for limb myogenesis in mouse), embryonic myoblasts differentiate to generate the primary muscle fibers that will serve as template fibers for subsequent myogenic development during the fetal phase of myogenesis (E14.5–P0) (Biressi, Molinaro, & Cossu, 2007). Fetal myoblasts fuse with primary fibers and can also fuse among themselves to generate additional secondary fibers. Rapid and extensive proliferation of postnatal myoblasts is responsible for further muscle growth and maturation observed during neonatal myogenesis (P0–P21) (White, Bierinx, Gnocchi, & Zammit, 2010). Notably, embryonic and fetal myoblasts have distinct features and can also be distinguished from postnatal myoblasts. Isolation of these different myoblast populations revealed that embryonic, fetal, and adult myoblasts exhibit differences in cellular morphology, proliferation, growth requirements, response to extrinsic molecular signals and drug treatments, and even in the morphology of their differentiated myofibers (Biressi, Molinaro, et al., 2007; Stockdale, 1992). Accordingly, microarray analysis of purified embryonic and fetal myoblasts demonstrated unique gene expression signatures with differences in transcription factor and cell surface receptor expression for each of these cell types (Biressi, Tagliafico, et al., 2007).

Different muscle groups exhibit distinct characteristics including anatomical location, structure and morphology, fiber type composition, innervation pattern, regenerative capacities, and biochemical properties. The diversities attributed to distinct muscle groups can be traced back to their differences in embryological origin. That different muscles have distinct developmental origins is best demonstrated by the varying phenotypes of the muscular dystrophies, a group of inherited skeletal muscle disorders that are categorized according to the affected muscle group (Emery, 2002). All skeletal muscle in vertebrates originates from the mesoderm, one of the three primary germ layers established in the early embryo. The majority of skeletal muscle, including muscles of the trunk and limb (the epaxial and hypaxial skeletal muscles), arise from the somites, segmented structures derived from paraxial mesoderm that form pair-wise along the anterior/posterior axis of the embryo (Christ & Ordahl, 1995). The dorsal portion of the somite matures into the dermomyotome, an epithelial structure that contains multipotent progenitor cells that will give rise to cells of the dermis as well as trunk and limb skeletal muscles. Maturation of progenitor cells at the dorsomedial and ventrolateral lips of the dermomyotome form the first primitive muscle tissue, the myotome. Subsequently, myogenic precursor cells, characterized by *Pax3* and *Pax7* expression, migrate from the central dermomyotome directly toward the myotome and undergo extensive proliferation to maintain a pool of progenitor cells and to undergo terminal differentiation to contribute toward growth of the myotome and eventual development of embryonic and fetal trunk muscles (Kassar-Duchossoy et al., 2005; Relaix, Rocancourt, Mansouri, & Buckingham, 2005).

Limb muscle originates from a population of somitic progenitor cells, which delaminate from the ventrolateral lip of the dermomyotome and migrate to the developing limb bud (Christ & Brand-Saberi, 2002). In mice, these long-range migratory precursor cells are dependent on *Pax3* expression and are initially Pax7 negative (in the chick embryo migrating progenitors co-express *Pax3* and *Pax7*) (Relaix, Rocancourt, Mansouri, & Buckingham, 2004). This requirement for Pax3 is consistent with the phenotype of *Pax3*-null mice, also known as splotch mice, which fail to develop limb musculature (Bober, Franz, Arnold, Gruss, & Tremblay, 1994). The cellular fate of these progenitor cells is decided upon arrival at the limb, as these multipotent progenitor cells give rise to either limb muscle or endothelial cells. In mouse progenitor cells, subsequent expression of *Pax7* specifies and commits these cells to the myogenic lineage (Hutcheson, Zhao, Merrell, Haldar, & Kardon, 2009).

In contrast to trunk and limb muscles, head muscles originate from multiple embryonic structures. Of the head muscle groups, only the tongue and certain muscles of the larynx and neck derive from the somites. Analogous to limb myogenesis, *Pax3*-expressing progenitor cells delaminate from the ventrolateral border of occipital somites and migrate to the branchial arches to create muscles of the tongue. Conversely, extraocular muscles develop from the prechordal and cranial–paraxial mesoderm, while the branchiomeric muscles are derived from the paraxial unsegmented mesoderm and the lateral splanchnic mesoderm (Harel et al., 2009; Sambasivan et al., 2009). Distinct from the progenitors that give rise to trunk, limb, and tongue satellite cells, which are characterized by *Pax3* expression, the progenitor cells that give rise to the majority of head muscles are characterized by expression of the transcription factor *MesP1* (Harel et al., 2009).

Throughout development, a population of uncommitted *Pax3* and *Pax7*-expressing cells that do not express any of the MRFs were found to persist in developing embryonic trunk and limb skeletal muscles and are thought to give rise to satellite cells (Kassar-Duchossoy et al., 2005). The first satellite cells in the mouse embryo are observed at day E15.5, when *Pax3/Pax7*-expressing cells are found aligned with embryonic muscle fibers in a sublaminar position (Relaix et al., 2005). Initial studies to address the origin of satellite cells were performed using chick–quail chimeras. Substituting quail somites into chick embryos revealed that cells of quail origin contributed to muscle fibers and satellite cells within the chick host, suggesting that satellite cells are somitic in origin (Armand, Boutineau, Mauger, Pautou, & Kieny, 1983). Recent efforts using *Pax3* and *Pax7* knock-in reporter alleles combined with Cre–LoxP-based lineage tracing to determine the origin of satellite cells have reinforced the notion that satellite cells from different muscle types originate from the same embryonic source as the muscle in which they reside. Multiple lines of evidence indicate that adult satellite cells from trunk and limb muscles originate from the dermomyotome (Gros, Manceau, Thome, & Marcelle, 2005; Kassar-Duchossoy et al., 2005; Relaix et al., 2005; Schienda et al., 2006). Limb satellite cells specifically arise from *Pax3*-expressing cells in the hypaxial somite, which migrate from the dermomyotome to the developing limbs (Schienda et al., 2006). Analogous to the somite-derived trunk and limb muscles, lineage-tracing experiments also suggest that satellite cells from various head muscles are derived from the same distinct embryonic origin as their host muscle (Harel et al., 2009).

The first 3 weeks of postnatal life in the mouse represent a period of intense growth, in which muscle mass increases nearly threefold. After P21, skeletal muscle reaches a steady-state level of homeostasis. From birth to P21, there is a significant decrease in satellite cell number from approximately 30% (percentage of total myonuclei per fiber) to less than 5%. During this growth period, a decrease in satellite cell numbers is correlated with an increase in myonuclear content of myofibers, a process termed myonuclear accretion (White et al., 2010). After postnatal muscle growth, satellite cells enter quiescence, which is essential for the maintenance of satellite cells throughout adulthood. The Notch signaling pathway has been implicated in the regulation of satellite cell entry into quiescence; the downstream Notch target genes, Hairy and Enhancer of Split-related 1 and 3 (*Hesr1* and *Hesr3*), are required during this process as *Hesr1/3* double knockout mice fail to generate undifferentiated quiescent satellite cells during postnatal development (Fukada et al., 2011).

Of note, the requirement for Pax7 in satellite cells after birth has remained a highly controversial issue within the satellite cell field. Numerous studies performed on *Pax7*-null mice have demonstrated that these mice undergo normal embryonic myogenesis, but exhibit significant and progressive loss of satellite cells after birth (Kuang et al., 2006; Oustanina et al., 2004; Relaix et al., 2006; Seale et al., 2000). These mice are unable to effectively regenerate muscle upon injury and generally fail to live past 2–3 weeks. In stark contrast, a study employing a *Pax7* floxed allele and tamoxifen to induce deletion of *Pax7* in adult mice (after P21) suggested that Pax7 is dispensable for muscle regeneration in the adult (Lepper, Conway, & Fan, 2009). While the authors reported that Pax7 is required for muscle regeneration in juvenile (P7–11) mice (consistent with *Pax7*-null mice), muscle regeneration appeared normal in adult mice, a phenotype that was not altered upon additional ablation of *Pax3*. The observed discrepancy in muscle regeneration abilities between conditional deletion of *Pax7* in the adult (Lepper et al., 2009) and the conventional *Pax7* knockout mice (Kuang et al., 2006) requires further investigation, especially since our studies have shown that loss of *Pax7* expression in satellite cells or myoblasts from any given age results in loss of proliferative capacity and reduction of *Myf5* expression (Kawabe, Wang, McKinnell, Bedford, & Rudnicki, 2012; McKinnell et al., 2008). As well, the results described by Lepper and colleagues are hard to grasp due to recent literature demonstrating that satellite cells, and thereby their proliferation, are essential for regeneration of skeletal muscle (Lepper et al., 2011; Murphy et al., 2011; Sambasivan et al., 2011).

3. ADULT SATELLITE CELLS: A HETEROGENEOUS POPULATION OF STEM CELLS AND COMMITTED PROGENITORS

During the final stages of fetal development, most satellite cells downregulate *Pax3* while maintaining *Pax7* expression (Kassar-Duchossoy et al., 2005). In the adult, all satellite cells express *Pax7*, while a subset of satellite cells, limited to the diaphragm and body wall muscles, also expresses *Pax3* (Relaix et al., 2006). Interestingly, differential expression of *Pax3* in adult satellite cells has not been correlated with differences in embryonic origin. Rather, a recent study proposed that *Pax3* expression is posttranscriptionally regulated by microRNA-206. Degradation of *Pax3* mRNA by miR-206 is dependent on the 3′-untranslated region of *Pax3* transcripts, which is subject to alternative polyadenylation in different muscle groups (Boutet et al., 2012). Additionally, while *Pax7* expression is characteristic of all satellite cells, the level of *Pax7* expression within the satellite cell population is not uniform. Using *Pax7–GFP* transgenic mice where expression of GFP correlates with Pax7 expression, quiescent satellite cells were isolated by fluorescence-activated cell sorting and categorized into *Pax7*-high and -low-expressing subpopulations. *Pax7*-high satellite cells were found to have lower metabolic activity, slower proliferation kinetics and were more resistant to differentiation than *Pax7*-low cells (Rocheteau, Gayraud-Morel, Siegl-Cachedenier, Blasco, & Tajbakhsh, 2012).

Quiescent satellite cells are characterized by the expression of *Pax7* and the absence of *MyoD* expression (Zammit et al., 2006). Satellite cell transition from quiescence to an activated state is reversibly regulated by the mitogen-activated protein kinase (MAPK) pathway involving the p38α/β kinases (Jones et al., 2005). Activated satellite cells robustly express *MyoD* and/or *Myf5*. Notably, in cultured myoblasts and satellite cells on cultured myofibers, a subset of Pax7+/MyoD+ cells, were found to downregulate *MyoD*, resist differentiation, and maintain a mitotically inactive state resembling quiescence (Olguin & Olwin, 2004; Zammit et al., 2004). This return of satellite cells to quiescence after muscle injury is dependent on Sprouty1 (*Spry1*), an inhibitor of receptor tyrosine kinase signaling. During muscle regeneration, *Spry1* is temporally expressed; quiescent satellite cells express robust levels of *Spry1*, which are subsequently downregulated in activated satellite cells and re-induced in satellite cells that reenter quiescence. Depletion of *Spry1* in satellite cells prevents satellite cells from reentering quiescence and results in a diminished satellite cell pool after regeneration (Shea et al., 2010).

Previously thought to be a dormant inactive state, increasing evidence demonstrates that quiescence is actively maintained and that satellite cells are poised for rapid activation and differentiation in response to injury. One recent study proposed that quiescent satellite cells transcribe *Myf5* mRNA but are able to maintain quiescence through the suppression of post-transcriptional regulation of *Myf5* by microRNA-31. Interestingly, miR-31 targeting of *Myf5* results in their sequestration into messenger ribonucleoprotein (mRNP) granules. The activation of satellite cells causes dissociation of mRNP granules and the release of *Myf5* transcripts, resulting in rapid translation of Myf5 protein and swift entry into the myogenic program (Crist, Montarras, & Buckingham, 2012). This model illustrates how satellite cells are primed for myogenesis and explains how satellite cells can quickly transition from a quiescent to activated state.

Satellite cells are stem cells that are capable of self-renewal and maintaining their population despite numerous rounds of muscle regeneration throughout the lifespan of the organism. Increasing evidence suggests that satellite cells represent a heterogeneous population of stem cells and committed progenitor cells, and that only a small proportion of satellite cells are "true" stem cells. Continuous labeling of rat satellite cells *in vivo* with bromodeoxyuridine (BrdU), a thymidine analogue, revealed differences in proliferation rates within the satellite cell population. The majority of satellite cells (around 80%) were quick to enter the cell cycle and contributed myonuclei to growing myofibers, while the remaining 20%, referred to as the "reserve population," exhibited much slower proliferation kinetics and resist differentiation (Schultz, 1996). Moreover, transplantation experiments of isolated satellite cells have demonstrated that only a small portion of transplanted cells are capable of repopulating the satellite cell niche (Collins et al., 2005).

As stem cells, satellite cells are able to self-renew via symmetric (generating two identical daughter cells) or asymmetric cell divisions (generating one self-renewing daughter cell and one daughter cell committed to differentiate). Using BrdU to pulse-chase label satellite cells, researchers have proposed that a subpopulation of satellite cells undergo nonrandom template DNA strand segregation. In accordance with the "immortal strand hypothesis," BrdU-label-retaining cells represent stem cell populations that retain original template DNA strands during cell division, a process that protects the stem cell genome from DNA replication errors and maintains epigenetic marks (Cairns, 1975; Conboy, Karasov, & Rando, 2007; Shinin, Gayraud-Morel, Gomes, & Tajbakhsh, 2006). Additionally, the cell fate determinant Numb was found to distribute asymmetrically and co-segregate with older template DNA in label-retaining cells following asymmetric cell division (Shinin et al., 2006). This finding suggests that inhibition of Notch signaling by Numb is important for satellite cell self-renewal. In contrast, another study found that during asymmetric cell divisions, Numb expression was detected in daughter cells that would undergo differentiation (Conboy & Rando, 2002). This observation is consistent with a proposed role for Notch signaling in maintaining satellite cell self-renewal (Kuang, Kuroda, Le Grand, & Rudnicki, 2007; Wen et al., 2012). Thus, the precise role for Notch signaling during asymmetric cell division and how it regulates cell fate determination of satellite cells remains unclear.

Additionally, the mechanisms that determine polarity and govern asymmetric cell divisions in satellite cells are not well understood. The evolutionarily conserved Par complex was recently described to establish asymmetry in dividing satellite cells by asymmetric activation of the p38α/β MAPK pathway (Troy et al., 2012). Asymmetric sequestration of p38α/β activates the MAPK pathway in only one daughter cell to induce *MyoD* expression and generates a proliferating myoblast. Absence of p38α/β in the other daughter cell returns the satellite cell to a quiescent state. Notably, the p38α/β kinases have also been described to regulate epigenetic silencing of *Pax7* via the Polycomb repressive complex 2 (PRC2) to influence satellite cell differentiation (Palacios et al., 2010).

Utilizing a genetic lineage-tracing approach in mice carrying *Myf5-Cre* and *Rosa26YFP* reporter alleles, our laboratory identified a population of adult satellite cells that have never expressed *Myf5* (Kuang et al., 2007). These Myf5 − cells represent a minor proportion of total *Pax7*-expressing satellite cells (around 10%) and demonstrate superior engraftment ability and niche repopulation as satellite cells compared to the more committed Myf5 + progenitor cells. Importantly, Myf5 − cells are capable of giving rise to Myf5 + cells by means of asymmetric cell division along the apical–basal axis perpendicular to the muscle fiber (Fig. 6.1). These Pax7 +/Myf5 − cells exhibit features of a less committed stem cell and are proposed to represent a "true" stem cell population. It is likely that Pax7 +/Myf5 − cells in adult muscle represent a lineage continuum of the described Pax+/MRF − embryonic progenitor cells that give rise to adult satellite cells. Induction of *Myf5* expression requires Pax7 recruitment of a histone methyltransferase complex to enhancer elements upstream of the *Myf5* promoter (McKinnell et al., 2008). Furthermore, coactivator-associated arginine methyltransferase 1 (Carm1) methylates Pax7 in order to regulate its transcriptional activity following asymmetric cell division of a Myf5 − satellite stem cell (Kawabe et al., 2012). The observation that Carm1/Pax7 interactions are present only in Myf5 + cells provides mechanistic insight to key differences in the active molecular pathways between Myf5 − satellite stem cells and Myf5 + satellite progenitor cells.

Homeostatic maintenance of a satellite stem cell pool is critical for preserving the regenerative potential of muscle. Our laboratory also found that noncanonical Wnt signaling drives expansion of Myf5 − satellite stem cells via symmetric cell division along the planar axis, parallel to the muscle fiber (Fig. 6.1). Wnt7a is upregulated during muscle regeneration and signals through its receptor Frizzled 7, which is expressed at higher levels in Myf5 − satellite stem cells compared to Myf5 + cells. In dividing satellite stem cells, Wnt7a induced polarized expression of Vangl2, an effector of the planar cell polarity pathway, which was required for Wnt7a-mediated satellite cell expansion (Le Grand, Jones, Seale, Scime, & Rudnicki, 2009). As well, we recently discovered that during muscle regeneration, satellite cells are capable of remodeling their niche in order to regulate their own expansion. Activated Myf5+ satellite cells express and contribute high levels of the extracellular matrix glycoprotein fibronectin to the satellite cell niche. We found that fibronectin enhanced Wnt7a activity in regenerating muscle, and that knockdown of fibronectin in prospectively isolated satellite cells abrogated the ability of these cells to repopulate the satellite cell niche upon transplantation (Bentzinger, Wang, von Maltzahn, Soleimani, et al., 2013). Together, these findings highlight the complexity of satellite cell regulation during muscle regeneration and the existence of cross talk between multiple extrinsic and intrinsic cellular pathways.

4. AGE AND DISEASE-RELATED DECLINE IN SATELLITE CELL FUNCTION

Progressive loss of muscle function occurs naturally with age and can be exacerbated by skeletal muscle diseases, also known as myopathies. The reduced regenerative potential associated with aged muscle is correlated with a decline in satellite cell function. While decreases in satellite cell numbers in aged muscle have been reported (Day, Shefer, Shearer, & Yablonka-Reuveni, 2010), the current thinking is that age-induced alterations of the satellite cell niche and microenvironment, which ultimately have an effect on satellite cell activity, are primarily responsible for the deterioration observed. A seminal study in the Rando laboratory employing parabiotic pairings between young and old mice, so that they share the same circulatory system, demonstrated that the systemic environment greatly influences satellite cell activity. Delivery of systemic factors from young mice demonstrated a remarkable ability to restore the proliferation and regenerative capacity of satellite cells in aged mice (Conboy et al., 2005). In aged muscle, upregulation of the Notch ligand Delta was significantly impaired in response to injury. Attenuation of Notch signaling in aged muscle results in defective satellite cell proliferation and an inability to generate myoblasts during regeneration (Conboy, Conboy, Smythe, & Rando, 2003). Additionally, in resting muscles of aged mice, higher expression levels of fibroblast growth factor (FGF) were observed. Increased FGF levels induce quiescent satellite cells to cycle and break quiescence, resulting in an overall loss of self-renewal ability (Chakkalakal, Jones, Basson, & Brack, 2012).

Among the diseases that affect skeletal muscle regeneration, the muscular dystrophies have been most studied. Muscular dystrophies are a group of inherited genetic disorders that are characterized by progressive muscle weakening and, ultimately, muscle wasting. There are approximately 40 different types of muscular dystrophy disorders, which vary in severity and are categorized by affected muscle distribution (Emery, 2002). Affecting 1 in 3500 male births, Duchenne muscular dystrophy (DMD) is the most common and most severe form (Emery, 1991). DMD is caused by mutations in the *DMD* gene that result in an absence of functional dystrophin protein. Dystrophin is a large (427 kDa) cytoskeletal protein and an essential component of the dystrophin glycoprotein complex (DGC), which connects the basal lamina of the muscle fiber to the extracellular matrix. Notably, several forms of muscular dystrophy are the result of primary mutations in genes that encode for proteins that constitute the DGC. Loss of dystrophin weakens the structural integrity of the sarcolemma, resulting in progressive damage with increased muscle contractile activity (Petrof, Shrager, Stedman, Kelly, & Sweeney, 1993). Constitutive muscle damage induces chronic cycles of muscle fiber degeneration and regeneration, thus forcing satellite cells to undergo continuous rounds of activation, proliferation, and differentiation. It is thought that the satellite cells of dystrophic mice progressively lose their regenerative capacity (Blau, Webster, & Pavlath, 1983). Eventual exhaustion of the satellite cell pool renders the muscle unable to repair the damaged muscle, which is gradually replaced by fibrotic and adipose tissue.

Intriguingly, *mdx* mice, which are deficient for dystrophin and are the most commonly studied mouse model for DMD, fail to fully recapitulate DMD disease progression in humans, displaying a much milder phenotype (Bulfield, Siller, Wight, & Moore, 1984). One proposed explanation for this discrepancy is the difference in the telomere lengths between mice and humans. Longer telomeres in mice were predicted to have a protective effect on the satellite stem cell reserve. Accordingly, the generation of dystrophic mice lacking telomerase activity resulted in severe muscular dystrophy that progressively worsened with age. The proliferative and regenerative capacity of satellite cells derived from these mice was significantly impaired (Sacco et al., 2010). This study provides further evidence that muscle weakening in DMD is the result of continuous decline in satellite cell function, leading to an inability to repair chronic muscle damage.

5. SATELLITE CELL TRANSPLANTATION THERAPY

That transplanted satellite cells can participate in muscle regeneration and contribute to the satellite cell population in host mice suggests that satellite cells are ideal candidates for cell-based transplantation therapy for muscle degenerative diseases. Early studies demonstrating that transplantation of satellite cell-derived myoblasts is capable of restoring dystrophin expression in *mdx* mice generated initial excitement in the field of regenerative medicine (Partridge, Morgan, Coulton, Hoffman, & Kunkel, 1989). Host immune rejection, poor survival, inability to self-renew, and defective migration of donor cells following transplantation, however, have impeded the therapeutic use of satellite cells within the clinic.

The immunogenic nature of transplanted allogenic cells and their resulting newly formed muscle tissue remain one of the major obstacles for satellite cell transplantation therapy necessitating the use of immunosuppressants. Treatment of genetic muscle disorders, such as DMD, requires healthy donor-derived cells since autologous transplantation would be unable to reinstate dystrophin expression. In order to avoid transplant rejection of donor cells, genetic correction of induced pluripotent stem cells (iPSCs) has been proposed as an alternative to allotransplantation of satellite cells. The advantages for using iPSCs include feasibility in obtaining cells from patients and the ability to perform autologous transplantation. A recent study generated myogenic precursor cells from both human embryonic stem cells (ESCs) and iPSCs by inducible lentiviral overexpression of *Pax7*. The authors demonstrated that transplantation of ESC/iPSC-derived myogenic progenitors into dystrophic muscle of *mdx* mice resulted in the engraftment of human myogenic progenitors and generation of dystrophin-positive myofibers (Darabi et al., 2012). While there are clear benefits for generating iPSCs from diseased patients to study mechanisms of disease biology, the use of iPSCs for transplantation therapy remains questionable. Risks associated with genetic reprogramming of somatic cells into iPSCs and the requirement for ectopic expression of *Pax7* to generate myogenic progenitors will need to be addressed before translation into the clinic.

As satellite cells represent a rare and minor population in skeletal muscle tissue, another restriction in satellite cell transplantation is the limited availability of cells for transplantation. Removal of satellite cells from their niche results in their activation and *in vitro* culturing of satellite cells in culture dishes leads to their commitment toward myogenic differentiation and conversion into myoblasts. While freshly isolated satellite cells transplanted into mice demonstrate proficient contribution to skeletal muscle regeneration, transplantation of cultured muscle progenitor cells yielded poor results (Bentzinger, Wang, von Maltzahn, & Rudnicki, 2013). Thus, the challenge lies in the ability to expand satellite cells in culture without loss of their regenerative and self-renewal properties. Of note, fewer numbers of satellite cells are required for transplantation compared to myoblasts due to their higher engraftment efficiency. Intriguingly, a single satellite cell was sufficient to generate new myofibers and satellite cells when transplanted into the muscle of mice (Sacco, Doyonnas, Kraft, Vitorovic, & Blau, 2008).

The use of bioengineering approaches to generate methods of culturing stem cells that mimic the *in vivo* niche shows great potential in allowing the *ex vivo* expansion of satellite cells. A recent study using a polyethylene glycol hydrogen platform that mimicked muscle tissue rigidity as a substrate for culturing satellite cells demonstrated that these cultured cells resist differentiation and are able to self-renew *in vitro*. Significantly, transplantation of cells cultured on hydrogel substrate demonstrated superior engraftment potential compared to cells cultured on plastic dishes, and engraftment was comparable to transplantation of freshly isolated satellite cells (Gilbert et al., 2010).

A comprehensive understanding of the complex heterogeneity that exists within the satellite cell population is essential for the advancement of satellite cell-based therapies. Earlier myoblast transplantation studies revealed that only a minor subpopulation was able to survive transplantation and generate new muscle, suggesting that this subpopulation exhibited distinct stem cell-like characteristics (Beauchamp, Morgan, Pagel, & Partridge, 1999). Moreover, in the transplantation studies of single satellite cells by Sacco and colleagues, engraftment occurred at a frequency of 4% (Sacco et al., 2008). Accordingly, our discovery of two populations of satellite cells based on *Myf5* expression has revealed distinct functional properties between satellite stem cells (Pax7+/Myf5− cells) and more committed myogenic progenitors (Pax7+/Myf5+ cells). Injection of Pax7+/Myf5+ cells resulted in precocious differentiation and limited migration upon transplantation into host muscles, while Pax7+/Myf5− cells contributed extensively to the satellite cell compartment throughout the muscle tissue (Kuang et al., 2007). As well, transplantation of satellite cells from different muscle groups revealed differential engraftment potential (Collins et al., 2005). In the transplantation studies performed by Collins et al., they found that EDL and soleus satellite cells generated larger clusters of muscle in a shorter period of time (3 weeks) compared to TA muscle (5 weeks). Thus, the ability to isolate satellite cells with the greatest potential for self-renewal and regeneration will significantly advance the therapeutic efficacy of satellite cell-based treatments.

6. CONCLUDING REMARKS

The muscle satellite cell and its myofiber niche together represent an excellent model for studying the biology of adult stem cells. The ability to visualize satellite cells on associated myofibers by *ex vivo* myofiber culturing methods has led to an enhanced understanding of the underlying biology and the influence of extrinsic factors on satellite cell divisions (Keire, Shearer, Shefer, & Yablonka-Reuveni, 2013; Pasut, Jones, & Rudnicki, 2013). The challenge now lies in being able to study satellite cells precisely at the time of myogenic commitment, which requires knowledge of how cell fate decisions are controlled. It is evident that satellite cells represent a heterogeneous population of stem cells and committed progenitors. Elucidating satellite cell hierarchy will require the study of satellite cells at the single cell level and tracking their behavior over time.

The ability of adult satellite cells to generate nonmyogenic progeny has also recently come into focus, thus challenging the paradigm that satellite cells give rise to cells solely within the myogenic lineage. Interestingly, our laboratory demonstrated that adult satellite cells can give rise to brown adipocytes, and that microRNA-133 targeting of *Prdm16*, a transcriptional regulator that specifies the brown fat lineage, regulates this cell fate decision (Yin et al., 2013). In another study, the Cossu laboratory revealed that skeletal myoblasts can be converted into pericytes, endothelial-associated mesenchymal cells that surround the vessels of the microvasculature. The switch from myoblasts to pericytes occurs via platelet-derived growth factor BB (PDGF-BB) and Notch ligand Delta-like 4 (Dll4), which are factors secreted by endothelial cells (Cappellari et al., 2013). Together, these studies imply that satellite cells are multipotent. Additional studies are required to determine the frequency of these cell fate decisions *in vivo*. It also remains unknown if all or only a unique subset of satellite cells/myoblasts exhibits the potential to adopt a nonmyogenic fate.

Finally, our knowledge of human satellite cell biology is severely lacking compared to what is known in mice. In particular, the concept of satellite cell heterogeneity remains largely unexplored in humans. Do the Pax7+/Myf5− and Pax7+/Myf5+ populations exist in humans, and if not, are there functionally equivalent cell populations? Ultimately, the success of muscle regenerative therapy is contingent upon a thorough understanding of satellite cell heterogeneity. Efforts are now shifting toward stimulating endogenous repair mechanisms that are intrinsic to stem cells as a more practical mode of regenerative therapy. With the current advances in high-throughput screening technologies and the increasing availability of small molecule drug libraries, hopefully such cures will soon be within reach.

ACKNOWLEDGMENTS

The authors thank Julia von Maltzahn and Melanie Lacaria for critical reading of the chapter. M. A. R. holds the Canada Research Chair in Molecular Genetics and N. C. C. is the recipient of fellowships from the Canadian Institutes of Health Research and the Ontario Stem Cell Initiative. Studies by the M. A. R. laboratory were supported by grants from the Canadian Institutes for Health Research, the National Institutes for Health, the Muscular Dystrophy Foundation, the Stem Cell Network, the Ontario Research Fund, and the Howard Hughes Medical Institute.

REFERENCES

Armand, O., Boutineau, A. M., Mauger, A., Pautou, M. P., & Kieny, M. (1983). Origin of satellite cells in avian skeletal muscles. *Archives d'Anatomie Microscopique et de Morphologie Expérimentale, 72*, 163–181.

Bajard, L., Relaix, F., Lagha, M., Rocancourt, D., Daubas, P., & Buckingham, M. E. (2006). A novel genetic hierarchy functions during hypaxial myogenesis: Pax3 directly activates Myf5 in muscle progenitor cells in the limb. *Genes & Development, 20*, 2450–2464.

Beauchamp, J. R., Morgan, J. E., Pagel, C. N., & Partridge, T. A. (1999). Dynamics of myoblast transplantation reveal a discrete minority of precursors with stem cell-like properties as the myogenic source. *The Journal of Cell Biology, 144*, 1113–1122.

Bentzinger, C. F., Wang, Y. X., von Maltzahn, J., & Rudnicki, M. A. (2013a). The emerging biology of muscle stem cells: Implications for cell-based therapies. *Bioessays, 35*, 231–241.

Bentzinger, C. F., Wang, Y. X., von Maltzahn, J., Soleimani, V. D., Yin, H., & Rudnicki, M. A. (2013b). Fibronectin regulates Wnt7a signaling and satellite cell expansion. *Cell Stem Cell, 12*, 75–87.

Biressi, S., Molinaro, M., & Cossu, G. (2007a). Cellular heterogeneity during vertebrate skeletal muscle development. *Developmental Biology, 308*, 281–293.

Biressi, S., Tagliafico, E., Lamorte, G., Monteverde, S., Tenedini, E., Roncaglia, E., et al. (2007b). Intrinsic phenotypic diversity of embryonic and fetal myoblasts is revealed by genome-wide gene expression analysis on purified cells. *Developmental Biology, 304*, 633–651.

Blau, H. M., Webster, C., & Pavlath, G. K. (1983). Defective myoblasts identified in Duchenne muscular dystrophy. *Proceedings of the National Academy of Sciences of the United States of America, 80*, 4856–4860.

Bober, E., Franz, T., Arnold, H. H., Gruss, P., & Tremblay, P. (1994). Pax-3 is required for the development of limb muscles: A possible role for the migration of dermomyotomal muscle progenitor cells. *Development, 120*, 603–612.

Boutet, S. C., Cheung, T. H., Quach, N. L., Liu, L., Prescott, S. L., Edalati, A., et al. (2012). Alternative polyadenylation mediates microRNA regulation of muscle stem cell function. *Cell Stem Cell, 10*, 327–336.

Bulfield, G., Siller, W. G., Wight, P. A., & Moore, K. J. (1984). X chromosome-linked muscular dystrophy (mdx) in the mouse. *Proceedings of the National Academy of Sciences of the United States of America, 81*, 1189–1192.

Cairns, J. (1975). Mutation selection and the natural history of cancer. *Nature, 255*, 197–200.

Cappellari, O., Benedetti, S., Innocenzi, A., Tedesco, F. S., Moreno-Fortuny, A., Ugarte, G., et al. (2013). Dll4 and PDGF-BB convert committed skeletal myoblasts to pericytes without erasing their myogenic memory. *Developmental Cell, 24*, 586–599.

Chakkalakal, J. V., Jones, K. M., Basson, M. A., & Brack, A. S. (2012). The aged niche disrupts muscle stem cell quiescence. *Nature, 490*, 355–360.

Christ, B., & Brand-Saberi, B. (2002). Limb muscle development. *The International Journal of Developmental Biology, 46*, 905–914.

Christ, B., & Ordahl, C. P. (1995). Early stages of chick somite development. *Anatomy and Embryology (Berlin), 191*, 381–396.

Collins, C. A., Olsen, I., Zammit, P. S., Heslop, L., Petrie, A., Partridge, T. A., et al. (2005). Stem cell function, self-renewal, and behavioral heterogeneity of cells from the adult muscle satellite cell niche. *Cell, 122*, 289–301.

Conboy, I. M., Conboy, M. J., Smythe, G. M., & Rando, T. A. (2003). Notch-mediated restoration of regenerative potential to aged muscle. *Science, 302*, 1575–1577.

Conboy, I. M., Conboy, M. J., Wagers, A. J., Girma, E. R., Weissman, I. L., & Rando, T. A. (2005). Rejuvenation of aged progenitor cells by exposure to a young systemic environment. *Nature, 433*, 760–764.

Conboy, M. J., Karasov, A. O., & Rando, T. A. (2007). High incidence of non-random template strand segregation and asymmetric fate determination in dividing stem cells and their progeny. *PLoS Biology, 5*, e102.

Conboy, I. M., & Rando, T. A. (2002). The regulation of Notch signaling controls satellite cell activation and cell fate determination in postnatal myogenesis. *Developmental Cell, 3*, 397–409.

Cooper, R. N., Tajbakhsh, S., Mouly, V., Cossu, G., Buckingham, M., & Butler-Browne, G. S. (1999). In vivo satellite cell activation via Myf5 and MyoD in regenerating mouse skeletal muscle. *Journal of Cell Science, 112*(Pt 17), 2895–2901.

Cornelison, D. D., & Wold, B. J. (1997). Single-cell analysis of regulatory gene expression in quiescent and activated mouse skeletal muscle satellite cells. *Developmental Biology, 191*, 270–283.

Crist, C. G., Montarras, D., & Buckingham, M. (2012). Muscle satellite cells are primed for myogenesis but maintain quiescence with sequestration of Myf5 mRNA targeted by microRNA-31 in mRNP granules. *Cell Stem Cell, 11*, 118–126.

Darabi, R., Arpke, R. W., Irion, S., Dimos, J. T., Grskovic, M., Kyba, M., et al. (2012). Human ES- and iPS-derived myogenic progenitors restore DYSTROPHIN and improve contractility upon transplantation in dystrophic mice. *Cell Stem Cell, 10*, 610–619.

Day, K., Shefer, G., Richardson, J. B., Enikolopov, G., & Yablonka-Reuveni, Z. (2007). Nestin-GFP reporter expression defines the quiescent state of skeletal muscle satellite cells. *Developmental Biology, 304*, 246–259.

Day, K., Shefer, G., Shearer, A., & Yablonka-Reuveni, Z. (2010). The depletion of skeletal muscle satellite cells with age is concomitant with reduced capacity of single progenitors to produce reserve progeny. *Developmental Biology, 340*, 330–343.

Emery, A. E. (1991). Population frequencies of inherited neuromuscular diseases—A world survey. *Neuromuscular Disorders, 1*, 19–29.

Emery, A. E. (2002). The muscular dystrophies. *Lancet, 359*, 687–695.

Fukada, S., Yamaguchi, M., Kokubo, H., Ogawa, R., Uezumi, A., Yoneda, T., et al. (2011). Hesr1 and Hesr3 are essential to generate undifferentiated quiescent satellite cells and to maintain satellite cell numbers. *Development, 138*, 4609–4619.

Gilbert, P. M., Havenstrite, K. L., Magnusson, K. E., Sacco, A., Leonardi, N. A., Kraft, P., et al. (2010). Substrate elasticity regulates skeletal muscle stem cell self-renewal in culture. *Science, 329*, 1078–1081.

Gros, J., Manceau, M., Thome, V., & Marcelle, C. (2005). A common somitic origin for embryonic muscle progenitors and satellite cells. *Nature*, *435*, 954–958.
Harel, I., Nathan, E., Tirosh-Finkel, L., Zigdon, H., Guimaraes-Camboa, N., Evans, S. M., et al. (2009). Distinct origins and genetic programs of head muscle satellite cells. *Developmental Cell*, *16*, 822–832.
Horst, D., Ustanina, S., Sergi, C., Mikuz, G., Juergens, H., Braun, T., et al. (2006). Comparative expression analysis of Pax3 and Pax7 during mouse myogenesis. *The International Journal of Developmental Biology*, *50*, 47–54.
Hu, P., Geles, K. G., Paik, J. H., DePinho, R. A., & Tjian, R. (2008). Codependent activators direct myoblast-specific MyoD transcription. *Developmental Cell*, *15*, 534–546.
Hutcheson, D. A., Zhao, J., Merrell, A., Haldar, M., & Kardon, G. (2009). Embryonic and fetal limb myogenic cells are derived from developmentally distinct progenitors and have different requirements for beta-catenin. *Genes & Development*, *23*, 997–1013.
Jones, N. C., Tyner, K. J., Nibarger, L., Stanley, H. M., Cornelison, D. D., Fedorov, Y. V., et al. (2005). The p38alpha/beta MAPK functions as a molecular switch to activate the quiescent satellite cell. *The Journal of Cell Biology*, *169*, 105–116.
Kassar-Duchossoy, L., Giacone, E., Gayraud-Morel, B., Jory, A., Gomes, D., & Tajbakhsh, S. (2005). Pax3/Pax7 mark a novel population of primitive myogenic cells during development. *Genes & Development*, *19*, 1426–1431.
Kawabe, Y., Wang, Y. X., McKinnell, I. W., Bedford, M. T., & Rudnicki, M. A. (2012). Carm1 regulates Pax7 transcriptional activity through MLL1/2 recruitment during asymmetric satellite stem cell divisions. *Cell Stem Cell*, *11*, 333–345.
Keire, P., Shearer, A., Shefer, G., & Yablonka-Reuveni, Z. (2013). Isolation and culture of skeletal muscle myofibers as a means to analyze satellite cells. *Methods in Molecular Biology*, *946*, 431–468.
Kuang, S., Charge, S. B., Seale, P., Huh, M., & Rudnicki, M. A. (2006). Distinct roles for Pax7 and Pax3 in adult regenerative myogenesis. *The Journal of Cell Biology*, *172*, 103–113.
Kuang, S., Kuroda, K., Le Grand, F., & Rudnicki, M. A. (2007). Asymmetric self-renewal and commitment of satellite stem cells in muscle. *Cell*, *129*, 999–1010.
Le Grand, F., Jones, A. E., Seale, V., Scime, A., & Rudnicki, M. A. (2009). Wnt7a activates the planar cell polarity pathway to drive the symmetric expansion of satellite stem cells. *Cell Stem Cell*, *4*, 535–547.
Lepper, C., Conway, S. J., & Fan, C. M. (2009). Adult satellite cells and embryonic muscle progenitors have distinct genetic requirements. *Nature*, *460*, 627–631.
Lepper, C., Partridge, T. A., & Fan, C. M. (2011). An absolute requirement for Pax7-positive satellite cells in acute injury-induced skeletal muscle regeneration. *Development*, *138*, 3639–3646.
Mauro, A. (1961). Satellite cell of skeletal muscle fibers. *The Journal of Biophysical and Biochemical Cytology*, *9*, 493–495.
McKinnell, I. W., Ishibashi, J., Le Grand, F., Punch, V. G., Addicks, G. C., Greenblatt, J. F., et al. (2008). Pax7 activates myogenic genes by recruitment of a histone methyltransferase complex. *Nature Cell Biology*, *10*, 77–84.
Murphy, M. M., Lawson, J. A., Mathew, S. J., Hutcheson, D. A., & Kardon, G. (2011). Satellite cells, connective tissue fibroblasts and their interactions are crucial for muscle regeneration. *Development*, *138*, 3625–3637.
Olguin, H. C., & Olwin, B. B. (2004). Pax-7 up-regulation inhibits myogenesis and cell cycle progression in satellite cells: A potential mechanism for self-renewal. *Developmental Biology*, *275*, 375–388.
Oustanina, S., Hause, G., & Braun, T. (2004). Pax7 directs postnatal renewal and propagation of myogenic satellite cells but not their specification. *The EMBO Journal*, *23*, 3430–3439.

Palacios, D., Mozzetta, C., Consalvi, S., Caretti, G., Saccone, V., Proserpio, V., et al. (2010). TNF/p38alpha/polycomb signaling to Pax7 locus in satellite cells links inflammation to the epigenetic control of muscle regeneration. *Cell Stem Cell*, *7*, 455–469.

Partridge, T. A., Morgan, J. E., Coulton, G. R., Hoffman, E. P., & Kunkel, L. M. (1989). Conversion of mdx myofibres from dystrophin-negative to -positive by injection of normal myoblasts. *Nature*, *337*, 176–179.

Pasut, A., Jones, A. E., & Rudnicki, M. A. (2013). Isolation and culture of individual myofibers and their satellite cells from adult skeletal muscle. *Journal of Visualized Experiments*, e50074.

Petrof, B. J., Shrager, J. B., Stedman, H. H., Kelly, A. M., & Sweeney, H. L. (1993). Dystrophin protects the sarcolemma from stresses developed during muscle contraction. *Proceedings of the National Academy of Sciences of the United States of America*, *90*, 3710–3714.

Relaix, F., Montarras, D., Zaffran, S., Gayraud-Morel, B., Rocancourt, D., Tajbakhsh, S., et al. (2006). Pax3 and Pax7 have distinct and overlapping functions in adult muscle progenitor cells. *The Journal of Cell Biology*, *172*, 91–102.

Relaix, F., Rocancourt, D., Mansouri, A., & Buckingham, M. (2004). Divergent functions of murine Pax3 and Pax7 in limb muscle development. *Genes & Development*, *18*, 1088–1105.

Relaix, F., Rocancourt, D., Mansouri, A., & Buckingham, M. (2005). A Pax3/Pax7-dependent population of skeletal muscle progenitor cells. *Nature*, *435*, 948–953.

Rocheteau, P., Gayraud-Morel, B., Siegl-Cachedenier, I., Blasco, M. A., & Tajbakhsh, S. (2012). A subpopulation of adult skeletal muscle stem cells retains all template DNA strands after cell division. *Cell*, *148*, 112–125.

Rudnicki, M. A., Le Grand, F., McKinnell, I., & Kuang, S. (2008). The molecular regulation of muscle stem cell function. *Cold Spring Harbor Symposia on Quantitative Biology*, *73*, 323–331.

Sacco, A., Doyonnas, R., Kraft, P., Vitorovic, S., & Blau, H. M. (2008). Self-renewal and expansion of single transplanted muscle stem cells. *Nature*, *456*, 502–506.

Sacco, A., Mourkioti, F., Tran, R., Choi, J., Llewellyn, M., Kraft, P., et al. (2010). Short telomeres and stem cell exhaustion model Duchenne muscular dystrophy in mdx/mTR mice. *Cell*, *143*, 1059–1071.

Sambasivan, R., Gayraud-Morel, B., Dumas, G., Cimper, C., Paisant, S., Kelly, R. G., et al. (2009). Distinct regulatory cascades govern extraocular and pharyngeal arch muscle progenitor cell fates. *Developmental Cell*, *16*, 810–821.

Sambasivan, R., Yao, R., Kissenpfennig, A., Van Wittenberghe, L., Paldi, A., Gayraud-Morel, B., et al. (2011). Pax7-expressing satellite cells are indispensable for adult skeletal muscle regeneration. *Development*, *138*, 3647–3656.

Schienda, J., Engleka, K. A., Jun, S., Hansen, M. S., Epstein, J. A., Tabin, C. J., et al. (2006). Somitic origin of limb muscle satellite and side population cells. *Proceedings of the National Academy of Sciences of the United States of America*, *103*, 945–950.

Schultz, E. (1996). Satellite cell proliferative compartments in growing skeletal muscles. *Developmental Biology*, *175*, 84–94.

Schultz, E., Gibson, M. C., & Champion, T. (1978). Satellite cells are mitotically quiescent in mature mouse muscle: An EM and radioautographic study. *The Journal of Experimental Zoology*, *206*, 451–456.

Seale, P., Sabourin, L. A., Girgis-Gabardo, A., Mansouri, A., Gruss, P., & Rudnicki, M. A. (2000). Pax7 is required for the specification of myogenic satellite cells. *Cell*, *102*, 777–786.

Shea, K. L., Xiang, W., LaPorta, V. S., Licht, J. D., Keller, C., Basson, M. A., et al. (2010). Sprouty1 regulates reversible quiescence of a self-renewing adult muscle stem cell pool during regeneration. *Cell Stem Cell*, *6*, 117–129.

Shinin, V., Gayraud-Morel, B., Gomes, D., & Tajbakhsh, S. (2006). Asymmetric division and cosegregation of template DNA strands in adult muscle satellite cells. *Nature Cell Biology*, *8*, 677–687.

Soleimani, V. D., Punch, V. G., Kawabe, Y., Jones, A. E., Palidwor, G. A., Porter, C. J., et al. (2012). Transcriptional dominance of Pax7 in adult myogenesis is due to high-affinity recognition of homeodomain motifs. *Developmental Cell*, *22*, 1208–1220.

Stockdale, F. E. (1992). Myogenic cell lineages. *Developmental Biology*, *154*, 284–298.

Troy, A., Cadwallader, A. B., Fedorov, Y., Tyner, K., Tanaka, K. K., & Olwin, B. B. (2012). Coordination of satellite cell activation and self-renewal by Par-complex-dependent asymmetric activation of p38alpha/beta MAPK. *Cell Stem Cell*, *11*, 541–553.

Wen, Y., Bi, P., Liu, W., Asakura, A., Keller, C., & Kuang, S. (2012). Constitutive Notch activation upregulates Pax7 and promotes the self-renewal of skeletal muscle satellite cells. *Molecular and Cellular Biology*, *32*, 2300–2311.

White, R. B., Bierinx, A. S., Gnocchi, V. F., & Zammit, P. S. (2010). Dynamics of muscle fibre growth during postnatal mouse development. *BMC Developmental Biology*, *10*, 21.

Yin, H., Pasut, A., Soleimani, V. D., Bentzinger, C. F., Antoun, G., Thorn, S., et al. (2013). MicroRNA-133 controls brown adipose determination in skeletal muscle satellite cells by targeting Prdm16. *Cell Metabolism*, *17*, 210–224.

Zammit, P. S., Golding, J. P., Nagata, Y., Hudon, V., Partridge, T. A., & Beauchamp, J. R. (2004). Muscle satellite cells adopt divergent fates: A mechanism for self-renewal? *The Journal of Cell Biology*, *166*, 347–357.

Zammit, P. S., Relaix, F., Nagata, Y., Ruiz, A. P., Collins, C. A., Partridge, T. A., et al. (2006). Pax7 and myogenic progression in skeletal muscle satellite cells. *Journal of Cell Science*, *119*, 1824–1832.

CHAPTER SEVEN

Neural Stem Cell of the Hippocampus: Development, Physiology Regulation, and Dysfunction in Disease

Chiara Rolando, Verdon Taylor[1]
Department of Biomedicine, University of Basel, Basel, Switzerland
[1]Corresponding author: e-mail address: verdon.taylor@unibas.ch

Contents

Abstract

The formation of the hippocampus is generated during embryonic development, but most neurons within the structure are produced after birth. The hippocampus is a primary region of neurogenesis within the adult mammalian brain. Adult-born neurons have to integrate into the established neural circuitry throughout life. Although the function of neurogenesis in the adult hippocampus, particularly in humans, remains unclear, experimental data suggest that adult-born neurons are involved in some forms of memory, as well as in diseases. Adult hippocampal neurogenesis is dynamic, responding to physiological and pathological stimuli that may promote brain function or contribute to diseases such as epilepsy. Here, we review some of the mechanisms and signaling pathways involved in the development of the hippocampus, as well as in adult

Current Topics in Developmental Biology, Volume 107
ISSN 0070-2153
http://dx.doi.org/10.1016/B978-0-12-416022-4.00007-X

neurogenesis. We discuss some recent findings suggesting heterogeneity within the hippocampal stem cell pool and the regulation of activation of quiescent stem cells. Finally, we discuss some of the issues relating neurogenesis to pathophysiology and aging.

1. INTRODUCTION

Formation of the central nervous system (CNS) requires a precise control of proliferation, cell fate determination, and differentiation. Here we review some aspects of development of the hippocampus and regulation of adult neurogenesis from adult neural stem cells (aNSCs) in the dentate gyrus (DG). Following determination of the neural ectoderm and formation of the neural tube, a process known as neurulation, neuroectodermal (neuroepithelial) cells make up the entire neural tube. Neuroepithelial cells are patterned along the dorsoventral and anterior–posterior axis. This patterning defines the regions of the CNS and spinal cord. As neurogenesis commences, neuroepithelial cells, which span the entire thickness of the neural tube, transform into radial glial cells (RGCs). Although a few neurons are generated directly from the neuroepithelium, RGCs continue to act as primary progenitors for both neurons and glial cells (Anthony, Klein, Fishell, & Heintz, 2004; Malatesta et al., 2003). Most neurons are generated during embryogenesis by RGCs that undergo self-renewing asymmetric cell divisions. The generation of neurons from RGCs usually progresses through an intermediate progenitor stage, which expands the number of cells generated. RGCs transform into parenchymal astrocytes and ependymal cells in the peri- and postnatal period, or continue to act as aNSCs in the walls of the lateral ventricles (subventricular zone, SVZ) and the subgranular zone (SGZ) of the hippocampal DG (Kriegstein & Alvarez-Buylla, 2009). Under physiological conditions, aNSCs display the structural and antigenic features of astrocytes (Doetsch, Caille, Lim, Garcia-Verdugo, & Alvarez-Buylla, 1999; Garcia, Doan, Imura, Bush, & Sofroniew, 2004; Seri, García-Verdugo, Collado-Morente, McEwen, & Alvarez-Buylla, 2004; Seri, Garcia-Verdugo, McEwen, & Alvarez-Buylla, 2001). They retain the ability to self-renew throughout life and continue to generate actively dividing cell intermediates that function as transit-amplifying progenitors (TAPs). The aNSCs and TAPs of the SVZ and SGZ have distinct features, fates and functions (Kriegstein & Alvarez-Buylla, 2009; Ming & Song, 2011). In the SVZ, the immature neuroblasts migrate in chains to the olfactory bulb and differentiate into multiple distinct neuronal subtypes (Doetsch, 2003; Kriegstein & Alvarez-Buylla, 2009; Merkle, Mirzadeh, & Alvarez-Buylla, 2007). In the hippocampus, a single neuron-type is generated, DG granule neurons (Seri et al., 2004). In addition, the growth factor requirements and the response of the aNSCs in the SVZ and DG differ markedly.

In this review, we focus on NSCs in the hippocampal DG. The hippocampus is part of the limbic system and plays important roles in the consolidation of information, as well as long and short term memory, and spatial navigation. The DG is the primary input into the hippocampal formation receiving connections from the entorhinal cortex. Neurogenesis in the DG of the hippocampus is prominent in adult rodents as well as primates and humans (Bergmann et al., 2012; Eriksson et al., 1998; Spalding et al., 2013). Adult neurogenesis in the DG is critical for some forms of learning and memory and is modulated by pathological conditions (Zhao, Deng, & Gage, 2008). NSC identity in the adult DG has not been fully elucidated. However, embryonic RGCs and NSCs of the adult DG do share some features that suggest similarities between these two populations. Here we discuss formation of the hippocampal DG and some of the molecular pathways involved in the formation of this brain region. We then review some of the mechanisms controlling neurogenesis in the adult DG, the effects of physiological stimuli, and pathological insults including epilepsy and depression, as well as during aging.

2. ORIGIN OF ADULT HIPPOCAMPAL NEURAL STEM CELLS: EMBRYONIC DEVELOPMENT OF THE DG

The DG is one of the two regions of on-going neuron production in the adult brain. DG neurogenesis is regulated by a specialized stem cell niche at the border of the granule cell layer (GCL) and the hilus, the SGZ (Kempermann, Jessberger, Steiner, & Kronenberg, 2004; Kempermann, Wiskott, & Gage, 2004). aNSCs in the DG are generated from stem cells in the embryonic germinal zone located in the subpallium region. These embryonic NSCs undergo a multistep development and migration to the SGZ (Fig. 7.1). The DG has a protracted developmental period that extends from embryogenesis into the postnatal period and is finely tuned by distinct developmental steps critical to correct patterning and organization. DG formation begins in mice around embryonic day 13 (E13.5) in the dorsomedial portion of the telencephalic vesicles (Li & Pleasure, 2005). The hippocampal

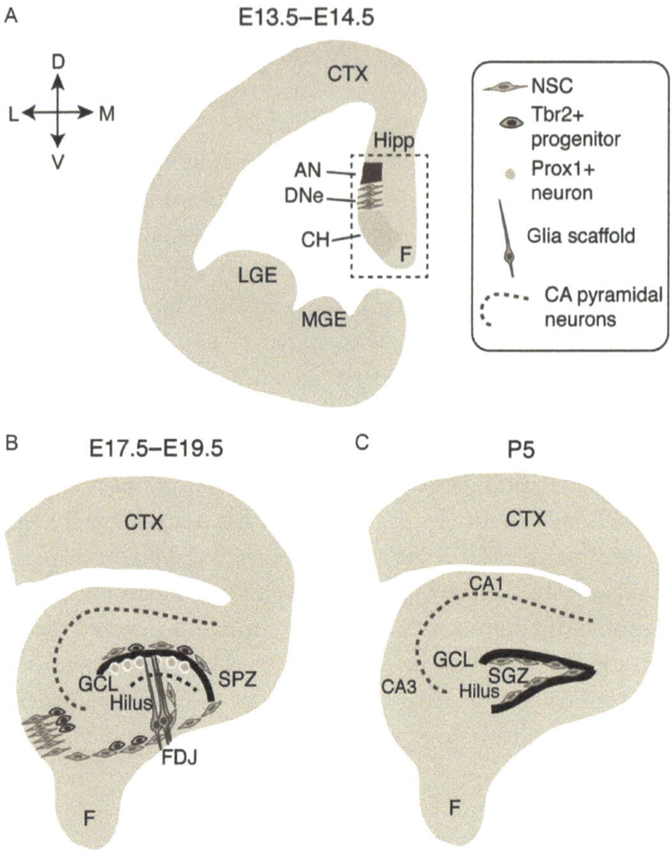

Figure 7.1 Embryonic and early postnatal hippocampal development. (A) Hippocampal primordium develops adjacent to the cortical hem (CH). The dentate neuroepithelium (DNe) contains NSCs that will generate the dentate gyrus. Conversely, the ammonic neuroepithelium (AN) will give rise to pyramidal neurons of the CA3 and CA1. (B) At later developmental stages, NSCs and Tbr2+ progenitors have already started to migrate tangentially and form the SPZ. The trans-hilar radial scaffold forms as progenitors become localized to the FDJ. At the same time, Prox1+ neurons have already populated the GCL. (C) During early postnatal development, progenitor cells that reside in the SPZ undergo redistribution to form the SGZ niche. LGE, lateral ganglionic eminence; MGE, medial ganglionic eminence; CTX, cortex; Hipp, hippocampus; F, fimbria; FJD, fimbriodentate junction; SPZ, subpial zone; GCL, granule cell layer; SGZ, subgranular zone. (See color plate.)

primordium is initially populated by Cajal–Retzius cells and RGCs responsible for the expansion of the tissue. Embryonic hippocampal RGCs proliferate in a domain of the ventricular zone called the dentate neuroepithelium (DNe), which is adjacent to the cortical hem (see later). These NSCs express the typical radial glial markers Nestin and Glast (Li, Kataoka, Coughlin, & Pleasure, 2009). Already, at embryonic time points, the hippocampal SVZ is populated by Tbr2+ committed neural progenitors that form a band of cells above the fimbria along the pial surface (Fig. 7.1). At E15.5, NSCs migrate through the dentate migratory stream toward a subpial region at the junction of the fimbria and the forming DG called the fimbriodentate junction (FDJ). The FDJ is a transient neurogenic zone critical for the maintenance of neural precursors in an undifferentiated state (Li et al., 2009). At E17.5, NSCs and Tbr2+ cells spread into the hilus and line the hippocampal fissure (HF) (Fig. 7.1). Persistent migration of NSCs and Tbr2+ cells in the perinatal period leads to the formation of the subpial zone (SPZ) lining the HF. At postnatal day 0 (P0), NSCs completely cover the dentate side of the HF as well as the subpial region of the future lower blade (Fig. 7.1). During these early phases of hippocampal development, the cortical hem functions as an organizing structure releasing instructive signals. Subsequently, the primary radial glia form the scaffold for the NSC and their derivatives to migrate and settle throughout the structure. Finally, at around P3, hippocampal NSCs undergo further reorganization to form the tertiary matrix and colonize the primordium from the hilus to the marginal zone of the GCL, whereas the Tbr2+ cells are mostly found in the nascent molecular layer (ML). At the end of the first postnatal week, the NSCs populate the SGZ where they will persist through life in the adult hippocampal neurogenic niche.

The classical view suggests that the aNSCs in the SGZ originate from the DNe at the equivalent longitudinal level; however, the exact embryonic origin of these aNSC has not been fully elucidated. It has recently been proposed that the aNSCs originate from the ventral hippocampus in the amygdala-hippocampus region (see later) (Li, Fang, Fernandez, & Pleasure, 2013). Therefore, precursors of the aNSCs might migrate from the ventral hippocampus along the longitudinal axis from the temporal to the septal poles before settling into the SGZ. This origin supports the hypothesis that the DG is a mosaic structure generated by stem cells with different embryonic origins.

2.1. Extracellular signaling driving embryonic hippocampal neurogenesis

A plethora of secreted factors and intrinsic signals have been identified that regulate DG development. In the following section, we outline two main molecular pathways that orchestrate embryonic hippocampal neurogenesis and that also have an important function in regulation of NSCs in the adult hippocampus.

2.1.1 Wnt/β-catenin signaling regulates hippocampus formation

The characteristic multistep development of the DG is orchestrated by different signals and molecular pathways. Wnt and Shh pathways have both been shown to be regulators of precursor behavior in the embryonic forebrain ventricular zone and SVZ, which give rise to the hippocampus (Galceran, Miyashita-Lin, Devaney, Rubenstein, & Grosschedl, 2000; Machold et al., 2003; Pozniak & Pleasure, 2006; Zhou, Zhao, & Pleasure, 2004). Furthermore, both of these pathways are known to be involved in the regulation of neural precursors in the postnatal and adult DG (Ahn & Joyner, 2005; Lai, Kaspar, Gage, & Schaffer, 2003; Lie et al., 2005).

Wnt/β-catenin function is crucial for the correct development of the hippocampal primordium, and some members of its pathway regulate its proper morphogenesis. Wnt signaling forms a medial-lateral gradient during mid-corticogenesis, with the DG being exposed to the highest Wnt activity. Early during embryonic development, Wnt3a is expressed by the cortical hem adjacent to the DNe. Wnt3a mutant mice display a severe reduction in hippocampal development because of an impaired cell proliferation in the ventricular zone, confirming that embryonic NSCs are Wnt-dependent (Lee, Tole, Grove, & McMahon, 2000). At early embryonic stages, the nuclear mediator of the Wnt pathway, Lef1, is widely expressed in the ventricular zone, including the DNe, and by a cell population that migrates out of the SVZ (Galceran et al., 2000). In addition, Lef1-deficient embryos display an impaired granule cell development. However, GFAP+ progenitors and cell migration are unaffected, which suggests a more specific effect on neuronal committed progenitors (i.e., Tbr2+ cells) than on NSCs (Galceran et al., 2000). Mice deficient in the Wnt coreceptor LPR6 display similar phenotypes and the number of Prox1- and Ngn2-expressing cells is decreased in the DG of these mutants, suggestive of hampered granule cell differentiation (Zhou et al., 2004). Interestingly, defective proliferation of ventricular zone progenitors is not observed outside the DG, suggesting a specific effect on progenitors already committed toward a DG granule cell fate. LPR6 mutant mice also have disorganized RGCs in the DG; however, it is unclear whether this is an effect specifically on structural scaffold-forming cells or on NSCs in the FJ (Zhou et al., 2004). In line with an important role for the pathway in hippocampal formation, ectopic expression of Wnt induces Prox1 + DG granule cells, suggesting that Wnt promotes DG neurogenesis (Machold et al., 2003).

2.1.2 Shh signaling orchestrates hippocampal development

Shh signaling is involved in DG development where it is essential for expanding granule neural progenitors during perinatal development to establish the aNSC population. Conditional deletion of Shh from Nestin-expressing progenitors does not lead to gross abnormalities during embryonic DG development but impairs late postnatal and adult DG neurogenesis (Machold et al., 2003). However, a recent report did show that conditional deletion of the Shh receptor Smoothened in hGFAP-promoter-expressing cells significantly reduces the number of BrdU-labeled progenitors in the DG and in the dentate migratory stream at E18.5 (Han et al., 2008). Smo receptor is localized to the primary cilium of cells, and it has been proposed that this localization is critical for its function and Shh signaling. In line with this, conditional ablation of the primary cilia kinesin-II motor protein Kif3a from GFAP-expressing RGCs results in defective proliferation in the late embryonic DG. This is followed by a dramatic decrease in the number of DCX neuroblasts and radial NSCs in the adult DG. In support of this, loss of the primary cilia in hGFAP::Cre;Kif3a$^{fl/fl}$ conditional mutant mice results in defective Shh signaling in the DG during late embryonic development (Han et al., 2008).

One source of Shh is the septum and this is responsible for establishing the SGZ (Machold et al., 2003). However, there is also a pallial source of Shh, which is critical for Hedgehog activity in the hippocampus (Li et al., 2013). Interestingly, Shh is a transcriptional target of high-mobility group transcription factor Sox2 in DG NSCs indicating that DG progenitors and their progeny produce Shh (Favaro et al., 2009). Conditional deletion of Shh from neurons using the NeuroD6::Cre allele abolished Hedgehog responding activity in the DG, which supports the hypothesis that neurons directly regulate NSC maintenance in the DG, partially through their secretion of Shh (Li et al., 2013).

Whether there are multiple populations of Shh-responsive progenitors in the developing hippocampus is an issue that needs to be addressed. Fate mapping analysis of Gli1::CreER^{T2}-expressing cells revealed that Shh-responsive cells are restricted to the ventral hippocampus at E17.5 and their progeny are found at all septotemporal levels in the postnatal SGZ. Moreover, Emx1::Cre-mediated conditional deletion of Smo demonstrates that SGZ development is dependent upon Hedgehog signaling. However, when

Shh is removed from the Emx1 expression domain, Hedgehog-responsive cells in the ventral hippocampus are unaffected and the SGZ develops, implying either that Shh is not the only Hedgehog involved in DG formation or that the source of ligand is not the EMX1 cells or their progeny (Li et al., 2013).

2.2. Intracellular mechanisms controlling developing NSCs

2.2.1 Transcriptional regulation of embryonic hippocampal development

Relatively little is known of the transcription factors regulating early stages of neurogenesis in the DG and, particularly, the generation and initial differentiation of the different populations of progenitors involved in the development of the DG. Proneural genes encode transcription factors of the basic helix loop helix (bHLH) class that initiate the development of neuronal lineages and promote the generation of committed progenitors. Proneural genes are activated downstream of signal cues and regulate neurogenic processes and the specification of progenitor cell identity (Hatakeyama et al., 2004).

The proneural transcription factor Ascl1 (Mash1) is expressed by hippocampal progenitors during development at different locations along the hippocampal primordium. The Ascl1+ population in the embryonic hippocampus comprises proliferating progenitors in the SVZ, in the dentate migratory stream, and in the tertiary matrix (Pleasure, Collins, & Lowenstein, 2000). However, despite this broad expression, Ascl1-null mice do not show major differences in the number of proliferating cells or differentiation marker expression in the hippocampus (Galichet, Guillemot, & Parras, 2008). The exact contribution of Ascl1 to embryonic hippocampal neurogenesis remains unclear. Interestingly, forced Ascl1 expression in adult DG progenitors promotes oligodendrocyte fate acquisition (Jessberger, Toni, Clemenson, Ray, & Gage, 2008).

The proneural gene *Ngn2* is also expressed by hippocampal progenitors and its expression overlaps that of Ascl1 to a major extent. Like Ascl1, Ngn2 is expressed by proliferating progenitors and is downregulated as the cells exit the cell cycle. Ngn2 knockout mice show a reduced number of progenitors in the DG, which results in the loss of a large fraction of dentate granule cells and a severe defect in DG morphogenesis because the lower blade fails to develop properly (Galichet et al., 2008). This granule cell defect reflects a unique role for Ngn2 in the developing DG. Other bHLH proteins are also present in the developing DG, but they are mostly absent from the embryonic neuroepithelium as their expression is restricted to postmitotic, differentiating neurons (Pleasure et al., 2000). One such factor, NeuroD1, is expressed by early postmitotic neurons intermingled with Ngn2+ progenitors. NeuroD1 plays a role in granule cell differentiation. NeuroD2, a related bHLH protein, is expressed by mature granule cells (Oldekamp, Kraemer, Alvarez-Bolado, & Skutella, 2004; Pleasure et al., 2000; Schwab et al., 2000).

The T-box transcription factor Tbr2 regulates glutamatergic neuron fate commitment in several brain regions including the DG (Hevner, Hodge, Daza, & Englund, 2006). Tbr2 orchestrates different aspects of hippocampal development spanning across neurogenesis to morphogenesis (Hodge et al., 2013, 2008, 2012). Tbr2-expressing intermediate progenitors are depleted in Tbr2-mutant mice, resulting in an impaired hippocampal neurogenesis. Moreover, conditional deletion of Tbr2 from Nestin-expressing progenitors results in an accumulation of Sox2+ cells that fail to differentiate into neuroblasts and neurons. Interestingly, Tbr2 directly binds to the promoter of *Sox2* and its overexpression in the developing hippocampus causes depletion of Sox2+ NSCs (Hodge et al., 2012). Recently, Tbr2 was shown to be expressed by Cajal–Retzius cells as well and to be required for their migration to the developing DG. Cajal–Retzius cells are important for establishing and maintaining the radial processes of RGCs and for migration of newborn neurons (Frotscher et al., 2003). Hence, loss of Tbr2 causes aberrant development of the radial glial scaffold in the hippocampus, which ultimately affects the proper formation of the GCL (Hodge et al., 2013).

Expression of the homeodomain protein Emx2 is also required for the proper growth of the hippocampus and for migration of DG progenitors (Backman et al., 2005; Oldekamp et al., 2004; Pellegrini, Mansouri, Simeone, Boncinelli, & Gruss, 1996). Emx2-mutant mice show a dramatic reduction in the DG and shrinkage of the entire hippocampus (Pellegrini et al., 1996). Interestingly, Emx2 loss of function in the DNe also influences the behavior of hippocampal progenitors. Proliferating progenitors in Emx2-null mice are confined to the VZ and only a few immature neurons are seen migrating away before their arrest in the migratory stream. This impaired neurogenesis leads to an absence of Ascl1 expression at later developmental stages, which supports defects in granule cell production. Interestingly, NSC markers as GFAP and Tenascin-C appear to be upregulated in the VZ of Emx2-null mice, suggesting an increase in progenitor and NSCs (Oldekamp et al., 2004).

Some transcription factors in the hippocampus may act differently depending on the developmental stage. The LIM-homeodomain transcription factor Lhx2 has temporally distinct roles in the regulation of neurogenesis in the developing hippocampus (Mangale et al., 2008; Subramanian et al., 2011). Therefore, in the dorsal telencephalon before E10.5, Lhx2 acts as a selector gene defining the cortex-hem boundary. Lhx2 conditional knockout causes cortical hem expansion at the expense of the neocortical epithelium, resulting in an ectopic expression of Wnt3a that promotes the development of ectopic hippocampi (Mangale et al., 2008). At later stages, Lhx2 controls the cell fate decisions made by hippocampal progenitors. Downregulation of Lhx2 in the embryonic hippocampus promotes astrogliogenesis at the expense of neurogenesis. In line with this, Lhx2 overexpression suppresses GFAP promoter activity, thereby prolonging the neurogenic phase (Subramanian et al., 2011).

2.2.2 miRNA regulation of embryonic hippocampal development

Posttranscriptional gene regulation mediated by microRNAs (miRNAs) plays an important role in the development of the CNS (Andersson et al., 2010; Kawase-Koga, Otaegi, & Sun, 2009). miRNAs are 22-nucleotide noncoding RNAs produced by the sequential activities of two RNAseIII complexes, one containing Drosha and the other containing Dicer. The miRNA Microprocessor (MP) contains Drosha and the RNA binding protein Dgcr8. The MP binds double-stranded RNA hairpins in the pre-miRNAs and cleaves them to release the pre-miRNA. The pre-miRNAs are exported from the nucleus and are processed into mature miRNAs by the Dicer complex. The leading strand of the mature miRNA is loaded onto the RISC complex, regulating the stability and translation of target mRNAs. miRNA-mediated temporal regulation of hippocampal neurogenesis has been demonstrated, opening a new concept of timely regulation of hippocampal neurogenesis (Davis et al., 2008; De Pietri Tonelli et al., 2008; Kawase-Koga et al., 2009; Li et al., 2011). Conditional ablation of Dicer from Emx1-expressing neural progenitors at early developmental stages causes severe defects in hippocampal development (Li et al., 2011). In particular, when miRNA biogenesis is blocked, Wnt signaling is reduced and Lhx2 expression decreased, suggesting that miRNA may regulate the correct level and domain of expression of hippocampal patterning signals (Li et al., 2011). Moreover, in Emx1-Dicer conditional knockout mice, DNe progenitors show decreased proliferation and start to differentiate precociously, resulting in a transient increase in Tbr1+ neurons. However, these Tbr1+ neurons die prematurely and do not differentiate into mature NeuN-positive granule cells. This contributes to a highly hypotrophic hippocampus in these mutant mice. Moreover, Dicer deletion from committed neurons using a Calmodulin Kinase II promoter-driven Cre-recombinase impairs axonal growth and spine formation in the developing hippocampus, highlighting a key role for Dicer in neural circuit formation (Davis et al., 2008).

It has been shown that the miRNA MP can regulate NSC in an miRNA-independent fashion. Indeed, during embryonic development, the MP maintains NSC in an undifferentiated state by directly reducing the mRNA stability of proneurogenic factors. Drosha loss of function in embryonic NSCs leads to a rapid increase in Ngn2 mRNA and premature differentiation, thus suggesting mechanisms through which neurogenesis can be regulated (Knuckles et al., 2012). Together, these data support that posttranscriptional mechanisms also play a crucial role in the maintenance of the NSC pool and development of the hippocampus.

3. ADULT HIPPOCAMPAL NEUROGENESIS: HETEROGENEOUS NEUROGENIC NSCs AND NICHE

3.1. Coexistence of active and quiescent adult NSCs in the adult DG

Adult DG NSCs are a heterogeneous population and include RGCs and nonradial cells that shuttle between active and quiescent states (Fig. 7.2). NSCs generate committed progenitors that differentiate but may divide to expand the number of neurons generated. The analysis of progenitors in the adult DG has moved from a retrospective analysis of *in vitro* properties of cells isolated from the hippocampus to *in vivo* labeling and lineage tracing. *In vivo* studies are uncovering an unprecedented range of cells that display stem cell character (Bonaguidi et al., 2011; Encinas et al., 2011; Knobloch et al., 2013; Lugert et al., 2010, 2011). Although *in vivo* identification and lineage tracing are important to study DG neurogenesis, results from different labs have led to different conclusions about stem cell behavior and potential (Bonaguidi et al., 2011; Encinas et al., 2011; Lugert et al., 2010).

It is generally accepted that aNSCs in the DG divide infrequently to generate more committed progeny. The mechanisms regulating activation of quiescent stem cells remain unclear but extrinsic and intrinsic pathways have been identified that play a role in the process *in vivo*. Notch signaling in the DG is a central niche-derived cue to control proliferation and differentiation

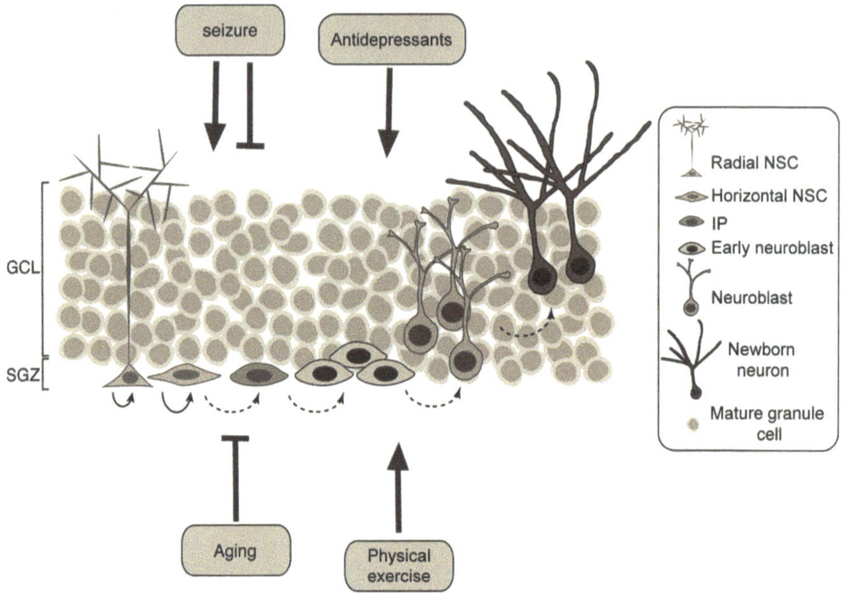

Figure 7.2 Adult neurogenesis in the hippocampal DG. Radial and horizontal NSCs populate the SGZ of the DG. Horizontal NSCs are the active pool that divides multiple times. NSCs originate intermediate progenitors (IPs) that then produce early mitotic neuroblasts. This pool will expand before generating postmitotic neuroblasts and then newborn neurons. In the scheme, different pathological conditions, physical exercise, and therapeutic treatments together with their effect on adult neurogenesis are also represented. GCL, granule cell layer; SGZ, subgranular zone. (See color plate.)

(Breunig, Silbereis, Vaccarino, Sestan, & Rakic, 2007; Ehm et al., 2010; Lugert et al., 2010). Both radial and nonradial Type1 NSCs display canonical Notch activity (Lugert et al., 2010, 2011). Although these two stem cell types seem to be able to self-renew and generate neurons, their mitotic characteristics and response to pathophysiology are decidedly different.

Quiescence of aNSCs is likely a general mechanism to preserve the stem cell pool and prevent exhaustion of the population under normal conditions. In addition, quiescent or dormant NSCs may be a mechanism by which the number of DNA mutations accumulated during cell division can be maintained at a minimum, reducing the production of potentially dangerous chromosomal aberrations. Interestingly, radial and nonradial stem cells in the adult DG respond selectively to inductive pathophysiological cues. For example, radial NSCs proliferate in response to physical exercise whereas nonradial cells do not (Fig. 7.2). Conversely, nonradial cells activate upon seizure-induced stimuli whereas radial cells do not (Lugert et al., 2010). Furthermore, aNSCs labeled based on Notch activity remain in the aged mouse brain but these mostly have a radial morphology, and the nonradial cells become inactive. Basket neuron activity also regulates aNSC activity through the release of GABA and activation of GABAA receptors (Song et al., 2012). Conversely, Nestin::GFP expressing cells are dramatically reduced in aged mice, suggesting loss of this population (Encinas et al., 2011).

3.2. Signaling pathways that regulate aNSC activity in the DG

Notch signaling regulates aNSC proliferation, maintenance, and differentiation, but how the same pathway controls these different processes remains unclear. Notch signaling blocks neurogenesis by suppressing the expression of the proneural transcription factors. This explains why active stem and progenitor cells are lost following ablation of canonical Notch activity as they differentiate into more committed intermediate progenitors that do not undergo long-term self-renewal. However, loss of Notch function results in the quiescent aNSCs becoming mitotically active, presumably entering the active NSC pool and then differentiating because of loss of neurogenic suppression. The mechanisms through which canonical Notch signals repress NSC activation and thus maintain the quiescent stem cell pool are less clear. It is possible that functional redundancy within the four Notch family members accounts for the differences in loss of quiescent cells seen by ablation of individual Notch receptors compared to deletion of the transcriptional effector of canonical Notch, RBP-Jk (Ables et al., 2010; Ehm et al., 2010; Lugert et al., 2010).

Other pathways control the activation of aNSCs in the DG. BMPs promote the exit of cells from the cell cycle and quiescence. Conditional inactivation of the BMP receptor BMPR-IA from aNSCs results in quiescent cells entering the cell cycle, a transient increase in mitotic progenitors, and a reduction in neurogenesis, presumably because the stem cell pool becomes exhausted (Mira et al., 2010). The repressor element 1-silencing transcription/neuron-restrictive silencer factor (REST/NRSF) is a negative regulator of transcription of many neuronal genes. REST recruits corepressors CoREST and Sin3a to target promoters preventing NSC differentiation. REST is expressed by aNSCs in the DG, and REST-deficient mice show a transient increase in neurogenesis at the expense of quiescent NSCs (Gao et al., 2011). Similarly, PTEN represses proliferation of aNSC in the DG. Deletion of PTEN results in aNSCs undergoing symmetric stem cell divisions at the expense of differentiation (Bonaguidi et al., 2011).

Wnt/β-catenin signaling is important during formation of the hippocampus and DG (see earlier). Wnt/β-catenin signaling is also crucial in adult neurogenesis in the DG. Wnt/β-catenin activity is strong in the SGZ and GCL of the DG and hippocampal progenitors express components of the Wnt pathway (Lie et al., 2005; Wexler, Paucer, Kornblum, Palmer, & Geschwind, 2009). Inhibition of Wnt activity *in vivo* results in reduced proliferation and neurogenesis whereas activation by expression of Wnt3a increases neurogenesis (Lie et al., 2005). Disrupted in schizophrenia 1 (Disc1) is a negative regulator of GSK3β, a key regulator of β-catenin destruction. Loss of Disc1 activity results in increased Wnt activity in DG progenitors, and increased proliferation and neurogenesis (Mao et al., 2009; Ming & Song, 2009). These Disc1 effects can be reversed by expression of a degradation-insensitive, stabilized form of β-catenin in DG progenitors (Mao et al., 2009). Similarly, negative regulators of the Wnt/β-catenin pathway Dickkopf1 (Dkk1) and sFRP3 regulate proliferation and differentiation in the adult DG (Jang et al., 2013; Seib et al., 2013). Dkk1 binds LRPs, Wnt coreceptors, preventing formation of a functional LRP/Fz Wnt receptor complex. Dkk1 deletion results in increased proliferation and neurogenesis consistent with increased Wnt activity in the DG (Seib et al., 2013). sFRP3 is a secreted antagonist of Wnt, and knockdown in the DG results in increased proliferation and Wnt pathway activation (Jang et al., 2013). The block in Wnt activity results in behavioral changes with impaired special memory and object recognition (Jessberger et al., 2009). Interestingly, during aging, the number of Wnt-expressing astrocytes in the DG decreases, supporting the hypothesis that astrocytes are a key source of Wnt and mediator of proliferation in the DG (Lie et al., 2005; Okamoto et al., 2011). Conversely, Dkk1 expression increases with age, presenting another potential explanation for the reduced proliferation and increased quiescence of aNSCs in the aged DG (Seib et al., 2013). Furthermore, sFRP levels seem to be reduced during exercise and following electroconvulsive treatment, both of which increase proliferation in the DG (Jang et al., 2013).

4. ADULT DG NSC IN EPILEPSY, AGING, AND DEPRESSION

Although studied extensively in rodents, the degree and function of neurogenesis in the DG of humans is unclear. A seminal study by Eriksson and colleagues demonstrated that adult neurogenesis also occurs in the DG of adult humans (Eriksson et al., 1998). More recently, Frisen and colleagues used elegant approaches to detect the formation of new neurons in the adult human brain by quantification of ^{14}C integrated into the DNA of dividing cells following a nuclear bomb test. This retrospective labeling of newborn cells and the birth date of newly generated hippocampal neurons revealed that up to one third of DG neurons turn over during postnatal life (Spalding et al., 2013). By contrast, quantification of ^{14}C levels in OB of humans revealed that the number of new neurons generated in the OB of adult humans is insignificant and below detection, indicating a major difference in the hippocampus (Bergmann et al., 2012).

The discovery of aNSC in the hippocampus of adult humans opened up the possibility that this plastic population could serve as an endogenous pool of progenitors for treating CNS disease. Similarly, the existence of NSCs and turnover of DG granule cells might be a key process in understanding the progressive cognitive decline associated with aging and brain pathologies. The generation of DG granule neurons is a dynamic process, and it is regulated in both physiological and pathological conditions in rodents (Fig. 7.2). Neurogenesis is controlled at the level of proliferation, differentiation, and survival of newly generated cells and can be modulated by pathologies (Zhao et al., 2008). Both physical activity and seizures strongly increase proliferation (Fabel & Kempermann, 2008; Parent & Murphy, 2008), whereas aging is associated with an exponential decrease in DG neurogenesis (Ben Abdallah, Slomianka, Vyssotski, & Lipp, 2010; Kempermann, Kuhn, & Gage, 1998; Kuhn, Dickinson-Anson, & Gage, 1996; Steiner, Zurborg, Horster, Fabel, & Kempermann, 2008). There is evidence that different neurogenic stimuli and pathological situations affect cells at different stages of neurogenesis (Kronenberg et al., 2003; Petrus et al., 2009; Steiner et al., 2008). Interestingly, distinct NSC populations seem to respond differently to these external stimuli; however, the mechanisms involved and whether they share common molecular pathways for maintenance and differentiation are unknown (Lugert et al., 2010).

4.1. Epilepsy

Neurogenesis in the adult brain is associated not only with cognitive functions. Evidence suggests that aberrant neurogenesis or stem cell activity could be involved in human pathology. A possible link exists between adult hippocampal neurogenesis and epilepsy (Fig. 7.2). Several studies demonstrate an association between seizures and increased neurogenesis. However, the nature of the relationship between the increased neuron production, the enhanced proliferation, and the disease state is far less clear. It remains to be clarified whether epilepsy is a cause, consequence, or a by-product of seizure-related disturbances in adult neurogenesis.

The altered neurogenesis following seizures contributes to several well-characterized cellular abnormalities including mossy fiber sprouting, ectopic dentate granule cell migration, and abnormal hilar basal dendrite development (Jessberger et al., 2007). In contrast, adult-born DG granule cells that integrate normally during epileptogenesis may serve a compensatory role to restore inhibition, thereby counteracting the epileptogenic stimuli.

Brain seizure-induced NSC proliferation and activation is a transient process and the increased neurogenesis in the DG returns to a normal level after 3–4 weeks (Jessberger et al., 2007; Parent & Helen, 2007). In addition, although radial and nonradial Type1 aNSCs are affected, all progenitor cell populations in the SGZ could be affected as well as the neuroblasts (Type-3 cells) (Huttmann et al., 2003; Jessberger, Romer, Babu, & Kempermann, 2005; Lugert et al., 2010; Steiner et al., 2008). Interestingly, kainic acid administration, a model for temporal lobe epilepsy, induces proliferation of Hes5::GFP+ NSCs subpopulations in the DG (Lugert et al., 2010). Hence, seizures increase the activation of Notch-dependent NSCs, recruiting quiescent cells to expand the active horizontal subpopulation. However, prolonged epileptic seizures markedly decrease adult neurogenesis, possibly causing exhaustion of the aNSC pool or modifying the neurogenic niche composition (Hattiangady, Rao, & Shetty, 2004). The mechanisms by which epileptic seizures alter adult neurogenesis and whether restoring normal aNSC behavior after such insults will attenuate the development of epilepsy or its complication remains to be determined.

4.2. Aging

Neurogenesis in the SGZ declines with age (Fig. 7.2). Thus, the total number of neurons generated with progressing age quickly decreases after a peak in early adulthood (Kronenberg et al., 2006). In line with this, the most dramatic decrease in the number of neuroblasts in the DG occurs during the first postnatal months in both mice and humans (Ben Abdallah et al., 2010; Knoth et al., 2010). However, one major difference in rodents is an approximately fourfold decline during adult life in humans compared to a 10-fold decrease in neurogenesis from 2 to 9 months of age in mice (Ben Abdallah et al., 2010; Spalding et al., 2013). The age-related decrease in neurogenesis could be linked to changes in the environment of the neurogenic niche as well as intrinsic changes in aNSC activity. One possible player in the regulation of DG neurogenesis in aged animals could be corticosteroids that normally inhibit neurogenesis. Corticosteroid levels increase during aging and the expression level of their receptor is subject to modification with age (Garcia, Steiner, et al., 2004).

In addition, intrinsic determinants of NSC fate that make them more susceptible to aging and reduce the mitotic activity need to be elucidated. However, alterations or loss of the NSC pool could also be a factor that contributes to reduced neurogenesis (Hattiangady & Shetty, 2010). Reduced proliferation of aNSCs has a major effect on hippocampal neurogenesis during aging, and the reduction in Notch-dependent NSCs is mostly attributable to loss of the active pool, while the quiescent pool is still present (Hattiangady, Rao, & Shetty, 2008; Lugert et al., 2010). Moreover, imbalance between maintenance and differentiation of NSCs might cause reduced neurogenesis even in the presence of canonical Notch signaling activity (Bonaguidi et al., 2011; Lugert et al., 2010). An alternative view is that NSCs differentiate into astrocytes during aging, thus exhausting the NSC pool (Encinas et al., 2011). The reality is likely to be a combination of both processes, with some NSCs being lost and others entering a dormant state.

Physical exercise could help to counteract the age-induced decline in cell proliferation. Several studies suggest that voluntary exercise could prevent the decreased proliferation and improve hippocampal-dependent cognitive performance (Kronenberg et al., 2006; Lugert et al., 2010; van Praag, Kempermann, & Gage, 1999). Moreover, growth factor administration (i.e., VEGF or IGF) stimulates cell proliferation in the DG, confirming the importance of factors in the extracellular environment as modulators of aNSC behavior during aging (Fabel et al., 2003; Trejo, Carro, & Torres-Aleman, 2001). On the whole, given the negative impact of aging on hippocampal neurogenesis and on cognitive functions, the stimulation of the endogenous aNSC pools might be able to restore memory and cognitive defects associated with age-related disorders and neurodegenerative diseases.

4.3. Mood-related disorders

Hippocampal neurogenesis has also been linked to depressive disorders, and current antidepressant treatments are thought to target neurogenesis (Sahay & Hen, 2007; Santarelli et al., 2003). In fact, growing evidence supports the concept that chronic treatment with clinical antidepressants enhances hippocampal neurogenesis (Perera et al., 2007). In particular, norepinephrine and antidepressants that block its reuptake can directly stimulate proliferation of quiescent progenitors in the adult hippocampus, leading to increased production of neurons (Jhaveri et al., 2010). Interestingly, the formation of new neurons seems to be critical for the beneficial effect of antidepressant treatment as the efficacy of treatment correlates with the time of onset of increased neurogenesis. Therefore, the behavioral effects of selective serotonin reuptake inhibitors and tricyclic antidepressants were blocked in two rodent models of anxiety/depression by radiological and genetic ablation of neurogenesis in the DG (David et al., 2009; Santarelli et al., 2003). It remains unclear how changes in hippocampal neurogenesis could affect anxiety and depression but this is potentially related to the connection of the hippocampus with subcortical structures. However, a specific effect on an aNSC subpopulation cannot be excluded. In fact, antidepressants that selectively block norepinephrine, but not serotonin, activate a subset of aNSC in the adult hippocampus. In this case, Hes5+ quiescent NSCs activate and generate more neurons both *in vivo* and *in vitro* (Jhaveri et al., 2010). Understanding molecular mechanisms that increase hippocampal neurogenesis may lead to building a substrate for powerful antidepressant therapies.

5. CONCLUSIONS

The hippocampus is critical for learning and memory and its formation is precisely regulated during embryonic and early postnatal development. The convergence of many signaling pathways, controlled movement, and migration of cells within the hippocampal primordium are important in establishing the compartmentalization of the hippocampus and DG. The DG is one of the few regions of the adult brain that continue to generate neurons throughout life, including in humans. Although the role of neurogenesis in adult humans remains to be determined, the links to disease, cognitive function, and aging in animal models need to be studied further. The potential functions of newborn neurons in the DG of humans are implied by the effects of some antidepressants in regulating neurogenesis. However, the need for neurogenesis in the efficacy of antidepressant treatment needs further investigation. This may lead to combinatorial strategies including approaches to specifically increase neuron production in the DG to treat depressive disorders in humans. The recent description of heterogenous stem cell-like populations in the DG and their selective response to pathophysiological stimuli suggest that the hippocampus and DG have retained selective mechanisms and pools of progenitors to allow adaptation to the animals' need. The precise lineage relationships and functions of these putative distinct aNSCs will require intense investigation in the future.

ACKNOWLEDGMENTS

We apologize to those colleagues who have contributed to the field but whose work we have been unable to cite because of space restrictions. We thank Dr. Claudio Giachino for his critical reading of the chapter. This work was supported by the Swiss National Science Foundation and the Deutsche Forschungsgemeinschaft (TA-310-3).

REFERENCES

Ables, J. L., DeCarolis, N. A., Johnson, M. A., Rivera, P. D., Gao, Z., Cooper, D. C., et al. (2010). Notch1 is required for maintenance of the reservoir of adult hippocampal stem cells. *The Journal of Neuroscience*, *30*(31), 10484–10492.

Ahn, S., & Joyner, A. L. (2005). In vivo analysis of quiescent adult neural stem cells responding to Sonic hedgehog. *Nature*, *437*, 894–897.

Andersson, T., Rahman, S., Sansom, S. N., Alsio, J. M., Kaneda, M., Smith, J., et al. (2010). Reversible block of mouse neural stem cell differentiation in the absence of dicer and microRNAs. *PLoS One*, *5*, e13453.

Anthony, T. E., Klein, C., Fishell, G., & Heintz, N. (2004). Radial glia serve as neuronal progenitors in all regions of the central nervous system. *Neuron*, *41*, 881–890.

Backman, M., Machon, O., Mygland, L., van den Bout, C. J., Zhong, W., Taketo, M. M., et al. (2005). Effects of canonical Wnt signaling on dorso-ventral specification of the mouse telencephalon. *Developmental Biology*, *279*, 155–168.

Ben Abdallah, N. M., Slomianka, L., Vyssotski, A. L., & Lipp, H. P. (2010). Early age-related changes in adult hippocampal neurogenesis in C57 mice. *Neurobiology of Aging*, *31*, 151–161.

Bergmann, O., Liebl, J., Bernard, S., Alkass, K., Yeung, M. S., Steier, P., et al. (2012). The age of olfactory bulb neurons in humans. *Neuron*, *74*, 634–639.

Bonaguidi, M. A., Wheeler, M. A., Shapiro, J. S., Stadel, R. P., Sun, G. J., Ming, G. L., et al. (2011). In vivo clonal analysis reveals self-renewing and multipotent adult neural stem cell characteristics. *Cell*, *145*, 1142–1155.

Breunig, J. J., Silbereis, J., Vaccarino, F. M., Sestan, N., & Rakic, P. (2007). Notch regulates cell fate and dendrite morphology of newborn neurons in the postnatal dentate gyrus. *Proceedings of the National Academy of Sciences of the United States of America*, *104*, 20558–20563.

David, D. J., Samuels, B. A., Rainer, Q., Wang, J.-W., Marsteller, D., Mendez, I., et al. (2009). Neurogenesis-dependent and -independent effects of fluoxetine in an animal model of anxiety/depression. *Neuron*, *62*, 479–493.

Davis, T. H., Cuellar, T. L., Koch, S. M., Barker, A. J., Harfe, B. D., McManus, M. T., et al. (2008). Conditional loss of dicer disrupts cellular and tissue morphogenesis in the cortex and hippocampus. *The Journal of Neuroscience*, *28*, 4322–4330.

De Pietri Tonelli, D., Pulvers, J. N., Haffner, C., Murchison, E. P., Hannon, G. J., & Huttner, W. B. (2008). miRNAs are essential for survival and differentiation of newborn neurons but not for expansion of neural progenitors during early neurogenesis in the mouse embryonic neocortex. *Development*, *135*, 3911–3921.

Doetsch, F. (2003). The glial identity of neural stem cells. *Nature Neuroscience*, *6*, 1127–1134.

Doetsch, F., Caille, I., Lim, D. A., Garcia-Verdugo, J. M., & Alvarez-Buylla, A. (1999). Subventricular zone astrocytes are neural stem cells in the adult mammalian brain. *Cell*, *97*, 703–716.

Ehm, O., Goritz, C., Covic, M., Schaffner, I., Schwarz, T. J., Karaca, E., et al. (2010). RBPJkappa-dependent signaling is essential for long-term maintenance of neural stem cells in the adult hippocampus. *Journal of Neuroscience*, *30*, 13794–13807.

Encinas, J. M., Michurina, T. V., Peunova, N., Park, J.-H., Tordo, J., Peterson, D. A., et al. (2011). Division-coupled astrocytic differentiation and age-related depletion of neural stem cells in the adult hippocampus. *Cell Stem Cell*, *8*, 566–579.

Eriksson, P. S., Perfilieva, E., Bjork-Eriksson, T., Alborn, A. M., Nordborg, C., Peterson, D. A., et al. (1998). Neurogenesis in the adult human hippocampus. *Nature Medicine*, *4*, 1313–1317.

Fabel, K., & Kempermann, G. (2008). Physical activity and the regulation of neurogenesis in the adult and aging brain. *Neuromolecular Medicine*, *10*, 59–66.

Fabel, K., Tam, B., Kaufer, D., Baiker, A., Simmons, N., Kuo, C., et al. (2003). VEGF is necessary for exercise-induced adult hippocampal neurogenesis. *European Journal of Neuroscience*, *18*, 2803–2812.

Favaro, R., Valotta, M., Ferri, A. L. M., Latorre, E., Mariani, J., Giachino, C., et al. (2009). Hippocampal development and neural stem cell maintenance require Sox2-dependent regulation of Shh. *Nature Neuroscience*, *12*, 1248–1256.

Frotscher, M., Haas, C. A., & Forster, E. (2003). Reelin controls granule cell migration in the dentate gyrus by acting on the radial glial scaffold. *Cerebral Cortex*, *13*(6), 634–640.

Galceran, J., Miyashita-Lin, E. M., Devaney, E., Rubenstein, J. L., & Grosschedl, R. (2000). Hippocampus development and generation of dentate gyrus granule cells is regulated by LEF1. *Development*, *127*, 469–482.

Galichet, C., Guillemot, F., & Parras, C. M. (2008). Neurogenin 2 has an essential role in development of the dentate gyrus. *Development*, *135*, 2031–2041.

Gao, Z., Ure, K., Ding, P., Nashaat, M., Yuan, L., Ma, J., et al. (2011). The master negative regulator REST/NRSF controls adult neurogenesis by restraining the neurogenic program in quiescent stem cells. *Journal of Neuroscience*, *31*, 9772–9786.

Garcia, A. D. R., Doan, N. B., Imura, T., Bush, T. G., & Sofroniew, M. V. (2004a). GFAP-expressing progenitors are the principal source of constitutive neurogenesis in adult mouse forebrain. *Nature Neuroscience*, *7*, 1233–1241.

Garcia, A., Steiner, B., Kronenberg, G., Bick-Sander, A., Kempermann, G., & Goetz, M. (2004b). Age-dependent expression of glucocorticoid and mineralocorticoid receptors on neural precursor cell populations in the adult murine hippocampus. *Aging cell*, *3*, 363–371.

Han, Y.-G., Spassky, N., Romaguera-Ros, M., Garcia-Verdugo, J.-M., Aguilar, A., Schneider-Maunoury, S., et al. (2008). Hedgehog signaling and primary cilia are required for the formation of adult neural stem cells. *Nature Neuroscience*, *11*, 277–284.

Hatakeyama, J., Bessho, Y., Katoh, K., Ookawara, S., Fujioka, M., Guillemot, F., et al. (2004). Hes genes regulate size, shape and histogenesis of the nervous system by control of the timing of neural stem cell differentiation. *Development*, *131*, 5539–5550.

Hattiangady, B., Rao, M. S., & Shetty, A. K. (2004). Chronic temporal lobe epilepsy is associated with severely declined dentate neurogenesis in the adult hippocampus. *Neurobiology of Disease*, *17*, 473–490.

Hattiangady, B., Rao, M. S., & Shetty, A. K. (2008). Plasticity of hippocampal stem/progenitor cells to enhance neurogenesis in response to kainate-induced injury is lost by middle age. *Aging Cell*, *7*, 207–224.

Hattiangady, B., & Shetty, A. K. (2010). Decreased neuronal differentiation of newly generated cells underlies reduced hippocampal neurogenesis in chronic temporal lobe epilepsy. *Hippocampus*, *20*(1), 97.

Hevner, R. F., Hodge, R. D., Daza, R. A. M., & Englund, C. (2006). Transcription factors in glutamatergic neurogenesis: Conserved programs in neocortex, cerebellum, and adult hippocampus. *Neuroscience Research*, *55*, 223–233.

Hodge, R. D., Garcia, A. J., Elsen, G. E., Nelson, B. R., Mussar, K. E., Reiner, S. L., et al. (2013). Tbr2 expression in Cajal-Retzius cells and intermediate neuronal progenitors is required for morphogenesis of the dentate gyrus. *The Journal of Neuroscience*, *33*, 4165–4180.

Hodge, R. D., Kowalczyk, T. D., Wolf, S. A., Encinas, J. M., Rippey, C., Enikolopov, G., et al. (2008). Intermediate progenitors in adult hippocampal neurogenesis: Tbr2 expression and coordinate regulation of neuronal output. *The Journal of Neuroscience*, *28*, 3707–3717.

Hodge, R. D., Nelson, B. R., Kahoud, R. J., Yang, R., Mussar, K. E., Reiner, S. L., et al. (2012). Tbr2 is essential for hippocampal lineage progression from neural stem cells to intermediate progenitors and neurons. *The Journal of Neuroscience*, *32*, 6275–6287.

Huttmann, K., Sadgrove, M., Wallraff, A., Hinterkeuser, S., Kirchhoff, F., Steinhauser, C., et al. (2003). Seizures preferentially stimulate proliferation of radial glia-like astrocytes in the adult dentate gyrus: Functional and immunocytochemical analysis. *European Journal of Neuroscience*, *18*, 2769–2778.

Jang, M. H., Bonaguidi, M. A., Kitabatake, Y., Sun, J., Song, J., Kang, E., et al. (2013). Secreted frizzled-related protein 3 regulates activity-dependent adult hippocampal neurogenesis. *Cell Stem Cell*, *12*, 215–223. http://dx.doi.org/10.1016/j.stem.2012.1011.1021.

Jessberger, S., Clark, R. E., Broadbent, N. J., Clemenson, G. D., Jr., Consiglio, A., Lie, D. C., et al. (2009). Dentate gyrus-specific knockdown of adult neurogenesis impairs spatial and object recognition memory in adult rats. *Learning and Memory*, *16*, 147–154.

Jessberger, S., Romer, B., Babu, H., & Kempermann, G. (2005). Seizures induce proliferation and dispersion of doublecortin-positive hippocampal progenitor cells. *Experimental Neurology*, *196*, 342–351.

Jessberger, S., Toni, N., Clemenson, G. D., Jr., Ray, J., & Gage, F. H. (2008). Directed differentiation of hippocampal stem/progenitor cells in the adult brain. *Nature Neuroscience*, *11*, 888–893.

Jessberger, S., Zhao, C., Toni, N., Clemenson, G. D., Li, Y., & Gage, F. H. (2007). Seizure-associated, aberrant neurogenesis in adult rats characterized with retrovirus-mediated cell labeling. *The Journal of Neuroscience*, *27*, 9400–9407.

Jhaveri, D. J., Mackay, E. W., Hamlin, A. S., Marathe, S. V., Nandam, L. S., Vaidya, V. A., et al. (2010). Norepinephrine directly activates adult hippocampal precursors via β≤3-adrenergic receptors. *The Journal of Neuroscience*, *30*, 2795–2806.

Kawase-Koga, Y., Otaegi, G., & Sun, T. (2009). Different timings of dicer deletion affect neurogenesis and gliogenesis in the developing mouse central nervous system. *Developmental Dynamics*, *238*, 2800–2812.

Kempermann, G., Jessberger, S., Steiner, B., & Kronenberg, G. (2004a). Milestones of neuronal development in the adult hippocampus. *Trends in Neurosciences*, *27*, 447–452.

Kempermann, G., Kuhn, H. G., & Gage, F. H. (1998). Experience-induced neurogenesis in the senescent dentate gyrus. *The Journal of Neuroscience*, *18*, 3206–3212.

Kempermann, G., Wiskott, L., & Gage, F. H. (2004b). Functional significance of adult neurogenesis. *Current Opinion in Neurobiology*, *14*, 186–191.

Knobloch, M., Braun, S. M., Zurkirchen, L., von Schoultz, C., Zamboni, N., Arauzo-Bravo, M. J., et al. (2013). Metabolic control of adult neural stem cell activity by Fasn-dependent lipogenesis. *Nature*, *493*, 226–230.

Knoth, R., Singec, I., Ditter, M., Pantazis, G., Capetian, P., Meyer, R. P., et al. (2010). Murine features of neurogenesis in the human hippocampus across the lifespan from 0 to 100 years. *PLoS One*, *5*, e8809.

Knuckles, P., Vogt, M. A., Lugert, S., Milo, M., Chong, M. M. W., Hautbergue, G. M., et al. (2012). Drosha regulates neurogenesis by controlling neurogenin 2 expression independent of microRNAs. *Nature Neuroscience, 15*, 962–969.

Kriegstein, A., & Alvarez-Buylla, A. (2009). The glial nature of embryonic and adult neural stem cells. *Annual Review of Neuroscience, 32*, 149–184.

Kronenberg, G., Bick-Sander, A., Bunk, E., Wolf, C., Ehninger, D., & Kempermann, G. (2006). Physical exercise prevents age-related decline in precursor cell activity in the mouse dentate gyrus. *Neurobiology of Aging, 27*, 1505–1513.

Kronenberg, G., Reuter, K., Steiner, B., Brandt, M. D., Jessberger, S., Yamaguchi, M. K., et al. (2003). Subpopulations of proliferating cells of the adult hippocampus respond differently to physiologic neurogenic stimuli. *Journal of Comparative Neurology, 467*, 455–463.

Kuhn, H. G., Dickinson-Anson, H., & Gage, F. H. (1996). Neurogenesis in the dentate gyrus of the adult rat: Age-related decrease of neuronal progenitor proliferation. *The Journal of Neuroscience, 16*, 2027–2033.

Lai, K., Kaspar, B. K., Gage, F. H., & Schaffer, D. V. (2003). Sonic hedgehog regulates adult neural progenitor proliferation in vitro and in vivo. *Nature Neuroscience, 6*, 21–27.

Lee, S. M., Tole, S., Grove, E., & McMahon, A. P. (2000). A local Wnt-3a signal is required for development of the mammalian hippocampus. *Development, 127*, 457–467.

Li, Q., Bian, S., Hong, J., Kawase-Koga, Y., Zhu, E., Zheng, Y., et al. (2011). Timing specific requirement of microRNA function is essential for embryonic and postnatal hippocampal development. *PLoS One, 6*, e26000.

Li, G., Fang, L., Fernandez, G., & Pleasure, S. J. (2013). The ventral hippocampus is the embryonic origin for adult neural stem cells in the dentate gyrus. *Neuron, 78*, 658–672.

Li, G., Kataoka, H., Coughlin, S. R., & Pleasure, S. J. (2009). Identification of a transient subpial neurogenic zone in the developing dentate gyrus and its regulation by Cxcl12 and reelin signaling. *Development, 136*, 327–335.

Li, G., & Pleasure, S. J. (2005). Morphogenesis of the dentate gyrus: What we are learning from mouse mutants. *Developmental Neuroscience, 27*, 93–99.

Lie, D.-C., Colamarino, S. A., Song, H.-J., Desire, L., Mira, H., Consiglio, A., et al. (2005). Wnt signalling regulates adult hippocampal neurogenesis. *Nature, 437*, 1370–1375.

Lugert, S., Basak, O., Knuckles, P., Haussler, U., Fabel, K., Gotz, M., et al. (2010). Quiescent and active hippocampal neural stem cells with distinct morphologies respond selectively to physiological and pathological stimuli and aging. *Cell Stem Cell, 6*, 445–456.

Lugert, S., Vogt, M., Tchorz, J. S., Muller, M., Giachino, C., & Taylor, V. (2011). Homeostatic neurogenesis in the adult hippocampus does not involve amplification of Ascl1 (high) intermediate progenitors. *Nature Communications, 3*, 670.

Machold, R., Hayashi, S., Rutlin, M., Muzumdar, M. D., Nery, S., Corbin, J. G., et al. (2003). Sonic hedgehog is required for progenitor cell maintenance in telencephalic stem cell niches. *Neuron, 39*, 937–950.

Malatesta, P., Hack, M. A., Hartfuss, E., Kettenmann, H., Klinkert, W., Kirchhoff, F., et al. (2003). Neuronal or glial progeny: Regional differences in radial glia fate. *Neuron, 37*, 751–764.

Mangale, V. S., Hirokawa, K. E., Satyaki, P. R. V., Gokulchandran, N., Chikbire, S., Subramanian, L., et al. (2008). Lhx2 selector activity specifies cortical identity and suppresses hippocampal organizer fate. *Science, 319*, 304–309.

Mao, Y., Ge, X., Frank, C. L., Madison, J. M., Koehler, A. N., Doud, M. K., et al. (2009). Disrupted in schizophrenia 1 regulates neuronal progenitor proliferation via modulation of GSK3beta/beta-catenin signaling. *Cell, 136*, 1017–1031.

Merkle, F. T., Mirzadeh, Z., & Alvarez-Buylla, A. (2007). Mosaic organization of neural stem cells in the adult brain. *Science, 317*, 381–384.

Ming, G. L., & Song, H. (2009). DISC1 partners with GSK3beta in neurogenesis. *Cell, 136*, 990–992.

Ming, G. L., & Song, H. (2011). Adult neurogenesis in the mammalian brain: Significant answers and significant questions. *Neuron, 70*, 687–702.

Mira, H., Andreu, Z., Suh, H., Lie, D. C., Jessberger, S., Consiglio, A., et al. (2010). Signaling through BMPR-IA regulates quiescence and long-term activity of neural stem cells in the adult hippocampus. *Cell Stem Cell, 7*, 78–89.

Okamoto, M., Inoue, K., Iwamura, H., Terashima, K., Soya, H., Asashima, M., et al. (2011). Reduction in paracrine Wnt3 factors during aging causes impaired adult neurogenesis. *The FASEB Journal, 25*, 3570–3582.

Oldekamp, J., Kraemer, N., Alvarez-Bolado, G., & Skutella, T. (2004). bHLH gene expression in the Emx2-deficient dentate gyrus reveals defective granule cells and absence of migrating precursors. *Cerebral Cortex, 14*, 1045–1058.

Parent, J. M. (2007). Adult neurogenesis in the intact and epileptic dentate gyrus. *Progress in Brain Research, 163*, 529–817.

Parent, J. M., & Murphy, G. G. (2008). Mechanisms and functional significance of aberrant seizure-induced hippocampal neurogenesis. *Epilepsia, 49*, 19–25.

Pellegrini, M., Mansouri, A., Simeone, A., Boncinelli, E., & Gruss, P. (1996). Dentate gyrus formation requires Emx2. *Development, 122*, 3893–3898.

Perera, T. D., Coplan, J. D., Lisanby, S. H., Lipira, C. M., Arif, M., Carpio, C., et al. (2007). Antidepressant-induced neurogenesis in the hippocampus of adult nonhuman primates. *The Journal of Neuroscience, 27*, 4894–4901.

Petrus, D. S., Fabel, K., Kronenberg, G., Winter, C., Steiner, B., & Kempermann, G. (2009). NMDA and benzodiazepine receptors have synergistic and antagonistic effects on precursor cells in adult hippocampal neurogenesis. *European Journal of Neuroscience, 29*, 244–252.

Pleasure, S. J., Collins, A. E., & Lowenstein, D. H. (2000). Unique expression patterns of cell fate molecules delineate sequential stages of dentate gyrus development. *The Journal of Neuroscience, 20*, 6095–6105.

Pozniak, C. D., & Pleasure, S. J. (2006). A tale of two signals: Wnt and Hedgehog in dentate neurogenesis. *Science's STKE, 319*, pe5.

Sahay, A., & Hen, R. (2007). Adult hippocampal neurogenesis in depression. *Nature Neuroscience, 10*, 1110–1115.

Santarelli, L., Saxe, M., Gross, C., Surget, A., Battaglia, F., Dulawa, S., et al. (2003). Requirement of hippocampal neurogenesis for the behavioral effects of antidepressants. *Science, 301*, 805–809.

Schwab, M. H., Bartholomae, A., Heimrich, B., Feldmeyer, D., Druffel-Augustin, S., Goebbels, S., et al. (2000). Neuronal basic helix-loop-helix proteins (NEX and BETA2/Neuro D) regulate terminal granule cell differentiation in the hippocampus. *Journal of Neuroscience, 20*, 3714–3724.

Seib, D. R., Corsini, N. S., Ellwanger, K., Plaas, C., Mateos, A., Pitzer, C., et al. (2013). Loss of Dickkopf-1 restores neurogenesis in old age and counteracts cognitive decline. *Cell Stem Cell, 12*, 204–214.

Seri, B., García-Verdugo, J. M., Collado-Morente, L., McEwen, B. S., & Alvarez-Buylla, A. (2004). Cell types, lineage, and architecture of the germinal zone in the adult dentate gyrus. *The Journal of Comparative Neurology, 478*, 359–378.

Seri, B., Garcia-Verdugo, J. M., McEwen, B. S., & Alvarez-Buylla, A. (2001). Astrocytes give rise to new neurons in the adult mammalian hippocampus. *The Journal of Neuroscience, 21*, 7153–7160.

Song, J., Zhong, C., Bonaguidi, M. A., Sun, G. J., Hsu, D., Gu, Y., et al. (2012). Neuronal circuitry mechanism regulating adult quiescent neural stem-cell fate decision. *Nature, 489*, 150–154.

Spalding, K. L., Bergmann, O., Alkass, K., Bernard, S., Salehpour, M., Huttner, H. B., et al. (2013). Dynamics of hippocampal neurogenesis in adult humans. *Cell, 153*, 1219–1227.

Steiner, B., Zurborg, S., Horster, H., Fabel, K., & Kempermann, G. (2008). Differential 24 h responsiveness of Prox1-expressing precursor cells in adult hippocampal neurogenesis to physical activity, environmental enrichment, and kainic acid-induced seizures. *Neuroscience, 154*, 521–529.

Subramanian, L., Sarkar, A., Shetty, A. S., Muralidharan, B., Padmanabhan, H., Piper, M., et al. (2011). Transcription factor Lhx2 is necessary and sufficient to suppress astrogliogenesis and promote neurogenesis in the developing hippocampus. *Proceedings of the National Academy of Sciences of the United States of America, 108*, 265–274.

Trejo, J. L., Carro, E., & Torres-Aleman, I. (2001). Circulating insulin-like growth factor I mediates exercise-induced increases in the number of new neurons in the adult hippocampus. *The Journal of Neuroscience, 21*, 1628–1634.

van Praag, H., Kempermann, G., & Gage, F. H. (1999). Running increases cell proliferation and neurogenesis in the adult mouse dentate gyrus. *Nature Neuroscience, 2*, 266–270.

Wexler, E. M., Paucer, A., Kornblum, H. I., Palmer, T. D., & Geschwind, D. H. (2009). Endogenous Wnt signaling maintains neural progenitor cell potency. *Stem Cells, 27*, 1130–1141.

Zhao, C., Deng, W., & Gage, F. H. (2008). Mechanisms and functional implications of adult neurogenesis. *Cell, 132*, 645–660.

Zhou, C.-J., Zhao, C., & Pleasure, S. J. (2004). Wnt signaling mutants have decreased dentate granule cell production and radial glial scaffolding abnormalities. *The Journal of Neuroscience, 24*, 121–126.

CHAPTER EIGHT

Lung Stem and Progenitor Cells in Tissue Homeostasis and Disease

Kristen T. Leeman[*,†,‡,§,1], Christine M. Fillmore[†,‡,§,1], Carla F. Kim[†,‡,§,2]

[*]Division of Newborn Medicine, Boston Children's Hospital, Harvard Medical School, Boston, Massachusetts, USA

[†]Stem Cell Program, Boston Children's Hospital, Boston, Massachusetts, USA

[‡]The Harvard Stem Cell Institute, Cambridge, Massachusetts, USA

[§]Department of Genetics, Harvard Medical School, Boston, Massachusetts, USA

[1]These authors contributed equally

[2]Corresponding author: e-mail address: carla.kim@childrens.harvard.edu

Contents

Abstract

The mammalian lung is a complex organ containing numerous putative stem/progenitor cell populations that contribute to region-specific tissue homeostasis and repair. In this review, we discuss recent advances in identifying and studying these cell populations in the context of lung homeostasis and disease. Genetically engineered mice now allow for lineage tracing of several lung stem and progenitor cell populations *in vivo* during different types of lung injury repair. Using specific sets of cell surface markers, these cells can also be isolated from murine and human lung and tested in 3D culture systems and *in vivo* transplant assays. The pathology of devastating lung diseases, including lung cancers, is likely in part due to dysregulation and dysfunction of lung stem cells. More precise characterization of stem cells with identification of new, unique markers; improvement in isolation and transplant techniques; and further development of functional assays will ultimately lead to new therapies for a host of human lung diseases. In particular, lung cancer biology may be greatly informed by findings in normal lung stem cell biology as evidence suggests that lung cancer is a disease that begins in, and may be driven by, neoplastic lung stem cells.

Current Topics in Developmental Biology, Volume 107
ISSN 0070-2153
http://dx.doi.org/10.1016/B978-0-12-416022-4.00008-1

1. INTRODUCTION

The mammalian respiratory system is a highly complex three-dimensional organ historically described as containing over 40 different cell types, each with specialized functions to maintain adequate gas exchange and protect against environmental exposures. During development, the primordial lung undergoes branching morphogenesis to form the proximal conducting airways and distal gas-exchanging alveolar space (Morrisey & Hogan, 2010). The adult murine lung contains several distinct epithelial cell populations with unique anatomical positions and specialized functions (Fig. 8.1). The proximal airway includes the cartilaginous trachea, lined by pseudostratified columnar epithelial cells with submucosal glands interspersed. Noncartilaginous bronchioles, lined with simple columnar epithelium, branch from the trachea in an organized pattern. Secretory Clara cells also line the basement membrane of the airway with ciliated, neuroendocrine, and goblet cell populations (Bertoncello & McQualter, 2013). Lung cell-type terminology is undergoing a transition as the name Clara cell is being replaced by club cell; this review will use the historic term Clara cell. Neuroendocrine cells are present individually as well as in clusters termed neuroendocrine bodies that may play a role in sensing stimuli within the airway lumen (Van Lommel, 2001). Terminal bronchioles lead to the distal alveolar space containing surfactant-producing alveolar type II (AT2) cells and gas-exchanging alveolar type I (AT1) cells (Rock & Hogan, 2011).

Diverse experimental approaches have provided evidence that different populations of lung stem/progenitor cells reside in distinct niches and act in region-specific homeostasis and injury repair. Murine mouse models of injury have been utilized to study stem cells because of the low baseline levels of lung cell turnover during homeostasis and the increased rate of proliferation to replace ablated tissue following injury (Rawlins & Hogan, 2006). For example, bleomycin injures the alveolar epithelium, and naphthalene specifically injures the bronchiolar epithelium (Rawlins & Hogan, 2006). For more proximal airway injury, sulfur dioxide inhalation damages the tracheal epithelium (Borthwick, Shahbazian, Krantz, Dorin, & Randell, 2001), while ozone and nitrogen dioxide damage airway epithelial cells (Evans, Johnson, Stephens, & Freeman, 1976; Evans, Shami, Cabral-Anderson, & Dekker, 1986). Using these region-specific epithelial injury mouse models, it is possible to study cellular proliferation and epithelial regeneration. Lineage tracing is another valuable *in vivo* tool that has been

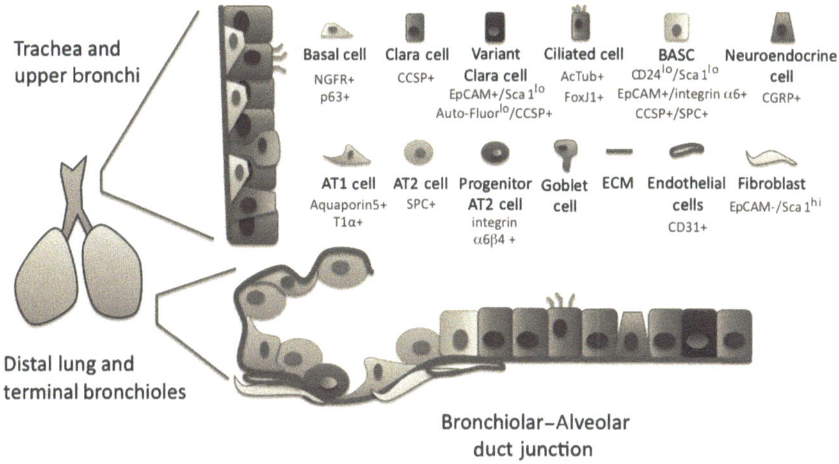

Figure 8.1 Cell types in the Lung. The proximal region of the murine respiratory system is lined by a pseudostratified epithelium containing secretory CCSP + Clara cells, mucus-producing goblet cells, and host-defending FoxJ1/Actub + ciliated cells. Variant Clara cells are thought to give rise to ciliated and Clara cell lineages after injuries such as naphthalene, and are enriched within the EpCAM^{hi}/Sca1^{lo}/Auto-fluorescence^{lo} cells. At the basal edge of the epithelium are the NGFR +/p63 + basal cells, which are thought to be able to give rise to Clara and ciliated cells during repair and in culture. In the more distal bronchioles, ciliated and Clara cells are interspersed with CGRP + neuroendocrine cells. Alveolar epithelial type 1 cells (AT1 cells), which express T1α and Aquaporin5, and SPC + alveolar epithelial type 2 cells (AT2 cells) line the alveolar space where gas exchange takes place. An alveolar progenitor cell has been identified that can give rise to AT2 and AT1 cells after injuries such as bleomycin, and is termed integrin α6β4 +. At the brochioalveolar duct junction, a rare cell population termed the brochioalveolar stem cells (BASC) coexpresses both CCSP and SPC, and is enriched in the CD24^{lo}/Sca1^{lo}/EpCAM +/integrin α6 + fraction of lung epithelial cells. BASCs are thought to be able to give rise to both Clara and AT2 cell lineages after injury. Alveolar epithelial cells and BASCs are closely associated with mesenchymal cells such as fibroblasts, the extracellular matrix (ECM), and CD31 + endothelial cells. Useful markers for FACS isolation of cell types are indicated. (See color plate.)

used to study stem cell populations and their role in lung injury and repair without removing them from the lung (Barkauskas et al., 2013; Raiser et al., 2008; Rawlins et al., 2009; Rock et al., 2011; Tropea et al., 2012). Investigators have created mouse models to label stem cell populations of choice, which coupled with the injuries mentioned earlier, allow for detection of the lineage label, which will remain in both the progenitor population and progeny cells after injury repair.

Three-dimensional (3D) culture systems have emerged as an important method of characterizing lung stem cell properties including proliferation, differentiation, and self-renewal (Lee et al., 2012; McQualter, Yuen, Williams, & Bertoncello, 2010; Rock et al., 2009). Fluorescence-activated cell sorting (FACS) can be used to isolate individual stem cell populations (Kim et al., 2005; Lee et al., 2013; McQualter et al., 2009; Rock et al., 2009; Summer, Fitzsimmons, Dwyer, Murphy, & Fine, 2007b; Teisanu, Lagasse, Whitesides, & Stripp, 2009; Zacharek et al., 2011), which can then be grown in clonal 3D assays in the presence of various microenvironmental factors such as Matrigel, nonepithelial cells, and growth factors to assess growth and differentiation properties. An ongoing challenge has been the ability to directly compare the functions of FACS-isolated lung stem cell populations with reparative cells *in situ*. Limited knowledge of markers distinguishing lung cell types *in vivo* has prevented precise concordance between the identity of lung cells with stem cell functions *in vitro* and *in vivo*. Transplantation assays have also been lacking in lung stem cell biology; currently there is no *in vivo* transplant assay for freshly sorted stem cells delivered to the lung. Recently, a kidney capsule transplantation model has been utilized as an alternative *in vivo* method of examining stem cell autonomous properties (Chapman et al., 2011). Subcutaneous injection of multipotent lung cells with Matrigel and the use of *ex vivo* decellularized lung models have also provided a new means to assess potential progenitor cell function (Longmire et al., 2012; Mou et al., 2012). The development of *in vivo* or *in vitro* assays to interrogate the function of stem cells at the single cell level, and the discovery of unique marker sets (rather than single markers) for each lung cell type are critical advances necessary to better identify lung stem cell populations and understand their relative contributions to tissue maintenance and repair.

The pathology of devastating lung diseases including lung cancers, chronic obstructive pulmonary disease, idiopathic pulmonary fibrosis (IPF), cystic fibrosis (CF), asthma, bronchopulmonary dysplasia (BPD), and others is likely caused, in part, by dysregulation and dysfunction of lung stem cells. Investigators continue to work to unlock the mystery of the lung stem cell landscape by first identifying the stem cell populations, examining their cell autonomous properties, and trying to better understand their microenvironmental interactions. While much progress has been made over recent years, important questions remain and the complete picture remains unclear. Lung stem cell studies offer a new avenue for developing treatment strategies for lung disease. In this review, we discuss the current knowledge of the lung stem field and the assays and tools used to dissect the complex biology of the lung in homeostasis and disease.

2. ENDOGENOUS LUNG STEM AND PROGENITOR CELLS

Endogenous adult lung stem/progenitor cells are regenerative cell populations important for epithelial cell maintenance and injury repair. In multiple adult organs, tissue-specific stem cells have been identified as multipotent cells with the capacity for long-term self-renewal and the ability to give rise to at least two distinct differentiated lineages. Tissue-specific stem cells are typically quiescent in normal conditions and proliferate during injury repair (Kolios & Moodley, 2013; Potten & Loeffler, 1990), although there are some exceptions such as the proliferative intestinal crypt stem cells (Clevers, 2013). Unipotent stem cells are cells with the ability to self-renew and differentiate into a single restricted lineage (e.g., spermatogonial stem cells). Progenitor cells may be multipotent or unipotent, and they are generally thought to have limited self-renewal capacity. Based on murine studies, it is believed that the lung contains multiple stem cell and progenitor cell populations, some of which reside in distinct regional niches (Bertoncello & McQualter, 2013). Current evidence suggests that different lung stem or progenitor cells may have distinct roles to play during the injury response (e.g., depending on the exact nature or extent of injury) and homeostasis. Microenvironmental signals likely play a key role in triggering specific regenerative functions (Hong, Reynolds, Giangreco, Hurley, & Stripp, 2001), yet they remain to be understood in detail. Perhaps unique from cells of some other organs, lung stem and progenitor cells also tend to have "daily" functional roles in tissue function other than a regenerative role. For example, some lung stem cells secrete proteins critical for gas exchange and become activated to proliferate for a reparative role after injury. The ability to assess self-renewal properties *in vivo* or *in vitro* in order to discern lung stem cells from lung progenitor cells is debated and may be influenced by microenvironmental cues. Herein we have attempted to refer to these important cell populations in the lung by the nomenclature used in the primary literature describing them.

In the proximal airway, basal cells have been identified as important stem cells of bronchiolar epithelium. Basal cells can be isolated based on their expression of the markers NGFR (nerve growth factor receptor), p63, cyto-keratin 5 (CK5), cyto-keratin 14 (CK14), and aquaporin 3 (Rock, Randell, & Hogan, 2010). FACS-isolated basal cells (p63+, NGFR+, CK5+) can be cultured in Matrigel and form clonal structures positive for markers of both ciliated and Clara cells. This 3D tracheosphere assay includes the use of growth medium containing EGF (epidermal growth factor), FGF (fibroblast growth factor), and bovine pituitary extract for colony growth. In mouse lineage tracing studies using a CK5–CreER transgenic mouse model, basal cells give rise to ciliated cells in the proximal airways both during homeostasis and after sulfur dioxide inhalation injury (Rock et al., 2009, 2010). Recently, a transient population of $p63^{+}/CK5^{+}$ cells were reported to migrate from the bronchiolar and peribronchiolar regions and proliferate to repair damaged alveolar lung tissue after H1N1 viral infection *in vivo* (Kumar et al., 2011). These cells were shown to expand locally into "pods" with an organized spherical shape, and then gain expression of alveolar-specific proteins. It is possible that the $p63^{+}/CK5^{+}$ migratory cells observed were basal cells, raising the possibility that basal cells can act as progenitors for alveolar epithelium during repair of particular injuries. Another proximal airway stem/progenitor cell population of tracheal submucosal gland duct cells has recently been identified to regenerate submucosal gland tubules, duct cells, and airway surface epithelium after severe hypoxic–ischemic injury (Hegab et al., 2011; Hegab, Nickerson, Ha, Darmawan, & Gomperts, 2012).

In the bronchioles of mouse lungs, two distinct niches have been identified for lung stem cells and progenitor cells. After nitrogen dioxide bronchiolar injury, Clara cells have been shown to give rise to ciliated cells; ciliated cells cannot self-renew (Evans, Cabral-Anderson, & Freeman, 1978; Rawlins & Hogan, 2008). Lineage tracing has shown that Clara cells can self-renew and generate ciliated cells during epithelial homeostasis (Rawlins et al., 2009). Lineage tracing studies labeling calcitonin gene-regulated peptide (CGRP)-expressing neuroendocrine cells showed their ability to self-renew in uninjured adult lung and to differentiate into Clara and ciliated cells after naphthalene injury (Song et al., 2012). The area surrounding the neuroendocrine bodies and the bronchioalveolar duct junction (BADJ) contains a population of Clara cells that are resistant to naphthalene injury. These cells, which have been termed variant Clara cells, can self-renew and differentiate into Clara cells to repair naphthalene bronchiolar epithelial injury (Giangreco, Reynolds, & Stripp, 2002; Reynolds, Giangreco, Power, & Stripp, 2000). The Stripp laboratory identified a population of $CD45^{neg}$ $CD31^{neg}$ $CD34^{neg}$ $Epcam^{pos}$ $Sca\text{-}1^{low}$ AutoFluorescence $(AF)^{low}$ cells that contained naphthalene-resistant progenitors (Teisanu et al., 2009). In contrast, the AF^{high} population contained the naphthalene-sensitive Clara cells (Teisanu et al., 2009). Also located at the BADJ is a population of cells identified as bronchioalveolar stem cells (BASCs) (Kim et al., 2005). BASCs are a rare population of epithelial cells with coexpression of both Clara cell secretory protein (CCSP, also known as Scgb1a1 or CC10) and AT2 cell marker prosurfactant protein C (SPC). BASCs increase in number after naphthalene bronchiolar injury and bleomycin alveolar injury, which suggests their role in repair (Kim et al., 2005). BASCs can be isolated by FACS for cells positive for the stem cell marker Sca-1 and negative for hematopoietic and endothelial cell markers. Recent studies have shown that a robust marker profile for enriching for BASCs utilizes $CD45^{neg}$ $CD31^{neg}$ $EpCAM^{pos}$ $Sca\text{-}1^{low}$ $CD24^{low}$ (Zacharek et al., 2011). BASCs self-renew over multiple passages *in vitro* and differentiate into both bronchiolar and alveolar colonies (Kim et al., 2005; Zacharek et al., 2011). Another similar population was recently isolated with the FACS signature of $CD45^{neg}$ $CD31^{neg}$ $EpCAM^{high}$ integrin $\alpha6^{pos}$ $CD104^{pos}$ $CD24^{low}$, and these cells produce colonies with airway, alveolar, or both airway and alveolar markers in a 3D coculture Matrigel assay with lung mesenchymal cells (McQualter et al., 2010). Work on improving specific cell surface marker expression profiles to better identify and isolate lung stem/progenitor cells is ongoing and necessary to isolate a truly homogenous population. In addition to phenotypic identification, functional assays including lineage tracing, single cell assays, and cell transplantation are necessary to better understand whether the roles of the many putative stem cells in the bronchioles, including Clara cells, variant Clara cells, and BASCs, are distinct during epithelial tissue repair *in vivo*.

In the distal lung, AT2 cells have long been proposed to be the progenitors of alveolar epithelium (Adamson & Bowden, 1974; Evans, Cabral, Stephens, & Freeman, 1975). Recent work from the Hogan laboratory showed evidence that AT2 cells are stem cells, through genetic lineage tracing experiments and 3D cultures (Barkauskas et al., 2013; Rock et al., 2010). After bleomycin injury, AT2 cells were previously shown to proliferate and were hypothesized to differentiate into AT2 cells and AT1 cells (Aso, Yoneda, & Kikkawa, 1976). Using an SPC-inducible CreER to fate-map AT2 cells, it has been shown that AT2 cells are not the major contributor to alveolar repair after bleomycin injury; instead, a new progenitor cell type has been suggested to be responsible for the majority of the AT2 cell regeneration (Chapman et al., 2011). This integrin $\alpha6\beta4$-positive cell population located in the alveolar epithelium has been identified as a progenitor cell population and shown to differentiate into mature SPC positive cells after *in vitro* colony expansion (Chapman et al., 2011). This cell type is capable of contributing to airway and alveolar structures when transplanted with embryonic lung cells under the kidney capsule *in vivo* (Chapman et al., 2011).

BASCs have also been identified as potential progenitors of alveolar epithelium because of their increased proliferation *in vivo* after bleomycin injury and their ability to differentiate into SPC-expressing cells in culture (Kim et al., 2005). Lineage tracing studies using a CCSP-CreER mouse model have shown that the percentage of lineage-labeled alveolar cells increased after bleomycin injury (Rock et al., 2011; Tropea et al., 2012). Specifically, after bleomycin lung injury and repair, the percentage of lineage-labeled AT2 and AT1 cells increased, indicating that a CCSP-expressing cell, such as a BASC or Clara cell contributed to alveolar epithelial repair (Rock et al., 2011; Tropea et al., 2012). In homeostasis or following hyperoxia injury, no increase in labeled alveolar cells was seen (Rawlins, Okubo, et al., 2009), suggesting that the CCSP-expressing cell contribution to alveolar repair may be injury-specific and may be dependent on specific microenvironmental factors. Ding et al. (2011) showed that unilateral pneumonectomy stimulates pulmonary capillary endothelial cells to produce angiocrine growth factors VEGFR2 and FGFR1, which are critical to initiate and sustain alveologenesis. This work supports the idea that signals from stromal cells are critical for lung stem cell functions. Together, these studies suggest that bronchiolar cells may be able to contribute to the alveolar lineage(s) under specific conditions and injury repair.

The studies outlined here provide evidence to support region-specific lung stem and progenitor cells in the murine lung; however, much less is known about stem cells in the human lung. Basal cells can be isolated from human lung using the markers NGFR and integrin $\alpha6$ (Rock et al., 2009) and human AT2 cells have been isolated for many years, most recently using the marker HTII-280 for FACS (Barkauskas et al., 2013). A putative population of c-kit positive human lung stem cells was reported to generate both epithelial and mesodermal lineages in culture (Kajstura et al., 2011). After mouse lung cryoinjury, injected human c-kit positive cells appear capable of contributing to airway, alveoli, and pulmonary vessel repair. These findings present a single multipotent lung stem cell, a concept not widely accepted in the current lung stem cell field. Further characterization and replication using vigorous techniques such as lineage tracing and functional assays are necessary. Recently, a new population of human E-Cad/Lgr6+ putative stem cells has been isolated, expanded, and shown to have self-renewal properties and the ability to differentiate *in vitro* and *in vivo* (Oeztuerk-Winder, Guinot, Ochalek, & Ventura, 2012). E-Cad/Lgr6+ single cell transplantations into the kidney capsule produce differentiated bronchioalveolar tissue and retain the ability to self-renew (Oeztuerk-

Winder et al., 2012). Human airway epithelial cell cultures have been developed as an important methodology to further study human lung progenitor cells (Fulcher, Gabriel, Burns, Yankaskas, & Randell, 2005). It is critical to continue to work toward linking the exciting mouse lung stem cell discoveries to the human lung and to begin to extrapolate findings to human disease.

3. THE OTHERS: LUNG MESENCHYMAL STROMAL CELLS AND LUNG ENDOTHELIAL PROGENITOR CELLS

Evidence continues to support the idea that adult mesenchymal stromal cells (MSCs) are an important element of epithelial stem/progenitor niches. Critical for regional specification of embryonic lung epithelium, MSCs at the distal tip of the branching epithelium are known to secrete FGF10, a critical component of the signaling network involving Bmp, Wnt, and sonic hedgehog pathways that is necessary for coordinating differentiation in the developing lung (Morrisey & Hogan, 2010). FGF10-positive mesenchymal cells have also been shown to give rise to smooth muscle cells (De Langhe, Carraro, Warburton, Hajihosseini, & Bellusci, 2006; Mailleux et al., 2005; Shan et al., 2008). MSCs have been shown to critically support the proliferation and differentiation of epithelial stem cells in coculture (McQualter et al., 2010). Supporting the importance of this microenvironmental relationship, recent *in vivo* work used the naphthalene injury model to show that parabronchial mesenchymal cells secrete FGF10 to support epithelial regeneration in the surviving progenitor cells (McQualter et al., 2010).

Several groups have identified resident lung mesenchymal stromal cells (LMSCs) that share some common characteristics with bone marrow-derived MSCs (Bentley et al., 2010; Hennrick et al., 2007; Hoffman et al., 2011; Martin et al., 2008; Summer, Fitzsimmons, Dwyer, Murphy, & Fine, 2007a). LMSCs fulfill the International Society of Cellular Therapy criteria for MSCs, including marker profile and differentiation abilities (Dominici et al., 2006), and are clonogenic when grown in Matrigel with bFGF or HGF (Hegab et al., 2011; Karoubi, Cortes-Dericks, Breyer, Schmid, & Dutly, 2009). Investigators have begun to dissect the role of LMSCs in injury showing that LMSC administration after elastase-induced emphysema resulted in decreased injury and increased survival, likely via a paracrine-mediated anti-inflammatory effect (Hoffman et al., 2011). Ongoing investigation is needed, including *in vivo* lineage tracing, to better characterize these cells in their *in vivo* state to test their potential role in homeostasis and injury.

An important cross talk is thought to exist between endothelial cells and epithelial cell progenitors. Angiogenic and angiocrine factor signaling are important in distal lung organogenesis and lung regeneration (Ding et al., 2011; Yamamoto et al., 2007). Coculture of human vascular endothelial cells with a human bronchiolar epithelial cell line in 3D culture has been shown to promote the generation of bronchioalveolar branching structures. This and other ongoing studies suggest a supportive role for endothelium in promoting branching of airway epithelium (Yamamoto et al., 2007). The role of the lung endothelial progenitor cell (EPC) in lung regeneration has begun to be examined, but it has proved difficult because of the inability to discriminate circulating EPCs from bone marrow and tissue resident EPCs. All three types of EPCs may contribute to endothelial cell regeneration and repair of lung injuries such as hyperoxia-induced injury that affect epithelial and endothelial cells (Balasubramaniam et al., 2010; Duong, Erzurum, & Asosingh, 2011). Ongoing work to better classify the *in vivo* roles of tissue resident EPCs is necessary to better understand their distinct role in development, homeostasis, and injury repair.

4. DIRECTING DIFFERENTIATION: EMBRYONIC STEM CELLS AND INDUCED PLURIPOTENT STEM CELLS

Pluripotent embryonic stem cells (ES cells) and induced pluripotent stem cells (iPS cells) hold great promise for regeneration of injured tissue and repair of disease states. ES cells are isolated from the inner cell mass of preimplantation blastocysts and under well-defined culture conditions, they can be maintained indefinitely in an undifferentiated state with the ability to give rise to cells of all three embryonic germ layers (Odorico, Kaufman, & Thomson, 2001). Takahashi and Yamanaka discovered that the transcription factors Klf4, Sox2, Oct4, and c-Myc, when introduced into mouse fibroblasts through retroviral transduction, led to clones with pluripotent properties (Takahashi & Yamanaka, 2006) and soon after, reprogrammed human iPS cells were first generated (Takahashi et al., 2007; Yu et al., 2007). The possibility of patient-specific iPS cells offers great hope for generating genetically matched patient-specific lung progenitor cells, with the opportunity for patient-specific drug screening and modeling of human disease.

ES cells resemble the early embryo and investigators have worked toward directed differentiation to lung lineages using a developmental approach. Definitive endoderm progenitor cells of the developing foregut give rise to the differentiated tissue of the thyroid, lung, liver, and pancreas. All lung epithelia must progress through a primordial progenitor stage defined by the onset of expression of homeodomain-containing transcription factor, Nkx2.1 (also known as TTF-1, TITF1), and downregulation of Sox2 along the dorsal–ventral axis of the gut tube (Lazzaro, Price, de Felice, & Di Lauro, 1991; Minoo, Su, Drum, Bringas, & Kimura, 1999; Que, Luo, Schwartz, & Hogan, 2009). Later in development, Sox2 expression increases in the area of the future lung trachea, bronchus, and bronchioles with the Nkx2.1-expressing cells of the embryonic lung giving rise to mature airway epithelium (Que et al., 2009). Sox9, FoxP2, and ID2 are expressed in the distal embryonic lung and mark a multipotent embryonic lung progenitor population (Perl, Kist, Shan, Scherer, & Whitsett, 2005; Rawlins, Clark, Xue, & Hogan, 2009; Shu et al., 2007). Nkx2.1 knock-out mice have lung and thyroid agenesis and malformations of forebrain (Kimura et al., 1996; Minoo et al., 1999). Initial attempts to generate functional lung epithelium from ES cells generated mixed cell populations with the risk of teratoma formation after transplantation because of the remaining undifferentiated pluripotent stem cells (Van Haute et al., 2009). Green et al. (2011) found that dual inhibition of TGF-β and bone morphogenic protein (BMP) signaling after specification of definitive endoderm from pluripotent cells highly enriches anterior foregut endoderm for the next steps in differentiation. Recently, Longmire et al. (2012) and Mou et al. (2012) published important advances and generation of putative lung progenitors from ES cells. Definitive endoderm was derived from mouse ES cells and converted first into foregut endoderm, and then into Nkx2.1+ lung endoderm using precisely-timed BMP, FGF, and WNT signaling. Nkx2.1+ lung endoderm was then converted into multipotent embryonic lung progenitor and airway progenitor cells with the formation of tracheospheres when subcutaneously transplanted into nude mice (Mou et al., 2012). Similarly, Longmire et al. (2012) purified and directed differentiation of ES cells via initial inhibition of TGF-β and BMP signaling, and then via stimulation of BMP and FGF signaling to form definitive endodermal precursors able to recellularize a 3D lung scaffold.

iPS cells can be induced in culture to develop into definitive endoderm, but it has also proved difficult to completely differentiate iPS into lung progenitor cells with phenotypic markers of differentiated epithelium. Several groups have produced patient-specific iPS cells to date, and Mou et al.

(2012) have produced human disease-specific immature airway epithelium derived from CF iPS cells. The Rossant laboratory has shown directed differentiation of human pluripotent stem cells into mature airway epithelium that express functional cystic fibrosis transmembrane conductance regulator protein (Wong et al., 2012). Work to better understand the details of ES cell- and iPS cell-directed differentiation in the pulmonary field is ongoing, showing great progress and promise.

5. MSCs: POTENTIAL ROLE OF CELL-BASED THERAPY IN LUNG DISEASE

MSCs have been recently examined as a potential cell-based therapy for lung disease. MSCs secrete growth factors and antimicrobial peptides, have immunomodulatory properties, and exhibit low immunogenicity (Lee, Fang, Krasnodembskaya, Howard, & Matthay, 2011). MSCs can respond, migrate, and facilitate repair of damaged tissue making them an attractive candidate for both prevention and treatment of lung disease. MSCs can be isolated from a variety of human tissues including bone marrow, adipose tissue, and placenta. The International Society of Cellular Therapy defined MSCs in 2006 by three standard criteria: (1) They must adhere to plastic; (2) they must express cell surface markers CD73, CD90, and CD105 and must not express CD45, CD34, CD14, and CD11b; and (3) they must have the capacity to differentiate into osteoblasts, adipocytes, and chondroblasts under *in vitro* conditions (Dominici et al., 2006). Paracrine-modulating factors secreted by MSCs appear to be responsible for injury repair rather than MSC engraftment (Lee et al., 2011). Recent evidence suggests that in addition to releasing soluble anti-inflammatory factors, the MSCs transfer microvesicles containing mitochondria, protein, and microRNA to other cells (Islam et al., 2012; Lee et al., 2012). MSC-based therapy has been studied in several important and devastating lung diseases, with promising results outlined here.

BPD is a disease with high morbidity and mortality that affects premature infants. BPD leads to distal airway epithelial cell simplification and vascular injury, and most current treatments are palliative (Ghanta, Leeman, & Christou, 2013). Bone marrow-derived MSCs and umbilical cord-derived MSCs have been shown to ameliorate lung injury in hyperoxia mouse models of BPD (Aslam et al., 2009; van Haaften et al., 2009; Vosdoganes, Lim, Moss, & Wallace, 2012). Low levels of MSC cell engraftment and an even greater preventative effect with delivery of cell-free MSC-conditioned media support the importance of a paracrine effect. A possible mechanism of action is stimulation of endogenous lung stem/progenitor cells by MSC-secreted factors, triggering their role in epithelial cell repair (Tropea et al., 2012). The overall mechanisms or specific factors responsible for amelioration of hyperoxia-induced lung injury remain active areas of investigation. No clinical trials have been completed in the United States, but at the time of writing, an open-label, single-center, phase 1 clinical study is underway to evaluate the safety and efficacy of umbilical cord-derived MSC treatment in premature infants with BPD. Ongoing preclinical work is necessary to better understand the mechanisms and to build upon current findings.

The therapeutic potential of MSCs has been tested and holds promise in studies of several other lung diseases such as pulmonary hypertension (PH), acute lung injury (ALI), and pulmonary fibrosis. PH is a progressive disease with ongoing endothelial dysfunction and vascular remodeling. In addition to EPCs as potential therapeutic cells for PH, the efficacy of MSC administration has been demonstrated in murine models of PH (Babar et al., 2007; Lee et al., 2012). MSCs genetically engineered to overexpress endothelial nitric oxide synthase, prostacyclin, or heme oxygenase 1 had even greater reversal effects on PH (Kanki-Horimoto et al., 2006; Liang et al., 2011). Several studies in animal models of ALI and also in a model of ALI in explanted human lungs have shown the beneficial effects of MSC administration, with increased survival, less lung damage, and decreased proinflammatory and increased anti-inflammatory cytokines (Gupta et al., 2007; Lee, Fang, Gupta, Serikov, & Matthay, 2009; Mei et al., 2007; Xu et al., 2007). IPF is a progressive disease currently without any effective therapy. MSC delivery to the bleomycin injury IPF mouse model has shown low levels of MSC engraftment, but significant improvement in lung injury (Aguilar et al., 2009; Kumamoto, Nishiwaki, Matsuo, Kimura, & Matsushima, 2009; Ortiz et al., 2003). Ongoing work continues to explore the possible benefits of treatment with MSCs for these lung diseases with high morbidity and mortality.

MSCs have been utilized in several clinical trials for immune- and inflammatory-related diseases and although efficacy has varied, no safety issues have arisen (Giordano, Galderisi, & Marino, 2007; Uccelli, Moretta, & Pistoia, 2008). Potential concerns of tumor or ectopic tissue formation have not been reported. A recent placebo-controlled, randomized, multicenter trial examined treatment with systemic MSCs for COPD and showed neither any adverse effects nor any differences in pulmonary function tests or quality of life indicators (Weiss, Casaburi, Flannery, Leroux-Williams, & Tashkin, 2013). This clinical trial showed important safety data, and it provides a basis for further clinical trials to test the efficacy of MSC treatment in respiratory diseases. Extensive work continues to examine the role of treatment with MSCs in lung disease in both mouse models of respiratory disease and human clinical trials.

6. STEM CELLS IN LUNG CANCER

Lung cancer is the leading cause of cancer related deaths in the United States and worldwide (Ettinger et al., 2013; Herbst, Heymach, & Lippman, 2008). In the United States alone, nearly 200,000 people will be diagnosed with lung cancer in 2013, and the predicted 5-year survival rate for all patients is a dismal 15% (Ettinger et al., 2013). Evidence suggests that lung cancer may originate from neoplastic lung stem/progenitor cell populations and that certain lung tumors contain cancer stem cells (CSCs). Through mouse models and molecular characterization of lung tumors, research aims to understand the molecular changes that lead to lung cancer development and to find novel ways to target lung cancer and lung CSCs.

Lung cancer is subdivided into two major groups: small cell and nonsmall cell lung cancers (Travis, Brambilla, & Riely, 2013). Small cell lung cancer (SCLC) accounts for ~15% of lung malignancies and is characterized by neuroendocrine cell morphology and gene expression. Nonsmall cell lung cancers (NSCLC) account for the remaining ~85% of lung malignances. Within the NSCLC group, at least three distinct histological subtypes exist: adenocarcinoma, squamous cell carcinoma, and large cell carcinoma. Adenocarcinoma is the predominant subtype, accounting for approximately 50% of NSCLCs. Adenocarcinomas have histology and biomarker expression consistent with distal lung, including expression of CK14, surfactant proteins, and amplification of the transcription factor Nkx2.1 (Herbst et al., 2008; Imielinski et al., 2012; Travis et al., 2013). Squamous cell carcinoma is the next most prevalent subtype, with ~40% of NSCLC patients receiving this diagnosis. Squamous cell carcinomas in the lung are characterized by squamous differentiation, which is most reminiscent of normal basal and secretory cell subtypes found in the more proximal lung and trachea (Herbst et al., 2008; TCGA, 2012; Travis et al., 2013; Wilkerson et al., 2010). Squamous cell carcinomas often express p63, CK5, and the transcription factor Sox2. Lastly, large cell carcinoma accounts for approximately 10% of NSCLCs. Some tumors have areas of histology that represent two or all three subtypes and, often, if the subtypes are commixed, the pathologist will term the lesion "adeno-squamous" (Travis et al., 2013). While smoking is associated with all types of lung tumors, it is most highly correlated with small cell and squamous cell carcinomas (Govindan et al., 2012; Sun, Schiller, & Gazdar, 2007).

Even within clonal tumor populations, it is believed that not all cells are equal in their ability to initiate or propagate disease. Both tumor recurrence following treatment and establishment of tumors at metastatic sites have been attributed to growth and survival of CSCs. Generally stated, the CSC hypothesis predicts that tumors contain a population of cells that are able to self-renew and differentiate, thus giving rise to the phenotypically distinct cells found within the tumor population (Nguyen, Vanner, Dirks, & Eaves, 2012). This hypothesis, coupled with emerging evidence that CSCs are relatively resistant to standard chemotherapies, has led to the theory that identification of CSCs in each type of tumor would enable development of treatments to specifically target the cells responsible for recurrent and metastatic disease (Li, Tiede, Massague, & Kang, 2007). Whether different oncogenic drivers, different cells of origin, or a combination of the two contribute to the various subtypes of lung tumors is unclear. The idea that stem and progenitor cells are tumor "cells of origin" is attractive because stem cells possess inherent self-renewal capacity and do not need to reawaken self-renewal programs during neoplastic transformation (Visvader, 2011). However, an equally plausible explanation for how tumor cells with the capacity to self-renew arise is that a more differentiated cell acquires the capacity for self-renewal through genetic and/or epigenetic mechanisms. Currently in debate is whether lung tumors are more likely to arise from stem or progenitor populations, and if CSCs, when present, maintain phenotypic resemblance to the cell of origin or the normal stem/progenitor cells of the lung.

CSCs are defined functionally by their ability to seed tumors in xenograft or allograft settings. Studies of human lung CSCs are hindered by genetic complexity and limited sample availability of fresh patient tissue. In efforts to expand viable cells for study, adherent lung cancer cell lines have been established from over 200 primary lung cancers (Gazdar, Girard, Lockwood, Lam, & Minna, 2010). The lines appear to recapitulate many features of the primary tumors from which they were derived, including amplification of genes such as Nkx2.1 and Sox2 (Bass et al., 2009; Tanaka et al., 2007; Wistuba et al., 1999). Many studies using human cell lines, or cell line-derived xenografts, have allowed for testing novel therapeutics, and tumor initiation and progression. Drug resistance was modeled efficiently in lung cancer cell lines and found to be mediated through selection of cells within the population that showed altered chromatin state and signaling properties (Sharma et al., 2010). In addition, cell lines have been utilized for several high-throughput synthetic lethal screens to identify novel druggable targets in defined genetic backgrounds (Barbie et al., 2009). However, the utility of cell lines for CSC assays remains in question; therefore, using primary human lung cancer samples remains the goal for many researchers. Patient-derived xenografts (PDXs), where a fresh or cryopreserved piece of resected lung tumor is embedded onto the side of an immunocompromised mouse, help to expand the numbers of cells available from a single tumor for analysis. One PDX study showed that cell surface marker CD166 and the gene *Lin28B* mark cells enriched for CSC activity in NSCLC; further characterization of this cell population is pending (Zhang et al., 2012). Other studies suggest that CD133 could enrich for human lung CSC activity, though reports on the utility of this marker are conflicting, (Bertolini et al., 2009; Eramo et al., 2008; Meng, Li, Wang, Wang, & Ma, 2009). Aldehyde dehydrogenase activity has also been shown to enrich for lung cancer cells with increased tumorigenic potential (Jiang et al., 2009; Sullivan et al., 2010).

The ability to produce transgenic mice has transformed the cancer research field by allowing for precise control of the genetic drivers of model tumors. The first genetically engineered mouse model (GEMM) of lung cancer, in which oncogenic Kras-G12D can be specifically induced in the murine lung, was described in 2001 (Jackson et al., 2001; Johnson et al., 2001). This model allows for activation of oncogenic Kras by Cre recombinase-mediated excision of the Lox-Stop-Lox cassette preceding the Kras oncogene. Lung specificity is obtained through administration of Cre-expressing viral vector directly to the cells of the airways through inhalation. The Kras-G12D mouse has been bred to many conditional knock-out allele backgrounds, including the tumor suppressor p53, yielding Kras-G12D/p53-null tumors (Jackson et al., 2005). These Kras-G12D/p53-null tumors are more advanced in stage than tumors from the Kras-G12D mice. In similar conditional systems, lung-specific promoters such as CCSP can be used to drive expression of transgenes in a doxycycline-inducible manner. This system has been particularly useful for expression of specific point mutants of epidermal growth factor receptor (EGFR) that were identified in patient samples (Ji et al., 2006; Politi et al., 2006).

GEMMs of lung cancer have allowed for numerous studies that would not have been possible to complete with patient samples. They have been used successfully for preclinical or coclinical trials of targeted therapies for lung cancer such as inhibition of EGFR (Ji et al., 2006; Zhou et al., 2009). GEMMs have also proved useful for studying metastatic spread of the disease (Carretero et al., 2010; Ji et al., 2007; Winslow et al., 2011). Winslow et al. demonstrated in the Kras-G12D/p53-null mice that primary tumors that downregulated Nkx2.1 expression were the tumors that seeded metastases in distant organs. Lastly, GEMMs have allowed for a comprehensive study of drug resistance mechanisms (Oliver et al., 2010). However, to date, GEMMs of lung cancer appear to predominantly represent the adenocarcinoma subtype of the disease. One exception is the Kras-G12D/LKB1-null mouse model, which appears to drive tumors of adenocarcinoma, squamous cell carcinoma, and large cell phenotypes (Ji et al., 2007). Recently, a transgenic mouse that expresses a kinase-dead IKKα transgene in all tissues was shown to develop spontaneous lung squamous cell carcinomas (Xiao et al., 2013). However, no lung-specific tumor suppressor loss or oncogene expression has been able to produce solely squamous cell carcinoma in the mouse.

Experimental data from GEMMs suggest that stem and progenitor cells are primary targets for transformation in the lung and that distinct lung cancer subtypes arise from different lung progenitor populations. The mouse model consisting of lung-specific inactivation of both p53 and Rb (p53-null/Rb-null) leads to highly penetrant SCLC (Park et al., 2011; Sutherland et al., 2011). Because SCLC has neuroendocrine features, it was long proposed that these tumors arise from neuroendocrine bodies within the lung, which are thought to contain their own stem cell population. In support of this cell of origin theory, in the p53-null/Rb-null model infected with Adeno-Cre, the majority of early lesions were composed of proliferating neuroendocrine cells (Park et al., 2011). Likewise, when a variety of cell type-specific Cre viruses were used, it was found that only a virus able to infect neuroendocrine or SPC-expressing cells efficiently drove small cell tumor formation (Sutherland et al., 2011). Furthermore, deletion of Rb and p53 by a Cre driven by the endogenous neuroendocrine gene CGRP was sufficient for SCLC development (Song et al., 2012). For models of adenocarcinoma, the SPC and CCSP-expressing BASCs were shown to be the first cells to proliferate in response to oncogenic Kras activation (Kim et al., 2005). However, lineage tracing studies also suggest that AT2 cells can be the cells of origin for Kras-expressing adenocarcionomas (Xu et al., 2012). For squamous cell carcinoma, It has been proposed that the basal cells of the trachea are the cells of origin (Lu et al., 2010), though the lack of GEMM to recapitulate this disease subtype prevents direct testing of this hypothesis. Expression of Sox2 directed by an inducible CCSP-driven Cre recombinase was able to induce lung tumors that expressed the marker p63, but histologically, these tumors were adenocarcinomas (Lu et al., 2010). This intriguing finding suggests that the CCSP-Cre allele targets a subset of cells, possibly variant Clara cells and BASCs, that are unable to produce a fully squamous phenotype despite expression of a squamous cell carcinoma transcription factor. If a transgenic mouse strain is designed that can specifically target basal cells, it may be possible to drive squamous cell carcinoma formation. Furthermore, derivation of BASC- or variant Clara cell-specific Cre mice will help resolve questions regarding cells of origin for adenocarcinomas.

GEMMs have also allowed the systematic study of CSC phenotypes by allowing for production of virtually unlimited genetically homogenous primary tumor samples. The gold standard for testing CSC potential involves injecting sorted, or otherwise enriched and depleted, tumor cell populations into immunocompromised or syngenic mouse recipients. In order to allow the transplanted tumor cells to engraft into the lung, they are often injected into the tail vein of the mouse, or delivered to the lung by intratracheal (IT) instillation. Using a serial IT transplantation assay, it was demonstrated that the BASC marker Sca-1 enriched for the tumor-propagating potential of the Kras-G12D/p53-null tumors (Fig. 8.2) (Curtis et al., 2010). However, in lung tumor models that rely only upon Kras or EGFR activation, this same sorting strategy did not show consistent CSC enrichment. This research exemplifies the importance of oncogenotype in determining the phenotype of CSC populations. It appears that genotype has a direct influence on whether CSCs retain cell-of-origin marker expression as the cancer evolves. These findings also exemplify the importance of functionally validating the tumor-propagating capacity of sorted populations in each new tumor type under study. Further experiments are needed to understand the importance of Sca-1 or other lung stem cell markers in defining tumor-propagating or metastatic tumor cell populations in each GEMM of lung cancer.

7. FUTURE DIRECTIONS

Identification and characterization of the various stem and progenitor cell populations in the lung will allow for the study of the mechanisms

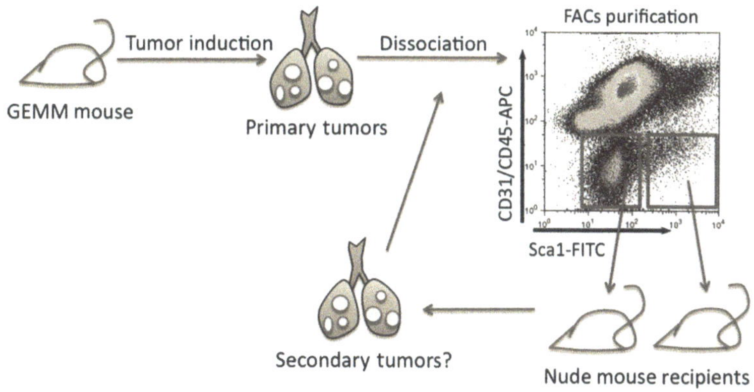

Figure 8.2 Orthotopic serial transplant assay for lung cancer stem cells. Using genetically engineered mouse models (GEMMs) of lung cancer, tumors with defined genetic characteristics can be induced. These primary tumors can be resected, enzymatically dissociated, and sorted, based on cell surface marker expression. Fluorescence-activated cell sorting (FACS) is typically used to separate tumor cell types. Antibodies against CD31 and CD45 can be used to exclude endothelial cells and hematopoietic cells, respectively, while the remaining tumor cell population can be fractionated using putative cancer stem cell markers such as Sca-1 (*x*-axis of representative FACS plot shown). FACS-isolated populations can then be introduced into the lungs of immunocompromised mice, such as nude mice, and cancer stem cell activity can be measured by secondary tumor growth. As cancer stem cells are thought to have self-renewal ability, a true cancer stem cell pool should be able to propagate tumors in a serial manner. (See color plate.)

through which these cells are maintained and stimulated. Because the lung is an essential organ, understanding these mechanisms and exploiting them therapeutically for lung regenerative medicine holds great promise for improving public health. Research will likely focus now on identification of unique cell surface markers that can be used to enrich for populations by FACS, as well as both *in vitro* and *in vivo* assays to fully examine the self-renewal and differentiation potential of each population. Extrapolation of findings in the mouse lung to the human, as well as the design of new mouse strains for lineage-specific gene control will be essential as this field matures. In parallel, the molecular characterization of lung stem cell markers in lung tumors will allow for testing of the CSC hypothesis in the various subtypes and genetic backgrounds of lung tumors. These advances will pave the way toward more personalized lung cancer treatment regimens and a deeper understanding of all lung disease.

ACKNOWLEDGMENTS

We thank members of our laboratory for their discussions and critical comments on the chapter. Work in our laboratory is supported by the Post-Doctoral Fellowship, PF-12-151-01-DMC, from the American Cancer Society (CMF), the Ikaria Advancing Newborn Medicine Grant (KTL), American Medical Association Foundation SEED Grant (KTL), 5 T32HD7466-15 (KTL), RO1 HL090136, U01 HL100402 RFA-HL-09-004, American Cancer Society Research Scholar Grant RSG-08-082-01-MGO, the V Foundation for Cancer Research, a Basil O'Conner March of Dimes Starter Award, and the Harvard Stem Cell Institute (CFK).

REFERENCES

Adamson, I., & Bowden, D. (1974). The type 2 cell as progenitor of alveolar epithelial regeneration. A cytodynamic study in mice after exposure to oxygen. *Laboratory Investigation, 30*, 35–42.

Aguilar, S., Scotton, C., McNutty, K., Nye, E., Stamp, G., Laurent, G., et al. (2009). Bone marrow stem cells expressing keratinocyte growth factor via an inducible lentivirus protects against bleomycin-induced pulmonary fibrosis. *PLoS One, 4*(11), e8013.

Aslam, M., Baveja, R., Liang, O., Fernandez-Gonzalez, A., Lee, C., Mitsialis, S., et al. (2009). Bone marrow stromal cells attenuate lung injury in a murine model of neonatal chronic lung disease. *American Journal of Respiratory and Critical Care Medicine, 180*(11), 1122–1130.

Aso, Y., Yoneda, K., & Kikkawa, Y. (1976). Morphologic and biochemical study of pulmonary changes induced by bleomycin in mice. *Laboratory Investigation, 35*, 558–568.

Baber, S., Deng, W., Master, R., Bunnell, B., Taylor, B., Murthy, S., et al. (2007). Intratracheal mesenchymal stem cell administration attenuates monocrotaline-induced pulmonary hypertension and endothelial dysfunction. *American Journal of Physiology. Heart and Circulatory Physiology, 292*(2), H1120–H1128.

Balasubramaniam, V., Ryan, S., Seedorf, G., Roth, E., Heumann, T., Yoder, M., et al. (2010). Bone marrow-derived angiogenic cells restore lung alveolar and vascular structure after neonatal hyperoxia in infant mice. *American Journal of Physiology. Lung Cellular and Molecular Physiology, 298*(3), L315–L323.

Barbie, D. A., Tamayo, P., Boehm, J. S., Kim, S. Y., Moody, S. E., Dunn, I. F., et al. (2009). Systematic RNA interference reveals that oncogenic KRAS-driven cancers require TBK1. *Nature, 462*(7269), 108–112.

Barkauskas, C., Cronce, M., Rackley, C., Bowie, E., Keene, D., Stripp, B., et al. (2013). Type 2 alveolar cells are stem cells in adult lung. *The Journal of Clinical Investigation, 123*(7), 3025–3036.

Bass, A. J., Watanabe, H., Mermel, C. H., Yu, S., Perner, S., Verhaak, R. G., et al. (2009). SOX2 is an amplified lineage-survival oncogene in lung and esophageal squamous cell carcinomas. *Nature Genetics, 41*(11), 1238–1242.

Bentley, J., Popova, A., Bozyk, P., Linn, M., Baek, A., Lei, J., et al. (2010). Ovalbumin sensitization and challenge increases the number of lung cells possessing a mesenchymal stromal cell phenotype. *Respiratory Research, 11*, 127.

Bertolini, G., Roz, L., Perego, P., Tortoreto, M., Fontanella, E., Gatti, L., et al. (2009). Highly tumorigenic lung cancer CD133+ cells display stem-like features and are spared by cisplatin treatment. *Proceedings of the National Academy of Sciences, 106*(38), 16281–16286.

Bertoncello, I., & McQualter, J. (2013). Lung stem cells: Do they exist? *Respirology, 18*(4), 587–595.

Borthwick, D., Shahbazian, M., Krantz, Q., Dorin, J., & Randell, S. (2001). Evidence for stem-cell niches in the tracheal epithelium. *American Journal of Respiratory Cell and Molecular Biology, 24*, 662–670.

Carretero, J., Shimamura, T., Rikova, K., Jackson, A. L., Wilkerson, M. D., Borgman, C. L., et al. (2010). Integrative Genomic and Proteomic Analyses Identify Targets for Lkb1-Deficient Metastatic Lung Tumors. *Cancer Cell, 17*(6), 547–559.

Chapman, H., Li, X., Alexander, J., Brumwell, A., Lorizio, W., Tan, K., et al. (2011). Integrin α6β4 identifies an adult distal lung epithelial population with regenerative potential in mice. *The Journal of Clinical Investigation, 121*(7), 2855–2862.

Clevers, H. (2013). Stem cells: A unifying theory for the crypt. *Nature, 495*(7439), 53–54.

Curtis, S. J., Sinkevicius, K. W., Li, D., Lau, A. N., Roach, R. R., Zamponi, R., et al. (2010). Primary tumor genotype is an important determinant in identification of lung cancer propagating cells. *Cell Stem Cell, 7*(1), 127–133.

De Langhe, S., Carraro, G., Warburton, D., Hajihosseini, M., & Bellusci, S. (2006). Levels of mesenchymal FGFR2 signaling modulate smooth muscle progenitor cell commitment in the lung. *Developmental Biology, 299*(1), 52–62.

Ding, B., Nolan, D., Guo, P., Babazadeh, A., Cao, Z., Rosenwaks, Z., et al. (2011). Endothelial-derived angiocrine signals induce and sustain regenerative lung alveolarization. *Cell, 147*(3), 539–553.

Dominici, M., Le Blanc, K., Mueller, I., Slaper-Cortenbach, I., Marini, F., Krause, D., et al. (2006). Minimal criteria for defining multipotent mesenchymal stromal cells. The International Society for Cellular Therapy position statement. *Cytotherapy, 8*(4), 315–317.

Duong, H., Erzurum, S., & Asosingh, K. (2011). Pro-angiogenic hematopoietic progenitor cells and endothelial colony-forming cells in pathological angiogenesis of bronchial and pulmonary circulation. *Angiogenesis, 14*, 411–422.

Eramo, A., Lotti, F., Sette, G., Pilozzi, E., Biffoni, M., Di Virgilio, A., et al. (2008). Identification and expansion of the tumorigenic lung cancer stem cell population. *Cell Death Differ, 15*(3), 504–514.

Ettinger, D. S., Akerley, W., Borghaei, H., Chang, A. C., Cheney, R. T., Chirieac, L. R., et al. (2013). Non-small cell lung cancer, version 2.2013. *Journal of the National Comprehensive Cancer Network, 11*(6), 645–653.

Evans, M., Cabral, L., Stephens, R., & Freeman, G. (1975). Transformation of alveolar type 2 cells to type 1 cells following exposure to NO2. *Experimental and Molecular Pathology, 22*(1), 142–150.

Evans, M., Cabral-Anderson, L., & Freeman, G. (1978). Role of the Clara cell in renewal of the bronchiolar epithelium. *Laboratory Investigation, 38*(6), 648–653.

Evans, M., Johnson, L., Stephens, R., & Freeman, G. (1976). Renewal of the terminal bronchiolar epithelium in the rat following exposure to NO2 or O3. *Laboratory Investigation, 35*(3), 246–257.

Evans, M., Shami, S., Cabral-Anderson, L., & Dekker, N. (1986). Role of nonciliated cells in renewal of the bronchial epithelium of rats exposed to NO2. *The American Journal of Pathology, 123*(1), 126–133.

Fulcher, M., Gabriel, S., Burns, K., Yankaskas, J., & Randell, S. (2005). Well-differentiated human airway epithelial cell cultures. *Methods in Molecular Medicine, 107*, 183–206.

Gazdar, A. F., Girard, L., Lockwood, W. W., Lam, W. L., & Minna, J. D. (2010). Lung cancer cell lines as tools for biomedical discovery and research. *Journal of the National Cancer Institute, 102*(17), 1310–1321.

Ghanta, S., Leeman, K., & Christou, H. (2013). An update on pharmacologic approaches to bronchopulmonary dysplasia. *Seminars in Perinatology, 37*(2), 115–123.

Giangreco, A., Reynolds, S., & Stripp, B. (2002). Terminal bronchioles harbor a unique airway stem cell population that localizes to the bronchioalveolar duct junction. *The American Journal of Pathology, 161*(1), 173–182.

Giordano, A., Galderisi, U., & Marino, I. (2007). From the laboratory bench to the patient's bedside: An update on clinical trials with mesenchymal stem cells. *Journal of Cellular Physiology*, *211*, 27–35.

Green, M., Chen, A., Nostro, M., d'Souza, S., Lemischka, I., Gouon-Evans, V., et al. (2011). Generation of anterior foregut endoderm from human embryonic and induced pluripotent stem cells. *Nature Biotechnology*, *29*(3), 267–272.

Govindan, R., Ding, L., Griffith, M., Subramanian, J., Dees, N. D., Kanchi, K. L., et al. (2012). Genomic Landscape of Non-Small Cell Lung Cancer in Smokers and Never-Smokers. *Cell*, *150*(6), 1121–1134.

Gupta, N., Su, X., Popov, B., Lee, J., Serikov, V., & Matthay, M. (2007). Intrapulmonary delivery of bone marrow-derived mesenchymal stem cells improves survival and attenuates endotoxin-induced acute lung injury in mice. *The Journal of Immunology*, *179*, 1855–1863.

Hegab, A., Ha, V., Gilbert, J., Zhang, K., Malkoski, S., Chon, A., et al. (2011). Novel stem/progenitor cell population from murine tracheal submucosal gland ducts with multipotent regenerative potential. *Stem Cells*, *29*(8), 1283–1293.

Hegab, A., Nickerson, D., Ha, V., Darmawan, D., & Gomperts, B. (2012). Repair and regeneration of tracheal surface epithelium and submucosal glands in a mouse model of hypoxic-ischemic injury. *Respirology*, *17*(7), 1101–1113.

Hennrick, K., Keeton, A., Nanua, S., Kijek, T., Goldsmith, A., Sajjan, U., et al. (2007). Lung cells from neonates show a mesenchymal stem cell phenotype. *American Journal of Respiratory and Critical Care Medicine*, *175*(11), 1158–1164.

Herbst, R. S., Heymach, J. V., & Lippman, S. M. (2008). Lung cancer. *New England Journal of Medicine*, *359*(13), 1367–1380.

Hoffman, A., Paxson, J., Mazan, M., Davis, A., Tyagi, S., Murthy, S., et al. (2011). Lung-derived mesenchymal stromal cell post-transplantation survival, persistence, paracrine expression, and repair of elastase-injured lung. *Stem Cells and Development*, *20*(10), 1779–1792.

Hong, K., Reynolds, S., Giangreco, A., Hurley, C., & Stripp, B. (2001). Clara cell secretory protein-expressing cells of the airway neuroepithelial body microenvironment include a label-retaining subset and are critical for epithelial renewal after progenitor cell depletion. *American Journal of Respiratory Cell and Molecular Biology*, *24*(6), 671–681.

Imielinski, M., Berger, A. H., Hammerman, P. S., Hernandez, B., Pugh, T. J., Hodis, E., et al. (2012). Mapping the Hallmarks of Lung Adenocarcinoma with Massively Parallel Sequencing. *Cell*, *150*(6), 1107–1120.

Islam, M., Das, S., Emin, M., Wei, M., Sun, L., Westphalen, K., et al. (2012). Mitochondrial transfer from bone marrow derived stromal cells to pulmonary alveoli protects against acute lung injury. *Nature Medicine*, *18*, 759–765.

Jackson, E. L., Willis, N., Mercer, K., Bronson, R. T., Crowley, D., Montoya, R., et al. (2001). Analysis of lung tumor initiation and progression using conditional expression of oncogenic K-ras. *Genes & Development*, *15*(24), 3243–3248.

Jackson, E. L., Olive, K. P., Tuveson, D. A., Bronson, R., Crowley, D., Brown, M., et al. (2005). The Differential Effects of Mutant p53 Alleles on Advanced Murine Lung Cancer. *Cancer Research*, *65*(22), 10280–10288.

Ji, H., Li, D., Chen, L., Shimamura, T., Kobayashi, S., McNamara, K., et al. (2006). The impact of human EGFR kinase domain mutations on lung tumorigenesis and in vivo sensitivity to EGFR-targeted therapies. *Cancer Cell*, *9*(6), 485–495.

Ji, H., Ramsey, M. R., Hayes, D. N., Fan, C., McNamara, K., Kozlowski, P., et al. (2007). LKB1 modulates lung cancer differentiation and metastasis. *Nature*, *448*(7155), 807–810.

Jiang, F., Qiu, Q., Khanna, A., Todd, N. W., Deepak, J., Xing, L., et al. (2009). Aldehyde dehydrogenase 1 is a tumor stem cell-associated marker in lung cancer. *Molecular Cancer Research*, 7(3), 330–338.

Johnson, L., Mercer, K., Greenbaum, D., Bronson, R. T., Crowley, D., Tuveson, D. A., et al. (2001). Somatic activation of the K-ras oncogene causes early onset lung cancer in mice. *Nature, 410*(6832), 1111–1116.

Kajstura, J., Rota, M., Hall, S. R., Hosoda, T., D'Amario, D., Sanada, F., et al. (2011). Evidence for human lung stem cells. *The New England Journal of Medicine, 364*(19), 1795–1806.

Kanki-Horimoto, S., Horimoto, H., Mieno, S., Kishida, K., Watanabe, F., Furuya, E., et al. (2006). Implantation of mesenchymal stem cells overexpressing endothelial nitric oxide synthase improves right ventricular impairments caused by pulmonary hypertension. *Circulation, 114*, I181–I185.

Karoubi, G., Cortes-Dericks, L., Breyer, I., Schmid, R., & Dutly, A. (2009). Identification of mesenchymal stromal cells in human lung parenchyma capable of differentiating into aquaporin 5-expressing cells. *Laboratory Investigation, 89*, 1100–1114.

Kim, C., Jackson, E., Woolfenden, A., Lawrence, S., Babar, I., Vogel, S., et al. (2005). Identification of bronchioalveolar stem cells in normal lung and lung cancer. *Cell, 121*(6), 823–835.

Kimura, S., Hara, Y., Pineau, T., Fernandez-Salguero, P., Fox, C., Ward, J., et al. (1996). The T/ebp null mouse: Thyroid-specific enhancer-binding protein is essential for the organogenesis of the thyroid, lung, ventral forebrain, and pituitary. *Genes & Development, 10*, 60–69.

Kolios, G., & Moodley, Y. (2013). Introduction to stem cells and regenerative medicine. *Respiration, 85*, 3–10.

Kumamoto, M., Nishiwaki, T., Matsuo, N., Kimura, H., & Matsushima, K. (2009). Minimally cultured bone marrow mesenchymal stem cells ameliorate fibrotic lung injury. *The European Respiratory Journal, 34*, 740–748.

Kumar, P. A., Hu, Y., Yamamoto, Y., Hoe, N. B., Wei, T. S., Mu, D., et al. (2011). Distal airway stem cells yield alveoli in vitro and during lung regeneration following H1N1 influenza infection. *Cell, 147*(3), 525–538.

Lazzaro, D., Price, M., de Felice, M., & Di Lauro, R. (1991). The transcription factor TTF-1 is expressed at the onset of thyroid and lung morphogenesis and in restricted regions of the foetal brain. *Development, 113*(4), 1093–1104.

Lee, J., Fang, X., Gupta, N., Serikov, V., & Matthay, M. (2009). Allogenic human mesenchymal stem cells for treatment of E. coli endotoxin-induced acute lung injury in the ex vivo perfused human lung. *Proceedings of the National Academy of Sciences of the United States of America, 106*(38), 16357–16362.

Lee, J., Fang, X., Krasnodembskaya, A., Howard, J., & Matthay, M. (2011). Concise review: Mesenchymal stem cells for acute lung injury: Role of paracrine soluble factors. *Stem Cells, 29*(6), 913–919.

Lee, J. H., Kim, J., Gludish, D., Roach, R. R., Saunders, A. H., Barrios, J., et al. (2013). Surfactant protein C chromatin-bound green fluorescence protein reporter mice reveal heterogeneity of surfactant protein C expressing lung cells. *American Journal of Respiratory Cell and Molecular Biology, 48*(3), 288–298.

Lee, C., Mitsialis, S., Aslam, M., Vitali, S., Vergadi, E., Konstantinou, G., et al. (2012). Exosomes mediate the cytoprotective action of mesenchymal stromal cells on hypoxia-induced pulmonary hypertension. *Circulation, 126*(22), 2601–2611.

Li, F., Tiede, B., Massague, J., & Kang, Y. (2007). Beyond tumorigenesis: Cancer stem cells in metastasis. *Cell Research, 17*(1), 3–14.

Liang, O., Mitsialis, S., Chang, M., Vergadi, E., Lee, C., Aslam, M., et al. (2011). Mesenchymal stromal cells expressing heme oxygenase-1 reverse pulmonary hypertension. *Stem Cells, 29*(1), 99–107.

Longmire, T., Ikonomou, L., Hawkins, F., Christodoulou, C., Cao, Y., Jean, J., et al. (2012). Efficient derivation of purified lung and thyroid progenitors from embryonic stem cells. *Cell Stem Cell, 10*(4), 398–411.

Lu, Y., Futtner, C., Rock, J. R., Xu, X., Whitworth, W., Hogan, B. L. M., et al. (2010). Evidence that Sox2 overexpression is oncogenic in the lung. *PLoS One*, *5*(6), e11022.

Mailleux, A., Kelly, R., Veltmaat, J., De Langhe, S., Zaffran, S., Thiery, J., et al. (2005). Fgf10 expression identifies parabronchial smooth muscle cell progenitors and is required for their entry into the smooth muscle cell lineage. *Development*, *132*(9), 2157–2166.

Martin, J., Helm, K., Ruegg, P., Varella-Garcia, M., Burnham, E., & Majka, S. (2008). Adult lung side populations have mesenchymal stem cell potential. *Cytotherapy*, *10*, 140–151.

McQualter, J., Brouard, N., Williams, B., Baird, B., Sims-Lucas, S., Yuan, K., et al. (2009). Endogenous fibroblastic progenitor cells in the adult mouse lung are highly enriched in the sca-1 positive cell fraction. *Stem Cells*, *27*(3), 623–633.

McQualter, J., Yuen, K., Williams, B., & Bertoncello, I. (2010). Evidence of an epithelial stem/progenitor cell hierarchy in the adult mouse lung. *Proceedings of the National Academy of Sciences of the United States of America*, *107*(4), 1414–1419.

Mei, S., McCarter, S., Deng, Y., Parker, C., Liles, W., & Stewart, D. (2007). Prevention of LPS-induced acute lung injury in mice by mesenchymal stem cells overexpressing angiopoietin 1. *PLoS Medicine*, *4*, e269.

Meng, X., Li, M., Wang, X., Wang, Y., & Ma, D. (2009). Both CD133+ and CD133− subpopulations of A549 and H446 cells contain cancer-initiating cells. *Cancer Science*, *100*(6), 1040–1046.

Minoo, P., Su, G., Drum, H., Bringas, P., & Kimura, S. (1999). Defects in tracheoesophageal and lung morphogenesis in Nkx2.1 (-/-) mouse embryos. *Developmental Biology*, *209*(1), 60–71.

Morrisey, E., & Hogan, B. (2010). Preparing for the first breath: Genetic and cellular mechanisms in lung development. *Developmental Cell*, *18*, 8–23.

Mou, H., Zhao, R., Sherwood, R., Ahfeldt, T., Lapey, A., Wain, J., et al. (2012). Generation of multipotent lung and airway progenitors from mouse ESCs and patient-specific cystic fibrosis iPSCs. *Cell Stem Cell*, *10*(4), 385–397.

Nguyen, L. V., Vanner, R., Dirks, P., & Eaves, C. J. (2012). Cancer stem cells: An evolving concept. *Nature Reviews. Cancer*, *12*(2), 133–143.

Odorico, J., Kaufman, D., & Thomson, J. (2001). Multilineage differentiation from human embryonic stem cell lines. *Stem Cells*, *19*, 193–204.

Oeztuerk-Winder, F., Guinot, A., Ochalek, A., & Ventura, J. (2012). Regulation of human lung alveolar multipotent cells by a novel p38alpha MAPK/mir-17-92 axis. *The EMBO Journal*, *31*(16), 3431–3441.

Oliver, T. G., Mercer, K. L., Sayles, L. C., Burke, J. R., Mendus, D., Lovejoy, K. S., et al. (2010). Chronic cisplatin treatment promotes enhanced damage repair and tumor progression in a mouse model of lung cancer. *Genes & Development*, *24*(8), 837–852.

Ortiz, L., Gambelli, F., McBride, C., Gaupp, D., Baddoo, M., Kaminski, N., et al. (2003). Mesenchymal stem cell engraftment in lung is enhanced in response to bleomycin exposure and ameliorates its fibrotic effects. *Proceedings of the National Academy of Sciences of the United States of America*, *100*, 8407–8411.

Park, K.-S., Liang, M.-C., Raiser, D. M., Zamponi, R., Roach, R. R., Curtis, S. J., et al. (2011). Characterization of the cell of origin for small cell lung cancer. *Cell Cycle*, *10*(16), 2806–2815.

Perl, A., Kist, R., Shan, Z., Scherer, G., & Whitsett, J. (2005). Normal lung development and function after Sox9 inactivation in the respiratory epithelium. *Genesis*, *41*, 23–32.

Politi, K., Zakowski, M. F., Fan, P.-D., Schonfeld, E. A., Pao, W., & Varmus, H. E. (2006). Lung adenocarcinomas induced in mice by mutant EGF receptors found in human lung cancers respond to a tyrosine kinase inhibitor or to down-regulation of the receptors. *Genes & Development*, *20*(11), 1496–1510.

Potten, C., & Loeffler, M. (1990). Stem cells: Attributes, cycles, spirals, pitfalls and uncertainties. Lessons for and from the crypt. *Development*, *110*, 1001–1020.

Que, J., Luo, X., Schwartz, R., & Hogan, B. (2009). Multiple roles for Sox2 in the developing and adult mouse trachea. *Development, 136*(11), 1899–1907.

Raiser, D., Zacharek, S., Roach, R., Curtis, S., Sinkevicius, K., Gludish, D., et al. (2008). Stem cell biology in the lung and lung cancers: Using pulmonary context and classic approaches. *Cold Spring Harbor Symposia on Quantitative Biology, 73*, 479–490.

Rawlins, E., Clark, C., Xue, Y., & Hogan, B. (2009). The Id2+ distal tip lung epithelium contains individual multipotent embryonic progenitor cells. *Development, 136*(22), 3741–3745.

Rawlins, E., & Hogan, B. (2006). Epithelial stem cells of the lung: Privileged few or opportunities for many? *Development, 133*(13), 2455–2465.

Rawlins, E., & Hogan, B. (2008). Ciliated epithelial cell lifespan in the mouse trachea and lung. *American Journal of Physiology. Lung Cellular and Molecular Physiology, 295*, L231–L234.

Rawlins, E., Okubo, T., Xue, Y., Brass, D., Auten, R., Hasegawa, H., et al. (2009). The role of Scgb1a1+ Clara cells in the long-term maintenance and repair of lung airway, but not alveolar epithelium. *Cell Stem Cell, 4*(6), 525–534.

Reynolds, S., Giangreco, A., Power, J., & Stripp, B. (2000). Neuroepithelial bodies of pulmonary airways serve as a reservoir of progenitor cells capable of epithelial regeneration. *The American Journal of Pathology, 156*(1), 269–278.

Rock, J., Barkauskas, C., Cronce, M., Xue, Y., Harris, J., Liang, J., et al. (2011). Multiple stromal populations contribute to pulmonary fibrosis without evidence for epithelial to mesenchymal transition. *Proceedings of the National Academy of Sciences of the United States of America, 108*(52), E1475–E1483.

Rock, J., & Hogan, B. (2011). Epithelial progenitor cells in lung development, maintenance, repair, and disease. *Annual Review of Cell and Developmental Biology, 27*, 493–512.

Rock, J., Onaitis, M., Rawlins, E., Lu, Y., Clark, C., Xue, Y., et al. (2009). Basal cells as stem cells of the mouse trachea and human airway epithelium. *Proceedings of the National Academy of Sciences of the United States of America, 106*(31), 12771–12775.

Rock, J., Randell, S., & Hogan, B. (2010). Airway basal stem cells: A perspective on their roles in epithelial homeostasis and remodeling. *Disease Models & Mechanisms, 3*(9–10), 545–556.

Shan, L., Subramaniam, M., Emanuel, R., Degan, S., Johnston, P., Tefft, D., et al. (2008). Centrifugal migration of mesenchymal cells in the embryonic lung. *Developmental Dynamics, 237*(3), 750–757.

Sharma, S. V., Lee, D. Y., Li, B., Quinlan, M. P., Takahashi, F., Maheswaran, S., et al. (2010). A chromatin-mediated reversible drug-tolerant state in cancer cell subpopulations. *Cell, 141*(1), 69–80.

Shu, W., Lu, M., Zhang, Y., Tucker, P., Zhou, D., & Morrisey, E. (2007). Foxp2 and Foxp1 cooperatively regulate lung and esophagus development. *Development, 134*, 1991–2000.

Song, H., Yao, E., Lin, C., Gacayan, R., Chen, M.-H., & Chuang, P.-T. (2012). Functional characterization of pulmonary neuroendocrine cells in lung development, injury, and tumorigenesis. *Proceedings of the National Academy of Sciences of the United States of America, 109*(43), 17531–17536.

Sullivan, J. P., Spinola, M., Dodge, M., Raso, M. G., Behrens, C., Gao, B., et al. (2010). Aldehyde dehydrogenase activity selects for lung adenocarcinoma stem cells dependent on notch signaling. *Cancer Research, 70*(23), 9937–9948.

Summer, R., Fitzsimmons, K., Dwyer, D., Murphy, J., & Fine, A. (2007). Isolation of an adult mouse lung mesenchymal progenitor cell population. *American Journal of Respiratory Cell and Molecular Biology, 37*, 152–159.

Sun, S., Schiller, J. H., & Gazdar, A. F. (2007). Lung cancer in never smokers - a different disease. *Nat Rev Cancer*, 7(10), 778–790.

Sutherland, K. D., Proost, N., Brouns, I., Adriaensen, D., Song, J.-Y., & Berns, A. (2011). Cell of origin of small cell lung cancer: Inactivation of Trp53 and Rb1 in distinct cell types of adult mouse lung. *Cancer Cell*, *19*(6), 754–764.

Takahashi, K., Tanabe, K., Ohnuki, M., Narita, M., Ichisaka, T., Tomoda, K., et al. (2007). Induction of pluripotent stem cells from adult human fibroblasts by defined factors. *Cell*, *131*(5), 861–872.

Takahashi, K., & Yamanaka, S. (2006). Induction of pluripotent stem cells from mouse embryonic and adult fibroblast cultures by defined factors. *Cell*, *126*, 663–676.

Tanaka, H., Yanagisawa, K., Shinjo, K., Taguchi, A., Maeno, K., Tomida, S., et al. (2007). Lineage-specific dependency of lung adenocarcinomas on the lung development regulator TTF-1. *Cancer Research*, *67*(13), 6007–6011.

TCGA, (2012). Comprehensive genomic characterization of squamous cell lung cancers. *Nature*, *489*(7417), 519–525.

Teisanu, R., Lagasse, E., Whitesides, J., & Stripp, B. (2009). Prospective isolation of bronchiolar stem cells based upon immunophenotypic and autofluorescence characteristics. *Stem Cells*, *27*(3), 612–622.

Travis, W. D., Brambilla, E., & Riely, G. J. (2013). New pathologic classification of lung cancer: Relevance for clinical practice and clinical trials. *Journal of Clinical Oncology*, *31*(8), 992–1001.

Tropea, K., Leder, E., Aslam, M., Lau, A., Raiser, D., Lee, J., et al. (2012). Bronchioalveolar stem cells increase after mesenchymal stromal cell treatment in a mouse model of bronchopulmonary dysplasia. *American Journal of Physiology. Lung Cellular and Molecular Physiology*, *302*(9), L829–L837.

Uccelli, A., Moretta, L., & Pistoia, V. (2008). Mesenchymal stem cells in health and disease. *Nature Reviews. Immunology*, *8*, 726–736.

van Haaften, T., Byrne, R., Bonnet, S., Rochefort, G., Akabutu, J., Bouchentouf, M., et al. (2009). Airway delivery of mesenchymal stem cells prevents arrested alveolar growth in neonatal lung injury in rats. *American Journal of Respiratory and Critical Care Medicine*, *180*(11), 1131–1142.

Van Haute, L., De Block, G., Liebaers, I., Sermon, K., & DeRycke, M. (2009). Generation of lung epithelial-like tissue from human embryonic stem cells. *Respiratory Research*, *5*(10), 105.

Van Lommel, A. (2001). Pulmonary neuroendocrine cells (PNEC) and neuroepithelial bodies (NEB): Chemoreceptors and regulators of lung development. *Paediatric Respiratory Reviews*, *2*, 171–176.

Visvader, J. E. (2011). Cells of origin in cancer. *Nature*, *469*(7330), 314–322.

Vosdoganes, P., Lim, R., Moss, T., & Wallace, E. (2012). Cell therapy: A novel treatment approach for bronchopulmonary dysplasia. *Pediatrics*, *130*, 727–737.

Weiss, D., Casaburi, R., Flannery, R., Leroux-Williams, M., & Tashkin, D. (2013). A placebo-controlled, randomized trial of mesenchymal stem cells in COPD. *Chest*, *143*(6), 1590–1598.

Wilkerson, M. D., Yin, X., Hoadley, K. A., Liu, Y., Hayward, M. C., Cabanski, C. R., et al. (2010). Lung squamous cell carcinoma mRNA expression subtypes are reproducible, clinically important, and correspond to normal cell types. *Clinical Cancer Research*, *16*(19), 4864–4875.

Winslow, M. M., Dayton, T. L., Verhaak, R. G. W., Kim-Kiselak, C., Snyder, E. L., Feldser, D. M., et al. (2011). Suppression of lung adenocarcinoma progression by Nkx2-1. *Nature*, *473*(7345), 101–104.

Wistuba, I. I., Bryant, D., Behrens, C., Milchgrub, S., Virmani, A. K., Ashfaq, R., et al. (1999). Comparison of features of human lung cancer cell lines and their corresponding tumors. *Clinical Cancer Research*, *5*(5), 991–1000.

Wong, A., Bear, C., Chin, S., Pasceri, P., Thompson, T., Huan, L., et al. (2012). Directed differentiation of human pluripotent stem cells into mature airway epithelia expressing functional CFTR protein. *Nature Biotechnology*, *9*, 876–882.

Xiao, Z., Jiang, Q., Willette-Brown, J., Xi, S., Zhu, F., Burkett, S., et al. (2013). The pivotal role of IKKα in the development of spontaneous lung squamous cell carcinomas. *Cancer Cell*, *23*(4), 527–540.

Xu, X., Rock, J. R., Lu, Y., Futtner, C., Schwab, B., Guinney, J., et al. (2012). Evidence for type II cells as cells of origin of K-Ras-induced distal lung adenocarcinoma. *Proceedings of the National Academy of Sciences of the United States of America*, *109*(13), 4910–4915.

Xu, J., Woods, C., Mora, A., Joodi, R., Brigham, K., Iyer, S., et al. (2007). Prevention of endotoxin-induced systemic response by bone marrow derived mesenchymal stem cells in mice. *American Journal of Physiology. Lung Cellular and Molecular Physiology*, *293*(1), L131–L141.

Yamamoto, H., Yun, E., Gerber, H., Ferrara, N., Whitsett, J., & Vu, T. (2007). Epithelial-vascular cross talk mediated by VEGF-A and HGF signaling directs primary septae formation during distal lung morphogenesis. *Developmental Biology*, *308*(1), 44–53.

Yu, J., Vodyanik, M., Smuga-Otto, K., Antosiewicz-Bourget, J., Frane, J., Tian, S., et al. (2007). Induced pluripotent stem cell lines derived from human somatic cells. *Science*, *318*(5858), 1917–1920.

Zacharek, S., Fillmore, C., Lau, A., Gludish, D., Chou, A., Ho, J., et al. (2011). Lung stem cell self-renewal relies on BMI1-dependent control of expression at imprinted loci. *Cell Stem Cell*, *9*(3), 272–281.

Zhang, W. C., Shyh-Chang, N., Yang, H., Rai, A., Umashankar, S., Ma, S., et al. (2012). Glycine decarboxylase activity drives non-small cell lung cancer tumor-initiating cells and tumorigenesis. *Cell*, *148*(1–2), 259–272.

Zhou, W., Ercan, D., Chen, L., Yun, C.-H., Li, D., Capelletti, M., et al. (2009). Novel mutant-selective EGFR kinase inhibitors against EGFR T790M. *Nature*, *462*(7276), 1070–1074.

CHAPTER NINE

Spermatogonial Stem Cell Functions in Physiological and Pathological Conditions

Qi-En Yang, Jon M. Oatley[1]

Center for Reproductive Biology, School of Molecular Biosciences, College of Veterinary Medicine, Washington State University, Pullman, Washington, USA

[1]Corresponding author: e-mail address: joatley@vetmed.wsu.edu

Contents

Abstract

Sperm have a vital role in the continuity of a species by contributing genetic information to the next generation. Production of these specialized gametes in numbers sufficient to confer normal fertility occurs via cycling of the spermatogenic lineage, a process referred to as spermatogenesis. Continuity relies on the activities of a self-renewing reservoir of spermatogonial stem cells (SSCs) from which progenitors will arise that transiently amplify in number before committing to a pathway of terminal differentiation. A primary population of SSCs is established during neonatal development from a pool of quiescent gonocyte precursors that forms in embryogenesis. Disruption of this process has dire consequences on maintenance of a cycling spermatogenic lineage in

Current Topics in Developmental Biology, Volume 107
ISSN 0070-2153
http://dx.doi.org/10.1016/B978-0-12-416022-4.00009-3

adulthood. At present, the molecular mechanisms underlying initial formation of the SSC pool are largely undefined. However, several transcription factors and posttranscriptional regulators have been identified as important regulators of SSC self-renewal from studies with mutant mouse models and experimental manipulation within primary cultures of mouse SSCs. Importantly, loss of function of these self-renewal factors may be underlying causes of infertility. Furthermore, disruption in the establishment of the SSC state within gonocytes or misregulation of self-renewal may manifest as testicular germ cell tumors in postnatal life.

1. INTRODUCTION

Germ cells play a vital role in the continuity of a species by transmitting genetic and epigenetic information to the next generation. Mature gametes are the product of a developmental pathway originating with an undifferentiated diploid state and ending with adoption of a specialized haploid state. In males, sperm are the end product of a process referred to as spermatogenesis which occurs within the seminiferous tubules of testes. Continual spermatogenesis is the cornerstone of male fertility and relies on the actions of a robust stem cell population. In the developed world, $\sim$5% of reproductive age men are infertile and there has been a rise in the incidence of testicular cancer over the past several decades (Bettocchi, Coker, Deacon, Parkinson, & Pryor, 1994; Hotaling & Walsh, 2009). A major cause of male infertility is insufficient number of spermatozoa in the ejaculate and most testicular cancers are of germ cell origin. Importantly, a vast majority of these pathologies are diagnosed as idiopathic and could be linked to aberrant regulation of male germline stem cell functions.

In general, stem cells are defined by a unique ability for self-renewal to sustain a foundational pool from which progenitors arise that are committed to a pathway of terminal differentiation (Jaenisch & Young, 2008; Oatley & Brinster, 2008, 2012). These actions support tissue homeostasis and provide regenerative capacity. For the male germline, stem cells are defined by the ability to support continual spermatogenesis during steady-state conditions and regenerate the spermatogenic lineage following cytotoxic insult or transplantation (Brinster & Zimmermann, 1994; de Ruiter-Bootsma, Kramer, & de Rooij, 1977). In mammals, it is well established that stem cell capacity resides within a subset of the spermatogonial population, thus prompting the label spermatogonial stem cell (SSC). Because SSCs are the ground state of spermatogenesis, disruption in their function has dire consequences on fertility. Impaired ability for self-renewal leads to complete loss of the germline and disruption in the ability of progenitors to arise from the SSC pool causes a block in spermatogenesis. Both deficiencies manifest as a condition referred to azoospermia in which spermatozoa are absent in the ejaculate. In addition to effects on fertility, misregulation of self-renewal and spermatogenic lineage commitment within SSCs could potentially result in the formation of testicular germ cell tumors (TGCTs).The majority of information about regulation of SSC functions has been gained from studies with mice. Importantly, several mutant mouse models have been generated over the past decade that emulate azoospermia and TGCT formation caused by aberrant SSC functions. In this review, we discuss the basic understanding of mechanisms regulating self-renewal of mouse SSCs and attempt to make connections between impairment of these and human testicular pathologies of azoospermia and TGCT formation.

2. SPERMATOGENESIS AND SSC BIOLOGY

2.1. Overview of spermatogenesis

In mammals, spermatogenesis occurs within seminiferous tubules that are bound by a basement membrane and lined along the outside by a layer of peritubular myoid cells which provide rigidity (Fig. 9.1A). Between tubules is interstitial tissue that is comprised of Leydig cells, blood vessels, and several immune cell types that contribute to the regulation of spermatogenesis via secretion of paracrine signaling factors (Fawcett, Neaves, & Flores, 1973). Within tubules is the seminiferous epithelium that is comprised of an array of germ cell types at different stages of maturation and a single somatic cell population termed Sertoli cells (Setchell, Voglmayr, & Waites, 1969). Because of their direct contact with germ cells, Sertoli cells have been regarded as "nurse cells" and are believed to play a central role in supporting all stages of germ cell development (Griswold, 1998). Testicular germ cells can be classified as three general populations: spermatogonia, spermatocyte, or spermatid (Fig. 9.1B). The spermatogonial population can be further distinguished as undifferentiated and differentiating subtypes (Fig. 9.1B).

For most, if not all, mammalian species the undifferentiated spermatogonial population consists of SSC and progenitor subtypes. Spermatogenesis initiates with the transition of progenitors from an undifferentiated to a differentiating state. The SSC population serves as a self-renewing reservoir from which a new cohort of progenitors will arise. In testes of rodents, the undifferentiated spermatogonial population consists of type A_{single} (A_s), A_{paired} (A_{pr}), and $A_{aligned}$ (A_{al}) subtypes. Classically, A_s spermatogonia

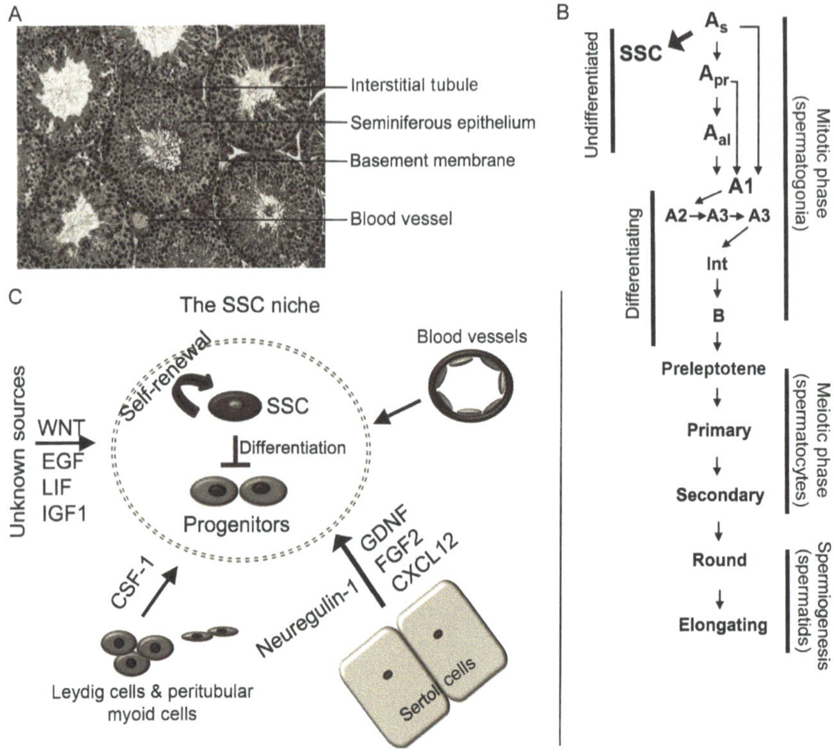

Figure 9.1 Spermatogenesis and the spermatogonial stem cell niche in mouse testes. (A) Representative image of a hematoxylin and eosin (H&E) stained cross-sections of seminiferous tubules from the testis of an adult mouse. Seminiferous tubules are bound by a basement membrane and lined along the outside by a layer of peritubular myoid cells which provide rigidity. Between tubules is interstitial tissue that is comprised of Leydig cells, blood vessels, and several immune cell types. The seminiferous epithelium consists of an array of germ cell types at different stages of maturation and a single somatic cell population termed Sertoli cells. (B) Schematic of the spermatogenic lineage. The undifferentiated spermatogonial population consists of type A_{single} (A_s), A_{paired} (A_{pr}), and $A_{aligned}$ (A_{al}) subtypes. The A_s spermatogonial population represents a self-renewing reservoir of SSCs from which A_{pr} arise that transiently amplify in number to produce A_{al} progenitors. At periodic intervals, the majority of A_{pr} and A_{al} spermatogonia transition to a differentiating state without a cell division becoming referred to as A1 spermatogonia. Another series of mitotic divisions yields A2, A3, A4, intermediate, and type B spermatogonia that initiate meiosis as preleptotene spermatocytes. Two meiotic divisions ensue with germ cells progressing as primary and then secondary spermatocytes. The end product of the spermatocyte stage is generation of haploid round spermatids that transform into elongated spermatids in a process referred to as spermiogenesis. (C) Current understanding of the SSC niche in mammalian testes. Sertoli cells serve as the major contributor via secreting GDNF, CXCL12, FGF2, and Neuregulin1 to control SSC activities. Leydig cells contribute via secretion of CSF-1 to promote SSC self-renewal. In addition, the vasculature is postulated to play a role but factors emanating from this source that influence SSC activities have not been described. (See color plate.)

have been considered the SSC population (Fig. 9.1B). Upon division, these cells can separate producing two A_s which represents a self-renewal event or remain connected by an intracellular bridge due to incomplete cytokinesis thereby becoming A_{pr}. The newly formed A_{pr} spermatogonia then undergo a series of mitotic divisions also with incomplete cytokinesis to form interconnected cohorts of 4, 8, 16, and sometimes 32 cells that constitute the A_{al} spermatogonia (Fig. 9.1B). At periodic intervals and under influence of retinoic acid signaling, the majority of A_{pr} and A_{al} spermatogonia transition to a differentiating state without a cell division becoming referred to as A1 spermatogonia (De Rooij & Lok, 1987). Another series of mitotic divisions yields A2, A3, A4, intermediate, and type B spermatogonia. Also during this period of development, the differentiating spermatogonia gain competence for meiotic initiation which occurs at the B spermatogonial stage resulting in formation of preleptotene spermatocytes. Two meiotic divisions ensue with germ cells progressing as primary and then secondary spermatocytes. The end product of the spermatocyte stage is generation of haploid round spermatids that transform into elongated spermatids in a process referred to as spermiogenesis. Finally, mature elongate spermatids are released as spermatozoa from the seminiferous epithelium into the lumen of tubules. Spermatogenesis occurs over a period of 35 days in mice (Franca, Ogawa, Avarbock, Brinster, & Russell, 1998), 52 days in rats (Franca et al., 1998), 61 days in cattle (Curtis & Amann, 1981), and 75 days in primates (Weathersbee, 1980).

In primates, the undifferentiated spermatogonial population includes A_{dark} and A_{pale} spermatogonial subtypes (Schulze, 1979). The mitotic activity of A_{dark} spermatogonia is reduced compared to A_{pale} spermatogonia, thus the populations are considered to be proliferating and quiescent subtypes, respectively. It has been proposed that A_{dark} spermatogonia represent the SSC pool; whereas, A_{pale} spermatogonia constitute the progenitor population (Ehmcke & Schlatt, 2006). Indeed, A_{pale} spermatogonia are known to give rise to differentiating type B spermatogonia, similar to A_{pr} and A_{al} spermatogonial subtypes in rodents. The type B spermatogonia then undergo a series of mitotic divisions before initiating meiotic prophase. At this point, germ cell maturation progresses in a similar manner to that in rodents.

2.2. Defining characteristics of SSCs

The foundation for continual spermatogenesis is provided by actions of SSCs. Self-renewal promotes maintenance of a foundational pool from which progenitors arise that transiently amplify in number before committing to a pathway of terminal differentiation (Oatley & Brinster, 2008). Similar to other tissue-specific stem cells, these activities confer regenerative capacity to reestablish spermatogenesis following cytotoxic insult that eliminates most germ cells or transplantation into seminiferous tubules lacking an endogenous spermatogenic lineage. Also, during steady-state conditions transition of progenitors to the differentiation pathway occurs at periodic intervals and the SSC pool serves as a reservoir from which the next cohort will arise. Thus, the defining characteristics of SSCs can be described as an ability to sustain and regenerate a cycling spermatogenic lineage.

Studies from the 1950s revealed that spermatogonia are the undifferentiated diploid cell population from which spermatogenesis originates (Clermont & Leblond, 1953). Later, results of histological analyses and experiments investigating the response to irradiation suggested existence of a regenerative stem cell pool within the spermatogonial population (Clermont, 1966; Clermont & Leblond, 1959). However, it was not until 1994 that seminal studies by Brinster and colleagues provided unequivocal demonstration of an SSC population within mouse testes based on identification of cells capable of regenerating spermatogenesis following transplantation into testes of sterile recipient mice (Brinster & Zimmermann, 1994). This fundamental development instigated a new generation of studies that significantly advanced the knowledge of molecular characteristics of SSCs and regulatory machinery influencing their functions.

In most cell lineages, the actual stem cell pool is rare in number compared to progenitor and differentiating populations and this is also the case in the male germline. Testes of adult mice contain ~330,000 undifferentiated spermatogonia with ~35,000 of these cells being A_s spermatogonia (Tegelenbosch & de Rooij, 1993). Interestingly, transplantation analyses revealed that testes of adult mice contain ~3000 cells with regenerative capacity (Nagano, 2003). Collectively, these findings suggest that SSCs constitute only 0.01% of total testis cells and 12% of the A_s spermatogonial population (de Rooij & Russell, 2000; Nagano & Yeh, 2013).Thus, A_s spermatogonia are a functionally heterogeneous population consisting of a small fraction of SSCs and the majority of cells are progenitors that lack regenerative capacity. Indeed, several studies have observed that some A_s spermatogonia convert directly to A1 spermatogonia without clonal expansion through A_{pr} and A_{al} stages (de Rooij & Russell, 2000; Nakagawa, Sharma, Nabeshima, Braun, & Yoshida, 2010). Furthermore, fragmentation of A_{al} spermatogonial cohorts to produce new A_s spermatogonia during regeneration of the spermatogenic lineage following cytotoxic insult has been observed in testes of adult mice (Nakagawa et al., 2010; van Beek, Davids, van de Kant, & de Rooij, 1984). However, whether these newly formed A_s spermatogonia function as SSCs or simply bolster the progenitor population that will transition to a differentiating state upon exposure to retinoic acid has not been determined. Regardless, the actual SSC pool in mammalian testes is a rare subset of the spermatogonial population.

Functional heterogeneity of the undifferentiated spermatogonial population including the A_s spermatogonial pool presents major challenges for identifying and studying SSCs specifically when analyses based on morphological features alone are utilized. At present, the only unequivocal means to assess SSC function is by transplantation into seminiferous tubules of recipient testes (Brinster & Zimmermann, 1994). Following microinjection into the lumen of seminiferous tubules, SSCs migrate to the basement membrane and the population expands via self-renewal before progenitors begin to arise and then initiate differentiation to reestablish spermatogenesis (Nagano, 2003). While SSCs are capable of reestablishing colonies of continual spermatogenesis, progenitor spermatogonia generate a single round of spermatogenesis and the lineage is summarily lost (Nagano, 2003; Nakagawa, Nabeshima, & Yoshida, 2007). Each colony of continual donor-derived spermatogenesis within recipient testes following transplantation is clonally derived from a single SSC (Nagano, Avarbock, & Brinster, 1999). Thus, quantification of colonies after a time frame covering multiple rounds of differentiation provides a direct measure of SSC number within the transplanted cell population. In addition, examination of colony length provides an assessment of the capacity for proliferative expansion (Dobrinski, Ogawa, Avarbock, & Brinster, 1999; Nagano et al., 1999). Thus, the defining characteristic of SSCs is the capacity to regenerate spermatogenesis and transplantation methodology provides a direct means of evaluation.

3. ORIGIN OF SSC POOL

In testes of mammals, the SSC pool is not present at birth and must become established during postnatal life from a precursor population referred to as gonocytes (McLean, Friel, Johnston, & Griswold, 2003; Orwig, Ryu, Avarbock, & Brinster, 2002). The gonocyte population arises from postmigratory primordial germ cells (PGCs) around embryonic day (ED) 15 (Western, Miles, van den Bergen, Burton, & Sinclair, 2008; Yoshioka, McCarrey, & Yamazaki, 2009). At this point, the cells enter quiescence and remain in this state until shortly after birth when cell cycle progression resumes and transition to the postnatal spermatogonial lineage occurs (Moreno, Dutrillaux, & Coffigny, 2001; van Dissel-Emiliani, de Boer-Brouwer, Spek, van der Donk, & de Rooij, 1993). Thus, establishment of the foundation SSC pool can be considered to begin with formation of the germ cell lineage during embryonic development and end at a specific time point in early postnatal life.

3.1. The embryonic period

In the mouse, germ cell specification occurs around ED 7.2 when PGCs arise from pluripotent epiblast and become distinguished from the surrounding somatic cells by strong alkaline phosphatase (AP) activity (Ginsburg, Snow, & McLaren, 1990). Shortly after formation, PGCs migrate from the posterior primitive streak to the genital ridge around ED 11.5 (Anderson, Copeland, Scholer, Heasman, & Wylie, 2000). This process is guided by factors secreted from neighboring somatic cells including KIT ligand (KITL) and chemokine C-X-C motif ligand 12 (CXCL12). PGCs express c-KIT and C-X-C chemokine receptor type 4 (CXCR4) which are transmembrane receptors for KITL and CXCL12, respectively. Activation of c-KIT signaling regulates survival and general motility of PGCs but does not serve as a chemoattractant (Godin et al., 1991). Activation of CXCR4 signaling provides a directional cue that guides PGC migration to the genital ridge (Ara et al., 2003; Molyneaux et al., 2003). In addition to secreted factors, adhesion molecules also influence PGC migration including integrin β1 and E-cadherin. In embryos deficient for integrin β1, migration of PGCs is limited but their formation and survival is not impaired (Anderson et al., 1999). The role of E-cadherin appears to be more extensive as deficiency disrupts both PGC specification and postmigratory survival (Di Carlo & De Felici, 2000; Okamura, Kimura, Nakano, & Matsui, 2003).

Coincident with PGC specification in a subset of epiblast cells is the loss of pluripotency. In mice, pluripotent stem cells can be isolated from inner cell mass of preimplantation blastocysts or epiblast of peri- or postimplantation embryos, cell populations that are referred to as embryonic stem cells or epiblast stem cells (EpiSCs), respectively (Leitch & Smith, 2013). Epiblast cells isolated from ED 6.5 embryos and PGCs isolated from ED 8.5 or 12.5 embryos are able to generate colonies of spermatogenesis following transplantation into seminiferous tubules of adult recipient mice (Chuma et al., 2005). But, teratomas arise from some transplanted epiblast cells coincident with others that produce spermatogenic colonies (Chuma et al., 2005). These observations demonstrate further a heterogeneous nature of the epiblast population in which a subset of cells adopts a germ cell fate while others remain pluripotent. Collectively, these findings indicate that adoption of a germ cell fate results in loss of pluripotent potential *in vivo*. However, some PGCs from ED 7 and ED 8.5 embryos regain pluripotent properties after isolation and maintenance *in vitro* in the presence of the cytokines fibroblast growth factor 2 (FGF2), leukemia inhibitory factor, and KITL (Matsui, Zsebo, & Hogan, 1992; Resnick, Bixler, Cheng, & Donovan, 1992). Although the efficiency is low, pluripotent cells can also be derived from testis cells isolated from neonatal and adult animals following an extended period of *in vitro* maintenance (Kanatsu-Shinohara et al., 2004; Ko et al., 2010). These finding should be interpreted with caution as the existence of cells with pluripotent potential within testes has not been demonstrated and adaptation to culture conditions following long-term maintenance is known to alter genomic stability that could influence spontaneous reprogramming (Kimura & Nakano, 2011; Leitch & Smith, 2013).

After reaching the developing gonad, signals from somatic cells instruct PGCs to adopt a male germ cell fate becoming gonocytes or commit to a female germ cell lineage initiating meiosis. At ED 13.5, female germ cells arrest at the diplotene stage of meiosis I. However, male germ cells follow a different path by remaining in mitosis until ED 16 at which point all have entered a period of quiescence and remain in this state until after birth (Ginsburg et al., 1990; Vergouwen, Jacobs, Huiskamp, Davids, & de Rooij, 1991; Western et al., 2008). The molecular control of this mitotic arrest remains to be defined. However, it is clear that dynamic changes in the expression of genes involved in the G1–S phase transition occur. In particular, retinoblastoma protein (RB) is hyperphosphorylated in proliferative PGCs and prospermatogonia until ED 12.5 and is then rapidly dephosphorylated in arrested cells that begin to form around ED 13.5 (Spiller, Wilhelm, & Koopman, 2010; Western et al., 2008). In addition, expression of several cell cycle inhibitors including p27, p21, and p57 is upregulated in arrested prospermatogonia at ED 14.5 but undetectable at ED 12.5 when germ cells are mitotically active (Western et al., 2008).

3.2. The postnatal period

At birth, gonocytes reside within the center of seminiferous cords and are in a state of mitotic arrest. Formation of the SSC pool requires that these cells migrate to the basement membrane and resume proliferative activity. Gonocytes that do not migrate undergo apoptosis and are cleared from the seminiferous epithelium (Roosen-Runge & Leik, 1968). In testes of mice, a primary undifferentiated spermatogonial population arises from a subset of prospermatogonia during postnatal days (PD) 3 and 8 (Drumond, Meistrich, & Chiarini-Garcia, 2011; Vergouwen et al., 1991). Importantly, this process includes formation of both a primary progenitor spermatogonial population and specification of a foundational SSC pool. These events are required to sustain the undifferentiated spermatogonia population throughout the reproductive lifespan of a male. Interestingly, another subset of gonocytes transition directly to a differentiating A1 spermatogonial state and these cells are the starting point for a first round of spermatogenesis in postnatal life (de Rooij, 1998; Kluin & de Rooij, 1981; Yoshida et al., 2006). All subsequent rounds of spermatogenesis in adulthood derive from undifferentiated progenitor spermatogonia that arose from the foundational SSC pool. It is important to note that this heterogeneity of gonocyte behavior has been observed in testes of mice but not other mammalian species. Regardless, two possible models can be devised for formation of the primary undifferentiated and differentiating spermatogonial populations (Fig. 9.2). The first being a subset of gonocytes transition to SSCs from which progenitors then arise and another subset give rise to differentiating spermatogonia. The second possibility is SSCs, progenitor spermatogonia, and differentiating spermatogonia derived from three separate subsets of gonocytes. At present, experimental evidence that favors a particular model is limited. However, results of recent studies demonstrated that both progenitor spermatogonia and differentiating spermatogonia arise from gonocyte precursors lacking expression of the tumor suppressor RB but the populations are not renewed after several rounds of spermatogenesis (Hu, de Rooij, & Page, 2013). These findings demonstrate impaired formation of an SSC pool and suggest that progenitor spermatogonia can arise from gonocytes without an SSC intermediate phase.

It is generally accepted that an SSC state in spermatogonia is attained after migration and resumption of cell cycle progression occurs in gonocytes during early postnatal development. However, indirect evidence suggests that SSCs may be predetermined in the embryonic gonad. Results of studies from Kluin and de Rooij revealed similarities in the morphological features and cell cycle kinetics of a subset of gonocytes at ED 13–15 and undifferentiated spermatogonia in testes of neonatal mice (Kluin & de Rooij, 1981). In that study, the authors hypothesized that subpopulations

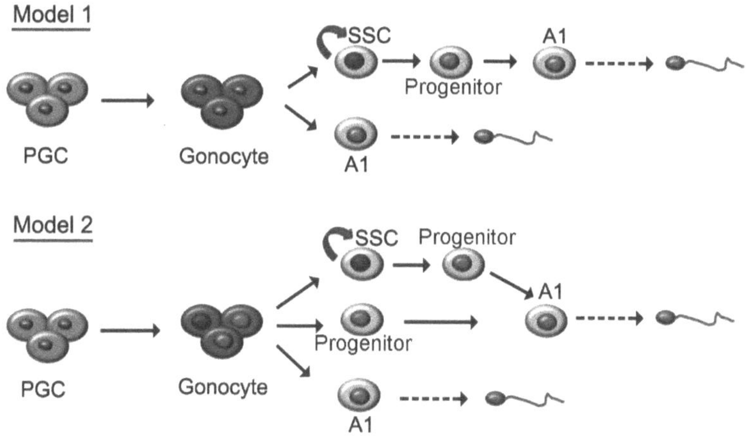

Figure 9.2 Two proposed models for the formation of the spermatogonial lineage during neonatal development. Model 1 dictates that a subset of prospermatogonia transition to SSCs from which progenitors then arise and another subset gives rise to differentiating spermatogonia directly. Model 2 proposes that the foundational SSC pool, initial progenitor spermatogonial population, and initial differentiating spermatogonial population derive individually from different subsets of gonocytes that are predetermined during embryogenesis. (See color plate.)

of migrating PGCs are primed for immediate differentiation by becoming A1 spermatogonia or transitioning to an undifferentiated spermatogonial state during neonatal development. This notion is supported by results of a recent lineage tracing study that demonstrated gonocytes form interconnected chains between ED 14.5 and 17.5 and then a subset undergo fragmentation before birth to produce shorter chains and single cells (Lei & Spradling, 2013). Clearly, the formation of different spermatogonial populations from gonocyte precursors during early postnatal life is required for establishment of a continually cycling spermatogenic lineage. However, the underlying regulatory mechanisms are largely undefined and will likely be a major area of emphasis in the field of male germ cell biology over the next decade.

4. MOLECULAR MECHANISMS REGULATING MAINTENANCE OF THE SSC POOL

After formation during early postnatal development, the foundational SSC pool is maintained via self-renewal activity. Whether the entire SSC pool is constantly in a proliferative phase similar to the stem cell population that maintains the intestinal epithelium (Barker et al., 2007) or proliferates rarely like stem cells in the bone marrow (Cheshier, Morrison, Liao, & Weissman, 1999) and hair follicle (Hsu, Pasolli, & Fuchs, 2011) is currently unknown. The progenitor spermatogonial population appears to be in a state of constant proliferation during steady-state spermatogenesis (Klein, Nakagawa, Ichikawa, Yoshida, & Simons, 2010), but proliferation kinetics of the SSC pool specifically are undefined. The relative rarity of SSC numbers within the spermatogenic lineage suggests the population is slow cycling similar to stem cell populations in most other developmental lineages.

Similar to other tissue-specific stem cells, the self-renewal of SSCs is influenced by signaling from extrinsic factors of a cognate niche environment that activate or suppress intracellular molecular mechanisms. The niche factors are contributions of somatic support cells that affect intracellular molecular pathways to guide stem cell activities. In addition, intrinsic mechanisms not influenced by extrinsic factors are known to govern the stem cell state. While evidence has started to emerge that factors secreted from interstitial cells influence SSC functions (Oatley, Oatley, Avarbock, Tobias, & Brinster, 2009), Sertoli cells are the key somatic cell population that dictate SSC niches (Oatley & Brinster, 2012).

Impairment of SSC maintenance due to deficiencies of the niche microenvironment or molecular mechanisms within SSCs manifests as loss of the spermatogenic lineage producing a condition referred to as Sertoli-cell-only (Fig. 9.3). In addition, the undifferentiated spermatogonial population of rodents can be maintained as a primary culture consisting of SSCs and progenitor spermatogonia. This methodology serves as an experimental platform to explore the influence of both extrinsic and intracellular regulators of SSC maintenance. When utilized in conjunction with transplantation analyses, it is possible to determine the impact of experimental manipulation of the microenvironment and molecular mechanisms within the cells on maintenance of the SSC pool within these cultures. The current knowledge base of mechanisms regulating SSC functions has been obtained largely from studies examining testes of mutant mice for the formation of seminiferous tubules with a Sertoli-cell-only phenotype and alteration of SSC content within primary cultures of undifferentiated spermatogonia (Fig. 9.3). In this review, we focus on transcription factors and posttranscriptional regulators within SSCs that influence self-renewal and survival. Furthermore, we consider only those for which there is direct experimental evidence that disrupted function leads to hallmarks of impaired maintenance of the SSC

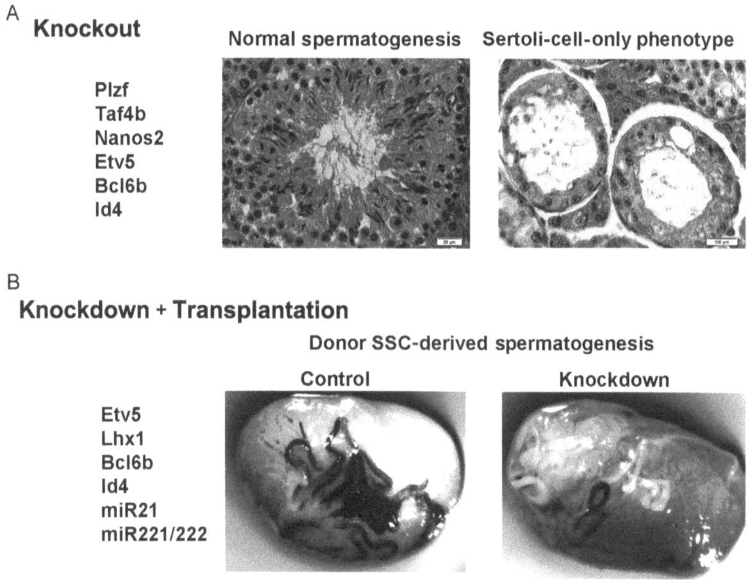

Figure 9.3 Transcription factors and posttranscriptional regulators that have been identified as important regulators of SSC self-renewal using functional criteria. (A) Transcription factors determined to influence maintenance of the SSC pool based on loss-of-function studies with mutant mice in which the germline is lost after several rounds of normal spermatogenesis resulting in formation of seminiferous tubules with a Sertoli-cell-only phenotype. Testes of mice harboring inactivation of Plzf, Taf4b, Nanos2, Bcl6b, Etv5, and Id4 contain seminiferous tubules with a Sertoli-cell-only phenotype and the percentage increases with advancing age. (B) Transcription factors and miRNAs determined to influence maintenance of the SSC pool based on knockdown of expression in primary cultures of undifferentiated spermatogonia followed by functional transplantation to determine alteration in SSC numbers. (See color plate.)

pool including formation of seminiferous tubules with a Sertoli-cell-only phenotype or loss of cells with regenerative potential in primary cultures of spermatogonia following transplantation (Fig. 9.3). For an up-to-date understanding of the niche, we encourage readers to refer to Fig. 9.1C and a recent review of the subject (Oatley & Brinster, 2012).

4.1. Influence of transcription factors

The fate decisions of most, if not all, stem cell populations are influenced greatly by transcription factors. Thus, there has been intense investigation of those that control SSC functions, most notably the capacity for self-renewal. The first niche factor identified to be obligatory for SSC maintenance was glial cell line-derived neurotropic factor (GDNF). Results of studies by Meng et al. (2000) revealed that male mice heterozygous for a null Gdnf allele possess testes with seminiferous tubules that are devoid of germ cells, resembling a Sertoli-cell-only phenotype. Moreover, supplementation of media with recombinant GDNF is required for the maintenance of the SSC pool within primary cultures of undifferentiated spermatogonia (Kubota, Avarbock, & Brinster, 2004a, 2004b). Based on this knowledge, several studies focused on identifying genes that are activated or suppressed by GDNF signaling within spermatogonia to identify key regulators of self-renewal. Furthermore, a number of genes not influenced by GDNF signaling have been identified that appear to play important roles in SSC maintenance. While an influence of GDNF signaling on expression of these genes has not been observed, regulation by other niche factors is possible and awaits discovery.

Using microarray-based gene expression profiling, Oatley, Avarbock, Telaranta, Fearon, and Brinster (2006) examined transcriptome changes within primary cultures of mouse undifferentiated spermatogonia in response to GDNF withdrawal and replacement (Oatley et al., 2006). Those experiments revealed that expression of the transcription factors Bcl6b, Etv5, and Lhx1 is induced by GDNF signaling. In addition, transplantation analyses revealed that maintenance of the SSC pool was impaired within primary cultures of undifferentiated spermatogonia following transient knockdown of their abundance using siRNA methodology. Furthermore, testes of Bcl6b and Etv5 null mice were found to contain seminiferous tubules with a Sertoli-cell-only phenotype thereby confirming an important role in regulating SSC functions (Oatley et al., 2006; Wu, Goodyear, Tobias, Avarbock, & Brinster, 2011). Interestingly, studies by Wu et al. (2011) revealed that Bcl6b gene expression is reduced in cultures of undifferentiated spermatogonia following transient reduction of ETV5 protein abundance. This finding suggests existence of a feed-forward regulatory loop for GDNF-regulated transcription factors within SSCs that influences self-renewal.

Another important GDNF-regulated factor is the transcription repressor inhibitor of DNA binding 4 (ID4). Expression of ID4 is localized to a subset of A_s spermatogonia in mouse testes and male mice homozygous for an Id4 null allele undergo aging-related increase in the percentage of seminiferous tubules with a Sertoli-cell-only phenotype (Oatley, Kaucher, Racicot, & Oatley, 2011). Moreover, transient reduction of ID4 abundance in primary cultures of undifferentiated spermatogonia using siRNA leads to impaired expansion of SSC numbers without affecting the progenitor cell population (Oatley et al., 2011). These findings suggest that ID4 is a specific marker of SSCs in the mouse germline and influences self-renewal to sustain the foundation pool.

In addition to the role of GDNF-regulated genes, several transcription factors not influenced by the niche are also indispensable for SSC maintenance. One of these is PLZF, which is known to regulate many developmental processes during embryonic and postnatal life. Mice that are homozygous for a Plzf null allele are viable but possess abnormities in limb patterning (Barna, Hawe, Niswander, & Pandolfi, 2000; Doulatov et al., 2009) and are infertile due to progressive loss of the germline upon aging (Buaas et al., 2004; Costoya et al., 2004). The root cause of this germline phenotype has been attributed to impaired self-renewal of SSCs because seminiferous tubules with a Sertoli-cell-only phenotype begin to appear around 2 months of age (Buaas et al., 2004; Costoya et al., 2004). However, primary cultures of progenitor spermatogonia can be established and expanded from testes of Plzf null mice indicating existence of a functional undifferentiated spermatogonial population (Hobbs, Seandel, Falciatori, Rafii, & Pandolfi, 2010). Interestingly, Plzf gene expression is not regulated by GDNF signaling but PLZF protein does interact with downstream components of the intracellular cascade to promote proliferation of the undifferentiated spermatogonial population (Hobbs et al., 2010). The receptor complex for GDNF signaling consists of RET and GFRa1 subunits and their expression is suppressed by the activity of mTORC1. Importantly, expression of the mTORC1 inhibitor REDD1 is induced by PLZF (Hobbs et al., 2010). Thus, spermatogonia from Plzf null mice possess elevated mTORC1 activity which attenuates expression of RET and GFRa1 thereby inducing a greater propensity for apoptosis and promoting premature differentiation compared to spermatogonia in testes of wild-type mice.

Another transcription factor expressed in spermatogonia that appears to influence SSC functions but for which regulation by niche factors has not been described is TAF4B. This molecule is a coactivator of the RNA Pol II transcriptional apparatus and is expressed specifically in the male germline (Falender et al., 2005). Male mice homozygous for a Taf4b null allele are of normal fertility at puberty demonstrating dispensability for spermatogenesis (Falender et al., 2005). However, seminiferous tubules with a Sertoli-cell-only phenotype are apparent at PD 0 and the percentage of these dramatically increases with advancing postnatal life (Falender et al., 2005). Thus, TAF4B appears to influence the survival of gonocyte precursors. As a result, deficiency of TAF4B in the germline likely impairs formation of the SSC pool during neonatal development which would lead to appearance of Sertoli-cell-only tubules early in life. The progressive increase in Sertoli-cell-only tubules with advancing age indicates that ongoing spermatogenesis is established and then lost due to disruption of SSC functions. Collectively, these findings indicate that TAF4B plays an important role both in the regulation of gonocyte survival and self-renewal of SSCs. However, the fact that TAF4B is detectable in most, if not all, spermatogonial subtypes and primary spermatocytes (Falender et al., 2005), suggests a general role in germ cell maintenance rather than specific influence on SSC functions.

The Foxo family of transcription factors influences the maintenance of various cell lineages via regulating cell cycle progression, apoptosis, and development (Huang & Tindall, 2007). While Foxo3 and Foxo4 are dispensable for continual spermatogenesis, Foxo1 appears to play an important role in maintenance of the spermatogonial population (Goertz, Wu, Gallardo, Hamra, & Castrillon, 2011). In mice with germline-specific inactivation of Foxo1, the number of gonocytes at birth is normal but proliferation of the general spermatogonial population during early postnatal development is severely impaired. At PD 21 when formation of the spermatogenic lineage is nearly complete, the majority of seminiferous tubules in testes of mice with germline deficiency of Foxo1 possess disrupted spermatogenesis including arrested differentiation of spermatogonia and appearance of a Sertoli-cell-only phenotype (Goertz et al., 2011). In contrast, Sertoli-cell-only tubules were not observed when these mice reached adulthood. Similar phenotypes were observed in mice with conditional germline inactivation of Foxo1/3/4. In these triple knockout animals, spermatogonial number is reduced as early as PD 7. Importantly, seminiferous tubules with a Sertoli-cell-only phenotype were observed at PD 21, but whether this phenotype is sustained into adulthood has not been described. Collectively, these findings imply deficiency of Foxo factors in the germline impairs multiple facets of spermatogonial function; however, the presence of germ cells in all tubules in adulthood indicates the presence of a functional SSC pool. Furthermore, the finding that spermatogonia are lacking at PD 21 but not later in life suggests a role for Foxo1 in control of differentiation but not self-renewal. The first round of spermatogenesis arises from a subset of prospermatogonial precursors that transition directly into a differentiating state (Yoshida et al., 2006). Thus, seminiferous tubules with a Sertoli-cell-only phenotype at PD 21 could result from impaired formation of these cells without corresponding defects in the SSC pool. If this were the case, these "empty" tubules would become repopulated with spermatogonia arising from the SSC pool later in life.

4.2. Influence of posttranscriptional regulators

While a major level of regulation for cell fate decisions is at the level of transcriptional control, emerging evidence over the last decade has revealed an influence of posttranscriptional mechanisms on maintenance of a stem cell state in various cell lineages. The regulatory molecules include epigenetic modifiers, RNA-binding proteins, and small noncoding RNAs. Epigenetic modifiers exert an influence on the transcriptome via altering specific histone residues and DNA methylation. The most well-described functions of RNA-binding proteins are to stabilize target mRNAs and repress translation. Similarly, small noncoding RNAs also influence the transcriptome by targeting mRNAs for degradation or translational repression. Here, we focus on the role of RNA-binding proteins and noncoding RNAs that have been reported to influence maintenance of the SSC pool in the mouse germline.

In the drosophila male germline, maintenance of the stem cell pool requires expression of the RNA-binding protein Nanos which interacts with Pumilio to repress translation of specific RNAs (Asaoka-Taguchi, Yamada, Nakamura, Hanyu, & Kobayashi, 1999). Of the three Nanos paralogues in the mouse genome, only Nanos2 and Nanos3 are expressed in the germline. Mice homozygous for a Nanos2 allele lack the germ cell lineage due to elimination of the PGC population during embryonic development. However, in testes of wild-type mice Nanos2 is detectable in A_s, A_{pr}, and A_{al} spermatogonia during postnatal life (Tsuda et al., 2003). Importantly, conditional inactivation in adulthood leads to progressive loss of the germline indicating exhaustion of SSC pool (Tsuda et al., 2003). Moreover, ectopic overexpression of Nanos2 causes impaired spermatogenesis in which most differentiating germ cells are lost resulting in an increased concentration of undifferentiated spermatogonia (Sada, Suzuki, Suzuki, & Saga, 2009). Interestingly, testes of Nanos2 overexpressing mice contain an increased frequency of A_s and A_{pr} spermatogonia within the undifferentiated population (Sada, Hasegawa, Pin, & Saga, 2012; Tsuda et al., 2003). Collectively, these findings indicate that in addition to having an essential role in promoting the germ cell state in general, Nanos2 influences maintenance of the stem cell state in postnatal spermatogonia. At present, the mRNA targets through which Nanos2 imparts an influence on maintenance of the SSC pool are undefined.

Another mechanism of posttranscriptional control for mRNAs is the influence of a diverse family of small noncoding RNAs that includes endogenous small interference RNAs (endo-siRNA, ~21 nucleotides), microRNAs (miRNAs, ~22 nucleotides), and Piwi-interacting RNAs (piRNAs, ~31 nucleotides). All of these regulatory RNA classes are known to play essential roles in multiple development processes during embryogenesis and postnatal life (Suh & Blelloch, 2011). However, the role of these in regulation of SSC functions is largely undefined. While enod-siRNAs exist within male germ cells, expression in SSCs or the undifferentiated spermatogonial population and a functional role has not been described (Song et al., 2011). Also, piRNAs and their interacting partners are known to be major regulators of meiosis (Zheng et al., 2010), but an influence on spermatogonial functions is unclear at present. In contrast, the role of miRNAs has begun to be uncovered over the last several years. Studies of Niu et al. (2011) used high-throughput sequencing of the miRNA population to identify enriched abundance of miR21 within the undifferentiated spermatogonial population of mouse testes (Niu et al., 2011). Results of additional studies revealed that inhibition of miR21 function in primary cultures of undifferentiated spermatogonia induces apoptosis leading to a reduction of SSC number (Niu et al., 2011). Recently, a role for the miR221/222 cluster in regulating SSC maintenance has been described (Yang, Racicot, Kaucher, Oatley, & Oatley, 2013). Inhibition of miR221/222 function in primary cultures of undifferentiated spermatogonia leads to loss of the SSC pool without effecting apoptosis. In yet another study utilizing primary cultures of testis cells, a role of miR146 in controlling the maintenance of the undifferentiated spermatogonial population has been suggested (Huszar & Payne, 2013). While studies of miR21 and miR221/222 utilized transplantation analyses to demonstrate an importance in maintenance of the SSC pool, similar analyses have not been reported for miR146. Collectively, these findings suggest that at least some miRNAs including miR21, miR221/222, and possibly miR146 influence SSC functions in the mouse germline. Whether, the role of these is to promote survival alone or to also influence the stem cell state remains to be defined as mRNA targets are revealed in future studies. It is important to note that the findings described earlier have been obtained from analysis of the undifferentiated spermatogonia *in vitro*. At present, the role of any miRNA in maintenance of the SSC pool *in vivo* has not been described.

4.3. Association of mechanisms defined in mice with infertility in humans

In humans, male factor infertility is a major health concern throughout the modern world. In the United States alone, ~10% of reproductive age couples experience issues with infertility and 50% of cases are attributed to a male factor (Hotaling & Walsh, 2009). Azoospermia is a pathological condition describing the absence of spermatozoa in the ejaculate and is a leading diagnosis of fertility disorders in men. Furthermore, ~10% of men diagnosed with azoospermia have testes that are devoid of germ cells, resembling a Sertoli-cell-only phenotype. While abnormalities in sex chromosomes are known causes, the majority of Sertoli-cell-only cases are idiopathic. Based on knowledge gained from mouse studies, impaired establishment and maintenance of the foundational SSC pool could be major causes of these pathologies.

Initial studies aimed at determining a genetic basis of azoospermia led to discovery of the azoospermia factor (Azf) region of the human Y chromosome (Tiepolo & Zuffardi, 1976). Further analyses revealed that a transcriptional unit located on the long arm of the Y chromosome termed deletion of the azoospermia factor (Daz) is absent in ~13% of men with non-obstructive azoospermia (Reijo, Alagappan, Patrizio, & Page, 1996). Multiple Daz isoforms have been detected in germ cells and all encode putative RNA-binding proteins with unknown functions (Shinka & Nakahori, 1996). A variety of phenotypes are observed in patients harboring Daz deletion including arrest of germ cell maturation in meiosis and seminiferous tubules with appearance of Sertoli-cell-only syndrome (Ferras et al., 2004). The Sertoli-cell-only phenotype could manifest from defects in SSC self-renewal or disruption in formation of the SSC pool due to deficiencies of the PGC or gonocyte precursor populations. While it is difficult to evaluate the role of Daz in human spermatogenesis directly, a mouse model harboring null alleles of the DAZ-like gene (Dazl) has been generated. Unlike Daz in the human genome, Dazl is located within an autosome in the mouse rather than the Y chromosome. However, mice homozygous for the Dazl null allele are infertile due to multiple defects in germ cell development including disruption of PGC survival, impaired spermatogonial differentiation, and arrest during meiosis in postnatal life (Lin & Page, 2005; Saunders et al., 2003). Thus, while Daz is clearly essential for spermatogenesis and loss of function is a direct cause of azoospermia in men, manifestation of this disease via impairment of SSC functions is unlikely.

Although linked to azoospermia, Y-chromosome abnormalities account for a small percentage of male infertility. In attempts to identify other genetic causes, mapping of single-nucleotide polymorphisms and genome wide association studies have been conducted with azoospermia patients. Mutations in the gene spermatogenesis and oogenesis helix-loop-helix 1 (Sohlh1) have been discovered in the genome of men with azoospermia (Choi et al., 2010). Importantly, mice homozygous for a Sohlh1 null allele are azoospermic due to impaired differentiation of spermatogonia. However, the undifferentiated spermatogonial population is present within Sohlh1 null mice indicating that the SSC pool is not disrupted. Another gene that is important for regulation of the spermatogonial population in mice which has been linked to infertility in men is Etv5. The frequency of a mutation in the Etv5 gene was found to be associated with homozygosity in men possessing a Sertoli-cell-only phenotype (O'Bryan et al., 2012). This finding indicates a conserved role of ETV5 for regulation of the SSC pool in mouse and man. While Nanos1 is dispensable for spermatogenesis in mice, mutations are correlated with azoospermia in humans (Kusz-Zamelczyk et al., 2013). An association of Nanos2, which is required for maintenance of the spermatogenic lineage in mice, with azoospermia in humans has not been described. Similarly, association of mutations in genes of other transcription factors and posttranscriptional regulators discovered to be essential for maintenance of the SSC pool in mice with infertility in humans has not been described. As we define further the molecular mechanisms regulating SSC self-renewal in mice, relationships between the key players and pathological conditions of azoospermia and Sertoli-cell-only syndrome in humans will undoubtedly be made.

5. TGCT FORMATION

5.1. Manifestation in human testes

Male infertility can be attributed to many factors that greatly impact quality of life but do not adversely influence overall health and longevity. However, formation of TGCTs is not only detrimental to spermatogenesis and fertility but can also manifest as a life-threatening disease. In humans, TGCTs are classified as seminomas, nonseminomas, or spermatocytic seminomas (Oosterhuis & Looijenga, 2005). Seminomas arise as a mass of cells with a morphological appearance resembling PGCs or gonocytes. Nonseminomas include embryonal carcinoma, teratoma, and yolk sac choriocarcinoma, all of which appear to lack defining features of the germ cell lineage (Fig. 9.4). Spermatocytic seminomas are rare benign tumors that share morphological similarities with both spermatogonia and spermatocytes. It is widely accepted that the common origin of human TGCT is carcinoma *in situ* (CIS) of the testis (Skakkebaek, 1972). CIS cells are defined as having undergone neoplastic transformation and are precursors to invasive cancer in a variety of tissues including lung (Tokai, 1975), breast (Foucar, 2013), bladder (Ranasinghe et al., 2013), and testis (Skakkebaek, 1972). These cells are

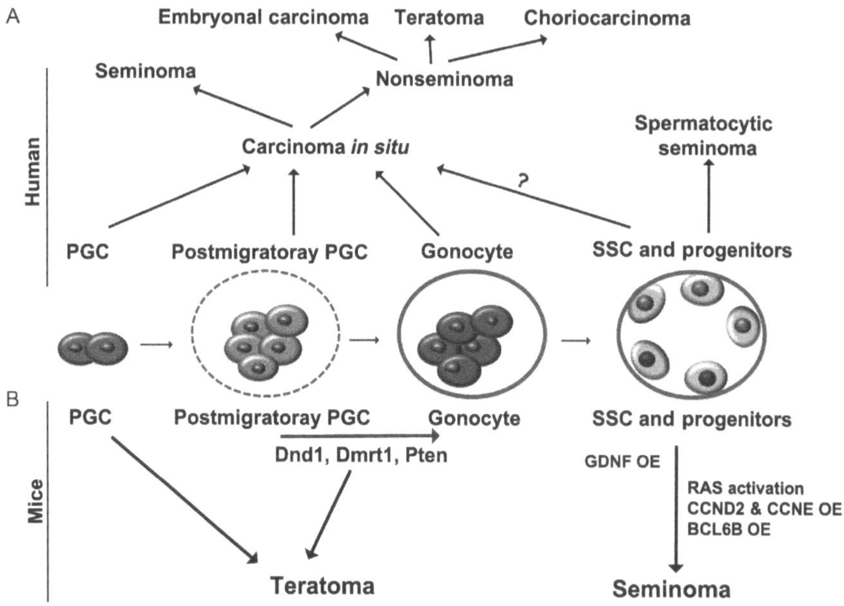

Figure 9.4 Impact of misregulated molecular mechanisms of spermatogonial lineage development on testicular germ cell tumor formation in humans and mice. (A) Schematic of sources for the variety of testicular germ cell tumors that form in humans. Carcinoma *in situ* (CIS) cells are thought to be the common source of seminoma and nonseminoma tumor types. CIS cells arise in postnatal life as a mass with morphological appearance resembling PGCs or gonocytes. Spermatocytic seminomas are rare benign tumors that share morphological similarities with both spermatogonia and spermatocytes. (B) Impact of misregulated molecular mechanisms influencing spermatogonial stem cell self-renewal on testicular germ cell tumor formation in mice. Teratomas arise from PGCs in which repression of pluripotency and cell cycle arrest is impaired. Seminomas can be induced by GDNF overexpression (OE) in a transgenic mouse line. Primary cultures of mouse undifferentiated spermatogonia engineered to express an activated form of RAS, or overexpress CCND2/CCNE or BCL6B give rise to seminomas following transplantation into recipient mouse seminiferous tubules. (See color plate.)

often detected in testes containing TGCTs and clinical observations have revealed that ~50% of patients diagnosed with CIS develop TGCTs within 5 years (Skakkebaek, 1978). In human testes, CIS cells reside as a single layer along the basement membrane of seminiferous tubules and do not display invasive properties. However, spermatogenesis is disrupted in tubules containing CIS cells.

At present, the defining characteristics of CIS cells and molecular etiology of their formation are largely undefined. From a morphological standpoint, CIS cells contain large dysplastic nucleoli and elevated amounts of cytoplasmic glycogen (Sigg & Hedinger, 1984; Skakkebaek, Berthelsen, Giwercman, & Muller, 1987). Among the currently accepted marker criteria of CIS cells is expression of placental AP, OCT3/4, and AP-2γ (van Casteren et al., 2009). While these criteria are applied to identifying CIS cells within testes, they are not causative agents of their formation. Several lines of evidence suggest the source of these cells is PGCs or gonocytes that fail in transition to the spermatogonial lineage during postnatal development. The expression of markers and global gene expression profile of CIS is more similar to PGCs and gonocytes than postnatal spermatogonia (Albrechtsen, Nielsen, Skakkebaek, & Wewer, 1982; Almstrup et al., 2005, 2004). In addition, CIS cells have been observed in prepubertal boys when gonocytes are the only germ cell type present (Muller, Skakkebaek, Nielsen, & Graem, 1984).

A current dogma in the TGCT field is that CIS cells possess the capacity to generate seminoma or retain plasticity to derive nonseminomas. Embryonal carcinomas are the most primitive nonseminomas which can differentiate into teratomas, choriocarcinomas, or yolk sac tumors (Rajpert-De Meyts et al., 2003). Classic pluripotent cells such as ES cells or EpiSCs can only form teratomas when transplanted into testes (Chuma et al., 2005; Reubinoff, Pera, Fong, Trounson, & Bongso, 2000). These cells do not appear to contribute to trophoblast lineages during normal embryonic development or when introduced into the testis environment (Reubinoff et al., 2000; Rossant & Cross, 2001). In contrast, choriocarcinomas, seminomas, and teratomas express markers of the trophoblast lineage (Lee, Vogt, Hsiao, & Osunkoya, 2012). These observations suggest CIS cells possess totipotent potential allowing for adoption of both embryonic and extraembryonic lineage fates (Fig. 9.4). Spermatocytic seminomas lack the expression of classical CIS markers. Because this class of TGCT possess similar morphological features and express several markers of undifferentiated spermatogonia (e.g., UTF-1, SALL4), SSCs and/or progenitor spermatogonia could be a primary source (Cao, Li, Guo, Allan, & Humphrey, 2009; Waheeb & Hofmann, 2011; Wang et al., 2010).

Another potential underlying mechanism of TGCT formation is aberrant regulation of cell cycle progression, in particular the G1–S transition. Expression of the cell cycle regulators RB, p19, and p18 is downregulated or absent in CIS cells (Bartkova et al., 2003; Bartkova, Thullberg, Rajpert-De Meyts, Skakkebaek, & Bartek, 2000). Also, expression of the cell cycle inhibitor p16 is often lacking in TGCTs (Chaubert et al., 1997). In addition, expression of Cyclin D2 (CCND2) is detectable in CIS cells and various

TGCTs but undetectable in human male germ cells during normal conditions (Houldsworth, Reuter, Bosl, & Chaganti, 1997). Furthermore, the Ras oncogene family (H-Ras, K-Ras, and N-Ras) is often mutated in human TGCTs including spermatocytic seminomas (Goriely et al., 2009; Moul, Theune, & Chang, 1992). Lastly, expression of the cell cycle repressor PTEN is not detectable in teratomas, 56% of seminomas, and 86% of embryonal carcinomas (Andreassen et al., 2013; Di Vizio et al., 2005). Collectively, these observations highlight the necessity of cell cycle checkpoint control during normal germ cell development. Misregulation of mitotic arrest during PGC-to-gonocyte transition and reentry of cell cycle progression during gonocyte-to-SSC transition may contribute directly to the formation of CIS cells in testes during fetal and neonatal development.

5.2. Impact of misregulated molecular mechanisms of SSC self-renewal on TGCT formation in mice

Understanding the molecular etiology of TGCT formation in humans has been hampered by lack of suitable mouse models. However, over the last decade several lines of mutant mice and experimental systems utilizing primary cultures of undifferentiated spermatogonia have been produced that emulate aspects of TGCT, in particular, is the formation of testicular teratocarcinoma. Groundbreaking studies of Leroy Stevens revealed a relatively high incidence of spontaneous teratoma formation in testes of the inbred 129 strains of mice (Stevens, 1958). Using this model, early studies determined that tumors begin to arise as early as E15.5 (Noguchi & Noguchi, 1985), but the underlying molecular cause of these manifestations remained unknown for many years. Later studies revealed that the Ter mutation within the gene Dnd1 confers tumorigenic potential (Cook, Munger, Nadeau, & Capel, 2011). In mice, Dnd1 encodes for an RNA-binding protein that influences the PGC-to-gonocyte transition via direct targeting of the cell cycle regulators p27 and p21 (Cook et al., 2011). Loss of Dnd1 function results in delayed cell cycle arrest and retention of pluripotent potential. Gonocytes harboring an inactivated Dnd1 gene retain expression of Nanog which is repressed in wild-type male germ cells (Heaney et al., 2012). The pluripotent state within ES cells requires Nanog activity and expression of this factor is detectable in human TGCTs (Clark et al., 2004). These observations suggest that reactivation of the normally suppressed core molecular machinery driving a pluripotent state in germ cells promotes teratoma formation. While activation of this program in gonocytes induces tumor formation, whether germ cells at other stages of development in embryonic and postnatal life are also susceptible is unknown.

Other models of testicular teratocarcinoma formation have been described including mice that are homozygous for a Dmrt1 (Krentz et al., 2009) or Pten (Kimura et al., 2003) null allele. Specification and migration of PGCs is not affected by Dmrt1 or Pten loss of function but some gonocytes continue to express pluripotency genes and these cells give rise to teratomas in adulthood. Importantly, mutations in both the *Pten* and *Drmt1* gene are highly associated with the development of metastatic TGCTs in humans (Turnbull et al., 2010). Interestingly, neither Dmrt1 nor Pten appears to function as a tumor suppressor in germ cells during postnatal life. Loss of Dmrt1 function after the gonocyte stage of development impairs maintenance of the spermatogonial state leading to premature initiation of meiosis; whereas, deficiency of Pten impairs germ cell survival (Goertz et al., 2011). Collectively, these results suggest that manifestation of testicular teratomas originates with impaired suppression of the pluripotent state in PGCs and/or gonocytes during fetal development.

At present, mouse models that mimic CIS transformation in humans have not been described. While transient overexpression of the transcription factor E2F1 leads to appearance of cells with a CIS-like phenotype in postnatal life, the propensity of these cells to form TGCT has not been described (Agger, Santoni-Rugiu, Holmberg, Karlstrom, & Helin, 2005). Interestingly, E2F1 is a target of the tumor suppressor RB and results of our recent studies suggest that conditional germline inactivation of RB in mice leads to appearance of cells with a CIS-like morphology in postnatal life (Yang et al., 2013).

An array of evidence has started to emerge over the last 5 years suggesting that misregulation of SSC self-renewal can induce TGCT formation, in particular development of seminomas (Fig. 9.4). Signaling from the cytokine GDNF is required for maintenance of the SSC pool in testes of mice (Meng et al., 2000). However, ubiquitous overexpression of GDNF in mice impairs spermatogonial differentiation resulting in the formation of clusters of undifferentiated spermatogonia that exhibit features of testicular seminomas (Meng, de Rooij, Westerdahl, Saarma, & Sariola, 2001). Also, ectopic expression of an activated form of RAS or overexpression of CCND2 and CCNE drives growth factor-independent proliferation of primary cultures of mouse undifferentiated spermatogonia (Lee et al., 2009). After transplantation into recipient testis, some of these transformed cells generate abnormal colonies of germ cells resembling seminomas.

Furthermore, cultured mouse undifferentiated spermatogonia engineered to overexpress the transcriptional repressor BCL6B, which influences maintenance of the SSC pool during normal conditions, derive seminoma-like tumors in testes of recipient mice following transplantation (Ishii, Kanatsu-Shinohara, Toyokuni, & Shinohara, 2012). Even though the histological and pathological features of these seminomas are not identical to those observed in humans, these findings indicate that aberrant regulation of self-renewal within SSCs could confer tumorigenic properties thereby inducing stem cell derived carcinoma. Clearly, additional research is required to determine whether such a possibility is an underlying cause of seminoma formation within human testes.

6. CONCLUDING REMARKS

Actions of the SSC pool provide the basis for continual cycling of the spermatogenic lineage. While a foundational SSC pool is established during neonatal development, emerging evidence suggests these cells are predetermined in a subset of the precursor gonocyte population that is formed during embryogenesis. In the mouse, we propose that three different lineages are predetermined in the gonocyte population that are independently responsible for formation of the foundational SSC pool, a primary progenitor population, and an initial differentiating spermatogonial population during neonatal development. At present, molecular markers that label these possible gonocyte subtypes have not been identified. However, several mouse models have been produced in which the SSC pool is not formed or is exhausted in early life coincident with the completion of several rounds spermatogenesis. These observations support the idea of predetermined lineage fates within the male germline.

Once formed, the foundational SSC pool must be sustained via self-renewal and several transcription factors and posttranscriptional regulators that influence this activity have been identified from studies of the mouse male germline. While impaired function of these factors leads to a Sertoli-cell-only phenotype and azoospermia in mice, the role of most in regulating the fertility of humans has not been described. Thus, future studies will be aimed at exploring the relationship between self-renewal factors described in mice and azoospermia in men. Lastly, origins of most, if not all, cancer in humans and the role of misregulated stem cell self-renewal in tumorigenesis remains largely unknown. The male germline is an attractive model for addressing these gaps in knowledge because of several unique features. Germ cells are specified directly from epiblast cells during embryogenesis and summarily lose pluripotent potential coincident with cell cycle arrest. Emerging evidence suggests that disruption in the mechanisms suppressing pluripotency and influencing cell cycle checkpoint control confer tumorigenic potential. Thus, defining the molecular etiology of TGCT formation may provide valuable information for the diagnosis and treatment of various cancers.

REFERENCES

Agger, K., Santoni-Rugiu, E., Holmberg, C., Karlstrom, O., & Helin, K. (2005). Conditional E2F1 activation in transgenic mice causes testicular atrophy and dysplasia mimicking human CIS. *Oncogene, 24*, 780–789.

Albrechtsen, R., Nielsen, M. H., Skakkebaek, N. E., & Wewer, U. (1982). Carcinoma in situ of the testis. Some ultrastructural characteristics of germ cells. *Acta Pathologica, Microbiologica, et Immunologica Scandinavica. Section A, 90*, 301–303.

Almstrup, K., Hoei-Hansen, C. E., Nielsen, J. E., Wirkner, U., Ansorge, W., Skakkebaek, N. E., et al. (2005). Genome-wide gene expression profiling of testicular carcinoma in situ progression into overt tumours. *British Journal of Cancer, 92*, 1934–1941.

Almstrup, K., Hoei-Hansen, C. E., Wirkner, U., Blake, J., Schwager, C., Ansorge, W., et al. (2004). Embryonic stem cell-like features of testicular carcinoma in situ revealed by genome-wide gene expression profiling. *Cancer Research, 64*, 4736–4743.

Anderson, R., Copeland, T. K., Scholer, H., Heasman, J., & Wylie, C. (2000). The onset of germ cell migration in the mouse embryo. *Mechanisms of Development, 91*, 61–68.

Anderson, R., Fassler, R., Georges-Labouesse, E., Hynes, R. O., Bader, B. L., Kreidberg, J. A., et al. (1999). Mouse primordial germ cells lacking beta1 integrins enter the germline but fail to migrate normally to the gonads. *Development, 126*, 1655–1664.

Andreassen, K. E., Kristiansen, W., Karlsson, R., Aschim, E. L., Dahl, O., Fossa, S. D., et al. (2013). Genetic variation in AKT1, PTEN and the 8q24 locus, and the risk of testicular germ cell tumor. *Human Reproduction, 28*, 1995–2002.

Ara, T., Nakamura, Y., Egawa, T., Sugiyama, T., Abe, K., Kishimoto, T., et al. (2003). Impaired colonization of the gonads by primordial germ cells in mice lacking a chemokine, stromal cell-derived factor-1 (SDF-1). *Proceedings of the National Academy of Sciences of the United States of America, 100*, 5319–5323.

Asaoka-Taguchi, M., Yamada, M., Nakamura, A., Hanyu, K., & Kobayashi, S. (1999). Maternal Pumilio acts together with Nanos in germline development in Drosophila embryos. *Nature Cell Biology, 1*, 431–437.

Barker, N., van Es, J. H., Kuipers, J., Kujala, P., van den Born, M., Cozijnsen, M., et al. (2007). Identification of stem cells in small intestine and colon by marker gene Lgr5. *Nature, 449*, 1003–1007.

Barna, M., Hawe, N., Niswander, L., & Pandolfi, P. P. (2000). Plzf regulates limb and axial skeletal patterning. *Nature Genetics, 25*, 166–172.

Bartkova, J., Lukas, C., Sorensen, C. S., Rajpert-De Meyts, E., Skakkebaek, N. E., Lukas, J., et al. (2003). Deregulation of the RB pathway in human testicular germ cell tumours. *The Journal of Pathology, 200*, 149–156.

Bartkova, J., Thullberg, M., Rajpert-De Meyts, E., Skakkebaek, N. E., & Bartek, J. (2000). Lack of p19INK4d in human testicular germ-cell tumours contrasts with high expression during normal spermatogenesis. *Oncogene, 19*, 4146–4150.

Bettocchi, C., Coker, C. B., Deacon, J., Parkinson, C., & Pryor, J. P. (1994). A review of testicular intratubular germ cell neoplasia in infertile men. *Journal of Andrology, 15*(Suppl), 14S–16S.

Brinster, R. L., & Zimmermann, J. W. (1994). Spermatogenesis following male germ-cell transplantation. *Proceedings of the National Academy of Sciences of the United States of America, 91*, 11298–11302.

Buaas, F. W., Kirsh, A. L., Sharma, M., McLean, D. J., Morris, J. L., Griswold, M. D., et al. (2004). Plzf is required in adult male germ cells for stem cell self-renewal. *Nature Genetics, 36*, 647–652.

Cao, D., Li, J., Guo, C. C., Allan, R. W., & Humphrey, P. A. (2009). SALL4 is a novel diagnostic marker for testicular germ cell tumors. *The American Journal of Surgical Pathology, 33*, 1065–1077.

Chaubert, P., Guillou, L., Kurt, A. M., Bertholet, M. M., Metthez, G., Leisinger, H. J., et al. (1997). Frequent p16INK4 (MTS1) gene inactivation in testicular germ cell tumors. *The American Journal of Pathology, 151*, 859–865.

Cheshier, S. H., Morrison, S. J., Liao, X., & Weissman, I. L. (1999). In vivo proliferation and cell cycle kinetics of long-term self-renewing hematopoietic stem cells. *Proceedings of the National Academy of Sciences of the United States of America, 96*, 3120–3125.

Choi, Y., Jeon, S., Choi, M., Lee, M. H., Park, M., Lee, D. R., et al. (2010). Mutations in SOHLH1 gene associate with nonobstructive azoospermia. *Human Mutation, 31*, 788–793.

Chuma, S., Kanatsu-Shinohara, M., Inoue, K., Ogonuki, N., Miki, H., Toyokuni, S., et al. (2005). Spermatogenesis from epiblast and primordial germ cells following transplantation into postnatal mouse testis. *Development, 132*, 117–122.

Clark, A. T., Rodriguez, R. T., Bodnar, M. S., Abeyta, M. J., Cedars, M. I., Turek, P. J., et al. (2004). Human STELLAR, NANOG, and GDF3 genes are expressed in pluripotent cells and map to chromosome 12p13, a hotspot for teratocarcinoma. *Stem Cells, 22*, 169–179.

Clermont, Y. (1966). Renewal of spermatogonia in man. *The American Journal of Anatomy, 118*, 509–524.

Clermont, Y., & Leblond, C. P. (1953). Renewal of spermatogonia in the rat. *The American Journal of Anatomy, 93*, 475–501.

Clermont, Y., & Leblond, C. P. (1959). Differentiation and renewal of spermatogonia in the monkey, Macacus rhesus. *The American Journal of Anatomy, 104*, 237–273.

Cook, M. S., Munger, S. C., Nadeau, J. H., & Capel, B. (2011). Regulation of male germ cell cycle arrest and differentiation by DND1 is modulated by genetic background. *Development, 138*, 23–32.

Costoya, J. A., Hobbs, R. M., Barna, M., Cattoretti, G., Manova, K., Sukhwani, M., et al. (2004). Essential role of Plzf in maintenance of spermatogonial stem cells. *Nature Genetics, 36*, 653–659.

Curtis, S. K., & Amann, R. P. (1981). Testicular development and establishment of spermatogenesis in Holstein bulls. *Journal of Animal Science, 53*, 1645–1657.

de Rooij, D. G. (1998). Stem cells in the testis. *International Journal of Experimental Pathology, 79*, 67–80.

De Rooij, D. G., & Lok, D. (1987). Regulation of the density of spermatogonia in the seminiferous epithelium of the Chinese hamster: II. Differentiating spermatogonia. *The Anatomical Record, 217*, 131–136.

de Rooij, D. G., & Russell, L. D. (2000). All you wanted to know about spermatogonia but were afraid to ask. *Journal of Andrology, 21*, 776–798.

de Ruiter-Bootsma, A. L., Kramer, M. F., & de Rooij, D. G. (1977). Survival of spermatogonial stem cells in the mouse after split-dose irradiation with fission neutrons of 1-MeV mean energy or 300-kV X rays. *Radiation Research, 71*, 579–592.

Di Carlo, A., & De Felici, M. (2000). A role for E-cadherin in mouse primordial germ cell development. *Developmental Biology*, *226*, 209–219.

Di Vizio, D., Cito, L., Boccia, A., Chieffi, P., Insabato, L., Pettinato, G., et al. (2005). Loss of the tumor suppressor gene PTEN marks the transition from intratubular germ cell neoplasias (ITGCN) to invasive germ cell tumors. *Oncogene*, *24*, 1882–1894.

Dobrinski, I., Ogawa, T., Avarbock, M. R., & Brinster, R. L. (1999). Computer assisted image analysis to assess colonization of recipient seminiferous tubules by spermatogonial stem cells from transgenic donor mice. *Molecular Reproduction and Development*, *53*, 142–148.

Doulatov, S., Notta, F., Rice, K. L., Howell, L., Zelent, A., Licht, J. D., et al. (2009). PLZF is a regulator of homeostatic and cytokine-induced myeloid development. *Genes & Development*, *23*, 2076–2087.

Drumond, A. L., Meistrich, M. L., & Chiarini-Garcia, H. (2011). Spermatogonial morphology and kinetics during testis development in mice: A high-resolution light microscopy approach. *Reproduction*, *142*, 145–155.

Ehmcke, J., & Schlatt, S. (2006). A revised model for spermatogonial expansion in man: Lessons from non-human primates. *Reproduction*, *132*, 673–680.

Falender, A. E., Freiman, R. N., Geles, K. G., Lo, K. C., Hwang, K., Lamb, D. J., et al. (2005). Maintenance of spermatogenesis requires TAF4b, a gonad-specific subunit of TFIID. *Genes & Development*, *19*, 794–803.

Fawcett, D. W., Neaves, W. B., & Flores, M. N. (1973). Comparative observations on intertubular lymphatics and the organization of the interstitial tissue of the mammalian testis. *Biology of Reproduction*, *9*, 500–532.

Ferras, C., Fernandes, S., Marques, C. J., Carvalho, F., Alves, C., Silva, J., et al. (2004). AZF and DAZ gene copy-specific deletion analysis in maturation arrest and Sertoli cell-only syndrome. *Molecular Human Reproduction*, *10*, 755–761.

Foucar, E. (2013). Case 8-2013: A woman with carcinoma in situ of the breast. *The New England Journal of Medicine*, *368*, 2439–2440.

Franca, L. R., Ogawa, T., Avarbock, M. R., Brinster, R. L., & Russell, L. D. (1998). Germ cell genotype controls cell cycle during spermatogenesis in the rat. *Biology of Reproduction*, *59*, 1371–1377.

Ginsburg, M., Snow, M. H., & McLaren, A. (1990). Primordial germ cells in the mouse embryo during gastrulation. *Development*, *110*, 521–528.

Godin, I., Deed, R., Cooke, J., Zsebo, K., Dexter, M., & Wylie, C. C. (1991). Effects of the steel gene product on mouse primordial germ cells in culture. *Nature*, *352*, 807–809.

Goertz, M. J., Wu, Z., Gallardo, T. D., Hamra, F. K., & Castrillon, D. H. (2011). Foxo1 is required in mouse spermatogonial stem cells for their maintenance and the initiation of spermatogenesis. *The Journal of Clinical Investigation*, *121*, 3456–3466.

Goriely, A., Hansen, R. M., Taylor, I. B., Olesen, I. A., Jacobsen, G. K., McGowan, S. J., et al. (2009). Activating mutations in FGFR3 and HRAS reveal a shared genetic origin for congenital disorders and testicular tumors. *Nature Genetics*, *41*, 1247–1252.

Griswold, M. D. (1998). The central role of Sertoli cells in spermatogenesis. *Seminars in Cell & Developmental Biology*, *9*, 411–416.

Heaney, J. D., Anderson, E. L., Michelson, M. V., Zechel, J. L., Conrad, P. A., Page, D. C., et al. (2012). Germ cell pluripotency, premature differentiation and susceptibility to testicular teratomas in mice. *Development*, *139*, 1577–1586.

Hobbs, R. M., Seandel, M., Falciatori, I., Rafii, S., & Pandolfi, P. P. (2010). Plzf regulates germline progenitor self-renewal by opposing mTORC1. *Cell*, *142*, 468–479.

Hotaling, J. M., & Walsh, T. J. (2009). Male infertility: A risk factor for testicular cancer. *Nature Reviews. Urology*, *6*, 550–556.

Houldsworth, J., Reuter, V., Bosl, G. J., & Chaganti, R. S. (1997). Aberrant expression of cyclin D2 is an early event in human male germ cell tumorigenesis. *Cell Growth & Differentiation*, *8*, 293–299.

Hsu, Y. C., Pasolli, H. A., & Fuchs, E. (2011). Dynamics between stem cells, niche, and progeny in the hair follicle. *Cell, 144*, 92–105.

Hu, Y. C., de Rooij, D. G., & Page, D. C. (2013). Tumor suppressor gene Rb is required for self-renewal of spermatogonial stem cells in mice. *Proceedings of the National Academy of Sciences of the United States of America, 110*, 12685–12690.

Huang, H., & Tindall, D. J. (2007). Dynamic FoxO transcription factors. *Journal of Cell Science, 120*, 2479–2487.

Huszar, J. M., & Payne, C. J. (2013). MicroRNA 146 (Mir146) modulates spermatogonial differentiation by retinoic acid in mice. *Biology of Reproduction, 88*, 15.

Ishii, K., Kanatsu-Shinohara, M., Toyokuni, S., & Shinohara, T. (2012). FGF2 mediates mouse spermatogonial stem cell self-renewal via upregulation of Etv5 and Bcl6b through MAP2K1 activation. *Development, 139*, 1734–1743.

Jaenisch, R., & Young, R. (2008). Stem cells, the molecular circuitry of pluripotency and nuclear reprogramming. *Cell, 132*, 567–582.

Kanatsu-Shinohara, M., Inoue, K., Lee, J., Yoshimoto, M., Ogonuki, N., Miki, H., et al. (2004). Generation of pluripotent stem cells from neonatal mouse testis. *Cell, 119*, 1001–1012.

Kimura, T., & Nakano, T. (2011). Induction of pluripotency in primordial germ cells. *Histology and Histopathology, 26*, 643–650.

Kimura, T., Suzuki, A., Fujita, Y., Yomogida, K., Lomeli, H., Asada, N., et al. (2003). Conditional loss of PTEN leads to testicular teratoma and enhances embryonic germ cell production. *Development, 130*, 1691–1700.

Klein, A. M., Nakagawa, T., Ichikawa, R., Yoshida, S., & Simons, B. D. (2010). Mouse germ line stem cells undergo rapid and stochastic turnover. *Cell Stem Cell, 7*, 214–224.

Kluin, P. M., & de Rooij, D. G. (1981a). A comparison between the morphology and cell kinetics of gonocytes and adult type undifferentiated spermatogonia in the mouse. *International Journal of Andrology, 4*, 475–493.

Ko, K., Arauzo-Bravo, M. J., Tapia, N., Kim, J., Lin, Q., Bernemann, C., et al. (2010). Human adult germline stem cells in question. *Nature, 465*, E1, discussion E3.

Krentz, A. D., Murphy, M. W., Kim, S., Cook, M. S., Capel, B., Zhu, R., et al. (2009). The DM domain protein DMRT1 is a dose-sensitive regulator of fetal germ cell proliferation and pluripotency. *Proceedings of the National Academy of Sciences of the United States of America, 106*, 22323–22328.

Kubota, H., Avarbock, M. R., & Brinster, R. L. (2004a). Culture conditions and single growth factors affect fate determination of mouse spermatogonial stem cells. *Biology of Reproduction, 71*, 722–731.

Kubota, H., Avarbock, M. R., & Brinster, R. L. (2004b). Growth factors essential for self-renewal and expansion of mouse spermatogonial stem cells. *Proceedings of the National Academy of Sciences of the United States of America, 101*, 16489–16494.

Kusz-Zamelczyk, K., Sajek, M., Spik, A., Glazar, R., Jedrzejczak, P., Latos-Bielenska, A., et al. (2013). Mutations of NANOS1, a human homologue of the Drosophila morphogen, are associated with a lack of germ cells in testes or severe oligo-astheno-teratozoospermia. *Journal of Medical Genetics, 50*, 187–193.

Lee, J., Kanatsu-Shinohara, M., Morimoto, H., Kazuki, Y., Takashima, S., Oshimura, M., et al. (2009). Genetic reconstruction of mouse spermatogonial stem cell self-renewal in vitro by Ras-cyclin D2 activation. *Cell Stem Cell, 5*, 76–86.

Lee, M. J., Vogt, A. P., Hsiao, W., & Osunkoya, A. O. (2012). CDX-2 expression in malignant germ cell tumors of the testes, intratubular germ cell neoplasia, and normal seminiferous tubules. *Tumour Biology, 33*, 2185–2188.

Lei, L., & Spradling, A. C. (2013). Mouse primordial germ cells produce cysts that partially fragment prior to meiosis. *Development, 140*, 2075–2081.

Leitch, H. G., & Smith, A. (2013). The mammalian germline as a pluripotency cycle. *Development, 140*, 2495–2501.

Lin, Y., & Page, D. C. (2005). Dazl deficiency leads to embryonic arrest of germ cell development in XY C57BL/6 mice. *Developmental Biology*, *288*, 309–316.

Matsui, Y., Zsebo, K., & Hogan, B. L. (1992). Derivation of pluripotential embryonic stem cells from murine primordial germ cells in culture. *Cell*, *70*, 841–847.

McLean, D. J., Friel, P. J., Johnston, D. S., & Griswold, M. D. (2003). Characterization of spermatogonial stem cell maturation and differentiation in neonatal mice. *Biology of Reproduction*, *69*, 2085–2091.

Meng, X., de Rooij, D. G., Westerdahl, K., Saarma, M., & Sariola, H. (2001). Promotion of seminomatous tumors by targeted overexpression of glial cell line-derived neurotrophic factor in mouse testis. *Cancer Research*, *61*, 3267–3271.

Meng, X., Lindahl, M., Hyvonen, M. E., Parvinen, M., de Rooij, D. G., Hess, M. W., et al. (2000). Regulation of cell fate decision of undifferentiated spermatogonia by GDNF. *Science*, *287*, 1489–1493.

Molyneaux, K. A., Zinszner, H., Kunwar, P. S., Schaible, K., Stebler, J., Sunshine, M. J., et al. (2003). The chemokine SDF1/CXCL12 and its receptor CXCR4 regulate mouse germ cell migration and survival. *Development*, *130*, 4279–4286.

Moreno, S. G., Dutrillaux, B., & Coffigny, H. (2001). Status of p53, p21, mdm2, pRb proteins, and DNA methylation in gonocytes of control and gamma-irradiated rats during testicular development. *Biology of Reproduction*, *64*, 1422–1431.

Moul, J. W., Theune, S. M., & Chang, E. H. (1992). Detection of RAS mutations in archival testicular germ cell tumors by polymerase chain reaction and oligonucleotide hybridization. *Genes, Chromosomes & Cancer*, *5*, 109–118.

Muller, J., Skakkebaek, N. E., Nielsen, O. H., & Graem, N. (1984). Cryptorchidism and testis cancer. Atypical infantile germ cells followed by carcinoma in situ and invasive carcinoma in adulthood. *Cancer*, *54*, 629–634.

Nagano, M. C. (2003). Homing efficiency and proliferation kinetics of male germ line stem cells following transplantation in mice. *Biology of Reproduction*, *69*, 701–707.

Nagano, M., Avarbock, M. R., & Brinster, R. L. (1999). Pattern and kinetics of mouse donor spermatogonial stem cell colonization in recipient testes. *Biology of Reproduction*, *60*, 1429–1436.

Nagano, M. C., & Yeh, J. R. (2013). The identity and fate decision control of spermatogonial stem cells: Where is the point of no return? *Current Topics in Developmental Biology*, *102*, 61–95.

Nakagawa, T., Nabeshima, Y., & Yoshida, S. (2007). Functional identification of the actual and potential stem cell compartments in mouse spermatogenesis. *Developmental Cell*, *12*, 195–206.

Nakagawa, T., Sharma, M., Nabeshima, Y., Braun, R. E., & Yoshida, S. (2010). Functional hierarchy and reversibility within the murine spermatogenic stem cell compartment. *Science*, *328*, 62–67.

Niu, Z., Goodyear, S. M., Rao, S., Wu, X., Tobias, J. W., Avarbock, M. R., et al. (2011). MicroRNA-21 regulates the self-renewal of mouse spermatogonial stem cells. *Proceedings of the National Academy of Sciences of the United States of America*, *108*, 12740–12745.

Noguchi, T., & Noguchi, M. (1985). A recessive mutation (ter) causing germ cell deficiency and a high incidence of congenital testicular teratomas in 129/Sv-ter mice. *Journal of the National Cancer Institute*, *75*, 385–392.

Oatley, J. M., Avarbock, M. R., Telaranta, A. I., Fearon, D. T., & Brinster, R. L. (2006). Identifying genes important for spermatogonial stem cell self-renewal and survival. *Proceedings of the National Academy of Sciences of the United States of America*, *103*, 9524–9529.

Oatley, J. M., & Brinster, R. L. (2008). Regulation of spermatogonial stem cell self-renewal in mammals. *Annual Review of Cell and Developmental Biology*, *24*, 263–286.

Oatley, J. M., & Brinster, R. L. (2012). The germline stem cell niche unit in mammalian testes. *Physiological Reviews*, *92*, 577–595.

Oatley, M. J., Kaucher, A. V., Racicot, K. E., & Oatley, J. M. (2011). Inhibitor of DNA binding 4 is expressed selectively by single spermatogonia in the male germline and regulates the self-renewal of spermatogonial stem cells in mice. *Biology of Reproduction, 85*, 347–356.

Oatley, J. M., Oatley, M. J., Avarbock, M. R., Tobias, J. W., & Brinster, R. L. (2009). Colony stimulating factor 1 is an extrinsic stimulator of mouse spermatogonial stem cell self-renewal. *Development, 136*, 1191–1199.

O'Bryan, M. K., Grealy, A., Stahl, P. J., Schlegel, P. N., McLachlan, R. I., & Jamsai, D. (2012). Genetic variants in the ETV5 gene in fertile and infertile men with non-obstructive azoospermia associated with Sertoli cell-only syndrome. *Fertility and Sterility, 98*(827–835), e821–e823.

Okamura, D., Kimura, T., Nakano, T., & Matsui, Y. (2003). Cadherin-mediated cell interaction regulates germ cell determination in mice. *Development, 130*, 6423–6430.

Oosterhuis, J. W., & Looijenga, L. H. (2005). Testicular germ-cell tumours in a broader perspective. *Nature Reviews. Cancer, 5*, 210–222.

Orwig, K. E., Ryu, B. Y., Avarbock, M. R., & Brinster, R. L. (2002). Male germ-line stem cell potential is predicted by morphology of cells in neonatal rat testes. *Proceedings of the National Academy of Sciences of the United States of America, 99*, 11706–11711.

Rajpert-De Meyts, E., Bartkova, J., Samson, M., Hoei-Hansen, C. E., Frydelund-Larsen, L., Bartek, J., et al. (2003). The emerging phenotype of the testicular carcinoma in situ germ cell. *APMIS, 111*, 267–278, discussion 278–269.

Ranasinghe, W. K., Attia, J., Oldmeadow, C., Lawrentschuk, N., Robertson, J., Ranasinghe, T., et al. (2013). Bladder carcinoma in situ (CIS) in Australia: A rising incidence for an under-reported malignancy. *BJU International*, http://dx.doi.org/10.1111/bju.12052.

Reijo, R., Alagappan, R. K., Patrizio, P., & Page, D. C. (1996). Severe oligozoospermia resulting from deletions of azoospermia factor gene on Y chromosome. *Lancet, 347*, 1290–1293.

Resnick, J. L., Bixler, L. S., Cheng, L., & Donovan, P. J. (1992). Long-term proliferation of mouse primordial germ cells in culture. *Nature, 359*, 550–551.

Reubinoff, B. E., Pera, M. F., Fong, C. Y., Trounson, A., & Bongso, A. (2000). Embryonic stem cell lines from human blastocysts: Somatic differentiation in vitro. *Nature Biotechnology, 18*, 399–404.

Roosen-Runge, E. C., & Leik, J. (1968). Gonocyte degeneration in the postnatal male rat. *The American Journal of Anatomy, 122*, 275–299.

Rossant, J., & Cross, J. C. (2001). Placental development: Lessons from mouse mutants. *Nature Reviews. Genetics, 2*, 538–548.

Sada, A., Hasegawa, K., Pin, P. H., & Saga, Y. (2012). NANOS2 acts downstream of glial cell line-derived neurotrophic factor signaling to suppress differentiation of spermatogonial stem cells. *Stem Cells, 30*, 280–291.

Sada, A., Suzuki, A., Suzuki, H., & Saga, Y. (2009). The RNA-binding protein NANOS2 is required to maintain murine spermatogonial stem cells. *Science, 325*, 1394–1398.

Saunders, P. T., Turner, J. M., Ruggiu, M., Taggart, M., Burgoyne, P. S., Elliott, D., et al. (2003). Absence of mDazl produces a final block on germ cell development at meiosis. *Reproduction, 126*, 589–597.

Schulze, C. (1979). Morphological characteristics of the spermatogonial stem cells in man. *Cell and Tissue Research, 198*, 191–199.

Setchell, B. P., Voglmayr, J. K., & Waites, G. M. (1969). A blood-testis barrier restricting passage from blood into rete testis fluid but not into lymph. *The Journal of Physiology, 200*, 73–85.

Shinka, T., & Nakahori, Y. (1996). The azoospermic factor on the Y chromosome. *Acta Paediatrica Japonica, 38*, 399–404.

Sigg, C., & Hedinger, C. (1984). Comparative ultrastructural and immunohistochemical investigations. *Virchows Archiv. A, Pathological Anatomy and Histopathology, 402*, 439–450.
Skakkebaek, N. E. (1972). Possible carcinoma-in-situ of the testis. *Lancet, 2*, 516–517.
Skakkebaek, N. E. (1978). Carcinoma in situ of the testis: Frequency and relationship to invasive germ cell tumours in infertile men. *Histopathology, 2*, 157–170.
Skakkebaek, N. E., Berthelsen, J. G., Giwercman, A., & Muller, J. (1987). Carcinoma-in-situ of the testis: Possible origin from gonocytes and precursor of all types of germ cell tumours except spermatocytoma. *International Journal of Andrology, 10*, 19–28.
Song, R., Hennig, G. W., Wu, Q., Jose, C., Zheng, H., & Yan, W. (2011). Male germ cells express abundant endogenous siRNAs. *Proceedings of the National Academy of Sciences of the United States of America, 108*, 13159–13164.
Spiller, C. M., Wilhelm, D., & Koopman, P. (2010). Retinoblastoma 1 protein modulates XY germ cell entry into G1/G0 arrest during fetal development in mice. *Biology of Reproduction, 82*, 433–443.
Stevens, L. C. (1958). Studies on transplantable testicular teratomas of strain 129 mice. *Journal of the National Cancer Institute, 20*, 1257–1275.
Suh, N., & Blelloch, R. (2011). Small RNAs in early mammalian development: From gametes to gastrulation. *Development, 138*, 1653–1661.
Tegelenbosch, R. A., & de Rooij, D. G. (1993). A quantitative study of spermatogonial multiplication and stem cell renewal in the C3H/101 F1 hybrid mouse. *Mutation Research, 290*, 193–200.
Tiepolo, L., & Zuffardi, O. (1976). Localization of factors controlling spermatogenesis in the nonfluorescent portion of the human Y chromosome long arm. *Human Genetics, 34*, 119–124.
Tokai, C. (1975). Letter: In situ lung carcinoma. *Acta Cytologica, 19*, 91–92.
Tsuda, M., Sasaoka, Y., Kiso, M., Abe, K., Haraguchi, S., Kobayashi, S., et al. (2003). Conserved role of nanos proteins in germ cell development. *Science, 301*, 1239–1241.
Turnbull, C., Rapley, E. A., Seal, S., Pernet, D., Renwick, A., Hughes, D., et al. (2010). Variants near DMRT1, TERT and ATF7IP are associated with testicular germ cell cancer. *Nature Genetics, 42*, 604–607.
van Beek, M. E., Davids, J. A., van de Kant, H. J., & de Rooij, D. G. (1984). Response to fission neutron irradiation of spermatogonial stem cells in different stages of the cycle of the seminiferous epithelium. *Radiation Research, 97*, 556–569.
van Casteren, N. J., de Jong, J., Stoop, H., Steyerberg, E. W., de Bekker-Grob, E. W., Dohle, G. R., et al. (2009). Evaluation of testicular biopsies for carcinoma in situ: Immunohistochemistry is mandatory. *International Journal of Andrology, 32*, 666–674.
van Dissel-Emiliani, F. M., de Boer-Brouwer, M., Spek, E. R., van der Donk, J. A., & de Rooij, D. G. (1993). Survival and proliferation of rat gonocytes in vitro. *Cell and Tissue Research, 273*, 141–147.
Vergouwen, R. P., Jacobs, S. G., Huiskamp, R., Davids, J. A., & de Rooij, D. G. (1991). Proliferative activity of gonocytes, Sertoli cells and interstitial cells during testicular development in mice. *Journal of Reproduction and Fertility, 93*, 233–243.
Waheeb, R., & Hofmann, M. C. (2011). Human spermatogonial stem cells: A possible origin for spermatocytic seminoma. *International Journal of Andrology, 34*, e296–e305, discussion e305.
Wang, P., Li, J., Allan, R. W., Guo, C. C., Peng, Y., & Cao, D. (2010). Expression of UTF1 in primary and metastatic testicular germ cell tumors. *American Journal of Clinical Pathology, 134*, 604–612.
Weathersbee, P. S. (1980). Early reproductive loss and the factors that may influence its occurrence. *The Journal of Reproductive Medicine, 25*, 315–318.
Western, P. S., Miles, D. C., van den Bergen, J. A., Burton, M., & Sinclair, A. H. (2008). Dynamic regulation of mitotic arrest in fetal male germ cells. *Stem Cells, 26*, 339–347.

Wu, X., Goodyear, S. M., Tobias, J. W., Avarbock, M. R., & Brinster, R. L. (2011). Spermatogonial stem cell self-renewal requires ETV5-mediated downstream activation of Brachyury in mice. *Biology of Reproduction*, *85*, 1114–1123.
Yang, Q. E., Gwost, I., Oatley, M. J., & Oatley, J. M. (2013). Retinoblastoma protein (RB1) controls fate determination in stem cells and progenitors of the mouse male germline. *Biology of Reproduction*, *89*, 113.
Yang, Q. E., Racicot, K. E., Kaucher, A. V., Oatley, M. J., & Oatley, J. M. (2013). MicroRNAs 221 and 222 regulate the undifferentiated state in mammalian male germ cells. *Development*, *140*, 280–290.
Yoshida, S., Sukeno, M., Nakagawa, T., Ohbo, K., Nagamatsu, G., Suda, T., et al. (2006). The first round of mouse spermatogenesis is a distinctive program that lacks the self-renewing spermatogonia stage. *Development*, *133*, 1495–1505.
Yoshioka, H., McCarrey, J. R., & Yamazaki, Y. (2009). Dynamic nuclear organization of constitutive heterochromatin during fetal male germ cell development in mice. *Biology of Reproduction*, *80*, 804–812.
Zheng, K., Xiol, J., Reuter, M., Eckardt, S., Leu, N. A., McLaughlin, K. J., et al. (2010). Mouse MOV10L1 associates with Piwi proteins and is an essential component of the Piwi-interacting RNA (piRNA) pathway. *Proceedings of the National Academy of Sciences of the United States of America*, *107*, 11841–11846.

CHAPTER TEN

The Origin, Biology, and Therapeutic Potential of Facultative Adult Hepatic Progenitor Cells

Soona Shin, Klaus H. Kaestner[1]
Department of Genetics and Institute for Diabetes, Obesity, and Metabolism, University of Pennsylvania School of Medicine, Philadelphia, Pennsylvania, USA
[1]Corresponding author: e-mail address: kaestner@mail.med.upenn.edu

Contents

Abstract

The liver plays an essential role in glucose and lipid metabolism, synthesis of plasma proteins, and detoxification of xenobiotics and other toxins. Chronic disease of this important organ is one of the leading causes of death in the United States. Following loss of tissue, liver mass can be restored by two mechanisms. Under normal conditions, or after massive loss of parenchyma by surgical resection, liver mass is maintained by division of hepatocytes. After chronic injury, or when proliferation of hepatocytes is

Current Topics in Developmental Biology, Volume 107
ISSN 0070-2153
http://dx.doi.org/10.1016/B978-0-12-416022-4.00010-X

impaired, facultative adult hepatic progenitor cells (HPCs) proliferate and differentiate into hepatocytes and cholangiocytes (biliary epithelial cells). HPCs are attractive candidates for cell transplantation because of their potential contribution to liver regeneration. However, until recently, the lack of highly specific markers has hampered efforts to better understand the origin and physiology of HPCs. Recent advances in cell isolation methods and genetic lineage tracing have enabled investigators to explore multiple aspects of HPC biology. In this review, we describe the potential origins of HPCs, the markers used to detect them, the contribution of HPCs to recovery, and the signaling pathways that regulate their biology. We end with an examination of the therapeutic potential of HPCs and their derivatives.

1. INTRODUCTION

The mammalian liver has a remarkable ability to regenerate its functional mass in response to tissue loss. In surgical models of liver resection, such as 70% partial hepatectomy (PH) in rat, the remaining uninjured hepatocytes proliferate and replace the parenchyma within 20 days (Martins, Theruvath, & Neuhaus, 2008). Under circumstances in which hepatocyte proliferation is blocked, as is the case after toxic liver injury, small cells that have scant cytoplasm and oval-shaped nuclei proliferate in the portal area and are thought to contribute to replacement of the parenchyma (Libbrecht & Roskams, 2002; Yovchev et al., 2008). These hepatic progenitor cells (HPCs), also called "oval cells" due to their morphology (Farber, 1956; Yovchev et al., 2008), can undergo bidirectional differentiation into hepatocytes and cholangiocytes (biliary epithelial cells) at least in experimental conditions (Fig. 10.1; Okabe et al., 2009; Shin et al., 2011). Hepatic progenitors are very different from tissue-resident stem cells in other epithelial tissue such as intestine and skin. In the latter tissues, stem and progenitor cells are required throughout life to replenish cells lost daily, and without continued replication of the tissue-resident stem cells, the epithelia of intestine and skin fail rapidly. In contrast, hepatic progenitors are "facultative," meaning (1) they are not needed to replenish liver tissue under normal, healthy conditions and (2) that many markers of HPCs are only expressed in the liver after injury and when the progenitor cells are activated.

The terms "HPCs," "oval cells," "liver progenitor cells," and "hepatic stem cells" describe the heterogeneous population of cells that have been proposed to maintain and regenerate liver during the repair process. To avoid confusion, in this review we will use the term HPCs to denote the epithelial component located within ductular reactions in the injured adult

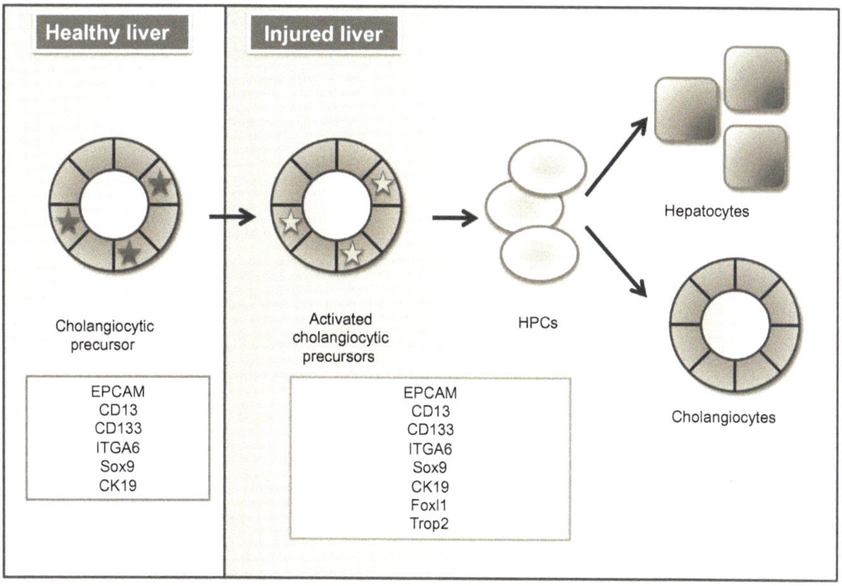

Figure 10.1 Adult hepatic progenitor cells. A subset of cholangiocytes is activated upon injury and gives rise to HPCs that can differentiate into hepatocytes and cholangiocytes. (See color plate.)

liver (Roskams et al., 2004). The "ductular reaction" is defined as the proliferation of apparent ductules that accompanies leukocyte infiltration and deposition of extracellular matrix (ECM) in response to liver injury (Roskams & Desmet, 1998; Roskams et al., 2004). The markers, origin, fate, and regenerative capability of these HPCs remain the subject of controversy in the field. In this review, we attempt to provide an overview of recent advances and to point out unanswered questions.

2. CONDITIONS THAT ACTIVATE HPC PROLIFERATION

The mechanism of ductular reaction initiation and proliferation of HPCs are poorly characterized. In humans, ductular reactions have been observed in multiple diseases such as fulminant hepatic failure, focal nodular hyperplasia, primary biliary cirrhosis, primary sclerosing cholangitis (Turanyi et al., 2010), cancer (Farber, 1956; Libbrecht & Roskams, 2002), pediatric nonalcoholic fatty liver disease (Nobili et al., 2012), genetic hemochromatosis, alcoholic liver disease, and chronic hepatitis C (Lowes, Brennan, Yeoh, & Olynyk, 1999). The number of HPCs correlates with the severity of liver diseases in human and dog (Lowes et al., 1999; Schotanus et al., 2009). In humans, a minimum of 50% hepatocyte loss is required for significant activation of the HPC compartment (Katoonizadeh, Nevens, Verslype, Pirenne, & Roskams, 2006), and there is an inverse correlation between the number of HPCs and the number of hepatocytes that express the proliferation marker Ki67 (Katoonizadeh et al., 2006). This suggests that a combination of hepatocyte loss and impaired hepatocyte proliferation is required to activate HPC proliferation. In rats, treatment with carbon tetrachloride (CCl_4) preferentially induced proliferation of hepatocytes, and proliferation of nonparenchymal cells required simultaneous administration of D-galactosamine, which blocks RNA and protein synthesis in hepatocytes (Dabeva, Alpini, Hurston, & Shafritz, 1993; Dabeva & Shafritz, 1993). Furthermore, in rats that had undergone PH, the expansion of HPCs was negligible (Dusabineza et al., 2012). Significant HPC induction was only observed when 2-acetaminofluorene (2-AAF) was added to block cell cycle progression in hepatocytes (Dusabineza et al., 2012; Trautwein et al., 1999). An inverse correlation between the activation of HPCs and hepatocyte proliferation was also observed in the canine liver (Schotanus et al., 2009), providing strong evidence that suppression of hepatocyte proliferation is required for activation of the HPC compartment. In mice, however, inhibition of the hepatocyte cell cycle might not be a definitive criterion for activation of the ductular reaction. Various treatments such as CCl_4 (Chu et al., 2011; Espanol-Suner et al., 2012; Malato et al., 2011; Rountree, Barsky, et al., 2007), choline-deficient, ethionine-supplemented diet (CDE) (Espanol-Suner et al., 2012; Knight & Yeoh, 2005; Tirnitz-Parker, Tonkin, Knight, Olynyk, & Yeoh, 2007), methionine choline-deficient diet supplemented with ethionine (MCDE) (Huch et al., 2013), bile duct ligation (BDL) (Chu et al., 2011; Malato et al., 2011; Sackett, Li, et al., 2009), PH (Malato et al., 2011), alpha-naphthylisothiocyanate (ANIT) (Rountree, Barsky, et al., 2007), high-fat diet combined with ethanol (Jung et al., 2008), and 3,5-diethoxycarbonyl-1,4-dihydrocollidine (DDC) (Carpentier et al., 2011; Malato et al., 2011; Rountree, Barsky, et al., 2007; Sackett, Li, et al., 2009; Shin et al., 2011) have been successfully used to induce proliferation of HPCs in mice in the absence of impaired hepatocyte proliferation. Rountree, Barsky, et al. (2007) tested multiple means of inducing HPC reaction in mouse models. CCl_4 preferentially induced proliferation of hepatocytes, ANIT preferentially induced proliferation of cholangiocytes, and DDC-induced cycling of both hepatocytes and cholangiocytes, and the models also induced proliferation of HPCs (Rountree, Barsky, et al., 2007). Moreover, another group reported that administration of CCl_4 or PH led to formation of new hepatocytes from HPCs (Malato et al., 2011). These results suggest that inhibition of hepatocyte proliferation is not necessarily a prerequisite for the activation of HPCs under all physiological conditions.

Studies using genetic mouse models have provided further insight. Endo and colleagues have demonstrated that deletion of *damaged DNA-binding protein 1* (*Ddb1*), an adaptor for Cullin4 ubiquitin ligase (Higa & Zhang, 2007), in hepatocytes led to the activation of HPC proliferation (Endo, Zhang, Yamaji, & Cang, 2012). Ablation of *Ddb1* caused only minor liver damage compared to the degree of injury induced by administration of DDC. Deletion of *Ddb1* also led to a blockade of hepatocyte proliferation, which was partially due to accumulation of p21. It is also worth noting that thermoreversible gelation polymer, a synthetic material that polymerizes when exposed to temperatures above a defined threshold, induced ductular reactions when injected directly into normal rat liver (Nagaya, Kubota, Suzuki, Akashi, & Mitaka, 2006). This suggests that ECM can provide cues for activation of the HPC compartment. In Section 5, we will delve more deeply into the signaling pathways and molecular mechanism that regulate the biology of HPCs.

3. ORIGINS AND MARKERS FOR HPCs

While there is little doubt about the presence of facultative HPCs in the injured liver, the existence and identity of HPC precursors in a healthy, uninjured liver remains controversial. For example, a study using incorporation of 5-bromo-2′-deoxyuridine identified multiple niches for label-retaining cells such as the proximal biliary branches, including the canal of Hering and intralobular bile ducts (Kuwahara et al., 2008). This review summarizes the ongoing debate about the possible origins of HPCs which include: (1) biliary epithelial cells, (2) progenitor cells persisting from the fetal liver, (3) bone marrow cells, (4) hepatic stellate cells, and (5) mature adult hepatocytes.

3.1. Biliary origin of facultative HPCs

Multiple lines of evidence suggest that cholangiocytes (biliary epithelial cells) serve as facultative stem cells in the liver. Thus, these differentiated cells may acquire stem cell-like characteristics under particular conditions (Yanger & Stanger, 2011). Recent work has demonstrated the utility of protein markers in yielding insights into HSC physiology. Table 10.1 summarizes markers

Table 10.1 Markers for HPCs

Marker	Diseases (models)	Species	References
A6	DDC, CDE, viral hepatitis, deletion of *Ddb1*	Mouse	Endo et al. (2012), Fan et al. (2012), and Tonkin, Knight, Curtis, Abraham, and Yeoh (2008)
AFP	Acute necrotizing hepatitis, HCV, primary biliary cirrhosis	Human	Spee et al. (2010)
	2-AAF/PH	Rat	Jensen et al. (2004) and Yovchev et al. (2008)
CD44	Acute necrotizing hepatitis, HCV, primary biliary cirrhosis	Human	Spee et al. (2010)
	2-AAF/PH	Rat	Yovchev et al. (2008)
	DDC, deletion of *Ddb1*	Mouse	Endo et al. (2012) and Shin et al. (2011)
CD133	Alcoholic hepatitis, acute necrotizing hepatitis, HCV, HBV, primary biliary cirrhosis	Human	Porretti et al. (2010), Sancho-Bru et al. (2012), and Spee et al. (2010)
	2-AAF/PH	Rat	Yovchev et al. (2008)
	DDC, ANIT, CCl_4	Mouse	Dorrell et al. (2011), Rountree, Barsky, et al. (2007), and Suzuki et al. (2008)
CK7	Pediatric nonalcoholic fatty liver disease, alcoholic hepatitis, HCV, primary biliary cirrhosis	Human	Nobili et al. (2012), Sancho-Bru et al. (2012), and Spee et al. (2010)
	2-AAF/PH	Rat	Yovchev et al. (2008)
	DDC	Mouse	Suzuki et al. (2008)
CK19	Acetaminophen-induced necrosis, genetic hemochromatosis, alcoholic liver disease, HCV, HBV, acute necrotizing hepatitis, primary biliary cirrhosis	Human	Feng et al. (2012), Lowes et al. (1999), Spee et al. (2010), and Theise et al. (1999)
	2-AAF/PH	Rat	Dusabineza et al. (2012) and Yovchev et al. (2008)
	DDC, CDE, deletion of *Ddb1*	Mouse	Carpentier et al. (2011), Endo et al. (2012), Fan et al. (2012), and Takase et al. (2013)

Table 10.1 Markers for HPCs—cont'd

Marker	Diseases (models)	Species	References
CLDN4	2-AAF/PH	Rat	Yovchev et al. (2008)
	Deletion of *Ddb1*	Mouse	Endo et al. (2012)
DLK1	Retrorsine, 2-AAF/PH	Rat	Jensen et al. (2004)
EPCAM	Alcoholic hepatitis, HCV, HBV	Human	Porretti et al. (2010) and Sancho-Bru et al. (2012)
	2-AAF/PH	Rat	Yovchev et al. (2008)
	DDC, deletion of *Ddb1*	Mouse	Endo et al. (2012), Fan et al. (2012), Okabe et al. (2009), Shin et al. (2011), and Takase et al. (2013)
Foxl1	BDL, DDC	Mouse	Sackett, Li, et al., (2009) and Shin et al. (2011)
GGT	2-AAF/PH, GalN	Rat	Dabeva and Shafritz (1993) and Yovchev et al. (2008)
ITGB4	AAF-PH	Rat	Yovchev et al. (2008)
	Deletion of *Ddb1*	Mouse	Endo et al. (2012)
ITGA6	HCV, HBV	Human	Porretti et al. (2010)
	2-AAF/PH	Rat	Yovchev et al. (2008)
Lgr5	CCl_4, DDC, MCDE	Mouse	Huch et al. (2013)
M2PK	Genetic hemochromatosis, alcoholic liver disease, HCV	Human	Lowes et al. (1999)
	CDE	Mouse	Davies, Knight, Tian, Yeoh, and Olynyk (2006)
NCAM	HCV, HBV, acute necrotizing hepatitis, primary biliary cirrhosis	Human	Porretti et al. (2010) and Spee et al. (2010)
OPN	DDC, CDE, CCl_4, PH	Mouse	Carpentier et al. (2011) and Espanol-Suner et al. (2012)
Sox9	DDC, CDE	Mouse	Carpentier et al. (2011)
Trop2	DDC, CDE	Mouse	Carpentier et al. (2011), Okabe et al. (2009), and Shin et al. (2011)

2-AAF, 2-acetaminofluorene; ANIT, alpha-naphthylisothiocyanate; CCl_4, carbon tetrachloride; CDE, choline-deficient, ethionine-supplemented diet; *Ddb1*, damaged DNA-binding protein 1; DDC, 3,5-diethoxycarbonyl-1,4-dihydrocollidine; GalN, D-galactosamine; HBV, Hepatitis B virus; HCV, Hepatitis C virus; MCDE, methionine choline-deficient diet supplemented with ethionine; PH, partial hepatectomy.

that are commonly used to isolate or detect HPCs in different injury models or disease contexts. Unfortunately, many of the widely used markers such as epithelial cell adhesion molecule (EPCAM) (Inada et al., 2008; Jozefczuk et al., 2010; Kamiya, Kakinuma, Yamazaki, & Nakauchi, 2009; Okabe et al., 2009; Schmelzer et al., 2007; Yovchev et al., 2008), CD133 (encoded by the *Prom1* gene) (Dorrell et al., 2011; Jozefczuk et al., 2010; Kamiya et al., 2009; Rountree, Barsky, et al., 2007; Suzuki et al., 2008), SRY (sex-determining region Y)-box 9 (Sox9) (Carpentier et al., 2011; Dorrell et al., 2011), and osteopontin (OPN) (Espanol-Suner et al., 2012) are found both in HPCs of the injured liver and in cholangiocytes and/or non-parenchymal cells of the healthy liver (Dorrell et al., 2008, 2011; Espanol-Suner et al., 2012; Furuyama et al., 2011; Inada et al., 2008; Rountree, Barsky, et al., 2007). Furthermore, facultative HPCs were reported to express the ductal markers cytokeratin 19 (CK19) (Espanol-Suner et al., 2012; Furuyama et al., 2011; Grozdanov, Yovchev, & Dabeva, 2006; Inada et al., 2008; Kamiya et al., 2009; Takase et al., 2013; Theise et al., 1999; Yovchev et al., 2008) and MIC1-1C3 (Dorrell et al., 2011).

The application of the above markers has been limited by their lack of cell-type specificity. Forkhead box L1 (*Foxl1*) (Shin et al., 2011), Trop2 (encoded by tumor-associated calcium signal transducer 2) (Okabe et al., 2009), and Lgr5 (leucine-rich-repeat-containing G-protein-coupled receptor 5) (Huch et al., 2013) are three newly described markers expressed exclusively by HPCs of the injured liver. Furthermore, *Foxl1*$^{+}$ cells and Trop2^{+} cells express EPCAM and CK19 (Okabe et al., 2009; Sackett, Li, et al., 2009; Shin et al., 2011), and organoids expressing Lgr5 could be grown from biliary duct fragments in culture (Huch et al., 2013), further supporting the concept that at least some types of HPCs can originate from cholangiocytes. Furthermore, gene expression analyses comparing *Foxl1*$^{+}$ HPCs, cholangiocytes, and hepatocytes demonstrate that the gene expression profile of *Foxl1*$^{+}$ HPCs resembles that of cholangiocytes rather than hepatocytes, supporting the idea that HPCs are derived from a cholangiocytic lineage (Shin et al., 2011).

Moreover, a number of reports have demonstrated that the proliferating HPC compartment is connected to preexisting ductules, an observation that strongly supports a ductal origin for HPCs (Alpini et al., 1992; Dusabineza et al., 2012; Theise et al., 1999). Interestingly, although CD133^{+} cells are found in both normal and injured liver, only CD133^{+} cells isolated from injured liver were self-renewable (Suzuki et al., 2008). Kamiya et al. (2009)

described the existence of "cholangiocytic precursor cells" that were present in the healthy liver. These cells expressed CD13 (also known as alanyl (membrane) aminopeptidase), integrin alpha 6 (ITGA6 or CD49f), CD133, and CK19 *in vivo*, and when these cells were cultured, they were self-renewable and began to express markers of HPCs (CD44) and hepatocytes (albumin). These results further support the idea that at least a subset of cholangiocytes can act as precursors of facultative HPCs. It has been difficult to ascertain precisely how activated cholangiocytes and HPCs differ from each other, as the two compartments share protein markers (Fig. 10.1; Kamiya et al., 2009). It is also worth mentioning that cholangiocytes comprise a heterogeneous population of cells (Roskams et al., 2004), and it would therefore be interesting to investigate whether cholangiocytic precursor cells are different from other cholangiocytes in terms of function, gene expression profile, or physiology.

In humans, Schmelzer, Wauthier, and Reid (2006) identified cells found in the normal liver from fetal and postnatal donors that are capable of being expanded in culture. These cells, termed human hepatic stem cells (hHpSCs), expressed EPCAM, CK19, neural cell adhesion molecule 1 (NCAM), claudin 3 (CLDN3), Kit, and aquaporin 4. The same group reported that EPCAM-expressing cells exist in the Canal of Hering in the livers of postnatal donors, and that these cells can repopulate the liver of immunodeficient mice (Schmelzer et al., 2007). Furthermore, the existence of an even more primitive stem population in the biliary tree has been reported. These stem cells are a heterogeneous population that give rise to hepatocytes, cholangiocytes, and pancreatic islets. At least a subset of these multipotent stem cells were positive for pancreatic and duodenal homeobox 1 (PDX1), SRY-box 17 (SOX17), and EPCAM (Carpino et al., 2012). At present, the relationship between the fetally derived hHpSCs and the facultative hepatic progenitors that appear in the adult liver following injury described earlier remains unresolved.

3.2. Embryonic origin of HPCs

Hepatoblasts are bipotential cells found in the fetal liver (between E9.5 and E14.5 in mice) that have the ability to differentiate into hepatocytes and cholangiocytes *in vivo*. Because both hepatoblasts and HPCs are proliferating and are capable of differentiating to hepatocytes and cholangiocytes, it has been suggested that fetal hepatoblasts give rise to HPCs (Schmelzer et al., 2007). The fact that hepatoblasts and HPCs share the same markers, such as delta-like 1 homolog (Jensen et al., 2004; Qiu, Hernandez, Dean, Rao, & Darlington, 2011; Yovchev et al., 2008) and alpha-fetoprotein (AFP) (Jensen et al., 2004; Yovchev et al., 2008), is cited as evidence that HPCs are derived from hepatoblasts. Fetal cells also share other characteristics of HPCs, such as clonogenicity and the ability to repopulate liver *in vivo*. For instance, the $CD13^{+}DLK^{+}$ fraction of cells from murine fetal liver has the ability to form colonies in culture (Kakinuma et al., 2009), and the $AFP^{+}CK19^{+}$ fraction from rat fetal liver has the ability to repopulate liver *in vivo* (Dabeva et al., 2000). As previously mentioned, the existence of hHpSCs that are different from hepatoblasts has also been reported in human fetal liver. These hHpSCs were observed in the ductal plate of the fetal liver and in the Canal of Hering of adult liver (Schmelzer et al., 2007). In mice, Carpentier et al. (2011) reported that the ductal plate of the fetal liver serves as the precursor for cholangiocytes, periportal hepatocytes, and HPCs in adult liver. However, one cannot rule out the possibility that the ductal plate gives rise to the fetal/neonatal adult cholangiocytes, which in turn give rise to facultative HPCs when the need arises. The main issue with this model, however, is that it requires that fetal hepatoblasts persist in the adult liver. Most studies have not observed either a $CD13^{+}DLK^{+}$ or a $AFP^{+}CK19^{+}$ fraction of cells in the healthy, uninjured liver, which argues against the continued presence of fetal hepatoblasts in the adult organ. Furthermore, the expression of AFP in the adult liver is not sufficient to prove descendancy from a persistent fetal hepatoblast, it could simply the case that the *Afp* promoter (or that of other fetally expressed genes) is reactivated in adult facultative HPCs. An analogous line of reasoning holds for hepatocellular carcinoma: merely because some tumors share characteristics with fetal hepatoblasts (such as rapid cycling and AFP expression) does not prove that the tumors originated from fetal hepatoblasts residing in the adult organ.

3.3. Bone marrow as the source of HPCs

In 1999, Petersen et al. (1999) demonstrated that when they transplanted bone marrow from male rats into female rats, 0.14% of hepatocytes carried the Y chromosome. They also reported that bone marrow-derived cells differentiated into hepatocytes, although the extent of their contribution to the parenchyma was less than 1% (Piscaglia, Shupe, Oh, Gasbarrini, & Petersen, 2007). The hypothesis that HPCs can originate from bone marrow is further supported by the fact that HPCs and hematopoietic stem cells express the same markers, such as Thy1 (Petersen, Goff, Greenberger, & Michalopoulos, 1998), CD34 (Omori et al., 1997), and Kit (Fujio, Evarts, Hu, Marsden, & Thorgeirsson, 1994). Subsequently, two independent groups transplanted bone marrow cells and detected donor-derived hepatic nodules in the livers of fumarylacetoacetate hydrolase (FAH)-deficient recipient mice (Vassilopoulos, Wang, & Russell, 2003; Wang et al., 2003). However, hepatocytes positive for FAH derived from transplanted bone marrow cells were shown subsequently to be the result of cell fusion rather than actual differentiation of bone marrow cells into HPCs or hepatocytes (Wang et al., 2003).

To address this discrepancy, two independent research groups transplanted GFP-expressing bone marrow cells or lacZ-transgenic bone marrow cells, respectively, and demonstrated that bone marrow cells played a negligible role in liver regeneration when recipient mice were subjected to CCl_4, ANIT, DDC, or CDE diet-mediated liver injury (Rountree, Wang, et al., 2007; Tonkin et al., 2008). It has also been demonstrated that bone marrow cells do not contribute to HPCs and hepatocytes in rats (Menthena et al., 2004). Finally, recent studies demonstrated that EPCAM-positive HPCs do not express Thy1, Kit, and CD34 (Okabe et al., 2009; Yovchev et al., 2008). Similarly, FACS-sorted $CD34^+$ and $Thy1^+$ cells do not express genes associated with HPCs, hepatocytes, and cholangiocytes (Rountree, Barsky, et al., 2007). In sum, these results strongly suggest that the bone marrow is not a major source of HPCs.

3.4. Hepatic stellate cells

A recent report proposed hepatic stellate cells as a type of HPCs. This study was based on the use of glial fibrillary acidic protein (GFAP), a marker of stellate cells, for genetic lineage tracing with the Cre-loxP technology (Fig. 10.2). *Gfap*-Cre-labeled cells contributed to the HPC compartment when mice were fed an MCDE diet (Yang et al., 2008). However, the *Gfap*-Cre driver employed is not only active in hepatic stellate cells, but also in a subset of cholangiocytes (Ueberham, Bottger, Ueberham, Grosche, & Gebhardt, 2010). Thus, the authors could not exclude the possibility that the facultative HPCs observed in their mice arose from cholangiocytes. In fact, a subsequent study by Scholten et al. (2010) provided strong evidence that stellate cells do not contribute to the HPC compartment. Given that stellate cells are derived from a different embryonic germ layer than are hepatocytes and cholantiocytes (from the mesoderm and endoderm, respectively), the transdifferentiation of stellate cells into HPCs might require too large a shift in the gene expression program to occur naturally.

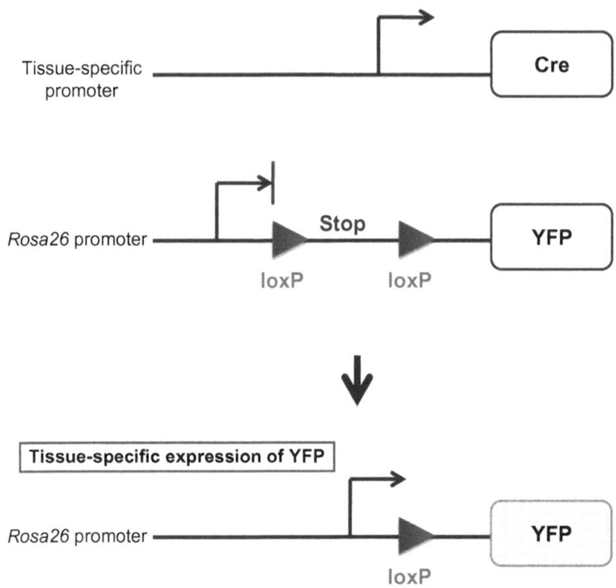

Figure 10.2 Lineage tracing using the Cre-LoxP approach. *Rosa26* promoter is ubiquitously active in mice. However, transcription of YFP reporter is blocked because of the transcriptional stop sequence (Srinivas et al., 2001). The Cre recombinase excises the sequence located between two loxP sites thus allowing for the expression of YFP in a specific cell type. (See color plate.)

3.5. Adult hepatocytes as facultative HPCs

Recently, Yanger and Stanger (2011) proposed that hepatocytes can undergo reprogramming into biliary lineage. They injected AAV8-TBG-Cre into $Rosa^{YFP}$ mice to lineage trace fully differentiated hepatocytes in the adult liver. Upon DDC treatment, cells that expressed both biliary maker A6 and hepatocyte marker HNF4α emerged. Forty-one percent of these $A6^+$ $HNF4\alpha^+$ cells were binucleated, and staining for YFP further confirmed that these cells originated from hepatocytes. They identified a substantial number of YFP^+ cells that acquired a biliary morphology and expressed OPN, A6, Sox9, or CK19. They also identified binucleated HPCs in the livers of DDC-fed mice but not in the livers of control mice. These results suggest that hepatocytes can differentiate into cholangiocytes and/or HPCs in response to liver injury.

3.6. Summary

The contradictory findings regarding the origins and location of HPCs may suggest multiple origins (Kuwahara et al., 2008) or multiple maturational stages. The existence of multiple stages of HPCs has been discussed (reviewed in Duncan, Dorrell, & Grompe, 2009; Turner et al., 2011), but conclusive evidence is lacking. Future work will be required to address the exact lineage relationships between different types of precursor cells reported in the literature over the past 20 years.

4. THE CONTRIBUTION OF HPCs TO LIVER REGENERATION

Expansion of proposed HPCs has been observed in various disease contexts in human as well as in rodents (Table 10.1). However, the contribution of HPCs to restoration of parenchymal architecture and liver function has been assumed, but not proven (Takase et al., 2013). Thus far, it has been demonstrated that HPCs form both typical and atypical ducts with poorly defined lumina (Libbrecht & Roskams, 2002; Yaswen, Hayner, & Fausto, 1984), and it remains unclear whether these ducts are fully functional. While formation of tube-like structure in culture has been used as indication of HPCs' ability to differentiate into cholangiocytes (Kamiya et al., 2009; Shin et al., 2011), this assay does not provide definitive proof that HPC-derived cholangiocytes are fully integrated into the biliary tree. Conclusive evidence to support the hypothesis that endogenous HPCs form functional ducts *in vivo* is still lacking.

The contribution of HPCs to hepatocytes was also the subject of debate until recently. Genetic lineage tracing using *Foxl1*-Cre; $Rosa^{LacZ}$ mice demonstrated that the number of β-gal$^+$/HNF4α$^+$ cells increased significantly 2 weeks after BDL, suggesting that *Foxl1*-Cre marked HPCs can form new hepatocytes (Sackett, Li, et al., 2009). In addition, a hepatocyte fate tracing using an adeno-associated viral vector AAV8-Ttr-Cre in combination with $Rosa^{YFP}$ mice was employed in multiple disease models such as PH, DDC, BDL, and CCl_4 (Malato et al., 2011). The authors found newly formed hepatocytes derived from progenitor cells, although the contribution was limited. Similarly, another group has reported the existence of rare CK19$^+$ or GSTP$^+$CK19$^-$ intermediate hepatocytes in the 2-AAF/PH rat model (Dusabineza et al., 2012). In humans, intermediate hepatocytes that express CK7 or CK19 have been observed as well (Katoonizadeh et al., 2006).

Sackett, Gao, et al. (2009) demonstrated the correlation between impaired ductular reactions and impaired liver regeneration in $Foxl1^{-/-}$ mice after BDL. The expansion of the CK19$^+$ ductular area was reduced

14 days after BDL in the *Foxl1*$^{-/-}$ livers, while the control livers exhibited a progressive increase in CK19^{+} ductular area. This effect correlated with elevated serum AST levels, delayed proliferation of hepatocytes, and increased parenchymal necrosis in *Foxl1*$^{-/-}$ mice compared to the control mice. These results strongly support the idea that HPCs contribute to liver regeneration. Similarly, Takase et al. (2013) demonstrated that forced expression of fibroblast growth factor 7 (FGF7) causes an increased HPC response in conjunction with attenuated liver damage and mortality following DDC-induced injury. FGF7, which is induced in the liver during the HPC response, was suggested to stimulate proliferation of HPCs (Takase et al., 2013). In this model, newly formed A6^{+}CK19^{-} hepatocytes were dramatically increased around A6^{+}CK19^{+} HPCs in *Fgf7*-transgenic mice but not in control mice. They demonstrated that *Fgf7*-deficient mice exhibit decreased HPC expansion as well as higher mortality and hepatic injury upon DDC treatment.

The small number of HPC-derived hepatocytes may be explained by the finding that ECM deposition during HPC activation prevents expansion and migration of progeny to parenchyma (Espanol-Suner et al., 2012). When mice were exposed to a CDE diet for 3 weeks and then to chow diet for 2 weeks, the number of hepatocytes originating from *Opn*-iCreERT2^{+} HPCs was significantly increased compared to the mice exposed to the CDE diet alone, indicating that HPCs might play a role during the recovery phase rather than during disease progression. In sum, HPCs contribute to parenchyma of the liver following injury, and although the extent to which they do so may be minor in the experimental models tested, expanded HPCs might still be targeted for the treatment of liver diseases.

5. SIGNALING PATHWAYS AND NICHES THAT REGULATE THE BEHAVIOR OF HPCs

The signaling pathways that promote proliferation and activation of the HPC compartment, the migration of HPCs, and subsequent cell fate specification have been extensively studied. Knowledge of these pathways is essential for targeting HPCs to promote expansion and differentiation in patients with liver disease using small molecules or growth factors.

5.1. The source of signaling molecules that interact with HPCs

The ductular reaction comprises not only the expansion of HPCs, but also that of mesenchymal and inflammatory cells that surround HPCs, as well as the subsequent deposition of ECM (Roskams & Desmet, 1998; Roskams et al., 2004). Macrophages and T-cells interact with HPCs to regulate proliferation, differentiation, and migration of HPCs (Boulter et al., 2012; Lorenzini et al., 2010; Van Hul, Abarca-Quinones, Sempoux, Horsmans, & Leclercq, 2009; Viebahn et al., 2010). Hepatic stellate cells or myofibroblasts have also been observed in close proximity of HPCs both in mouse (Sackett, Li, et al., 2009) and in patients with liver diseases (Schotanus et al., 2009), suggesting that these cells may interact with HPCs. In addition, two reports demonstrated that Thy1$^+$ mesenchymal cells can act as the niche for HPCs (Takase et al., 2013; Yovchev et al., 2008). In addition, the ECM can also be a factor that regulates HPC activity and function (Carpentier et al., 2011; Lozoya et al., 2011; Van Hul et al., 2009).

5.2. Signaling pathways that regulate activation, survival, and proliferation of HPCs

Two growth factors, FGF7 and granulocyte-colony stimulating factor (G-CSF), have been proposed to act as mitogens for HPCs (Piscaglia et al., 2007; Takase et al., 2013). FGF7 promotes the expansion of HPCs *in vivo*, and Thy1$^+$ cells can act as a niche for activation of HPCs (Takase et al., 2013). G-CSF also has been shown to stimulate proliferation and migration of HPCs in rats (Piscaglia et al., 2007). Additionally, mice deficient of Bmi1, a component of the polycomb repressive complex 1 (Wu & Yang, 2011), exhibited reduced expansion of HPCs in response to DDC injury (Fan et al., 2012). Tirnitz-Parker and colleagues demonstrated that injection of "tumor necrosis factor-like weak inducer of apoptosis" to CDE-fed animals led to an increased number of HPCs and invasion of HPCs into the parenchyma. This effect correlated with activation of NF-κB (Tirnitz-Parker et al., 2010). Small molecule inhibitors of signaling pathways have also proven useful for the investigation of HPC biology. For example, SC-236, an inhibitor of cyclooxygenase 2, attenuated the CDE-induced HPC response (Davies et al., 2006). Treatment of CDE-fed mice with the PPAR gamma agonist ciglitazone resulted in reduction of HPC response as well (Knight, Yeap, Yeoh, & Olynyk, 2005).

The EPCAM$^+$ cells containing HPCs isolated from human fetal livers are enriched for components of hedgehog signaling pathway, specifically Indian hedgehog and patched 1. Cyclopamine, an inhibitor of hedgehog signaling, induced cell death of EPCAM$^+$ cells in culture, implying that hedgehog signaling plays a role in maintaining survival of this cell population (Sicklick et al., 2006). In addition, the Rho-associated protein kinase inhibitor Y-27632 significantly improved clonogenicity of HPCs in culture (Kamiya et al., 2009; Shin et al., 2011). Thus, multiple signaling pathways act in concert to effect the activation and proliferation of HPCs, and understanding their specific contributions will require further study.

Multiple cytokines are involved in the activation of the ductular reaction and proliferation of HPCs. These cytokines include tumor necrosis factor (Knight & Yeoh, 2005; Viebahn et al., 2010), lymphotoxin alpha (Knight & Yeoh, 2005), lymphotoxin beta (Akhurst et al., 2005), interferon gamma (Akhurst et al., 2005), interleukin-22 (Feng et al., 2012), and interleukin 6 (Yeoh et al., 2007). Signal transducer and activator of transcription 3 signaling has been suggested as the central mediator of activation of the HPC response to interleukins and interferon gamma (Akhurst et al., 2005; Feng et al., 2012; Yeoh et al., 2007). These findings suggest that the inflammatory response to liver injury plays a key role in the activation of HPC response.

5.3. Signaling pathways that regulate the specification of HPCs

There is increasing evidence to suggest that the Wnt and Notch signaling pathways are key regulators of fate determination of HPCs in rodents and humans (Boulter et al., 2012; Spee et al., 2010). In humans, acute necrotizing hepatitis causes activation of the Wnt signaling pathway (Spee et al., 2010), and Notch signaling was activated in HPCs in patients with primary biliary cirrhosis. These results imply that the Wnt pathway is involved in directing HPCs toward the hepatocytic fate, while the Notch pathway promotes differentiation of HPCs toward the biliary lineage (Spee et al., 2010). Similarly, in mice, myofibroblasts were identified as a source of Jagged 1, which activates Notch signaling in HPCs and promotes specification of cholangiocytes. In contrast, Wnt3a expressed by macrophages led to specification of HPCs to hepatocytes (Boulter et al., 2012).

Interaction between HPCs and the ECM may also influence the fate of HPCs. In the liver of CDE-fed mice, deposition of ECM occurs prior to the expansion of HPCs (Van Hul et al., 2009). Inhibition of transforming growth factor beta-induced accumulation of ECM resulted in an increased number of HPC-derived hepatocytes, suggesting that the ECM is also involved in fate determination of HPCs (Espanol-Suner et al., 2012).

Interestingly, a recent report demonstrated that HPCs can also express resistin and glucagon-like peptide-1, and their expression levels correlated with the degree of fibrosis and steatosis in pediatric nonalcoholic fatty liver disease, implying existence of signaling from HPCs to the surrounding cells (Nobili et al., 2012). Further investigation of the crosstalk between HPCs and their surrounding niche will be required in order to effectively direct first proliferation and then differentiation of these cells in culture so that they can be tested as a cell therapeutic.

6. CELL THERAPY

The utilization of HPCs for cell therapy is an attractive alternative to organ transplantation for the treatment of liver disease. There are several advantages to using HPCs instead of primary hepatocytes or whole organ transplantation. First, a single HPC can be expanded in culture without losing its bidirectional differentiation potential (Shin et al., 2011; Suzuki et al., 2008). Second, the fact that HPCs are found in the liver with various types of disorders (Lowes et al., 1999; Nobili et al., 2012; Turanyi et al., 2010) suggests that HPCs can be isolated from diseased livers that cannot be used for organ transplantation. Third, it might be possible to isolate HPCs from a patient who has liver disease, expand them in culture, and transplant back to the patient. This autologous transplant protocol would obviate the need for immunosuppression after transplantation. Finally, HPCs are smaller than hepatocytes, and it has been suggested that smaller cells might have less of a tendency to cause portal hypertension following injection into the portal vein (Sandhu, Petkov, Dabeva, & Shafritz, 2001).

Multiple laboratories have demonstrated that transplanted HPCs have the ability to differentiate into hepatocytes *in vivo* (Dorrell et al., 2011; Endo et al., 2012; Kamiya et al., 2009; Suzuki et al., 2008; Yovchev et al., 2008). Currently, preconditioning of the recipient liver is required for engraftment and expansion of transplanted cells. For example, engraftment of hHpSCs was enhanced when the recipient mice were treated with CCl_4 (Schmelzer et al., 2007). The *Fah*-deficient mice, a model of tyrosinemia (Grompe et al., 1993), are the most widely used recipient mice for transplantation of cells into liver in numerous studies (Dorrell et al., 2011; Huch et al., 2013; Qiu et al., 2011). FAH is a critical enzyme for tyrosine catabolism (Davern, 2001), and disruption of this enzyme results in the accumulation of toxic metabolites and cell-autonomous destruction of host hepatocytes. 2-(2-Nitro-4-trifluoromethylbenzoyl)-1,3-cyclohexanedione (NTBC) is an inhibitor of 4-hydroxyphenylpyruvate dioxygenase (an enzyme upstream of FAH in the tyrosine catabolic pathway) (Grompe et al., 1993; Holme & Lindstedt, 1998), and it is added to the drinking water of *Fah*-deficient mice to maintain their liver function before transplantation. Withdrawal of NTBC favors engraftment of transplanted cells because host hepatocytes are quickly lost.

However, efficient engraftment and restoration of full liver function following transplantation of HPCs remains difficult. For reasons that are not well understood, the engraftment efficiency of HPCs is very low compared to that of primary hepatocytes (Dorrell et al., 2011; Huch et al., 2013). When HPCs were injected intrasplenically into *Fah*-deficient mice, engraftment was observed only in two out of 20 transplanted mice (Dorrell et al., 2011). Another report also demonstrated that injection of clonal HPCs expanded from a single cell to immunodeficient *Fah*-deficient mice ($Fah^{-/-}Rag2^{-/-}Il2rg^{-/-}$ mice) resulted in less than 0.1% of engraftment of the liver (Huch et al., 2013), suggesting that the low engraftment efficiency is not due to rejection of transplanted cells by the host immune system. The low engraftment efficiency may be improved by optimizing the differentiation protocol in culture (Dorrell et al., 2011; Huch et al., 2013). Current differentiation protocols fail to differentiate HPCs into fully mature hepatocytes in culture (Shin et al., 2011). Another potential problem of cell therapy is that injected cells can be dispersed to the ectopic sites throughout the body. Turner et al. (2013) demonstrated that mixing cells with specific biomaterials can improve targeting of HPCs to the liver.

The route of cell transplantation is another subject that requires investigation. While intrasplenic injection is the most widely used method for injection of hepatocytes, it has been demonstrated that small cells are more likely to become entrapped in the spleen (Cheng et al., 2009), suggesting that intrasplenic injection may not be an optimal method for the delivery of HPCs. In addition, although cell therapy possesses great potential, the injury of the host liver associated with cell transplantation should not be underestimated. Intrasplenic injection of hepatocytes or nonparenchymal epithelial cells caused disruption of gap junctions and ischemic liver injury (Gupta et al., 2000). Whether HPCs can be tumorigenic is also a concern, although it has been shown that HPCs were nontumorigenic when transplanted subcutaneously into nude mice (Tirnitz-Parker et al., 2007). In sum, although HPC-based therapies pose significant advantages compared to organ or primary hepatocytes which cannot be expanded in culture, further optimization of the technology is required before clinical application can become a reality.

7. CONCLUDING REMARKS

Recent progress in molecular biology has enabled investigation of the origin, physiology, and therapeutic potential of HPCs. However, additional studies are required to reconcile conflicting findings from different research groups based on different injury models, species, and use of markers for detecting and isolating HPCs. For example, studies revealed the heterogeneity of HPCs between rat, mouse, and human (Bisgaard et al., 2002; Jelnes et al., 2007). In addition, protocols for the differentiation and transplantation of HPCs must be optimized before this exciting cell population can be translated into clinical application.

ACKNOWLEDGMENTS

Work in the Kaestner lab is supported by the 2012 Liver Scholar Award to S. S. from American Association for the Study of Liver Diseases/American Liver Foundation and by the P01-DK049210 grant to K. H. K. from National Institute of Diabetes and Digestive and Kidney Diseases.

REFERENCES

Akhurst, B., Matthews, V., Husk, K., Smyth, M. J., Abraham, L. J., & Yeoh, G. C. (2005). Differential lymphotoxin-beta and interferon gamma signaling during mouse liver regeneration induced by chronic and acute injury. *Hepatology*, *41*, 327–335.

Alpini, G., Aragona, E., Dabeva, M., Salvi, R., Shafritz, D. A., & Tavoloni, N. (1992). Distribution of albumin and alpha-fetoprotein mRNAs in normal, hyperplastic, and preneoplastic rat liver. *The American Journal of Pathology*, *141*, 623–632.

Bisgaard, H. C., Holmskov, U., Santoni-Rugiu, E., Nagy, P., Nielsen, O., Ott, P., et al. (2002). Heterogeneity of ductular reactions in adult rat and human liver revealed by novel expression of deleted in malignant brain tumor 1. *The American Journal of Pathology*, *161*, 1187–1198.

Boulter, L., Govaere, O., Bird, T. G., Radulescu, S., Ramachandran, P., Pellicoro, A., et al. (2012). Macrophage-derived Wnt opposes Notch signaling to specify hepatic progenitor cell fate in chronic liver disease. *Nature Medicine*, *18*, 572–579.

Carpentier, R., Suner, R. E., van Hul, N., Kopp, J. L., Beaudry, J. B., Cordi, S., et al. (2011). Embryonic ductal plate cells give rise to cholangiocytes, periportal hepatocytes, and adult liver progenitor cells. *Gastroenterology*, *141*, 1432–1438, 1438, e1431-1434.

Carpino, G., Cardinale, V., Onori, P., Franchitto, A., Berloco, P. B., Rossi, M., et al. (2012). Biliary tree stem/progenitor cells in glands of extrahepatic and intraheptic bile ducts: An anatomical in situ study yielding evidence of maturational lineages. *Journal of Anatomy*, *220*, 186–199.

Cheng, K., Benten, D., Bhargava, K., Inada, M., Joseph, B., Palestro, C., et al. (2009). Hepatic targeting and biodistribution of human fetal liver stem/progenitor cells and adult hepatocytes in mice. *Hepatology*, *50*, 1194–1203.

Chu, A. S., Diaz, R., Hui, J. J., Yanger, K., Zong, Y., Alpini, G., et al. (2011). Lineage tracing demonstrates no evidence of cholangiocyte epithelial-to-mesenchymal transition in murine models of hepatic fibrosis. *Hepatology*, *53*, 1685–1695.

Dabeva, M. D., Alpini, G., Hurston, E., & Shafritz, D. A. (1993). Models for hepatic progenitor cell activation. *Proceedings of the Society for Experimental Biology and Medicine*, *204*, 242–252.

Dabeva, M. D., Petkov, P. M., Sandhu, J., Oren, R., Laconi, E., Hurston, E., et al. (2000). Proliferation and differentiation of fetal liver epithelial progenitor cells after transplantation into adult rat liver. *The American Journal of Pathology*, *156*, 2017–2031.

Dabeva, M. D., & Shafritz, D. A. (1993). Activation, proliferation, and differentiation of progenitor cells into hepatocytes in the D-galactosamine model of liver regeneration. *The American Journal of Pathology*, *143*, 1606–1620.

Davern, T. J. (2001). Molecular therapeutics of liver disease. *Clinics in Liver Disease*, *5*, 381–414, vi.

Davies, R. A., Knight, B., Tian, Y. W., Yeoh, G. C., & Olynyk, J. K. (2006). Hepatic oval cell response to the choline-deficient, ethionine supplemented model of murine liver injury is attenuated by the administration of a cyclo-oxygenase 2 inhibitor. *Carcinogenesis*, *27*, 1607–1616.

Dorrell, C., Erker, L., Lanxon-Cookson, K. M., Abraham, S. L., Victoroff, T., Ro, S., et al. (2008). Surface markers for the murine oval cell response. *Hepatology*, *48*, 1282–1291.

Dorrell, C., Erker, L., Schug, J., Kopp, J. L., Canaday, P. S., Fox, A. J., et al. (2011). Prospective isolation of a bipotential clonogenic liver progenitor cell in adult mice. *Genes & Development*, *25*, 1193–1203.

Duncan, A. W., Dorrell, C., & Grompe, M. (2009). Stem cells and liver regeneration. *Gastroenterology*, *137*, 466–481.

Dusabineza, A. C., Van Hul, N. K., Abarca-Quinones, J., Starkel, P., Najimi, M., & Leclercq, I. A. (2012). Participation of liver progenitor cells in liver regeneration: Lack of evidence in the AAF/PH rat model. *Laboratory Investigation; A Journal of Technical Methods and Pathology*, *92*, 72–81.

Endo, Y., Zhang, M., Yamaji, S., & Cang, Y. (2012). Genetic abolishment of hepatocyte proliferation activates hepatic stem cells. *PloS One*, 7, e31846.

Espanol-Suner, R., Carpentier, R., Van Hul, N., Legry, V., Achouri, Y., Cordi, S., et al. (2012). Liver progenitor cells yield functional hepatocytes in response to chronic liver injury in mice. *Gastroenterology*, *143*(1564–1575), e1567.

Fan, L., Xu, C., Wang, C., Tao, J., Ho, C., Jiang, L., et al. (2012). Bmi1 is required for hepatic progenitor cell expansion and liver tumor development. *PloS One*, 7, e46472.

Farber, E. (1956). Similarities in the sequence of early histological changes induced in the liver of the rat by ethionine, 2-acetylamino-fluorene, and 3′-methyl-4-dimethylaminoazobenzene. *Cancer Research*, *16*, 142–148.

Feng, D., Kong, X., Weng, H., Park, O., Wang, H., Dooley, S., et al. (2012). Interleukin-22 promotes proliferation of liver stem/progenitor cells in mice and patients with chronic hepatitis B virus infection. *Gastroenterology*, *143*(188–198), e187.

Fujio, K., Evarts, R. P., Hu, Z., Marsden, E. R., & Thorgeirsson, S. S. (1994). Expression of stem cell factor and its receptor, c-kit, during liver regeneration from putative stem cells in adult rat. *Laboratory Investigation; A Journal of Technical Methods and Pathology*, *70*, 511–516.

Furuyama, K., Kawaguchi, Y., Akiyama, H., Horiguchi, M., Kodama, S., Kuhara, T., et al. (2011). Continuous cell supply from a Sox9-expressing progenitor zone in adult liver, exocrine pancreas and intestine. *Nature Genetics*, *43*, 34–41.

Grompe, M., al-Dhalimy, M., Finegold, M., Ou, C. N., Burlingame, T., Kennaway, N. G., et al. (1993). Loss of fumarylacetoacetate hydrolase is responsible for the neonatal hepatic dysfunction phenotype of lethal albino mice. *Genes & Development*, 7, 2298–2307.

Grozdanov, P. N., Yovchev, M. I., & Dabeva, M. D. (2006). The oncofetal protein glypican-3 is a novel marker of hepatic progenitor/oval cells. *Laboratory Investigation; A Journal of Technical Methods and Pathology*, *86*, 1272–1284.

Gupta, S., Rajvanshi, P., Malhi, H., Slehria, S., Sokhi, R. P., Vasa, S. R., et al. (2000). Cell transplantation causes loss of gap junctions and activates GGT expression permanently in host liver. *American Journal of Physiology. Gastrointestinal and Liver Physiology*, *279*, G815–G826.

Higa, L. A., & Zhang, H. (2007). Stealing the spotlight: CUL4-DDB1 ubiquitin ligase docks WD40-repeat proteins to destroy. *Cell Division*, *2*, 5.

Holme, E., & Lindstedt, S. (1998). Tyrosinaemia type I and NTBC (2-(2-nitro-4-trifluoromethylbenzoyl)-1,3-cyclohexanedione). *Journal of Inherited Metabolic Disease*, *21*, 507–517.

Huch, M., Dorrell, C., Boj, S. F., van Es, J. H., Li, V. S., van de Wetering, M., et al. (2013). In vitro expansion of single Lgr5+ liver stem cells induced by Wnt-driven regeneration. *Nature*, *494*, 247–250.

Inada, M., Benten, D., Cheng, K., Joseph, B., Berishvili, E., Badve, S., et al. (2008). Stage-specific regulation of adhesion molecule expression segregates epithelial stem/progenitor cells in fetal and adult human livers. *Hepatology International*, *2*, 50–62.

Jelnes, P., Santoni-Rugiu, E., Rasmussen, M., Friis, S. L., Nielsen, J. H., Tygstrup, N., et al. (2007). Remarkable heterogeneity displayed by oval cells in rat and mouse models of stem cell-mediated liver regeneration. *Hepatology*, *45*, 1462–1470.

Jensen, C. H., Jauho, E. I., Santoni-Rugiu, E., Holmskov, U., Teisner, B., Tygstrup, N., et al. (2004). Transit-amplifying ductular (oval) cells and their hepatocytic progeny are characterized by a novel and distinctive expression of delta-like protein/preadipocyte factor 1/fetal antigen 1. *The American Journal of Pathology*, *164*, 1347–1359.

Jozefczuk, J., Stachelscheid, H., Chavez, L., Herwig, R., Lehrach, H., Zeilinger, K., et al. (2010). Molecular characterization of cultured adult human liver progenitor cells. *Tissue Engineering. Part C, Methods*, *16*, 821–834.

Jung, Y., Brown, K. D., Witek, R. P., Omenetti, A., Yang, L., Vandongen, M., et al. (2008). Accumulation of hedgehog-responsive progenitors parallels alcoholic liver disease severity in mice and humans. *Gastroenterology*, *134*, 1532–1543.

Kakinuma, S., Ohta, H., Kamiya, A., Yamazaki, Y., Oikawa, T., Okada, K., et al. (2009). Analyses of cell surface molecules on hepatic stem/progenitor cells in mouse fetal liver. *Journal of Hepatology*, *51*, 127–138.

Kamiya, A., Kakinuma, S., Yamazaki, Y., & Nakauchi, H. (2009). Enrichment and clonal culture of progenitor cells during mouse postnatal liver development in mice. *Gastroenterology*, *137*, 1114–1126, 1126 e1111–1114.

Katoonizadeh, A., Nevens, F., Verslype, C., Pirenne, J., & Roskams, T. (2006). Liver regeneration in acute severe liver impairment: A clinicopathological correlation study. *Liver International: Official Journal of the International Association for the Study of the Liver*, *26*, 1225–1233.

Knight, B., Yeap, B. B., Yeoh, G. C., & Olynyk, J. K. (2005). Inhibition of adult liver progenitor (oval) cell growth and viability by an agonist of the peroxisome proliferator activated receptor (PPAR) family member gamma, but not alpha or delta. *Carcinogenesis*, *26*, 1782–1792.

Knight, B., & Yeoh, G. C. (2005). TNF/LTalpha double knockout mice display abnormal inflammatory and regenerative responses to acute and chronic liver injury. *Cell and Tissue Research*, *319*, 61–70.

Kuwahara, R., Kofman, A. V., Landis, C. S., Swenson, E. S., Barendswaard, E., & Theise, N. D. (2008). The hepatic stem cell niche: Identification by label-retaining cell assay. *Hepatology*, *47*, 1994–2002.

Libbrecht, L., & Roskams, T. (2002). Hepatic progenitor cells in human liver diseases. *Seminars in Cell & Developmental Biology*, *13*, 389–396.

Lorenzini, S., Bird, T. G., Boulter, L., Bellamy, C., Samuel, K., Aucott, R., et al. (2010). Characterisation of a stereotypical cellular and extracellular adult liver progenitor cell niche in rodents and diseased human liver. *Gut*, *59*, 645–654.

Lowes, K. N., Brennan, B. A., Yeoh, G. C., & Olynyk, J. K. (1999). Oval cell numbers in human chronic liver diseases are directly related to disease severity. *The American Journal of Pathology*, *154*, 537–541.

Lozoya, O. A., Wauthier, E., Turner, R. A., Barbier, C., Prestwich, G. D., Guilak, F., et al. (2011). Regulation of hepatic stem/progenitor phenotype by microenvironment stiffness in hydrogel models of the human liver stem cell niche. *Biomaterials*, *32*, 7389–7402.

Malato, Y., Naqvi, S., Schurmann, N., Ng, R., Wang, B., Zape, J., et al. (2011). Fate tracing of mature hepatocytes in mouse liver homeostasis and regeneration. *The Journal of Clinical Investigation*, *121*, 4850–4860.

Martins, P. N., Theruvath, T. P., & Neuhaus, P. (2008). Rodent models of partial hepatectomies. *Liver International: Official Journal of the International Association for the Study of the Liver*, *28*, 3–11.

Menthena, A., Deb, N., Oertel, M., Grozdanov, P. N., Sandhu, J., Shah, S., et al. (2004). Bone marrow progenitors are not the source of expanding oval cells in injured liver. *Stem Cells*, *22*, 1049–1061.

Nagaya, M., Kubota, S., Suzuki, N., Akashi, K., & Mitaka, T. (2006). Thermoreversible gelation polymer induces the emergence of hepatic stem cells in the partially injured rat liver. *Hepatology*, *43*, 1053–1062.

Nobili, V., Carpino, G., Alisi, A., Franchitto, A., Alpini, G., De Vito, R., et al. (2012). Hepatic progenitor cells activation, fibrosis, and adipokines production in pediatric nonalcoholic fatty liver disease. *Hepatology*, *56*, 2142–2153.

Okabe, M., Tsukahara, Y., Tanaka, M., Suzuki, K., Saito, S., Kamiya, Y., et al. (2009). Potential hepatic stem cells reside in EpCAM+ cells of normal and injured mouse liver. *Development*, *136*, 1951–1960.

Omori, N., Omori, M., Evarts, R. P., Teramoto, T., Miller, M. J., Hoang, T. N., et al. (1997). Partial cloning of rat CD34 cDNA and expression during stem cell-dependent liver regeneration in the adult rat. *Hepatology*, *26*, 720–727.

Petersen, B. E., Bowen, W. C., Patrene, K. D., Mars, W. M., Sullivan, A. K., Murase, N., et al. (1999). Bone marrow as a potential source of hepatic oval cells. *Science*, *284*, 1168–1170.

Petersen, B. E., Goff, J. P., Greenberger, J. S., & Michalopoulos, G. K. (1998). Hepatic oval cells express the hematopoietic stem cell marker Thy-1 in the rat. *Hepatology*, *27*, 433–445.

Piscaglia, A. C., Shupe, T. D., Oh, S. H., Gasbarrini, A., & Petersen, B. E. (2007). Granulocyte-colony stimulating factor promotes liver repair and induces oval cell migration and proliferation in rats. *Gastroenterology*, *133*, 619–631.

Porretti, L., Cattaneo, A., Colombo, F., Lopa, R., Rossi, G., Mazzaferro, V., et al. (2010). Simultaneous characterization of progenitor cell compartments in adult human liver. *Cytometry Part A: The Journal of the International Society for Analytical Cytology*, *77*, 31–40.

Qiu, Q., Hernandez, J. C., Dean, A. M., Rao, P. H., & Darlington, G. J. (2011). CD24-positive cells from normal adult mouse liver are hepatocyte progenitor cells. *Stem Cells and Development*, *20*, 2177–2188.

Roskams, T., & Desmet, V. (1998). Ductular reaction and its diagnostic significance. *Seminars in Diagnostic Pathology*, *15*, 259–269.

Roskams, T. A., Theise, N. D., Balabaud, C., Bhagat, G., Bhathal, P. S., Bioulac-Sage, P., et al. (2004). Nomenclature of the finer branches of the biliary tree: Canals, ductules, and ductular reactions in human livers. *Hepatology*, *39*, 1739–1745.

Rountree, C. B., Barsky, L., Ge, S., Zhu, J., Senadheera, S., & Crooks, G. M. (2007). A CD133-expressing murine liver oval cell population with bilineage potential. *Stem Cells*, *25*, 2419–2429.

Rountree, C. B., Wang, X., Ge, S., Barsky, L., Zhu, J., Gonzales, I., et al. (2007). Bone marrow fails to differentiate into liver epithelium during murine development and regeneration. *Hepatology*, *45*, 1250–1260.

Sackett, S. D., Gao, Y., Shin, S., Esterson, Y. B., Tsingalia, A., Hurtt, R. S., et al. (2009). Foxl1 promotes liver repair following cholestatic injury in mice. *Laboratory Investigation, 89*, 1387–1396.

Sackett, S. D., Li, Z., Hurtt, R., Gao, Y., Wells, R. G., Brondell, K., et al. (2009). Foxl1 is a marker of bipotential hepatic progenitor cells in mice. *Hepatology, 49*, 920–929.

Sancho-Bru, P., Altamirano, J., Rodrigo-Torres, D., Coll, M., Millan, C., Jose Lozano, J., et al. (2012). Liver progenitor cell markers correlate with liver damage and predict short-term mortality in patients with alcoholic hepatitis. *Hepatology, 55*, 1931–1941.

Sandhu, J. S., Petkov, P. M., Dabeva, M. D., & Shafritz, D. A. (2001). Stem cell properties and repopulation of the rat liver by fetal liver epithelial progenitor cells. *The American Journal of Pathology. 159*, 1323–1334.

Schmelzer, E., Wauthier, E., & Reid, L. M. (2006). The phenotypes of pluripotent human hepatic progenitors. *Stem Cells, 24*, 1852–1858.

Schmelzer, E., Zhang, L., Bruce, A., Wauthier, E., Ludlow, J., Yao, H. L., et al. (2007). Human hepatic stem cells from fetal and postnatal donors. *The Journal of Experimental Medicine, 204*, 1973–1987.

Scholten, D., Osterreicher, C. H., Scholten, A., Iwaisako, K., Gu, G., Brenner, D. A., et al. (2010). Genetic labeling does not detect epithelial-to-mesenchymal transition of cholangiocytes in liver fibrosis in mice. *Gastroenterology, 139*, 987–998.

Schotanus, B. A., van den Ingh, T. S., Penning, L. C., Rothuizen, J., Roskams, T. A., & Spee, B. (2009). Cross-species immunohistochemical investigation of the activation of the liver progenitor cell niche in different types of liver disease. *Liver International: Official Journal of the International Association for the Study of the Liver, 29*, 1241–1252.

Shin, S., Walton, G., Aoki, R., Brondell, K., Schug, J., Fox, A., et al. (2011). Foxl1-Cre-marked adult hepatic progenitors have clonogenic and bilineage differentiation potential. *Genes & Development, 25*, 1185–1192.

Sicklick, J. K., Li, Y. X., Melhem, A., Schmelzer, E., Zdanowicz, M., Huang, J., et al. (2006). Hedgehog signaling maintains resident hepatic progenitors throughout life. *American Journal of Physiology. Gastrointestinal and Liver Physiology, 290*, G859–G870.

Spee, B., Carpino, G., Schotanus, B. A., Katoonizadeh, A., Vander Borght, S., Gaudio, E., et al. (2010). Characterisation of the liver progenitor cell niche in liver diseases: Potential involvement of Wnt and Notch signalling. *Gut, 59*, 247–257.

Srinivas, S., Watanabe, T., Lin, C. S., William, C. M., Tanabe, Y., Jessell, T. M., et al. (2001). Cre reporter strains produced by targeted insertion of EYFP and ECFP into the ROSA26 locus. *BMC Developmental Biology, 1*, 4.

Suzuki, A., Sekiya, S., Onishi, M., Oshima, N., Kiyonari, H., Nakauchi, H., et al. (2008). Flow cytometric isolation and clonal identification of self-renewing bipotent hepatic progenitor cells in adult mouse liver. *Hepatology, 48*, 1964–1978.

Takase, H. M., Itoh, T., Ino, S., Wang, T., Koji, T., Akira, S., et al. (2013). FGF7 is a functional niche signal required for stimulation of adult liver progenitor cells that support liver regeneration. *Genes & Development, 27*, 169–181.

Theise, N. D., Saxena, R., Portmann, B. C., Thung, S. N., Yee, H., Chiriboga, L., et al. (1999). The canals of Hering and hepatic stem cells in humans. *Hepatology, 30*, 1425–1433.

Tirnitz-Parker, J. E., Tonkin, J. N., Knight, B., Olynyk, J. K., & Yeoh, G. C. (2007). Isolation, culture and immortalisation of hepatic oval cells from adult mice fed a choline-deficient, ethionine-supplemented diet. *The International Journal of Biochemistry & Cell Biology, 39*, 2226–2239.

Tirnitz-Parker, J. E., Viebahn, C. S., Jakubowski, A., Klopcic, B. R., Olynyk, J. K., Yeoh, G. C., et al. (2010). Tumor necrosis factor-like weak inducer of apoptosis is a mitogen for liver progenitor cells. *Hepatology, 52*, 291–302.

Tonkin, J. N., Knight, B., Curtis, D., Abraham, L. J., & Yeoh, G. C. (2008). Bone marrow cells play only a very minor role in chronic liver regeneration induced by a choline-deficient, ethionine-supplemented diet. *Stem Cell Research*, *1*, 195–204.

Trautwein, C., Will, M., Kubicka, S., Rakemann, T., Flemming, P., & Manns, M. P. (1999). 2-acetaminofluorene blocks cell cycle progression after hepatectomy by p21 induction and lack of cyclin E expression. *Oncogene*, *18*, 6443–6453.

Turanyi, E., Dezso, K., Csomor, J., Schaff, Z., Paku, S., & Nagy, P. (2010). Immunohistochemical classification of ductular reactions in human liver. *Histopathology*, *57*, 607–614.

Turner, R., Lozoya, O., Wang, Y., Cardinale, V., Gaudio, E., Alpini, G., et al. (2011). Human hepatic stem cell and maturational liver lineage biology. *Hepatology*, *53*, 1035–1045.

Turner, R. A., Wauthier, E., Lozoya, O., McClelland, R., Bowsher, J. E., Barbier, C., et al. (2013). Successful transplantation of human hepatic stem cells with restricted localization to liver using hyaluronan graftsdagger. *Hepatology*, *57*, 775–784.

Ueberham, E., Bottger, J., Ueberham, U., Grosche, J., & Gebhardt, R. (2010). Response of sinusoidal mouse liver cells to choline-deficient ethionine-supplemented diet. *Comparative Hepatology*, *9*, 8.

Van Hul, N. K., Abarca-Quinones, J., Sempoux, C., Horsmans, Y., & Leclercq, I. A. (2009). Relation between liver progenitor cell expansion and extracellular matrix deposition in a CDE-induced murine model of chronic liver injury. *Hepatology*, *49*, 1625–1635.

Vassilopoulos, G., Wang, P. R., & Russell, D. W. (2003). Transplanted bone marrow regenerates liver by cell fusion. *Nature*, *422*, 901–904.

Viebahn, C. S., Benseler, V., Holz, L. E., Elsegood, C. L., Vo, M., Bertolino, P., et al. (2010). Invading macrophages play a major role in the liver progenitor cell response to chronic liver injury. *Journal of Hepatology*, *53*, 500–507.

Wang, X., Willenbring, H., Akkari, Y., Torimaru, Y., Foster, M., Al-Dhalimy, M., et al. (2003). Cell fusion is the principal source of bone-marrow-derived hepatocytes. *Nature*, *422*, 897–901.

Wu, K. J., & Yang, M. H. (2011). Epithelial-mesenchymal transition and cancer stemness: The Twist1-Bmi1 connection. *Bioscience Reports*, *31*, 449–455.

Yang, L., Jung, Y., Omenetti, A., Witek, R. P., Choi, S., Vandongen, H. M., et al. (2008). Fate-mapping evidence that hepatic stellate cells are epithelial progenitors in adult mouse livers. *Stem Cells*, *26*, 2104–2113.

Yanger, K., & Stanger, B. Z. (2011). Facultative stem cells in liver and pancreas: Fact and fancy. *Developmental Dynamics: An Official Publication of the American Association of Anatomists*, *240*, 521–529.

Yaswen, P., Hayner, N. T., & Fausto, N. (1984). Isolation of oval cells by centrifugal elutriation and comparison with other cell types purified from normal and preneoplastic livers. *Cancer Research*, *44*, 324–331.

Yeoh, G. C., Ernst, M., Rose-John, S., Akhurst, B., Payne, C., Long, S., et al. (2007). Opposing roles of gp130-mediated STAT-3 and ERK-1/2 signaling in liver progenitor cell migration and proliferation. *Hepatology*, *45*, 486–494.

Yovchev, M. I., Grozdanov, P. N., Zhou, H., Racherla, H., Guha, C., & Dabeva, M. D. (2008). Identification of adult hepatic progenitor cells capable of repopulating injured rat liver. *Hepatology*, *47*, 636–647.

CHAPTER ELEVEN

Nephron Progenitor Cells: Shifting the Balance of Self-Renewal and Differentiation

Raphael Kopan[*,†,1], **Shuang Chen**[*], **Melissa Little**[*,‡,1]

[*]Department of Developmental Biology, Washington University, St. Louis, Missouri, USA
[†]Division of Developmental Biology, Department of Pediatrics, Cincinnati Children's Hospital, Cincinnati, Ohio, USA
[‡]Institute for Molecular Bioscience, The University of Queensland, St. Lucia, Queensland, Australia
[1]Corresponding author: e-mail address: raphael.kopan@cchmc.org; m.little@imb.uq.edu.au

Contents

Abstract

Within the developing mammalian kidney, several populations of progenitors form the discrete cellular components of the final organ. Fate mapping experiments revealed the cap mesenchyme (CM) to be the progenitor population for all nephron epithelial cells, whereas the neighboring stromal mesenchyme gives rise to mesangial, pericytic, renin-producing and interstitial cells. The collecting ducts are derived from a population of progenitors at the ureteric bud (UB) tip and a proportion of the endothelium is also derived from a dedicated mesenchymal progenitor. The stroma, CM, and UB interact to create spatially defined niches at the periphery of the developing organ. While the UB tip population persist, the CM represents a transient progenitor population that

Current Topics in Developmental Biology, Volume 107
ISSN 0070-2153
http://dx.doi.org/10.1016/B978-0-12-416022-4.00011-1

is exhausted to set the final organ size. The timing of CM exhaustion, and hence the final organ structure, is sensitive to disruptions such as premature birth. Here we will discuss our current understanding of the molecular processes allowing these populations to balance cell survival, self-renewal, support of branching, and maintain capacity to commit to differentiation.

1. INTRODUCTION

The permanent postnatal kidney in the mammal, containing in excess of 20 distinct cell types each with specific functional attributes, arises during embryonic development from a much smaller number of progenitors. The ureteric epithelium contributes to the collecting ducts whereas a mesenchymal progenitor population, the cap mesenchyme (CM), gives rise to all other specialized nephron epithelial cell types via a mesenchymal-to-epithelial transition (MET). In order for the CM population to generate the 300,000–1,800,000 nephrons present in the adult human kidney (Bertram, Douglas-Denton, Diouf, Hughson, & Hoy, 2011), the CM must balance commitment to differentiation with self-renewal/population expansion. Both of these activities are controlled via signaling from adjacent cell populations, notably the adjacent ureteric epithelial compartment and the surrounding stromal mesenchyme (SM). In addition, ureteric epithelial branching is regulated by reciprocal signals from the CM. Indeed, the periphery of the developing organ consists of a nephrogenic zone composed of nephrogenic niches, each consisting of a ureteric tip surrounded by CM itself surrounded by SM (Fig. 11.1). This close interrelationship is key to normal kidney development and consequently responsible for the considerable variation observed in organ size and final nephron number (Boyle et al., 2008; Costantini, 2006; Kobayashi et al., 2008; Sequeira Lopez & Gomez, 2011). While lineage tracing has established the cell types derived from these initial progenitor populations, the nature of the reciprocal signaling between UB and CM remains incompletely understood. Even less is known about the ability of newly formed nephrons to regulate the formation of subsequent nephrons or the origin or role of vascular, neural, and lymphatic progenitors within the developing kidney. In this review, we will focus on the nephron progenitor population, the CM. This population, with input from the UB and the SM, appears to be the major driver of organogenesis in the kidney; hence an understanding of this population will be critical for revealing the basis of developmental variation and anomaly.

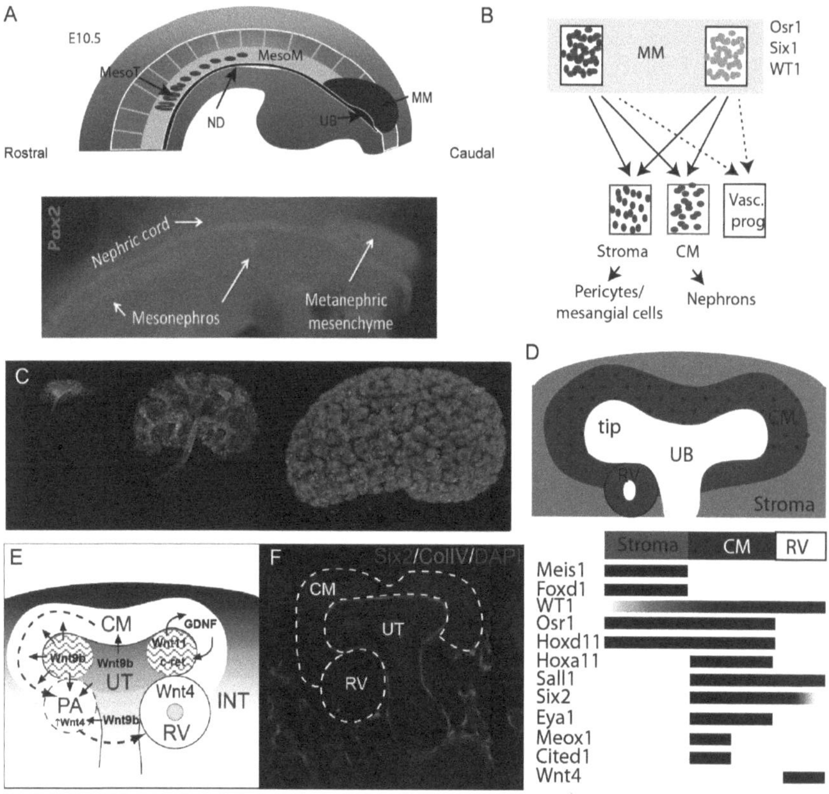

Figure 11.1 Development of the mammalian urinary system. (A) Diagram of the developing urinary system in the E10.5 mouse embryo (top) with wholemount Pax2 expression (bottom). ND, nephric duct/nephric cord; MM, metanephric mesenchyme; MesoM, mesonephric mesenchyme; MesoT, mesonephric tubules; UB, point of ureteric bud outgrowth. (B) Two models of the derivation of key kidney progenitor populations from the MM. The MM may represent a heterogeneous field containing several different mesenchymal progenitors (top left), including the stroma, CM and vascular progenitors (vasc. prog.). Alternatively, the MM may represent a homogeneous population of precursors (top right) all co-expressing characteristic MM marker genes and able to differentiate into distinct downstream progenitors. Solid lines indicate what the MM gives rise to. Dotted lines indicate the assumption that this also gives rise to vascular progenitors although renal vasculature is derived from vasculogenesis and angiogenesis. (C) Optical projection series of the developing mouse kidney at E11.5, E13.5 and E15.5 showing ureteric epithelium (white; cytokeratin), cap mesenchyme (green, Six2), and mitosing cells (red, pHH3). (D) Diagram of the nephrogenic niche within the developing metanephros, showing the CM, UB, stroma (SM) and the location of the forming renal vesicles (Stage 1 nephrons; RV). Gene expression within the mesenchymal compartment and derivatives are indicated below this diagram. Shaded regions within CM relate to the assumed position of the uninduced CM based upon expression of *Cited1*. (E) Diagram of the niche indicating the GDNF-c-ret-Wnt11 forward feedback loop driving branching

(*Continued*)

2. SETTING THE SCENE: THE DEVELOPING URINARY SYSTEM OF THE MAMMAL

The mammalian urogenital system, including the kidneys, is derived entirely from the intermediate mesoderm (IM). The dorsal IM forms the nephric duct (ND, also called the Wolffian duct (Saxen, 1987)), whereas the ventral IM remains undifferentiated as a nephrogenic cord of mesenchyme running the length of the ND and competent to respond to signals emanating from it (Fig. 11.1A). In a rostral to caudal temporal wave, ND-derived signals induce the formation of primitive renal tubules along the embryonic trunk forming firstly the pronephric and then the mesonephric kidneys. The developmental potential of the rostral IM becomes progressively restricted while a caudal bean-shaped region of the nephrogenic cord commits to forming the metanephric mesenchyme (MM) at the level of the hind limb (Fig. 11.1A). The MM will form the final permanent kidney: the metanephric kidney or metanephros.

By E10, the MM contains the mesenchymal progenitors feted for differentiation into the nephron epithelial (Boyle et al., 2008; Kobayashi et al., 2008), stromal (Humphreys et al., 2010), and some vascular (Sequeira Lopez & Gomez, 2011) components of the metanephric (adult) kidneys (Fig. 11.1B). While a number of genes mark MM, it is unclear whether this represents a homogeneous multipotent population or a heterogeneous population of already committed progenitor types (Fig. 11.1B). Nevertheless, the MM gene expression is very similar to that of the mesonephric mesenchyme (Georgas et al., 2011). The pronephros or mesonephros arises via the simple induction of adjacent tubes that then fuse with the ND. In contrast, the metanephric field induces a distinct, singular response from the adjacent ND. Secretion of GDNF, FGF10, and other inductive signals by the MM (possibly an exclusive property of the nephron epithelial progenitors, although this has not been proven), activates receptor tyrosine kinases (RTKs) in the ND, triggering pseudostratification and evagination of a single ureteric bud (UB; Costantini, 2012; Costantini & Kopan, 2010). This

Figure 11.1—Cont'd and the Wnt signaling involved in nephron formation and CM survival, (F) Section immunohistochemistry of a single nephrogenic niche at E15.5 showing Six2 in the CM (red), Collagen IV marking all basement membranes (green), and DAPI identifying all nuclei (blue). UT, ureteric tip; CM, cap mesenchyme; RV, renal vesicle. *Panel (A): Adapted from Georgas, Chiu, Lesieur, Rumballe, and Little (2011). Panel (B): From Sequeira Lopez and Gomez (2011). Panel (C): Adapted from Rumballe et al. (2011).* (See color plate.)

UB evagination is led by tip stem cells (ureteric epithelial progenitors) expressing high levels of the receptor Ret (Chi et al., 2009; Shakya, Watanabe, & Costantini, 2005). In the absence of GDNF, or its receptors, (Ret and GFR1α), the ND initiates a UB (Chi et al., 2009) that does not grow (Costantini & Shakya, 2006; Skinner, Safford, Reeves, Jackson, & Freemerman, 2008). In contrast, in the absence of the MM transcription factor *Osr1*, no UB initiates (see Figure 7 in Mugford, Sipila, McMahon, & McMahon, 2008) suggesting an *Osr1*-regulated signal is required for this initial event with GDNF then promoting outgrowth via cell proliferation, cell sorting and migration within the ureteric epithelium (Chi et al., 2009; Mugford et al., 2008). The elongating UB extends toward the source of GDNF, the MM, with the two cell populations meeting in the mouse at E11–11.5. This meeting triggers the commencement of branching morphogenesis, the engine driving the formation of the metanephros. The previously unbranched UB forms an initial T-shaped epithelial structure (Fig. 11.1C). From this time until the completion of organogenesis, the UB undergoes dichotomous branching events to form the characteristic metanephric ureteric tree (Fig. 11.1C). The continued upregulation of Ret in response to mesenchymal GDNF signaling drives this repeated UB branching thereby forming the future collecting system with the initial extension from the ND forming the ureter. It is around this branched epithelial "tree" that the nephrons, vasculature, lymphatic, neural, and other interstitial elements differentiate.

3. THE NEPHROGENIC NICHE

As noted earlier, the periphery of the developing metanephros contains a rim of nephrogenic niches that represent the site of both branching and nephron formation (Fig. 11.1D–F). When the UB reaches the MM field, MM cells sort and organize into a cohesive domain of cells called the CM around UB tips. Differential gene expression distinguishes the CM (expressing *Six2* initially and *Six2, Cited1* after E12.5) and the SM (expressing *Foxd1*; Fig. 11.1D). During most of kidney development, the SM forms a thin layer along the periphery of the kidney, superficial to the CM, but also between the fields of CM around each UB tip. Hence, the CM is "sandwiched" between UB and SM with these populations controlling the balance between CM self-renewal and differentiation (Carroll & Das, 2013; Fetting et al., 2013; Hatini, Huh, Herzlinger, Soares, & Lai, 1996; Levinson & Mendelsohn, 2003).

Spatial analysis of gene expression across niche compartments has substantially improved our biological understanding of kidney development. Lineage tracing using either the *Six2* or *Cited1* promoter has established that CM cells form all the epithelial cell types within the final nephron, but do not contribute to the UB-derived collecting ducts (Boyle et al., 2008; Kobayashi et al., 2008). Nephron formation commences when a few CM cells at a new branch tip undergo MET to form a renal vesicle (RV) (Fig. 11.1D–F). Once formed, the distal RV fuses with the UB tip closest to its point of initiation (Georgas et al., 2009). RVs form at characteristic positions relative to the UB tip leaving an undifferentiated, GDNF-secreting CM surrounding the leading edge of the tip, thereby allowing branching to continue. Differentiation of each of the distinct segments of the final nephron involves cellular proliferation, extensive tubular elongation and intercalation (Fischer et al., 2006; Karner et al., 2009), and proximal-distal patterning driven by processes still being identified. *Notch2* signaling is required for podocyte and proximal tubule differentiation, the latter also requiring *Hnf1b,* however, additional ligand/receptor signals and transcriptional regulators will most likely emerge (Cheng et al., 2007; Heliot et al., 2013; Massa et al., 2013; Surendran et al., 2010). While UB branching and nephrogenesis via MET are continuing, the surrounding stromal and endothelial progenitors join the nascent S-shape bodies where they differentiate to form the mesangial and endothelial components (respectively) of the glomerulus located at the most proximal end of the patterning nephrons.

4. A SUBDIVISION WITHIN THE NEPHRON PROGENITOR POPULATION: EVIDENCE FOR MORE THAN ONE STATE

The CM is a unique mesenchymal progenitor/stem cell population and a critical driver of branching morphogenesis and nephrogenesis. The CM is characterized at the molecular level by expression of key transcription factors (Table 11.1), chief among these being *Sall1/4* and *Six1/2*. In the absence of *Six2*, CM cells differentiate as soon as the UB enters the CM (Self et al., 2006). In *Sall1/4* transheterozygotes, Six2 levels are greatly reduced and renal agenesis is frequently observed (Sakaki-Yumoto et al., 2006). Spatial gene expression patterns suggest that the CM is heterogeneous (Mugford, Yu, Kobayashi, & McMahon, 2009). Most notably, expression of Cbp/p300-interacting transactivator 1, *Cited1*, is restricted to the region of the CM closest to the periphery of the kidney (Boyle et al., 2007; Mugford et al., 2009). It is thought that the *Six2*$^{+}$*Cited1*$^{+}$ CM cells represent the

Table 11.1 Genes known to contribute to the CM formation and/or maintenance, ? is uncertain

	Gene	Role	References	Kidney LOF phenotype
Key transcription factors	Wt1	Upstream of FGF, Sal1, Bmp7, pax2,PTIP	Hartwig et al. (2010), Kreidberg (2010)	Agenesis
	Pax2	Epigenetic regulation in the stem cells, upstream of Six2	Dressler (2008, 2009), Fang et al. (2009), Kim, Patel, Xiao, and Dressler (2009), Lefevre, Patel, Kim, Tessarollo, and Dressler (2010), Patel, Bhumbra, Paknikar, and Dressler (2012), Wu, Prindle, Dressler, and Yu, (2009)	Agenesis
	MTA/NuRD	Interacts with Sal1, chromatin modifier	Denner and Rauchman (2013)	Loss of CM, apoptosis
	Eya1	Regulates Six2	Gong, Yallowitz, Sun, Dressler, and Wellik (2007), Wellik, Hawkes, and Capecchi, (2002)	
	Hoxc10,9	Upstream of Six2 uncertain		
	Hoxd11	Upstream of Six2	Wellik et al. (2002)	Agenesis
	Hoxa9	Upstream of Six2 uncertain		
	Eya4	Upstream of Six2 uncertain		

Continued

Table 11.1 Genes known to contribute to the CM formation and/or maintenance, ? is uncertain—cont'd

	Gene	Role	References	Kidney LOF phenotype
	Six1	Stem cell maintenance	Xu et al. (2003)	
	Six2	Stem cell maintenance	Brunskill, Lai, Jamison, Potter, and Patterson (2011), Kobayashi et al. (2008), Self et al. (2006)	Agenesis, premature differentiation
	Sall1	Stem cell maintenance	Sakaki-Yumoto et al. (2006)	Agenesis (premature differentiation?)
	Sall4	Stem cell maintenance	Sakaki-Yumoto et al. (2006)	Agenesis (premature differentiation?)
	Etv5/4	Downstream of FGFR	Lu et al. (2009), Barak et al. (2012)	Agenesis (premature differentiation?)
Genes contributing to the niche, maintenance	GDNF	Signaling to the UB	Durbec et al. (1996), Moore et al. (1996), Pepicelli, Kispert, Rowitch, and McMahon (1997), Sainio et al. (1997), Trupp et al. (1996)	Agenesis
	β-Catenin	Downstream of Wnt9b, Wnt4	Karner et al. (2011)	Few Six2 cell maintained, agenesis
	Kif26b	Regulates adhesion in CM, downstream of Sall1	Uchiyama et al. (2010)	Agenesis

Npnt	Nephronectin is a BM component bound by Itga8	Brandenberger et al. (2001), Fujiwara et al. (2011), Linton, Martin, and Reichardt (2007)	Agenesis
Itga8	Needed to bind to nephronectin to establish CM	Brandenberger et al. (2001), Fujiwara et al. (2011), Linton et al. (2007)	Agenesis (early deletion)
Fgfr1/2	Reception of niche signals	Poladia et al, 2006	Agenesis
Fgf9	Autocrine/paracrine Niche signal	Barak et al. (2012)	Agenesis with FGF20
Fgf20	Autocrine niche signal	Barak et al. (2012)	Agenesis in humans, with FGF9 in mice
Pod1	Expressed in MM and stroma	Acharya, Baek, Banfi, Eskiocak, and Tallquist (2011), Maezawa et al. (2012)	Die in the perinatal period with multiple renal defects (Cui et al., 2004; Cui, Schwartz, & Quaggin, 2003)
Cxcr4	CXCL12/SDF-1 receptor, proliferation?	Haege et al. (2012), Sims-Lucas et al. (2012), Takabatake et al. (2009)	Smaller
BMP7	CM aggregation, progenitors maintenance, differentiation	Brown et al. (2013), Dudley, Godin, and Robertson (1999), Dudley, Lyons, and Robertson (1995), Dudley and Robertson (1997)	Smaller kidneys

Continued

Table 11.1 Genes known to contribute to the CM formation and/or maintenance, ? is uncertain—cont'd

Gene	Role	References	Kidney LOF phenotype
Dcn	Antagonist of BMP7	Fetting et al. (2013)	Rescues FoxD1 mutants
Ppp3Ca (calcineurin α)	Proliferation (downstream of Wnt?) Differentiation?	Gooch, Toro, Guler, and Barnes (2004)	Smaller Kidneys, unclear role in CM
Dlg1 Cask	Act together to scaffold signaling at membrane	Ahn, Kim, Kim, Swat, and Miner (2013)	When both are removed, progenitors are rapidly depleted
cMyc	Growth	Couillard and Trudel (2009)	Hypoplasia
Fat4	Signals from stroma to CM to regulate Yap/Taz (Das et al., 2013) or PCP (Mao et al., 2011; Saburi et al., 2008; Saburi, Hester, Goodrich, & McNeill, 2012)	Das et al. (2013), Mao et al. (2011), Saburi et al. (2008, 2012)	Hypoplastic kidneys, cysts (all) Progenitor expansion in LOF (Das et al., 2013) Branching morphogenesis (Mao et al., 2011)
Yap	Hippo effector, Wnt antagonist (Barry et al., 2013)	Barry et al. (2013), Das et al. (2013), Reginensi et al. (2013)	Loss of CITED1 (all) Differentiation defects (Reginensi et al., 2013) Progenitor expansion in LOF (Das et al., 2013)
FoxD1	Required for stroma	Levinson et al. (2005)	Progenitor expansion in LOF
Pbx1	Required for stroma	Schnabel et al (2003)	Progenitor expansion in LOF (partner of FoxD1?)

Genes expressed in CM/niche without known function	Cadm1	Adhesion	Yamada et al. (2006)	None
	Capn6	Calpain		KOMP ES, no mice
	Pla2g7	Marker	Karner et al. (2011)	KOMP ES, no mice
	Tafa5/Fam19a5	Marker	Karner et al. (2011)	KOMP ES, no mice
	Lrig3	Regulates EGF, Wnt in development	Abraira et al. (2008), Abraira, Satoh, Fekete, and Goodrich (2010), Gur et al. (2004), Zhao, Tanegashima, Ro, and Dawid (2008)	None
	Meox1	TF, Expressed specifically in cap	Mankoo et al. (1999), Stamataki, Kastrinaki, Mankoo, Pachnis, and Karagogeos (2001)	None
	Meox2	TF	Mankoo et al. (1999), Stamataki et al. (2001)	None
	Vcan	Versican, adhesion	Dours-Zimmermann et al. (2009)	None
	Crym	Expressed specifically in cap	Imai et al. (2010), Suzuki et al. (2007)	None
	A430107O13Rik	CPED1, a cadherin domain protein		Unknown

Continued

Table 11.1 Genes known to contribute to the CM formation and/or maintenance, ? is uncertain—cont'd

Gene	Role	References	Kidney LOF phenotype
CITED1	*Important marker of uncommitted progenitors*	Boyle et al. (2008), Boyle, Shioda, Perantoni, and de Caestecker (2007), Brown et al. (2013), Murphy et al. (2012), Plisov et al. (2005)	None
Sobp	Unknown	Chen et al. (2008)	Jc/Jc mice are null, no kidney defects
Hmcn1	Unknown		KOMP ES, no mice
Cdh11	Adhesion	Horikawa, Radice, Takeichi, and Chisaka (1999)	None

uninduced, self-renewing population (Brown et al., 2013; Mugford et al., 2009) whereas a *Six2*$^{+}$*Cited1*$^{-}$ population, located closer to the region of nephron formation, is thought to have been induced (committed to differentiation and to MET, Fig. 11.3). Mutations in a number of genes result in shifts in CM molecular signature suggesting that the balance between induced and uninduced states can change and/or that the functions of this domain are molecularly separable. However, null alleles for many transcripts constituting these CM molecular subdomains do not reveal a function (i.e. *Cited1*, *Meox1/2*, *Crym*, Table 11.1) rendering them useful only as markers. Whether the CM is actually divided into distinct subpopulations, or is populated by cells oscillating between self-renewal and differentiation-ready states (similar to neural stem cells, Imayoshi, Sakamoto, Yamaguchi, Mori, & Kageyama, 2010; Kageyama, Niwa, Shimojo, Kobayashi, & Ohtsuka, 2010), is not clear. The difficulty in recapitulating CM behaviors in culture further confounds the definitive interrogation of CM heterogeneity.

5. ROLES PLAYED BY THE CM

The CM displays multiple activities, including survival, self-renewal, support/induction of UB branching, and a capacity to differentiate into the nephron epithelia. Here we will address what is known about how the interplay between these roles is regulated.

5.1. Support of ureteric branching

The CM initiates initial UB outgrowth. Without this event, renal agenesis occurs. Reciprocally, if no UB arises at E10.5, the CM will undergo apoptosis within 1–2 days (Kreidberg et al., 1993). Once the UB extends to the renal mesenchymal field, the continued secretion of GDNF by the CM ensures continued UB branching, a key determinant of overall kidney size and, indirectly, of nephron number. As long as a signaling loop between the CM and the tips of the UB is maintained, UB branching and differentiation will continue. The regulation of GDNF expression and function can be deduced from a rich collection of human and mouse mutations that cause renal agenesis. In this way, several transcription factors (e.g. *Sall1*, *Eya1*, *Pax2*, *Hox11* paralogs) and upstream ligands or receptors (e.g. *Nephronectin/a8b1 integrin*, *Gdf-11*) have been identified as required for normal GDNF expression (Bouchard, 2004; Boyle & de Caestecker, 2006). In the absence of GDNF, UB outgrowth and branching can be rescued by the simultaneous deletion of *Sprouty (Spry)* with *GDNF* or *Ret,* revealing a redundant role for the FGF10/FGFR pathway in UB outgrowth and branching (Fig. 11.2, Michos et al., 2010). This observation indicates shared signaling components downstream of Ret and FGFR control these processes. This convergence involves the Ras/Erk MAP kinase, PLCγ-Ca^{+} and PI3K-Akt signaling pathways (reviewed by Takahashi, 2001), all of which contribute to kidney development (Fisher, Michael, Barnett, & Davies, 2001; Jain, Encinas, Johnson, & Milbrandt, 2006; Tang, Cai, Tsai, Wang, & Dressler, 2002; Wong et al., 2005). GDNF-activated Ret upregulates *Wnt11*, a secreted positive regulator of GDNF expression from the MM, as well as *Spry1,* a negative regulator of GDNF and FGF signal reception in the UB (Basson et al., 2005) in an incoherent feed-forward loop (Kim, Kwon, & Cho, 2008; Muller, Obeyesekere, Mills, & Ram, 2008). Ret signaling also upregulates the RTK signaling inhibitors *Dusp6* and *Spred2* (Lu et al., 2009). Two ETS domain factors, *Etv4* and *Etv5*, act downstream of Ret and FGFR. Both are co-expressed in the UB tips, where they are upregulated by GNDF/Ret signaling via PI3K. In the MM, *Etv4* and *Etv5* are upregulated by FGFR signaling and lost when FGFR signaling is reduced (Barak et al., 2012). *Etv4* and *Etv5* are redundant; double homozygotes never develop kidneys (Lu et al., 2009). This network controls renal branching morphogenesis and is sensitive to reduction in signal strength (Fig. 11.2A). Hence, as long as there remains sufficient undifferentiated CM to provide GDNF to the ureteric tips, branching should continue. As each branching event occurs, the surrounding CM is effectively divided between the newly formed branches, however, there is also a corresponding reduction in tip size (M.L., unpublished observations). Nevertheless, the balance between CM self-renewal, which will replenish the GDNF source, and exit via commitment to differentiation, ultimately determines the capacity of this domain to support UB branching.

5.2. Differentiation to form nephrons

Classical embryological experiments identified the UB and several other tissues as sources of signal(s) that could support MM differentiation (Saxen, 1987). Subsequently, investigators realized that all these tissues produced "canonical" Wnt ligands and identified Wnt9b as the *bona fide* UB differentiation factor (Carroll, Park, Hayashi, Majumdar, & McMahon, 2005). In response to canonical Wnt signaling, cells within the CM cluster form *Cdh4/6*-expressing pretubular aggregates (PTA) which upregulate the expression of *Wnt4*, *Lhx1*, *Fgf8*, and *Fgfrl*. While Wnt9b is the primary signal

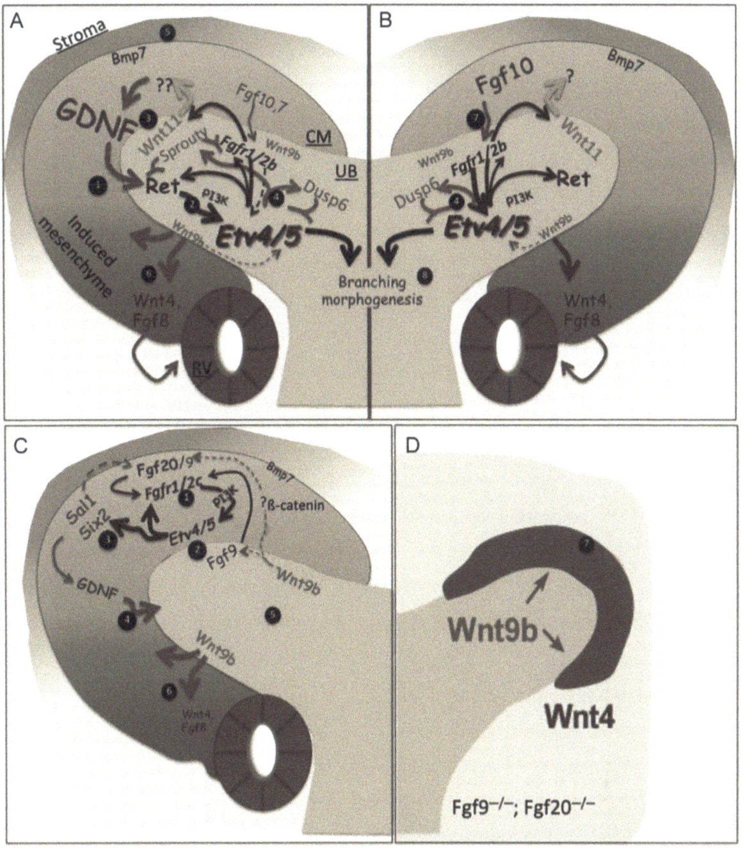

Figure 11.2 Genetic network controlling ureteric bud branching and MM maintenance. (A) Ret controls UB tip stem cells (for details see Costantini & Kopan, 2010). Bold arrows mark major effects; dotted arrows, relationships that are week or uncertain. Black arrows summarize complex, multistep processes. (1) GDNF signal to Ret and FgfR in the UB to activate Etv4/5 via PI3K. (2) Etv4/5 activates Ret, FgfR, and Wnt11. (3) Wnt11 signals to the CM to activate GDNF via an undetermined process. (4) At the same time, Etv4/5 also activates Sprouty, an inhibitor of both Ret and Fgfr, and Dusp6, an inhibitor of Ret and Fgfr signaling. This forms an incoherent feed-forward loop. The stroma cells produce (5) BMP7, which promotes aggregation and sorting of CM cells. The UB secretes Wnt9b, the main differentiation-inducing signal in the metanephros (6). Wnt9b-induced mesenchyme undergoes mesenchymal to epithelial transition and produces Wnt4 and FGF8, the latter will induce proliferation in the renal vesicle (RV). (B) Loss of GDNF leads to loss of Ret signaling in the UB tip cells; leading to failure in maintenance of Etv4/5 in the UB and loss of branching/tip cell identity. Mesenchymal Fgf7,10 *can* signal via Fgfr and PI3K to activate Etv4/5 to rescue loss of GDNF but will only do so if the inhibitor *spry* is removed (Costantini & Kopan, 2010) (7). The net output of the Ret/Fgfr signals in the UB is branching morphogenesis, which can be observed in the absence of the CM as long as GDNF is provided. Disruption of this network at any point eliminates

(Continued)

initiating CM differentiation into PTA via the stabilization of β-catenin within the CM (Park, Valerius, & McMahon, 2007), RV formation also requires Wnt4. Interestingly, sustained high levels of β-catenin signaling appear to retard the completion of the epithelialization event (Park et al., 2007). The induction of *Wnt4* expression within the PTA results in an auto-inductive response that completes MET (Kispert, Vainio, & McMahon, 1998; Stark, Vainio, Vassileva, & McMahon, 1994). Unlike Wnt9b, the Wnt4 signal also involves a noncanonical Ca^{+}-mediated autocrine signal (Burn et al., 2011; Tanigawa et al., 2011), which may help explain why *Wnt4*$^{-/-}$ kidneys fail to branch (see below for further discussion of this point).

Fgf8 expression in PTA acts in an auto-inductive fashion to drive proliferation of epithelial cells within the RV and its derivatives (the comma and S-shaped body; Grieshammer et al., 2005; Perantoni et al., 2005). The PTA also expresses *Fgfrl1*, a receptor mimic lacking an intracellular kinase domain. *Fgfrl1* is required for RV formation (Trueb, Amann, & Gerber, 2013), most likely by restricting FGF8 activity where it cannot interfere with nephrogenesis. While FGF8 is not thought to be involved in signaling directly to the CM (Barak et al., 2012), this does not eliminate the possibility that RV-derived factors do affect tip identity or proliferation in a feedback loop.

While Wnt9b is regarded as the initial trigger for CM differentiation, isolated MM *in vitro* can be induced to form epithelial structures in response to *Wnt4* alone (Osafune, Takasato, Kispert, Asashima, & Nishinakamura, 2006). Wnt4 may also act via stabilized β-catenin, or the Six2^{+} expressing CM is permissive and poised to undergo MET in response to a variety of stimuli. Indeed, in mice lacking both *Wnt9b* and *Wnt4*, the CM will respond to ectopic Notch activation by forming renal epithelia (Boyle, Kim, Valerius, McMahon, & Kopan, 2011). This is a property of CM cells and not of activated Notch proteins as stromal cells did not convert to epithelia, and Notch signals are used by stem cells in many nonepithelial lineages (Liu, Sato, Cerletti, & Wagers, 2010).

Figure 11.2—Cont'd progenitor identity in both the MM and the UB tip and renal agenesis. (C) The intra-CM Fgf signaling loop involves Fgf/920 and FgfR1/2 (1) which activate Etv4/5 in the MM via PI3K. (2). Etv4/5 activates Fgfr, and (3) CM maintenance factors Six2 and Sal1 which indirectly ensure GDNF expression. (4) GDNF induces branching. The UB cells produce (5) Wnt9b, which promotes maintenance of CM targets either directly or indirectly, perhaps by elevating Fgf9 and/or Fgf20. Wnt9 also induces differentiation (6) which takes over the entire CM when Fgf9/20 levels are critically reduced (D, (7)). (See color plate.)

5.3. Maintenance of a self-renewing progenitor population

The initial number of CM cells appears to be defined at E11 (~10,000 cells) with no further contribution from other compartments (Kobayashi et al., 2008). The promiscuity of the CM to respond to differentiation signals argues for the presence of strong counter survival/proliferation signals ensuring sufficient CM self-renewal to drive continued nephrogenesis throughout kidney development. Although genetic deletion of *Cited1* and *Cited2* did not result in any detectable phenotype (Boyle et al., 2007), Cited1$^+$ cells within the CM appear refractory to differentiation by β-catenin. BMP7-mediated SMAD signaling converted this population into a *Cited1*$^-$ Wnt9b-responsive population (Brown et al., 2013). While provocative, these observations were made on isolated, dissociated renal cortex and await confirmation *in vivo*. Because Cited1 is but a marker for this cell population, some other factor(s) must provide protection from the differentiation cue provided by Wnt9b and available to most if not all CM cells.

Six2 is part of the mechanism protecting the CM from differentiation: in its absence, a global loss of self-renewal results in the formation of a few renal epithelial structures in a rudimentary organ. The collapse of the CM is associated with loss UB branching and severe hypoplasia (Self et al., 2006). Because Six2 seemed to prevent the *en masse* differentiation of CM progenitors exposed to Wnt9b (Self et al., 2006), and because it acts as a transcriptional repressor, it was proposed to act as a nuclear Wnt antagonist. However, reanalysis of the Wnt9b knockout mice suggests that there is not only a loss of markers of nephron formation in these kidneys, but also the loss of expression of uninduced CM markers (Karner et al., 2011). This suggests that Wnt9b not only plays a role in triggering commitment to epithelialize but is also required for CM survival/self-renewal. Such a role for canonical signaling is also indicated by the disruption to CM structure seen with mesenchyme-wide deletion of canonical Wnt signaling (Karner et al., 2011) and the deletion of tankyrase (Karner et al., 2010).

Although Six2 and β-catenin had been proposed to antagonize each other (Kobayashi et al., 2008; Self et al., 2006), comprehensive analysis of chromatin binding by Six2 and β-catenin by ChIP within isolated CM either untreated or treated with a canonical Wnt agonist revealed a complex interrelationship between Six2 and β-catenin (Park et al., 2012), leading to the suggestion that Six2 and β-catenin may cooperate in regulating expression of enhancers with overlapping Six2/TCF sites. Support to this idea comes from the observation that Six2 can bind TCF *in vitro* (Park et al., 2012).

Alternatively, Six2 and β-catenin binding at these sites may be mutually exclusive (Park et al., 2012). While these alternatives and their functional outcomes remain to be resolved, a plausible emerging picture suggests that in $Six2^{HI}$ CM, Six2 and β-catenin cooperate to regulate both self-renewal and competence (Karner et al., 2011; Park et al., 2012), whereas in $Six2^{LO}$ cells, β-catenin promotes differentiation directly and/or with help from a stromal signal (see below). Notably, retention of *Six2* expression was not sufficient to retain competence to differentiate *in vitro* (Lusis et al., 2010). Indeed, *Six2* expression is not restricted to the CM but persists in the forming PTA and RV, albeit at low levels of expression (Brunskill, Georgas, Rumballe, Little, & Potter, 2011; Rumballe et al., 2011).

The identity of exogenous factors capable of maintaining a CM progenitor was first suggested when a combination of bFGF (Fgf2) and Bmp7 was seen to promote the growth of an isolated, competent renal progenitor population (Dudley et al., 1999). *Bmp7* is expressed both in the CM and in the adjacent UB tips (Dudley et al., 1995) and a global deletion of this growth factor results in a loss of nephron formation and a likely depletion of CM (Dudley et al., 1995; Luo et al., 1995). Of note, BMP7 is not autoregulatory and it cannot be induced or maintained by EGF or bFGF (Godin, Takaesu, Robertson, & Dudley, 1998). Thus, expression of *Bmp7* in the CM must require a different inductive signal, most likely a proteoglycan or Wnt signaling. Because a conditional ablation of *Bmp7* in the CM alone has not been performed, it is also unclear whether BMP7 may be provided by other sources.

BMP7 appears to elicit both BMP-SMAD and MAPK/JNK signaling within the CM acting through receptors yet to be identified (Blank, Brown, Adams, Karolak, & Oxburgh, 2009). The ablation of the Type 1 BMP receptor Alk3 from the mesenchyme with Rarb2-Cre resulted in hypomorphic kidneys, however, the defect was less severe than in the global BMP7 knockout (Di Giovanni, Alday, Chi, Mishina, & Rosenblum, 2011). Using an approach in which MM is isolated and cultured *in vitro*, it has been shown that proliferation of the $Cited1^{+}$ CM population is enhanced in response to BMP7-mediated MAPK/JNK signaling, and enhanced further by EGF (Brown et al., 2011). Conversely, phosphoSMAD1/5/8 appears to facilitate the commitment of these same uninduced CM cells to nephron formation in response to Wnt signaling (Brown et al., 2013). Similarly, BMP7 signaling via robust phosphoSMAD1/5/8 accumulation strongly supports epithelial regeneration in the postnatal kidney after acute injury (Lund, Davies, & Hruska, 2002; Manson, Niederhoff, Hruska, & Austin, 2011). These observations suggest a pro-differentiation role for BMP7 when acting via SMAD and a pro-progenitor role when acting via MAPK/JNK signaling (Brown et al., 2011, 2013), synergizing with FGF (Barak et al., 2012). The eventual outcome possibly depends upon dose, receptor identities, and the presence of other growth factors.

CM-specific inactivation of both FGF receptors 1 and 2 (FGFR1/2) arrests kidney development (Poladia et al., 2006; Sims-Lucas et al., 2011), indicating a critical role for a ligand, but the identity of signals involved in CM survival *in vivo* long remained obscure. UB-derived, epithelial FGF ligand(s) were sought because in most tissues, FGF ligands secreted from epithelial cells signal to the mesenchymal ("c") splice forms of FGFR, whereas ligands secreted from mesenchymal cells signal to epithelial ("b") splice forms of FGFRs (Zhang et al., 2006). FGF8, expressed in the RV, and FGF7/10, expressed by the CM, did not support survival of isolated MM cells (Barak et al., 2012; Brown et al., 2011). Targeted deletion of *Fgf7* does lead to a reduction in ureteric branch numbers and, secondarily, to fewer nephrons (Bates, 2011; Qiao et al., 1999), as expected from the CM/UB feedback loop. However, FGF9 and FGF20, two related paralogs shown to signal from within the UB (Fgf9) and the CM (Fgf9/20) in an unusual autocrine (mesenchyme to mesenchyme) manner, were found to provide the missing signal. *Fgf9*$^{-/-}$; *Fgf20*$^{-/-}$ mouse embryos and *FGF20*$^{-/-}$ human fetuses specify a transient CM cell population but lack kidneys (Barak et al., 2012). *In vitro*, addition of the carrier heparin with Fgf9/20 supported survival of MM cells without any evident epithelial differentiation for up to 6 days (Barak et al., 2012; Brown et al., 2011). Interestingly, aggregation (Barak et al., 2012) and both PI3K and MAPK activation (but not ERK signaling) appear critical for progenitor survival *in vitro* (Brown et al., 2011).

It is unclear if PI3K-activated CM cells retain the competence to differentiate. However, both isolated MM (which contains the CM and stromal progenitors) and sorted Six2^{+} progenitors maintained with Fgf9/20 retained the ability to differentiate when a Wnt source was added (Barak et al., 2012). Progenitor aggregation in culture was further promoted by BMP7, which improved overall survival and increased competence to respond to Wnt. Importantly, purified populations of Six2^{+} cells could not be maintained indefinitely in Fgf9-containing media. Within a few days, a rapidly dividing, Six2^{-} cell population emerged from sorted Six2^{+} cells; Six2^{+} cells dispersed and by 7–8 days were no longer detectable (Barak et al., 2012). Whether this suggests the requirement for other signals or a window of time within which a differentiation signal is required is yet to be resolved.

It is unclear what activates *Fgf9/20* expression in the CM, but Wnt proteins are strong candidates based on their activity elsewhere (Chamorro et al., 2005). Accordingly, when differentiation was induced in purified $Six2^+$ cells after 3 days in culture, Wnt4-expressing renal epithelia were surrounded with $Six2^+$ cells for at least 8 days (Barak et al., 2012), perhaps due to induction of endogenous Fgf9/20. *Fgf20* was among the genes that were downregulated in sorted MM cells, and its expression was restored by a pharmacological activator of canonical Wnt signaling (Park et al., 2012). However, CHIP-seq experiments could not identify β-catenin binding within 1 Mb of the *Fgf20* promoter. Instead, Six2 binds near *Fgf20* and may contribute to its expression (Park et al., 2012). *Fgf9* regulation appears more perplexing. *Fgf9* mRNA is absent from isolated E11.5 $Wnt9b^{-/-}$ MM, suggesting it is downstream of Wnt9b (Karner et al., 2011). Although *Fgf9* was bound by both Six2 and β-catenin (Park et al., 2012), it was not induced after activation of canonical Wnt signaling in sorted MM cells.

Fgf9 in the CM may represent a target for the cooperative regulation of Six2/β-catenin, as described earlier. However, *Fgf9* expression is reduced in $Wnt4^{-/-}$ MM which still receives ample Wnt9b signals (Karner et al., 2011). Because $Wnt9b^{-/-}$ embryos do not express *Wnt4*, this could mean that Wnt4, and not Wnt9b, activates *Fgf9* (but not *Fgf20*) in a β-catenin dependent manner. Much more remains to be done to uncover what may be yet another incoherent feed-forward loop involving Six2, β-catenin, Fgf9/20, and Etv4/5 (Fig. 11.2C).

Since *Fgf20* is produced by CM cells, why can't they survive on their own? Some evidence that they may be able to self-sustain was obtained when MM cells were maintained in media containing heparin. Wild-type MM cells survived longer and proliferated better than MM isolated from $Fgf20^{-/-}$ embryos, however, Six2 expression was eventually lost (H. Barak and R. Kopan, personal communication). This result suggests that improvements in isolation protocols may be needed or that additional factors are necessary for survival (Carroll & Das, 2013). One important consideration is that isolated MM contains stromal cells. As will be discussed below, stromal cells provide signals that promote differentiation over survival.

Mice with reduced amounts of Fgf9/20 ligands contain many branch tips in which $Six2^+$ CM cells were prematurely replaced by mature nephrons. *In vitro*, branching of $Fgf9^{+/-}$; $Fgf20^{-/-}$ metanephroi stalls when too few $Six2^+$ cells are present near a branch tip (Barak et al., 2012). Consistent with the idea that the CM and the UB are coupled in a feedback loop, stalled branching could be restarted by addition of GDNF to $Fgf9^{+/-}$; $Fgf20^{-/-}$

metanephroi *in vitro*. Importantly, resumed branching resulted often in an increased Six2$^+$ population (supplemental information, Barak et al., 2012). The later observation suggests that either GDNF elevated the levels of Fgf9 production from the UB or other UB factors can compensate for Fgf9/20 loss if Ret activation levels are sufficiently elevated.

The proliferation rates in the CM population are low: CM cells are estimated to only go through five or fewer divisions between E11.5 (the start of nephrogenesis) and P3 (cessation of nephrogenesis, Kobayashi et al., 2008). *In vitro*, selection for Six2$^+$ populations that self-renew effectively resulted in loss of competence to differentiate (Lusis et al., 2010); conversely, retention of competence to differentiate with BMP7/FGF9 was also associated with low proliferation rates of Six2$^+$ cells (Barak et al., 2012). Yap and Taz, the effectors of the Hippo pathway known for its role in regulating organ size and proliferation (Bossuyt et al., 2013), are both dispensable for the self-renewing capacity of CM cells (Reginensi et al., 2013). In contrast, loss of Wnt9b or β-catenin was associated with severe reduction in the number of Six2$^+$ cells, an indication of a role for Wnt signaling in stimulating proliferation. Indeed, pharmacological and genetic analysis showed that Wnt9b-null CM cells were correctly specified but had greatly reduced rates of proliferation, suggesting Wnt9b acts to maintain the size of progenitor pool (Karner et al., 2011; Park et al., 2012).

While showing mild phenotypes as single gene deletions, simultaneous removal of the cytoskeletal scaffolding proteins *Dlg1* (*disks-large homolog 1*) and *CASK* (*calcium/calmodulin-dependent serine protein kinase*) results in a depletion of CM via reduced proliferation and increased apoptosis (Ahn et al., 2013). This is a cell autonomous process as it can be recapitulated by deletion specifically within the CM. In mutants, the remaining CM was poorly aggregated and showed reduced Ras pathway activity, including reduced phospho(p)-Raf, p-P38, and p-JNK. Dlg1 and CASK link receptors and scaffolding proteins at intracellular junctions and can play roles in cell–cell interactions, cell motility, and cell signaling. The reduction in Ras pathway activity in the CM of these mutants implicates these proteins in FGF signaling. The authors propose a link with FGF8, but this is quite restricted to the PTA and hence may have declined due to a concomitant reduction in nephrogenesis (Ahn et al., 2013). FGF9/20 expression was not investigated. The association with ERK is unclear, but the reduction in p-JNK may relate to a reduction in Bmp7 expression in the CM.

CM cells are clearly mesenchymal, but they are distinguishable histologically from the surrounding mesenchyme and have long been described as forming via aggregation around the tip. What processes regulate the stereotypical localization of the forming nephrons is not known, and a role for cellular adhesion has not been fully explored. It is possible that the CM has a degree of motility such that induced CM sort out to the region where MET occurs. Since Dlg1 is also known to play an important role in cell migration, the mutant phenotype in *Dlg1:CASK* compound mutants may result from aberrant CM cell movement, either in the initial aggregation phase or during turnover (Ahn et al., 2013).

6. BALANCING SELF-RENEWAL AND DIFFERENTIATION: SIGNALS FROM THE STROMA

Until recently, focus has been placed primarily on the interactions between the CM and UB and how this balances branching, CM self-renewal, and nephron initiation. CM cells are also in contact with other key cellular components likely to regulate this balance. Recent studies have begun to delve into the role of the surrounding Foxd1$^+$ SM. Loss of SM by either deletion of *FoxD1* (Hatini et al., 1996; Levinson & Mendelsohn, 2003) or by overexpression of an activated Notch allele in the FoxD1-lineage (Boyle et al., 2011) resulted in an impressive expansion of Six2$^+$ populations (Boyle et al., 2011; Levinson et al., 2005) which retained an ability to differentiate, albeit with a considerable delay (Levinson et al., 2005). This data suggest that SM actively participates in balancing CM self-renewal and differentiation, restricting the former and promoting the latter. Recent observations noted the upregulation of five genes in *FoxD1*-deficient cells in cell culture (Fetting et al., 2013). One such gene encoded the small Leucine-rich proteoglycan decorin (*Dcn*), a modulator of growth factor signaling (Iozzo & Schaefer, 2010) expressed in the interstitium. Dcn has been proposed to bind BMP7, reducing its ability to trigger SMAD phosphorylation and thus preserving self-renewing uninduced CM. Supporting this hypothesis is the observation that Dcn antagonized BMP7 *in vitro*, and in *Dcn*$^{-/-}$ *FoxD1*$^{-/-}$ mutants CM differentiation was partially rescued, restoring some *Lef1* expression and reducing the numbers of Cited1$^+$Six2$^+$ cells to near wild-type levels (Fig. 11.3; Fetting et al., 2013).

According to this model, the expansion of CM in *FoxD1* mutants should also be reversed by overexpression of BMP ligands. Indeed, the loss of *FoxD1* leads to the formation of an aberrant vascular system and elevated expression of *Bmp4* by cells from outside the SM lineage (Levinson et al., 2005). We can speculate that the eventual differentiation of nephrons in

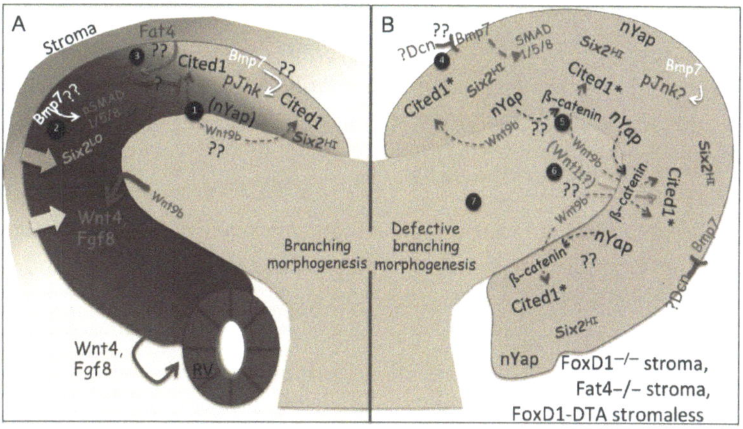

Figure 11.3 SM contribution to the niche. (A) The CM is divided into uninduced *Six2*^{HI},*Cited1*$^{+}$ cells (light green) that are refractory to Wnt differentiation signals (1), and *Six2*^{LO}, *Cited1*$^{-}$ cells (dark green) that can respond to Wnt and become competent to differentiate (purple) (2). A Wnt signal activates *Cited1* and additional markers; *Cited1* is not required for renal development and thus labeled in the figure with an asterisk. BMP7 signals through unknown receptor(s) to activate JNK in uninduced CM. Activation of pSMAD1/5/8 is thought to condition the CM and make it competent to respond to Wnt. The stroma also expresses Fat4 (3) that through an unknown kinase phosphorylate Yap. pYap is either unable to cooperate with β-catenin to maintain uninduced CM or interacts with Disheveled to block β-catenin from maintaining uninduced CM. (4). Alternatively, in the absence of FoxD1 or stroma, Decorin (DCN) is overexpressed in the interstitium to either inhibit SMAD phosphorylation or all BMP7 activity, failing to condition CM which remains refractory to Wnt. Without Fat4 Yap is nuclear (nYap). nYap may somehow impact β-catenin to maintain uninduced CM (and the marker *Cited1*) (5) or, in the absence of pYap, weak Wnt ligands like Wnt11 gain the ability to activate β-catenin and maintain uninduced CM and *Cited1* (6). Branching is affected without induced mesenchyme (7). (See color plate.)

FoxD1-deficient kidneys (Levinson et al., 2005) resumes when BMP4 reaches a level sufficient to condition the cells such that they regain an ability to respond to Wnt9b. Based on this view, the loss of stroma regulates CM by modulating the availability of BMP7, but may not be actively promoting CM differentiation. However, as loss of *Dcn* did not fully restore kidney morphogenesis, additional SM signals are involved.

Another possible SM signal emerged from studies of *Foxd1Cre;RosaDTA* embryos (Das et al., 2013). This strategy eliminated the SM and all its derivatives whereas loss of *FoxD1* still permits differentiation of PDGFRb^{+} cells (Fetting et al., 2013). As in the other models, the CM expands and differentiation is reduced. To ask if the retention of CM cells required Wnt signals, the authors used Wnt inhibitors in organ culture, discovering to their surprise that β-catenin dependent but Wnt-independent retention of β-catenin targets occurs in *Foxd1Cre;RosaDTA* (stromaless) CM. The authors noted further that Yap was unphosphorylated and exclusively nuclear in stromaless CM, but cytoplasmic and phosphorylated in the wild-type uninduced CM. This observation led to the proposal that nuclear Yap cooperates with β-catenin to maintain expression of progenitor gene expression and that Fat4 in the stroma regulates Yap localization in the adjacent CM, maintaining it in the cytoplasm and thus facilitating differentiation. As noted earlier, another study deleted Yap from the CM and reported no effect on the balance of CM self-renewal/differentiation (Reginensi et al., 2013). However, both Reginensi et al. (2013) and Das et al. (2013) agree that in the absence of Yap, the expression of many Wnt9b-dependent transcripts is reduced, including markers of the uninduced CM such as *Meox2* and *Cited1*. Both studies conclude that planar cell polarity (PCP) signaling is not involved in maintaining the CM niche.

Consistent with a possible Yap/β-catenin interaction, removal of *Fat4* in *Wnt9b*$^{-/-}$ CM cells allowed nuclear Yap accumulation and restored the expression of uninduced CM markers, including *Cited1*, without a change in pSMAD1/5/8 levels (Das et al., 2013). This is consistent with a role for Yap in regulating these transcripts via an unknown mechanism that does not require a direct interaction with β-catenin or binding near β-catenin targets. In this context, it is worth noting that in the intestinal stem cell, cytoplasmic Yap binds to Disheveled, acting as a Wnt signaling antagonist (Barry et al., 2013). If this later function was conserved in the kidney, cytoplasmic pYap would act as a Wnt antagonist whereas nuclear Yap would act as a Wnt agonist, while loss of pYap may allow weak Wnt ligands (Wnt11?) to maintain uninduced CM via β-catenin (Fig. 11.3B).

Based on these and other observations, Das et al. (2013) propose a model in which the SM plays an *active* role in promoting differentiation of CM cells, and Fat4, Yap/Taz, and β-catenin play a critical role in balancing differentiation versus proliferation. Their model predicts that removal of Yap/Taz will unleash rapid exhaustion of CM progenitors via premature differentiation, which in turn will lead to loss of branching. However, only a mild/late CM exhaustion phenotype was described and this was subtle enough to escape notice by Reginensi et al. (2013). A prior Fat4 deletion also did not report a CM expansion; however, this is evident in their supplemental information (Saburi et al., 2008) where *WT1* expression in the CM at postnatal day (P) 0 indicates loss of CM differentiation. The model also predicts that differentiating cells will always be adjacent to the stroma and uninduced progenitors will contain nYap and be closer to the UB, a prediction inconsistent with the expression domain of *Cited1* and the evidence that the *Cited1*-expressing cells are refractory to the differentiation-promoting Wnt9b signals (Brown et al., 2013), despite their contacts with Fat4-expressing SM cells. These various inconsistencies will need to be addressed and are likely to arise as a result of complex feedback systems yet to be identified. However, all three studies agreed that the SM plays an important role in the balance of differentiation/self-renewal, and that several potential SM signals are at work.

7. INFLUENCES FROM OTHER CELLULAR COMPARTMENTS

Similar to the SM, we predict that other cellular compartments adjacent to the nephrogenic niches may also provide feed-forward and feedback signals to regulate final nephron and ureteric tip number. Other non-Foxd1-derived interstitial components, such as the endothelium, also reside close to the CM. Signals from the newly formed nephrons themselves are also likely to influence the niche. Indeed, the distal RV most closely abutting the ureteric epithelium strongly expresses ligands for the Notch, FGF, Wnt, and BMP pathways. The possibility that some of these ligands influence the CM has not been fully ruled out. *Wnt4* expression in this region is regarded as being a cell autonomous signal for the process of epithelialization (Burn et al., 2011; Tanigawa et al., 2011). Notably, the loss of *Wnt4* arrests branching (Stark et al., 1994), suggesting a direct role on the tip or in a supporting a $GDNF^{+}$ self-renewing CM. Cells in the distal RV that invade the ureteric epithelium to form a patent connecting lumen (Georgas et al., 2009; Kao, Vasilyev, Miyawaki, Drummond, & McMahon, 2012) also express *Bmp2* and *Dll1*. Any given ureteric tip becomes physically attached to nascent nephrons while continuing to branch, with the maturing nephrons altering the spatial relationship of CM and tip. This alone could change the signaling environment within the niche across developmental time. A more comprehensive analysis of such effects will require temporospatial analyses in 3D in whole organs *in vivo*.

8. TRANSCRIPTIONAL REGULATION OF THE CM

Until this point, we have focused primarily on signaling pathways. Table 11.1 summarizes the transcription factors known to be expressed in the CM. Rather than each playing a discrete function in the CM, many appear to play complex roles in regulation of the CM state (and often also the formation of the MM). This is because many factors form protein–protein complexes with each other and with components regulating chromatin structure and DNA methylation. For example, Sall1 interacts with the NuRD complex (Denner & Rauchman, 2013) and Six2 may interact with β-catenin to regulate the balance between self-renewal and differentiation (Park et al., 2012). The deletion of *WT1* results in renal agenesis, suggesting a role in CM survival. WT1 also contributes to MET via direct regulation of *Wnt4* (Sim et al., 2002) with distinct activities likely to result from context-dependent interactions with other proteins. One of the earliest transcription factors expressed in the MM, and later becoming restricted to the CM, is Pax2. In the absence of *Pax2*, the MM does not secrete GDNF, and isolated *Pax2*$^{-/-}$ cells fails to respond to differentiation cues (Brophy, Ostrom, Lang, & Dressler, 2001). Pax2 can interact with GRG4 or PTIP to recruit either chromatin silencing or activating machinery, respectively, to target loci (Patel et al., 2012). What these targets are and whether Pax2 has distinct functions in CM versus UB versus the differentiating nephron is not yet clear. The fact that Pax2 can be phosphorylated by JNK may place Pax2 under feedback regulation by Bmp7 or by noncanonical Wnt4 signals, perhaps explaining why *Wnt4* loss impacts branching.

Removal of other MM/CM transcription factors, including *Hoxd11/a11*, *Osr1*, *Six1*, and *Sall1/4*, also results in loss of CM. Osr1 regulates *Six2, Pax2*, and *Eya1* and there is evidence for protein–protein complex formation including Pax2:Six1:Eya1 and Eya1:Hoxa11:Pax2 complexes (Gong et al., 2007; reviewed in Hendry & Little, 2012). Hence, it is interesting to speculate that the CM state can be recreated in an adult tubular epithelial cell via the reactivation of a few key regulatory genes that includes *Six1*, *Six2*, *Eya1*, *Hoxa11*, *Osr1*, and *Snai2* (Hendry et al., 2013). Fully understanding the complex regulatory network of the CM will require additional CHiP-Seq-based analyses such as those commenced for Six2 (Park et al., 2012).

9. RENAL HYPOPLASIA: TIME TO REASSESS WHAT WE KNOW

A disruption to the balance of self-renewal and differentiation clearly result in aberrant kidney development. While a severe early defect is easy to score, hypoplasia ("small kidney") has been associated with >150 genes/loci (www.ncbi.nlm.nih.gov/omim/). Moreover, the dysplastic spectrum is very broad and the degree of disruption to kidney organogenesis is often not well characterized. While reduced nephron number is linked with small kidneys at birth, subsequent hypertrophy within such kidneys can frequently result in kidneys of apparently normal weight. This phenomenon is undoubtedly going to mask many more subtle developmental defects and is likely to explain the large variability (10-fold) seen in nephron number in the human population (Bertram et al., 2011). Since our ability to identify the molecular basis of such variation emerged only after the relatively recent development of inducible Cre tools, many gene-deletion models generated without such tools, such as *Pax2* and *Bmp7*, need to be reexamined via selective deletion in the CM. A second major barrier is our lack of quantitative imaging capabilities in both mice and man. Nephron number cannot be readily quantified except at postmortem while total ureteric tip number, niche number, and timing of cessation of nephrogenesis are never comprehensively evaluated. The development of quantitative tools for such analyses will undoubtedly improve the phenotyping of various mutations quantifying the morphometric alterations in response to genetic and/or environmental perturbation during development. Imaging approaches such as optical confocal microscopy have begun to tackle this issue (Short, Hodson, & Smyth, 2013), allowing comprehensive 3D imaging of the entire kidney out to E15.5 and facilitating the accurate comparison of ureteric branching between wild-type and subtle mutant lines. In humans, better contrast agents are being developed to facilitate the counting of nephrons in living individuals (Beeman et al., 2011). Combined, these strategies will help identify the gene networks regulating nephron endowment.

Under normal conditions there is little evidence for apoptosis within the CM (Georgas et al., 2009), but it could be a factor in hypoplasia. If the UB fails to reach the metanephric region, as described in the *WT1* gene knockout, the MM forms but undergoes apoptosis and is lost within 48 h (Kreidberg et al., 1993). The trigger for this apoptotic induction is not clear, but loss of FGF signaling can cause apoptosis among $Six2^{+}$ cells (Barak et al., 2012; Poladia et al., 2006). Isolated MM cells, which include CM and SM, also undergo apoptosis, indicating that factors provided by adjacent tissues are important for survival. Such survival factors may be distinct from those pathways regulating self-renewal. Several antiapoptotic genes, including *Bcl2*, are expressed in the CM. Loss of *Bcl2* in the developing kidney has been reported to result in severe hypoplasia due to increased apoptosis in the nephrogenic region (Hammerman, 1998; Sorenson, Rogers, Korsmeyer, & Hammerman, 1995). Removal of p53, either globally or specifically in the CM, results in hypoplasia with an apparent CM depletion (Saifudeen et al., 2012). In this model, there is both a reduction in proliferation and an increase in apoptosis. The p53 phenotype may include the consequences of reduced *Pax2*, a hypothesis supported by reporter assays and ChIP studies showing that p53 can positively regulate *Pax2* expression and regulate *Bcl2* expression (Saifudeen et al., 2012). Deletion of both p53 alleles also worsens the hypoplastic phenotype of $Pax2^{+/-}$ mice (Saifudeen et al., 2012). CM also expresses *c-myc* and *N-myc*, both of which may regulate the turnover rate of this population. Deletion of *c-myc* in the MM (using a Bmp7Cre rather than the more commonly used Six2Cre) was reported to result in hypoplasia (Couillard & Trudel, 2009). Global deletion of *N-myc* did not have a kidney phenotype (Bates, Kharzai, Erwin, Rossant, & Parada, 2000); however, this requires further investigation. The regulation of expression of these genes in the CM is not understood.

10. CESSATION OF NEPHROGENESIS

The relationship between branching and nephron number is generally viewed as a simple reciprocal set of signals between the ureteric tip and the CM that continues stably throughout organogenesis. However, kidney development is not uniform. The majority of branch generations are formed within the first half of metanephric development (Cebrian, Borodo, Charles, & Herzlinger, 2004) while the majority of nephrons form in the immediate postnatal period (Hartman, Lai, & Patterson, 2007). The $Six2^+$ CM is lost and the final nephrons form at approximately P3 in the mouse (Rumballe et al., 2011). Although less thoroughly investigated in humans, the literature suggests that nephron formation in the human kidney ceases prior to birth (~36 weeks; Hinchliffe, Sargent, Howard, Chan, & van Velzen, 1991).

The molecular basis for the exhaustion of the CM at the cessation of nephrogenesis is not fully understood. While renal injury can trigger reexpression of a number of CM markers within damaged tubular epithelium, including *Sall1* (Abedin et al., 2011) and *Pax2* (Imgrund et al., 1999), the CM marker Six2 is not expressed in the postnatal kidney even in response to injury and all newly formed cells were derived from dedifferentiated adult epithelia (Humphreys et al., 2008). The dependence of the developing kidney on incoherent feed-forward loops may provide clues to how nephrogenesis comes to an end. RTK/Etv4/5 cross-activation amplifies signal reception but is also very sensitive to signal levels. Cells experiencing lower levels of Ret signaling relative to their neighbors will be demoted from tip to stalk and lose the ability to promote GDNF expression and maintain the CM (Chi et al., 2009; Shakya et al., 2005). As the number of CM cells surrounding any given tip declines, this may reduce GDNF levels below a threshold required for tip cell identity resulting in a termination of branching. Analyses of gene expression in the immediate postnatal period, however, suggests that the loss of CM markers precedes the loss of ureteric tip markers with the remaining CM rapidly committing to nephron formation (Rumballe et al., 2011). The ureteric epithelium appears to retain a competent tip progenitor population able to continue branching in response to isolated embryonic CM (Hartman et al., 2007).

The observed shift in the balance between CM self-renewal and differentiation could occur as a result of elevated levels of or access to Wnt9b, loss of Fgf9/20, or a shift to SMAD1/5/8 activation by BMP7. Mutations that prematurely deplete the CM population, such as loss of FGF reception in MM cells (Barak et al., 2012; Poladia et al., 2006; Sims-Lucas et al., 2012) or loss of Sox8/9 from the UB (Reginensi et al., 2011) appear to show regions of premature cessation of nephron formation with these instances representing a more stochastic "exhaustion" of individual niches as opposed to the global pattern of differentiation seen at birth. Indeed, the global depletion of all remaining CM after birth suggests an active trigger. This is supported by the absence of mutants in which there is evidence for prolonged nephron formation. The observed upregulation of glycolytic pathway genes during the final wave of nephron formation led to the proposal that the active trigger may involve changes in metabolism, possibly resulting from a shift in tissue oxygen tension. Such a parturition-based event is inconsistent with the proposed timing of cessation of nephrogenesis in humans (Hinchliffe et al., 1991); however, more recent data questions this conclusion arguing that nephron formation may continue postnatally in term and preterm humans (Faa et al., 2010). The association between prematurity and low nephron in humans is consistent with a parturition-based trigger (Faa et al., 2010). In the mouse, the final nephrons form in a distinct location with respect to the ureteric tip, the latter also showing an altered arrangement with respect to the edge of the kidney (the capsule) (Rumballe et al., 2011). This is coupled with the depletion of surrounding SM, the maturation of vasculature, and growing number of maturing nephrons. This dramatic change in niche topology may be sufficient to tip the balance of inductive/repressive signals, leading to cessation of nephrogenesis. The observation that premature cessation of nephrogenesis can be observed demands a reconciliation of these options.

Only once the mechanism driving cessation is identified can we better predict how one might influence the final nephron number in a positive way.

11. CONCLUSION

The existence of a cap mesenchymal progenitor population, able to give rise to all the mature tubular epithelial cell types within the nephron, makes it a desirable cell type to maintain or recreate for the purposes of generating more nephrons. While this embryonic progenitor population is not present in the postnatal mammal, preliminary data suggests the feasibility of recreating the CM state in an adult epithelial cell via the forced reexpression of key CM transcription factors (Hendry et al., 2013). The utility of such a population were it to be reintroduced into an adult organ is unclear; however, the data from Hartman et al. (2007) suggests that the ureteric compartment, at least in the early postnatal period, can still respond to an embryonic CM. Indeed, in the zebrafish it has been shown that the serial transplantation of nephron progenitors does result in the appropriate fusion with existing ND structures to form new nephrons (Diep et al., 2011). While the receptiveness of the adult mammalian collecting duct remains to be assessed, re-derived CM populations may prove valuable for the generation of mature renal cell types for bioengineering purposes. However, the complexity of the interrelationship between this progenitor population and the surrounding cellular compartments in the nephrogenic niches of the developing kidney suggests we have a lot left to learn before we can recapitulate what the embryo so beautifully regulates *in vivo*.

ACKNOWLEDGMENTS

M. L. is an NHMRC Senior Principal Research Fellow (APP1042093). R. K. and S. C. are both supported by Washington University and the NIH (DK066408). We thank Tom Carroll and Leif Oxburgh for communicating unpublished results. We also thank Adler Ju, Alex Combes, Hila Barak, Bree Rumballe, and Kylie Georgas for providing images.

REFERENCES

Abraira, V. E., Del Rio, T., Tucker, A. F., Slonimsky, J., Keirnes, H. L., & Goodrich, L. V. (2008). Cross-repressive interactions between Lrig3 and netrin 1 shape the architecture of the inner ear. *Development*, *135*, 4091–4099.

Abraira, V. E., Satoh, T., Fekete, D. M., & Goodrich, L. V. (2010). Vertebrate Lrig3-ErbB interactions occur in vitro but are unlikely to play a role in Lrig3-dependent inner ear morphogenesis. *PLoS One*, *5*, e8981.

Acharya, A., Baek, S. T., Banfi, S., Eskiocak, B., & Tallquist, M. D. (2011). Efficient inducible Cre-mediated recombination in Tcf21 cell lineages in the heart and kidney. *Genesis*, *49*, 870–877.

Abedin, M. J., Imai, N., Rosenberg, M. E., & Gupta, S. (2011). Identification and characterization of Sall1-expressing cells present in the adult mouse kidney. *Nephron Experimental nephrology, 119*(4), e75–82.

Ahn, S. Y., Kim, Y., Kim, S. T., Swat, W., & Miner, J. H. (2013). Scaffolding proteins DLG1 and CASK cooperate to maintain the nephron progenitor population during kidney development. *Journal of the American Society of Nephrology, 24*, 1127–1138.

Barak, H., Huh, S.-H., Chen, S., Jeanpierre, C., Martinovic, J., Parisot, M., et al. (2012). FGF9 and FGF20 maintain the stemness of nephron progenitors in mice and man. *Developmental Cell, 22*, 1191–1207.

Barry, E. R., Morikawa, T., Butler, B. L., Shrestha, K., de la Rosa, R., Yan, K. S., et al. (2013). Restriction of intestinal stem cell expansion and the regenerative response by YAP. *Nature, 493*, 106–110.

Basson, M. A., Akbulut, S., Watson-Johnson, J., Simon, R., Carroll, T. J., Shakya, R., et al. (2005). Sprouty1 is a critical regulator of GDNF/RET-mediated kidney induction. *Developmental Cell, 8*, 229–239.

Bates, C. M. (2011). Role of fibroblast growth factor receptor signaling in kidney development. *Pediatric Nephrology, 26*, 1373–1379.

Bates, C. M., Kharzai, S., Erwin, T., Rossant, J., & Parada, L. F. (2000). Role of N-myc in the developing mouse kidney. *Developmental Biology, 222*, 317–325.

Beeman, S. C., Zhang, M., Gubhaju, L., Wu, T., Bertram, J. F., Frakes, D. H., Cherry, B. R., & Bennett, K. M. (2011). Measuring glomerular number and size in perfused kidneys using MRI. *American journal of physiology Renal physiology, 300*(6), F1454–7.

Bertram, J. F., Douglas-Denton, R. N., Diouf, B., Hughson, M. D., & Hoy, W. E. (2011). Human nephron number: Implications for health and disease. *Pediatric Nephrology, 26*, 1529–1533.

Blank, U., Brown, A., Adams, D. C., Karolak, M. J., & Oxburgh, L. (2009). BMP7 promotes proliferation of nephron progenitor cells via a JNK-dependent mechanism. *Development, 136*, 3557–3566.

Bossuyt, W., Chen, C. L., Chen, Q., Sudol, M., McNeill, H., Pan, D., et al. (2013). An evolutionary shift in the regulation of the Hippo pathway between mice and flies. *Oncogene*, http://dx.doi.org/10.1038/onc.2013.82, ePublication.

Bouchard, M. (2004). Transcriptional control of kidney development. *Differentiation, 72*, 295–306.

Boyle, S., & de Caestecker, M. (2006). Role of transcriptional networks in coordinating early events during kidney development. *American Journal of Physiology. Renal Physiology, 291*, F1–F8.

Boyle, S. C., Kim, M., Valerius, M. T., McMahon, A. P., & Kopan, R. (2011). Notch pathway activation can replace the requirement for Wnt4 and Wnt9b in mesenchymal-to-epithelial transition of nephron stem cells. *Development, 138*, 4245–4254.

Boyle, S., Misfeldt, A., Chandler, K. J., Deal, K. K., Southard-Smith, E. M., Mortlock, D. P., et al. (2008). Fate mapping using Cited1-CreERT2 mice demonstrates that the cap mesenchyme contains self-renewing progenitor cells and gives rise exclusively to nephronic epithelia. *Developmental Biology, 313*, 234–245.

Boyle, S., Shioda, T., Perantoni, A. O., & de Caestecker, M. (2007). Cited1 and Cited2 are differentially expressed in the developing kidney but are not required for nephrogenesis. *Developmental Dynamics, 236*, 2321–2330.

Brandenberger, R., Schmidt, A., Linton, J., Wang, D., Backus, C., Denda, S., et al. (2001). Identification and characterization of a novel extracellular matrix protein nephronectin that is associated with integrin alpha8beta1 in the embryonic kidney. *The Journal of Cell Biology, 154*, 447–458.

Brophy, P. D., Ostrom, L., Lang, K. M., & Dressler, G. R. (2001). Regulation of ureteric bud outgrowth by Pax2-dependent activation of the glial derived neurotrophic factor gene. *Development, 128*, 4747–4756.

Brown, A. C., Adams, D., de Caestecker, M., Yang, X., Friesel, R., & Oxburgh, L. (2011). FGF/EGF signaling regulates the renewal of early nephron progenitors during embryonic development. *Development*, *138*, 5099–5112.

Brown, A. C., Muthukrishnan, S. D., Guay, J. A., Adams, D. C., Schafer, D. A., Fetting, J. L., et al. (2013). Role for compartmentalization in nephron progenitor differentiation. *Proceedings of the National Academy of Sciences of the United States of America*, *110*, 4640–4645.

Brunskill, E. W., Georgas, K., Rumballe, B., Little, M. H., & Potter, S. S. (2011). Defining the molecular character of the developing and adult kidney podocyte. *PLoS One*, *6*, e24640.

Brunskill, E. W., Lai, H. L., Jamison, D. C., Potter, S. S., & Patterson, L. T. (2011). Microarrays and RNA-Seq identify molecular mechanisms driving the end of nephron production. *BMC Developmental Biology*, *11*, 15.

Burn, S. F., Webb, A., Berry, R. L., Davies, J. A., Ferrer-Vaquer, A., Hadjantonakis, A. K., et al. (2011). Calcium/NFAT signalling promotes early nephrogenesis. *Developmental Biology*, *352*, 288–298.

Carroll, T. J., & Das, A. (2013). Defining the signals that constitute the nephron progenitor niche. *Journal of the American Society of Nephrology*, *24*, 873–876.

Carroll, T. J., Park, J. S., Hayashi, S., Majumdar, A., & McMahon, A. P. (2005). Wnt9b plays a central role in the regulation of mesenchymal to epithelial transitions underlying organogenesis of the mammalian urogenital system. *Developmental Cell*, *9*, 283–292.

Cebrian, C., Borodo, K., Charles, N., & Herzlinger, D. A. (2004). Morphometric index of the developing murine kidney. *Developmental Dynamics*, *231*, 601–608.

Chamorro, M. N., Schwartz, D. R., Vonica, A., Brivanlou, A. H., Cho, K. R., & Varmus, H. E. (2005). FGF-20 and DKK1 are transcriptional targets of beta-catenin and FGF-20 is implicated in cancer and development. *The EMBO Journal*, *24*, 73–84.

Chen, Z., Montcouquiol, M., Calderon, R., Jenkins, N. A., Copeland, N. G., Kelley, M. W., et al. (2008). Jxc1/Sobp, encoding a nuclear zinc finger protein, is critical for cochlear growth, cell fate, and patterning of the organ of corti. *The Journal of Neuroscience*, *28*, 6633–6641.

Cheng, H. T., Kim, M., Valerius, M. T., Surendran, K., Schuster-Gossler, K., Gossler, A., et al. (2007). Notch2, but not Notch1, is required for proximal fate acquisition in the mammalian nephron. *Development*, *134*, 801–811.

Chi, X., Michos, O., Shakya, R., Riccio, P., Enomoto, H., Licht, J. D., et al. (2009). Ret-dependent cell rearrangements in the Wolffian duct epithelium initiate ureteric bud morphogenesis. *Developmental Cell*, *17*, 199–209.

Costantini, F. (2006). Renal branching morphogenesis: Concepts, questions, and recent advances. *Differentiation*, *74*, 402–421.

Costantini, F. (2012). Genetic controls and cellular behaviors in branching morphogenesis of the renal collecting system. *Wiley Interdisciplinary Reviews: Developmental Biology*, *1*, 693–713.

Costantini, F., & Kopan, R. (2010). Patterning a complex organ: Branching morphogenesis and nephron segmentation in kidney development. *Developmental Cell*, *18*, 698–712.

Costantini, F., & Shakya, R. (2006). GDNF/Ret signaling and the development of the kidney. *Bioessays*, *28*, 117–127.

Couillard, M., & Trudel, M. (2009). C-myc as a modulator of renal stem/progenitor cell population. *Developmental Dynamics*, *238*, 405–414.

Cui, S., Ross, A., Stallings, N., Parker, K. L., Capel, B., & Quaggin, S. E. (2004). Disrupted gonadogenesis and male-to-female sex reversal in Pod1 knockout mice. *Development*, *131*, 4095–4105.

Cui, S., Schwartz, L., & Quaggin, S. E. (2003). Pod1 is required in stromal cells for glomerulogenesis. *Developmental Dynamics*, *226*, 512–522.

Das, A., Tanigaua, S., Karner, C. M., Xin, M., Lum, M., Chen, C., et al. (2013). Stromal-epithelial crosstalk regulates kidney progenitor cell differentiation. *Nature Cell Biology*, *15*, 1035–1044.
Denner, D. R., & Rauchman, M. (2013). Mi-2/NuRD is required in renal progenitor cells during embryonic kidney development. *Developmental Biology*, *375*, 105–116.
Diep, C. Q., Ma, D., Deo, R. C., Holm, T. M., Naylor, R. W., Arora, N., et al. (2011). Identification of adult nephron progenitors capable of kidney regeneration in zebrafish. *Nature*, *470*, 95–100.
Di Giovanni, V., Alday, A., Chi, L., Mishina, Y., & Rosenblum, N. D. (2011). Alk3 controls nephron number and androgen production via lineage-specific effects in intermediate mesoderm. *Development*, *138*, 2717–2727.
Dours-Zimmermann, M. T., Maurer, K., Rauch, U., Stoffel, W., Fassler, R., & Zimmermann, D. R. (2009). Versican V2 assembles the extracellular matrix surrounding the nodes of ranvier in the CNS. *The Journal of Neuroscience*, *29*, 7731–7742.
Dressler, G. R. (2008). Epigenetics, development, and the kidney. *Journal of the American Society of Nephrology*, *19*, 2060–2067.
Dressler, G. R. (2009). The specification and maintenance of renal cell types by epigenetic factors. *Organogenesis*, *5*, 73–82.
Dudley, A. T., Godin, R. E., & Robertson, E. J. (1999). Interaction between FGF and BMP signaling pathways regulates development of metanephric mesenchyme. *Genes & Development*, *13*, 1601–1613.
Dudley, A. T., Lyons, K. M., & Robertson, E. J. (1995). A requirement for bone morphogenetic protein-7 during development of the mammalian kidney and eye. *Genes & Development*, *9*, 2795–2807.
Dudley, A. T., & Robertson, E. J. (1997). Overlapping expression domains of bone morphogenetic protein family members potentially account for limited tissue defects in BMP7 deficient embryos. *Developmental Dynamics*, *208*, 349–362.
Durbec, P., Marcos-Gutierrez, C. V., Kilkenny, C., Grigoriou, M., Wartiowaara, K., Suvanto, P., et al. (1996). GDNF signalling through the Ret receptor tyrosine kinase. *Nature*, *381*, 789–793.
Faa, G., Gerosa, C., Fanni, D., Nemolato, S., Locci, A., Cabras, T., et al. (2010). Marked interindividual variability in renal maturation of preterm infants: Lessons from autopsy. *The Journal of Maternal-Fetal & Neonatal Medicine*, *23*(Suppl. 3), 129–133.
Fang, M., Ren, H., Liu, J., Cadigan, K. M., Patel, S. R., & Dressler, G. R. (2009). Drosophila ptip is essential for anterior/posterior patterning in development and interacts with the PcG and trxG pathways. *Development*, *136*, 1929–1938.
Fetting, J. L., Guay, J. A., Karolak, M. J., Iozzo, R. V., Adams, D. C., Brown, A. C., et al. (2013). FOXD1 promotes nephron progenitor differentiation by repressing Decorin in the embryonic kidney http://dx.doi.org/10.1242/dev.089078. Epub 2013 Nov 27.
Fischer, E., Legue, E., Doyen, A., Nato, F., Nicolas, J. F., Torres, V., et al. (2006). Defective planar cell polarity in polycystic kidney disease. *Nature Genetics*, *38*, 21–23.
Fisher, C. E., Michael, L., Barnett, M. W., & Davies, J. A. (2001). Erk MAP kinase regulates branching morphogenesis in the developing mouse kidney. *Development*, *128*, 4329–4338.
Fujiwara, H., Ferreira, M., Donati, G., Marciano, D. K., Linton, J. M., Sato, Y., et al. (2011). The basement membrane of hair follicle stem cells is a muscle cell niche. *Cell*, *144*, 577–589.
Georgas, K. M., Chiu, H. S., Lesieur, E., Rumballe, B. A., & Little, M. H. (2011). Expression of metanephric nephron-patterning genes in differentiating mesonephric tubules. *Developmental Dynamics*, *240*, 1600–1612.
Georgas, K., Rumballe, B., Valerius, M. T., Chiu, H. S., Thiagarajan, R. D., Lesieur, E., et al. (2009). Analysis of early nephron patterning reveals a role for distal RV proliferation in fusion to the ureteric tip via a cap mesenchyme-derived connecting segment. *Developmental Biology*, *332*, 273–286.

Godin, R. E., Takaesu, N. T., Robertson, E. J., & Dudley, A. T. (1998). Regulation of BMP7 expression during kidney development. *Development, 125*, 3473–3482.

Gong, K. Q., Yallowitz, A. R., Sun, H., Dressler, G. R., & Wellik, D. M. (2007). A Hox–Eya–Pax complex regulates early kidney developmental gene expression. *Molecular and Cellular Biology, 27*, 7661–7668.

Gooch, J. L., Toro, J. J., Guler, R. L., & Barnes, J. L. (2004). Calcineurin A-alpha but not A-beta is required for normal kidney development and function. *American Journal of Pathology, 165*, 1755–1765.

Grieshammer, U., Cebrian, C., Ilagan, R., Meyers, E., Herzlinger, D., & Martin, G. R. (2005). FGF8 is required for cell survival at distinct stages of nephrogenesis and for regulation of gene expression in nascent nephrons. *Development, 132*, 3847–3857.

Gur, G., Rubin, C., Katz, M., Amit, I., Citri, A., Nilsson, J., et al. (2004). LRIG1 restricts growth factor signaling by enhancing receptor ubiquitylation and degradation. *The EMBO Journal, 23*, 3270–3281.

Haege, S., Einer, C., Thiele, S., Mueller, W., Nietzsche, S., Lupp, A., et al. (2012). CXC chemokine receptor 7 (CXCR7) regulates CXCR4 protein expression and capillary tuft development in mouse kidney. *PLoS One*, 7, e42814.

Hammerman, M. R. (1998). Regulation of cell survival during renal development. *Pediatric Nephrology, 12*, 596–602.

Hartman, H. A., Lai, H. L., & Patterson, L. T. (2007). Cessation of renal morphogenesis in mice. *Developmental Biology, 310*, 379–387.

Hartwig, S., Ho, J., Pandey, P., Macisaac, K., Taglienti, M., Xiang, M., et al. (2010). Genomic characterization of Wilms' tumor suppressor 1 targets in nephron progenitor cells during kidney development. *Development, 137*, 1189–1203.

Hatini, V., Huh, S. O., Herzlinger, D., Soares, V. C., & Lai, E. (1996). Essential role of stromal mesenchyme in kidney morphogenesis revealed by targeted disruption of Winged Helix transcription factor BF-2. *Genes & Development, 10*, 1467–1478.

Heliot, C., Desgrange, A., Buisson, I., Prunskaite-Hyyrylainen, R., Shan, J., Vainio, S., et al. (2013). HNF1B controls proximal-intermediate nephron segment identity in vertebrates by regulating Notch signalling components and Irx1/2. *Development, 140*, 873–885.

Hendry, C. E., & Little, M. H. (2012). Reprogramming the kidney: A novel approach for regeneration. *Kidney International, 82*, 138–146.

Hendry, C. E., Vanslambrouck, J. M., Ineson, J., Suhaimi, N., Takasato, M., Rae, F., et al. (2013). Direct transcriptional reprogramming of adult cells to embryonic nephron progenitors. *Journal of the American Society of Nephrology, 24*, 1424–1434.

Hinchliffe, S. A., Sargent, P. H., Howard, C. V., Chan, Y. F., & van Velzen, D. (1991). Human intrauterine renal growth expressed in absolute number of glomeruli assessed by the disector method and Cavalieri principle. *Laboratory Investigation, 64*, 777–784.

Horikawa, K., Radice, G., Takeichi, M., & Chisaka, O. (1999). Adhesive subdivisions intrinsic to the epithelial somites. *Developmental Biology, 215*, 182–189.

Humphreys, B. D., Lin, S. L., Kobayashi, A., Hudson, T. E., Nowlin, B. T., Bonventre, J. V., et al. (2010). Fate tracing reveals the pericyte and not epithelial origin of myofibroblasts in kidney fibrosis. *American Journal of Pathology, 176*, 85–97.

Humphreys, B. D., Valerius, M. T., Kobayashi, A., Mugford, J. W., Soeung, S., Duffield, J. S., et al. (2008). Intrinsic epithelial cells repair the kidney after injury. *Cell Stem Cell, 2*, 284–291.

Imai, H., Ohta, K., Yoshida, A., Suzuki, S., Hashizume, K., Usami, S., et al. (2010). μ-Crystallin, new candidate protein in endotoxin-induced uveitis. *Investigative Ophthalmology & Visual Science, 51*, 3554–3559.

Imayoshi, I., Sakamoto, M., Yamaguchi, M., Mori, K., & Kageyama, R. (2010). Essential roles of Notch signaling in maintenance of neural stem cells in developing and adult brains. *The Journal of Neuroscience, 30*, 3489–3498.

Imgrund, M., Grone, E., Grone, H. J., Kretzler, M., Holzman, L., Schlondorff, D., & Rothenpieler, U. W. (1999). Re-expression of the developmental gene Pax-2 during experimental acute tubular necrosis in mice 1. *Kidney Int, 56*(4), 1423–1431.

Iozzo, R. V., & Schaefer, L. (2010). Proteoglycans in health and disease: Novel regulatory signaling mechanisms evoked by the small leucine-rich proteoglycans. *FEBS Journal, 277*, 3864–3875.

Jain, S., Encinas, M., Johnson, E. M., Jr., & Milbrandt, J. (2006). Critical and distinct roles for key RET tyrosine docking sites in renal development. *Genes & Development, 20*, 321–333.

Kageyama, R., Niwa, Y., Shimojo, H., Kobayashi, T., & Ohtsuka, T. (2010). Ultradian oscillations in Notch signaling regulate dynamic biological events. *Current Topics in Developmental Biology, 92*, 311–331.

Kao, R. M., Vasilyev, A., Miyawaki, A., Drummond, I. A., & McMahon, A. P. (2012). Invasion of distal nephron precursors associates with tubular interconnection during nephrogenesis. *Journal of the American Society of Nephrology, 23*, 1682–1690.

Karner, C. M., Chirumamilla, R., Aoki, S., Igarashi, P., Wallingford, J. B., & Carroll, T. J. (2009). Wnt9b signaling regulates planar cell polarity and kidney tubule morphogenesis. *Nature Genetics, 41*, 793–799.

Karner, C. M., Das, A., Ma, Z., Self, M., Chen, C., Lum, L., et al. (2011). Canonical Wnt9b signaling balances progenitor cell expansion and differentiation during kidney development. *Development, 138*, 1247–1257.

Karner, C. M., Merkel, C. E., Dodge, M., Ma, Z., Lu, J., Chen, C., et al. (2010). Tankyrase is necessary for canonical Wnt signaling during kidney development. *Developmental Dynamics, 239*, 2014–2023.

Kim, D., Kwon, Y. K., & Cho, K. H. (2008). The biphasic behavior of incoherent feed-forward loops in biomolecular regulatory networks. *Bioessays, 30*, 1204–1211.

Kim, D., Patel, S. R., Xiao, H., & Dressler, G. R. (2009). The role of PTIP in maintaining embryonic stem cell pluripotency. *Stem Cells, 27*, 1516–1523.

Kispert, A., Vainio, S., & McMahon, A. P. (1998). Wnt-4 is a mesenchymal signal for epithelial transformation of metanephric mesenchyme in the developing kidney. *Development, 125*, 4225–4234.

Kobayashi, A., Valerius, M. T., Mugford, J. W., Carroll, T. J., Self, M., Oliver, G., et al. (2008). Six2 defines and regulates a multipotent self-renewing nephron progenitor population throughout mammalian kidney development. *Cell Stem Cell, 3*, 169–181.

Kreidberg, J. A. (2010). WT1 and kidney progenitor cells. *Organogenesis, 6*, 61–70.

Kreidberg, J. A., Sariola, H., Loring, J. M., Maeda, M., Pelletier, J., Housman, D., et al. (1993). WT-1 is required for early kidney development. *Cell, 74*, 679–691.

Lefevre, G. M., Patel, S. R., Kim, D., Tessarollo, L., & Dressler, G. R. (2010). Altering a histone H3K4 methylation pathway in glomerular podocytes promotes a chronic disease phenotype. *PLoS Genetics, 6*, e1001142.

Levinson, R. S., Batourina, E., Choi, C., Vorontchikhina, M., Kitajewski, J., & Mendelsohn, C. L. (2005). Foxd1-dependent signals control cellularity in the renal capsule, a structure required for normal renal development. *Development, 132*, 529–539.

Levinson, R., & Mendelsohn, C. (2003). Stromal progenitors are important for patterning epithelial and mesenchymal cell types in the embryonic kidney. *Seminars in Cell and Developmental Biology, 14*, 225–231.

Linton, J. M., Martin, G. R., & Reichardt, L. F. (2007). The ECM protein nephronectin promotes kidney development via integrin alpha8beta1-mediated stimulation of Gdnf expression. *Development, 134*, 2501–2509.

Liu, J., Sato, C., Cerletti, M., & Wagers, A. (2010). Notch signaling in the regulation of stem cell self-renewal and differentiation. *Current Topics in Developmental Biology, 92*, 367–409.

Lu, B. C., Cebrian, C., Chi, X., Kuure, S., Kuo, R., Bates, C. M., et al. (2009). Etv4 and Etv5 are required downstream of GDNF and Ret for kidney branching morphogenesis. *Nature Genetics*, *41*, 1295–1302.

Lund, R. J., Davies, M. R., & Hruska, K. A. (2002). Bone morphogenetic protein-7: An anti-fibrotic morphogenetic protein with therapeutic importance in renal disease. *Current Opinion in Nephrology and Hypertension*, *11*, 31–36.

Luo, G., Hofmann, C., Bronckers, A. L., Sohocki, M., Bradley, A., & Karsenty, G. (1995). BMP-7 is an inducer of nephrogenesis, and is also required for eye development and skeletal patterning. *Genes & Development*, *9*, 2808–2820.

Lusis, M., Li, J., Ineson, J., Christensen, M. E., Rice, A., & Little, M. H. (2010). Isolation of clonogenic, long-term self renewing embryonic renal stem cells. *Stem Cell Research*, *5*, 23–39.

Maezawa, Y., Binnie, M., Li, C., Thorner, P., Hui, C. C., Alman, B., et al. (2012). A new Cre driver mouse line, Tcf21/Pod1-Cre, targets metanephric mesenchyme. *PLoS One*, *7*, e40547.

Mankoo, B. S., Collins, N. S., Ashby, P., Grigorieva, E., Pevny, L. H., Candia, A., et al. (1999). Mox2 is a component of the genetic hierarchy controlling limb muscle development. *Nature*, *400*, 69–73.

Manson, S. R., Niederhoff, R. A., Hruska, K. A., & Austin, P. F. (2011). Endogenous BMP-7 is a critical molecular determinant of the reversibility of obstruction-induced renal injuries. *American Journal of Physiology. Renal Physiology*, *301*, F1293–F1302.

Mao, Y., Mulvaney, J., Zakaria, S., Yu, T., Morgan, K. M., Allen, S., et al. (2011). Characterization of a Dchs1 mutant mouse reveals requirements for Dchs1-Fat4 signaling during mammalian development. *Development*, *138*, 947–957.

Massa, F., Garbay, S., Bouvier, R., Sugitani, Y., Noda, T., Gubler, M. C., et al. (2013). Hepatocyte nuclear factor 1beta controls nephron tubular development. *Development*, *140*, 886–896.

Michos, O., Cebrian, C., Hyink, D., Grieshammer, U., Williams, L., D'Agati, V., et al. (2010). Kidney development in the absence of Gdnf and Spry1 requires Fgf10. *PLoS Genetics*, *6*, e1000809.

Moore, M. W., Klein, R. D., Farinas, I., Sauer, H., Armanini, M., Phillips, H., et al. (1996). Renal and neuronal abnormalities in mice lacking GDNF. *Nature*, *382*, 76–79.

Mugford, J. W., Sipila, P., McMahon, J. A., & McMahon, A. P. (2008). Osr1 expression demarcates a multi-potent population of intermediate mesoderm that undergoes progressive restriction to an Osr1-dependent nephron progenitor compartment within the mammalian kidney. *Developmental Biology*, *324*, 88–98.

Mugford, J. W., Yu, J., Kobayashi, A., & McMahon, A. P. (2009). High-resolution gene expression analysis of the developing mouse kidney defines novel cellular compartments within the nephron progenitor population. *Developmental Biology*, *333*, 312–323.

Muller, M., Obeyesekere, M., Mills, G. B., & Ram, P. T. (2008). Network topology determines dynamics of the mammalian MAPK1,2 signaling network: Bifan motif regulation of C-Raf and B-Raf isoforms by FGFR and MC1R. *The FASEB Journal*, *22*, 1393–1403.

Murphy, A. J., Pierce, J., de Caestecker, C., Taylor, C., Anderson, J. R., Perantoni, A. O., et al. (2012). SIX2 and CITED1, markers of nephronic progenitor self-renewal, remain active in primitive elements of Wilms' tumor. *Journal of Pediatric Surgery*, *47*, 1239–1249.

Osafune, K., Takasato, M., Kispert, A., Asashima, M., & Nishinakamura, R. (2006). Identification of multipotent progenitors in the embryonic mouse kidney by a novel colony-forming assay. *Development*, *133*, 151–161.

Park, J.-S., Ma, W., O'Brien, L. L., Chung, E., Guo, J.-J., Cheng, J.-G., et al. (2012). Six2 and Wnt regulate self-renewal and commitment of nephron progenitors through shared gene regulatory networks. *Developmental Cell*, *23*.

Park, J. S., Valerius, M. T., & McMahon, A. P. (2007). Wnt/β-catenin signaling regulates nephron induction during mouse kidney development. *Development, 134*, 2533–2539.

Patel, S. R., Bhumbra, S. S., Paknikar, R. S., & Dressler, G. R. (2012). Epigenetic mechanisms of Groucho/Grg/TLE mediated transcriptional repression. *Molecular Cell, 45*, 185–195.

Pepicelli, C. V., Kispert, A., Rowitch, D. H., & McMahon, A. P. (1997). Gdnf induces branching and increased cell proliferation in the ureter of the mouse. *Developmental Biology, 192*, 193–198.

Perantoni, A. O., Timofeeva, O., Naillat, F., Richman, C., Pajni-Underwood, S., Wilson, C., et al. (2005). Inactivation of FGF8 in early mesoderm reveals an essential role in kidney development. *Development, 132*, 3859–3871.

Plisov, S., Tsang, M., Shi, G., Boyle, S., Yoshino, K., Dunwoodie, S. L., et al. (2005). Cited1 is a bifunctional transcriptional cofactor that regulates early nephronic patterning. *Journal of the American Society of Nephrology, 16*, 1632–1644.

Poladia, D. P., Kish, K., Kutay, B., Hains, D., Kegg, H., Zhao, H., et al. (2006). Role of fibroblast growth factor receptors 1 and 2 in the metanephric mesenchyme. *Developmental Biology, 291*, 325–339.

Qiao, J., Uzzo, R., Obara-Ishihara, T., Degenstein, L., Fuchs, E., & Herzlinger, D. (1999). FGF-7 modulates ureteric bud growth and nephron number in the developing kidney. *Development, 126*, 547–554.

Reginensi, A., Clarkson, M., Neirijnck, Y., Lu, B., Ohyama, T., Groves, A. K., et al. (2011). SOX9 controls epithelial branching by activating RET effector genes during kidney development. *Human Molecular Genetics, 20*, 1143–1153.

Reginensi, A., Scott, R. P., Gregorieff, A., Bagherie-Lachidan, M., Chung, C., Lim, D.-S., et al. (2013). Yap- and Cdc42-dependent nephrogenesis and morphogenesis during mouse kidney development. *PLoS Genetics, 9*, e1003380.

Rumballe, B. A., Georgas, K. M., Combes, A. N., Ju, A. L., Gilbert, T., & Little, M. H. (2011). Nephron formation adopts a novel spatial topology at cessation of nephrogenesis. *Developmental Biology, 360*, 110–122.

Saburi, S., Hester, I., Fischer, E., Pontoglio, M., Eremina, V., Gessler, M., et al. (2008). Loss of Fat4 disrupts PCP signaling and oriented cell division and leads to cystic kidney disease. *Nature Genetics, 40*, 1010–1015.

Saburi, S., Hester, I., Goodrich, L., & McNeill, H. (2012). Functional interactions between Fat family cadherins in tissue morphogenesis and planar polarity. *Development, 139*, 1806–1820.

Saifudeen, Z., Liu, J., Dipp, S., Yao, X., Li, Y., McLaughlin, N., et al. (2012). A p53-Pax2 pathway in kidney development: Implications for nephrogenesis. *PLoS One, 7*, e44869.

Sainio, K., Suvanto, P., Davies, J., Wartiovaara, J., Wartiovaara, K., Saarma, M., et al. (1997). Glial-cell-line-derived neurotrophic factor is required for bud initiation from ureteric epithelium. *Development, 124*, 4077–4087.

Sakaki-Yumoto, M., Kobayashi, C., Sato, A., Fujimura, S., Matsumoto, Y., Takasato, M., et al. (2006). The murine homolog of SALL4, a causative gene in Okihiro syndrome, is essential for embryonic stem cell proliferation, and cooperates with Sall1 in anorectal, heart, brain and kidney development. *Development, 133*, 3005–3013.

Saxen, L. (1987). Organogenesis of the kidney. In P. B. G. P. W. Barlow & C. C. Wylie (Eds.), *Developmental and cell biology series*. Cambridge: Cambridge University Press.

Schnabel, C. A., Godin, R. E., & Cleary, M. L. (2003). Pbx1 regulates nephrogenesis and ureteric branching in the developing kidney. *Dev Biol, 254*(2), 262–276.

Self, M., Lagutin, O. V., Bowling, B., Hendrix, J., Cai, Y., Dressler, R. D., et al. (2006). Six2 is required for suppression of nephrogenesis and progenitor renewal in the developing kidney. *The EMBO Journal, 25*, 5214–5228.

Sequeira Lopez, M. L., & Gomez, R. A. (2011). Development of the renal arterioles. *Journal of the American Society of Nephrology, 22*, 2156–2165.

Shakya, R., Watanabe, T., & Costantini, F. (2005). The role of GDNF/Ret signaling in ureteric bud cell fate and branching morphogenesis. *Developmental Cell, 8*, 65–74.

Short, K., Hodson, M., & Smyth, I. (2013). Spatial mapping and quantification of developmental branching morphogenesis. *Development, 140*, 471–478.

Sim, E. U., Smith, A., Szilagi, E., Rae, F., Ioannou, P., Lindsay, M. H., et al. (2002). Wnt-4 regulation by the Wilms' tumour suppressor gene, WT1. *Oncogene, 21*, 2948–2960.

Sims-Lucas, S., Cusack, B., Baust, J., Eswarakumar, V. P., Masatoshi, H., Takeuchi, A., et al. (2011). Fgfr1 and the IIIc isoform of Fgfr2 play critical roles in the metanephric mesenchyme mediating early inductive events in kidney development. *Developmental Dynamics, 240*, 240–249.

Sims-Lucas, S., Di Giovanni, V., Schaefer, C., Cusack, B., Eswarakumar, V. P., & Bates, C. M. (2012). Ureteric morphogenesis requires Fgfr1 and Fgfr2/Frs2alpha signaling in the metanephric mesenchyme. *Journal of the American Society of Nephrology, 23*, 607–617.

Skinner, M. A., Safford, S. D., Reeves, J. G., Jackson, M. E., & Freemerman, A. J. (2008). Renal aplasia in humans is associated with RET mutations. *The American Journal of Human Genetics, 82*, 344–351.

Sorenson, C. M., Rogers, S. A., Korsmeyer, S. J., & Hammerman, M. R. (1995). Fulminant metanephric apoptosis and abnormal kidney development in bcl-2-deficient mice. *The American Journal of Physiology, 268*, F73–F81.

Stamataki, D., Kastrinaki, M., Mankoo, B. S., Pachnis, V., & Karagogeos, D. (2001). Homeodomain proteins Mox1 and Mox2 associate with Pax1 and Pax3 transcription factors. *FEBS Letters, 499*, 274–278.

Stark, K., Vainio, S., Vassileva, G., & McMahon, A. P. (1994). Epithelial transformation of metanephric mesenchyme in the developing kidney regulated by Wnt-4. *Nature, 372*, 679–683.

Surendran, K., Boyle, S., Barak, H., Kim, M., Stromberski, C., McCright, B., et al. (2010). The contribution of Notch1 to nephron segmentation in the developing kidney is revealed in a sensitized Notch2 background and can be augmented by reducing Mint dosage. *Developmental Biology, 337*, 386–395.

Suzuki, S., Suzuki, N., Mori, J., Oshima, A., Usami, S., & Hashizume, K. (2007). micro-Crystallin as an intracellular 3,5,3'-triiodothyronine holder in vivo. *Molecular Endocrinology, 21*, 885–894.

Takabatake, Y., Sugiyama, T., Kohara, H., Matsusaka, T., Kurihara, H., Koni, P. A., et al. (2009). The CXCL12 (SDF-1)/CXCR4 axis is essential for the development of renal vasculature. *Journal of the American Society of Nephrology, 20*, 1714–1723.

Takahashi, M. (2001). The GDNF/RET signaling pathway and human diseases. *Cytokine & Growth Factor Reviews, 12*, 361–373.

Tang, M. J., Cai, Y., Tsai, S. J., Wang, Y. K., & Dressler, G. R. (2002). Ureteric bud outgrowth in response to RET activation is mediated by phosphatidylinositol 3-kinase. *Developmental Biology, 243*, 128–136.

Tanigawa, S., Wang, H., Yang, Y., Sharma, N., Tarasova, N., Ajima, R., et al. (2011). Wnt4 induces nephronic tubules in metanephric mesenchyme by a non-canonical mechanism. *Developmental Biology, 352*, 58–69.

Trueb, B., Amann, R., & Gerber, S. D. (2013). Role of FGFRL1 and other FGF signaling proteins in early kidney development. *Cellular and Molecular Life Sciences, 70*, 2505–2518.

Trupp, M., Arenas, E., Fainzilber, M., Nilsson, A. S., Sieber, B. A., Grigoriou, M., et al. (1996). Functional receptor for Gdnf encoded by the C-Ret proto-oncogene. *Nature, 381*, 785–789.

Uchiyama, Y., Sakaguchi, M., Terabayashi, T., Inenaga, T., Inoue, S., Kobayashi, C., et al. (2010). Kif26b, a kinesin family gene, regulates adhesion of the embryonic kidney mesenchyme. *Proceedings of the National Academy of Sciences of the United States of America, 107*, 9240–9245.

Wellik, D. M., Hawkes, P. J., & Capecchi, M. R. (2002). Hox11 paralogous genes are essential for metanephric kidney induction. *Genes & Development, 16*, 1423–1432.

Wong, A., Bogni, S., Kotka, P., de Graaff, E., D'Agati, V., Costantini, F., et al. (2005). Phosphotyrosine 1062 is critical for the in vivo activity of the Ret9 receptor tyrosine kinase isoform. *Molecular and Cellular Biology, 25*, 9661–9673.

Wu, J., Prindle, M. J., Dressler, G. R., & Yu, X. (2009). PTIP regulates 53BP1 and SMC1 at the DNA damage sites. *The Journal of Biological Chemistry, 284*, 18078–18084.

Xu, P. X., Zheng, W., Huang, L., Maire, P., Laclef, C., & Silvius, D. (2003). Six1 is required for the early organogenesis of mammalian kidney. *Development, 130*, 3085–3094.

Yamada, D., Yoshida, M., Williams, Y. N., Fukami, T., Kikuchi, S., Masuda, M., et al. (2006). Disruption of spermatogenic cell adhesion and male infertility in mice lacking TSLC1/IGSF4, an immunoglobulin superfamily cell adhesion molecule. *Molecular and Cellular Biology, 26*, 3610–3624.

Zhang, X., Ibrahimi, O. A., Olsen, S. K., Umemori, H., Mohammadi, M., & Ornitz, D. M. (2006). Receptor specificity of the fibroblast growth factor family. The complete mammalian FGF family. *The Journal of Biological Chemistry, 281*, 15694–15700.

Zhao, H., Tanegashima, K., Ro, H., & Dawid, I. B. (2008). Lrig3 regulates neural crest formation in Xenopus by modulating Fgf and Wnt signaling pathways. *Development, 135*, 1283–1293.

CHAPTER TWELVE

Adult Stem Cell Niches: Cellular and Molecular Components

Amélie Rezza[*,†], Rachel Sennett[*,†], Michael Rendl[*,†,‡,§,1]
[*]Black Family Stem Cell Institute, Icahn School of Medicine at Mount Sinai, New York, USA
[†]Department of Developmental and Regenerative Biology, Icahn School of Medicine at Mount Sinai, New York, USA
[‡]Department of Dermatology, Icahn School of Medicine at Mount Sinai, New York, USA
[§]Graduate School of Biomedical Sciences, Icahn School of Medicine at Mount Sinai, New York, USA
[1]Corresponding author: e-mail address: michael.rendl@mssm.edu

Contents

Abstract

As stem cells (SCs) in adult organs continue to be identified and characterized, it becomes clear that their survival, quiescence, and activation depend on specific signals in their microenvironment, or niche. Although adult SCs of diverse tissues differ by their developmental origin, cycling activity, and regenerative capacity, there appear to be conserved similarities regarding the cellular and molecular components of the SC niche. Interestingly, many organs house both slow-cycling and fast-cycling SC populations, which rely on the coexistence of quiescent and inductive niches for proper regulation. In this review we present a general definition of adult SC niches in the most studied mammalian systems. We further focus on dissecting their cellular organization and on highlighting recently identified key molecular regulators. Finally, we detail the potential involvement of the SC niche in tissue degeneration, with a particular emphasis on aging and cancer.

Current Topics in Developmental Biology, Volume 107
ISSN 0070-2153
http://dx.doi.org/10.1016/B978-0-12-416022-4.00012-3

1. INTRODUCTION

Stem cells (SCs) have been studied for decades and tremendous advances have been made regarding SC functions and potential applications in different biological and biomedical domains, such as organ physiology, cancer pathology, aging, and regenerative medicine. The concept of a specific microenvironment or "niche" that is required for SCs to function emerged more than three decades ago (Schofield, 1978). Numerous studies have since confirmed in multiple invertebrate and mammalian organ systems that adult SCs reside in specific local microenvironments providing structural support and molecular signals to regulate SC quiescence, self-renewal, and activation for tissue maintenance.

Early and ongoing studies utilized Drosophila and *Caenorhabditis Elegans* to investigate adult SC niches, as the identification of SC reservoirs and accompanying niche structures are relatively straightforward in these model organisms (Byrd & Kimble, 2009; de Cuevas & Matunis, 2011). In mammalian systems a great deal of work has explored three archetypal examples of adult SCs and their corresponding niches: the hematopoietic system in the bone marrow, the small intestine, and hair follicles (HFs) in the skin. Numerous studies have identified markers to distinguish and isolate distinct SC populations, in addition to exploring the cells and signals constituting the niche within these model systems. All three organs are characterized by a high cell turnover and, under homeostatic conditions, rely on the steady activity of fast-cycling SC populations that self-renew and differentiate into all the different lineages necessary to maintain these complex tissues (Barker et al., 2007; Greco et al., 2009; Sangiorgi & Capecchi, 2008; Takeda et al., 2011; Takizawa, Regoes, Boddupalli, Bonhoeffer, & Manz, 2011; Wilson et al., 2008; Zhang, Cheong, Ciapurin, McDermitt, & Tumbar, 2009). Importantly, these rapidly renewing tissues also contain separate slow-cycling SCs that are activated to replenish the cycling "workhorse" SCs or to regenerate damaged tissue upon injury, and therefore can be considered "back up" SCs. It is therefore an appealing concept to consider the coexistence of quiescent and inductive niches in these organs.

In contrast, organs that undergo slow rates of cell turnover, such as the brain (Beckervordersandforth et al., 2010; Coskun et al., 2008; Lugert et al., 2010; Nam & Benezra, 2009), muscle (Kuang, Gillespie, & Rudnicki, 2008; Pannerec, Marazzi, & Sassoon, 2012; Relaix & Zammit,

2012), and liver (Furuyama et al., 2011; Tanaka & Miyajima, 2012) contain only slow-cycling reserve SCs that maintain the tissue and can be activated following injury. Although specific SC markers have been identified in these organs, the exact cell types and specific signals composing the niche are just beginning to be uncovered in these systems.

The different functions of distinct SC populations appear to depend largely on extrinsic influences. Sustained signals from the environment are important for maintaining the quiescence, self-renewal, and continued survival of "reserve" SCs, while frequently different signals from unique niches are generated to induce SC activation for tissue turnover and repair. *In vitro* maintenance and/or activation of SCs requires addition of specific factors to the culture medium (Blanpain, Lowry, Geoghegan, Polak, & Fuchs, 2004; Roobrouck, Vanuytsel, & Verfaillie, 2011; Sato et al., 2009), confirming the importance of external inputs in regulating SC function. Interestingly, there is also accumulating evidence that this "medium-dependent" function of SCs exists for embryonic SCs (Nichols & Smith, 2012; Roobrouck et al., 2011).

In this review, we describe structural and cellular features of major adult SC niches in the mammalian system, mostly from well-characterized mouse models and where applicable also from human tissues. We further comprehensively discuss the molecular components of the niche that regulate SCs, distinguishing quiescent and activating signals. Lastly, we highlight the influence of the niche in nonphysiological conditions, such as during aging and in cancer.

2. CELLULAR ORGANIZATION OF ADULT SC NICHES

Along with characterization of SC reservoirs in adult tissues comes the exploration of their microenvironment. In most systems, the cellular components of SC niche can be classified into four categories (Fig. 12.1): (1) the SCs and progeny themselves, as they provide autocrine and paracrine regulation, respectively, within their own lineage; (2) neighboring mesenchymal or stromal cells providing paracrine signals; (3) extracellular matrix (ECM) or cell–cell contacts involving adhesion molecules; and (4) external cues from distant sources within the tissue or outside the tissue, such as from blood vessels, neurons, or immune cells. The concomitant expression and/or secretion of specific factors by these components creates a discretely

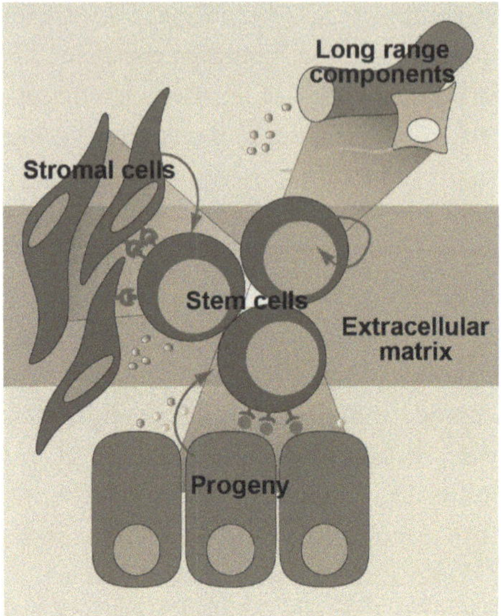

Figure 12.1 Common features of adult stem cell niches. Stem cells (SCs) reside in a specific microenvironment called the niche, composed of multiple components: SCs themselves (blue) and their progeny (purple), as well as close surrounding cells from the mesenchyme or stroma (green), more distant cells such as endothelial cells or neurons providing long-range signals (red and orange), and the extracellular matrix/basement membrane (ECM; gray). Each component expresses adhesion molecules, specific receptors, or secreted factors that signal to the SCs (gradients), creating a unique localized niche for SC maintenance and/or activation. (See color plate.)

localized niche, allowing modulation of SC activity (Fig. 12.1). In this chapter we will detail the localization of fast- and slow-cycling SC reservoirs and point out conserved niche components within several adult tissues.

2.1. Autocrine/intrinsic SC regulation and feedback from direct progeny

Cellular contributions to the niche, especially in the form of secreted signals, are foremost and center in many adult SC systems. Important cellular input is generated by the SC pool itself and its direct progeny (blue and purple cells in Fig. 12.1; reviewed in Hsu & Fuchs, 2012). The role of SCs in adult hematopoiesis has been extensively studied, generating paradigms by which we consider the mechanisms behind SC function in other adult tissues. Hematopoietic stem cells (HSCs) proliferate and differentiate to give rise to all blood cell lineages, and the behavior of distinct HSC subpopulations is influenced by niche stimuli present within the bone marrow (Fig. 12.2A). Traditionally, the localization of HSCs is thought to determine proliferative status, with quiescent SCs residing at the endosteum while activated SCs self-renew and differentiate in the perivascular space to maintain homeostasis, although more recent work has suggested the separation of these two niches is not as discrete as originally described (Lo Celso et al., 2009; Xie et al., 2009). A great deal of work has identified intrinsic regulation of HSC function and an important role of several stromal cell compartments within the HSC niche (discussed below), but recent studies are starting to highlight the important contributions of SC progeny as well. When macrophages are specifically targeted for ablation in mice, HSCs enter the peripheral bloodstream, suggesting that these myeloid progeny are important for maintaining HSC retention in the niche (Chow et al., 2011; Winkler et al., 2010). Further investigation revealed a circuitous mechanism, whereby macrophages secrete signals that act on other niche cells within the marrow to ultimately influence HSC localization. A separate study interrogated the role of lymphoid progeny in the niche by depleting regulatory T cells within the marrow of transplant recipient mice, before seeding donor HSCs. In the absence of Treg supportive stimuli, fewer transplanted HSCs survived (Fujisaki et al., 2011). Finally, neutrophil clearance from the bone marrow can also impact HSC activity (Casanova-Acebes et al., 2013).

Similarly, SCs and progeny in hair follicles (HFs) generate important niche signals. In adult skin, HFs cycle through periods of apoptotic destruction and regeneration (Fig. 12.2B) (Blanpain & Fuchs, 2009; Sennett & Rendl, 2012). Slow-cycling hair follicle stem cells (HFSCs) reside in a specialized "bulge" pocket, located near the top of the follicle that remains intact throughout the destruction phase of the hair cycle. During the subsequent regeneration phase, fast-cycling SC progeny in the HF germ multiply and reconstitute a new follicle, followed by proliferation and differentiation of direct SC progeny to continuously supply the growing HF until the next hair cycle. The distinct behavior of slow- and fast-cycling HFSCs is thought to depend on differential exposure to quiescent and activating niche signals. While housed in the relatively isolated bulge pocket, HFSCs are exposed only to quiescence-maintaining signals believed to be generated by the SCs themselves and possibly by nearby blood vessels and neurons (Blanpain et al., 2004; Greco et al., 2009; Tumbar et al., 2004). After the first hair cycle round, differentiated SC progeny also provide important niche input to maintain quiescence among bulge HFSCs (Hsu,

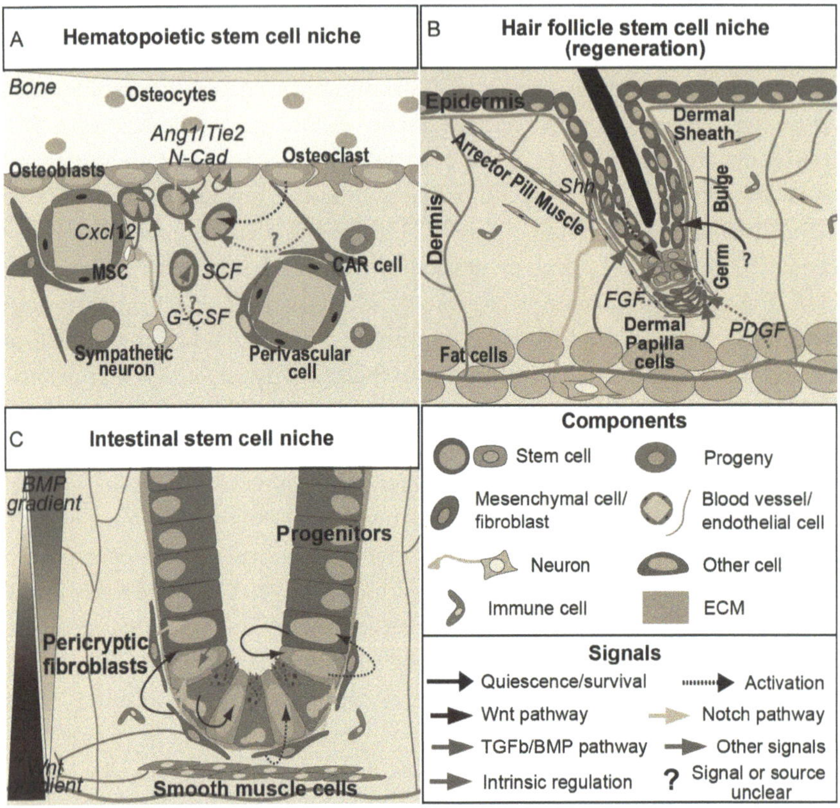

Figure 12.2 Cellular and molecular components of three fast-cycling adult stem cell niches: the hematopoietic, the hair follicle, and the intestinal systems. (A) The hematopoietic stem cell (HSC) niche includes osteoblasts (green), osteoclasts (orange), endothelial cells (red), perivascular cells (light brown), HSC progeny (purple), CXCL12-abundant reticular (CAR; dark green) cells, and mesenchymal stem cells (MSC; light green). Many signals regulate HSCs (blue): SCs autoregulate their cell cycle to prevent exhaustion (green arrow). TGFb from neural cells (red arrow), Notch signaling (yellow arrow), Ang1 from osteoblasts, perivascular SCF, and CXCL12 from MSC (brown arrows) are important for HSC maintenance. Activating signals are Wnts (still debated; dashed blue arrow), secreted factors from CAR cells and G-CSF (dashed brown arrows). (B) Hair follicle SCs (HFSCs) during regeneration reside in a niche composed of mesenchymal dermal papilla (DP) cells (dark green), fat cells (light brown), and dermis populated by blood vessels (red), immune cells (pink), loose fibroblasts and dermal sheath fibroblasts (light green), arrector pili muscle cells (dark brown), and neurons (yellow). Two populations of SCs are distinguishable during regeneration: bulge SCs (dark blue) and germ SCs (light blue). Maintenance signals to bulge SCs come from different niche components: HFSCs control quiescence through CDKi, Runx1/p21, Tbx1, NFATc1, Lhx2, and Sox9 (green arrow); subcutaneous fat cells express BMP ligands (red arrows); canonical Wnt signals from an unknown source (dark blue arrow); and nerve-secreted Shh (brown arrow). Activating signals include TGFb2

Pasolli, & Fuchs, 2011; Takeda et al., 2013). HFSCs can also act as niche for neighboring melanocyte SCs (Chang et al., 2013; Tanimura et al., 2011).

As opposed to HFs, constant SC activation is required for maintaining homeostasis in the intestine because of rapid cell turnover (Fig. 12.2C). Most studies have focused on the small intestine, where the epithelium is organized into crypts that house fast- and slow-cycling intestinal SCs (ISCs) together with transit-amplifying progeny. These crypts are continuous with villous extensions that consist of terminally differentiated absorptive and secreting cell progeny (Bjerknes & Cheng, 2006). Intriguingly, despite years of study, there is currently intense ongoing debate about the origins and relationship of the two ISC populations that are characterized by different marker expression and cell cycling activity (Rizk & Barker, 2012; Clevers, 2013). Lgr5+ ISCs are interspersed with differentiated Paneth cell progeny at the crypt base, cycle quickly under homeostatic conditions, and have been shown through lineage tracing to self-renew and give rise to all cell types within a crypt/villus unit (Barker et al., 2007). Because of their localization they are also known as crypt basal cells, or CBCs. Separate studies have characterized Bmi1+ SCs in the +4 position counting from the base, which similarly demonstrate both the ability to self-renew and give rise to all crypt cell lineages under homeostatic conditions (Sangiorgi & Capecchi, 2008). These +4 SCs were considered quiescent (Chwalinski, Potten, & Evans, 1988), although more recent studies contradict this hypothesis and suggest that both ISC populations are cycling under homeostatic conditions (Cambuli, Rezza, Nadjar, & Platerot, 2013; Sangiorgi & Capecchi, 2008). These cells can repair the intestinal epithelium in the absence of Lgr5+ cells, and even give rise to new Lgr5+ cells after targeted ablation (Tian et al., 2011). However, complementary experiments revealed that CBCs are also capable of giving rise to +4 SCs (Takeda et al., 2011).

(dashed red arrow) and FGF7 (dashed brown arrow) expressed by DP cells, Wnt ligands expressed by epithelial cells (dashed dark blue arrow), and PDGF from the fat probably acting through DP cells (dashed brown arrow). (C) The intestinal SC niche is composed of ISC progeny Paneth cells (purple), pericryptic fibroblasts (green), and smooth muscle cells (brown). Blood vessels (red) and immune cells (pink) are present in the pericryptic mesenchyme. Two SC pools can be distinguished: Lgr5+ Columnar Basal Cells (light blue) and +4 SCs (dark blue); and many signals regulate their behavior: Ascl2 and Lrig1 control ISC maintenance (green arrow), Wnt signaling regulates SC survival through Wnt ligands from Paneth cells and surrounding fibroblasts (plain and dashed blue arrows), Notch is involved in cell proliferation and differentiation (plain and dashed yellow arrows), BMP signaling is repressed by antagonists from pericryptic fibroblasts and smooth muscle cells (red blunt arrows). (See color plate.)

Even more recently, a study demonstrated the existence of label-retaining progenitors that only contribute to crypt repair after injury (Buczacki et al., 2013). These appear to exist as a subpopulation of Lgr5+ cells, but their relation to the +4 SC pool is unclear. Attempts to link the activity and origins of these distinct SC populations have generated controversy within the field, and explorations of important niche influences have been similarly ambiguous. Although SC progeny Paneth cells were highlighted as an important constituent of the niche (Sato et al., 2011), the physiological relevance of this finding has since been challenged (Kim, Escudero, & Shivdasani, 2012), and SCs themselves, their epithelial progeny and surrounding pericryptal fibroblasts are thought to secrete important factors for SC maintenance (Fig. 12.2C) (Farin, Van Es, & Clevers, 2012; Sakamori et al., 2012).

Quiescent niche signals are thought to prevail in adult muscle tissue, which has little proclivity for homeostatic regeneration and limited capacity for wound repair. While satellite SCs were identified within adult muscle many years ago (Mauro, 1961), the muscle fiber progeny was only recently recognized as an important source of niche signaling (Fig. 12.3A; Ratajczak et al., 2003; Sherwood et al., 2004; Tatsumi et al., 2006; Wozniak & Anderson, 2007). Satellite SCs reside along the edge of these fibers, nestled underneath an encompassing basement membrane, and their nuclei comprise only a small percent of those within and along the muscle fiber (Wagers & Conboy, 2005). Recently, studies found that satellite SCs are heterogeneous: some proliferate more frequently, presumably in order to achieve normal tissue homeostasis, whereas others cycle less frequently and are considered reserve quiescent SCs (Ono, Boldrin, Knopp, Morgan, & Zammit, 2010; Ono et al., 2012; Relaix & Zammit, 2012). The influence of differential niche stimuli on these two populations has not yet been precisely described.

Likewise, adult neurogenesis by SC activation is a young field currently under great investigation. Based on label incorporating studies it was first determined that two regions of the adult brain undergo active proliferation to generate new cells: the subventricular zone (SVZ) bordering the lateral ventricles, and the subgranular zone (SGZ) of the dentate gyrus within the hippocampus (Ming & Song, 2011). In both regions, glial fibrillary acidic protein-expressing radial-glia like cells are the putative SCs that proliferate and can generate differentiated neurons, but are largely quiescent until stimulated following tissue injury. Some evidence exists for a relatively more "active" SC population that exists in the SGZ, characterized by expression of Sox2 (Suh et al., 2007). Based on lineage tracing experiments,

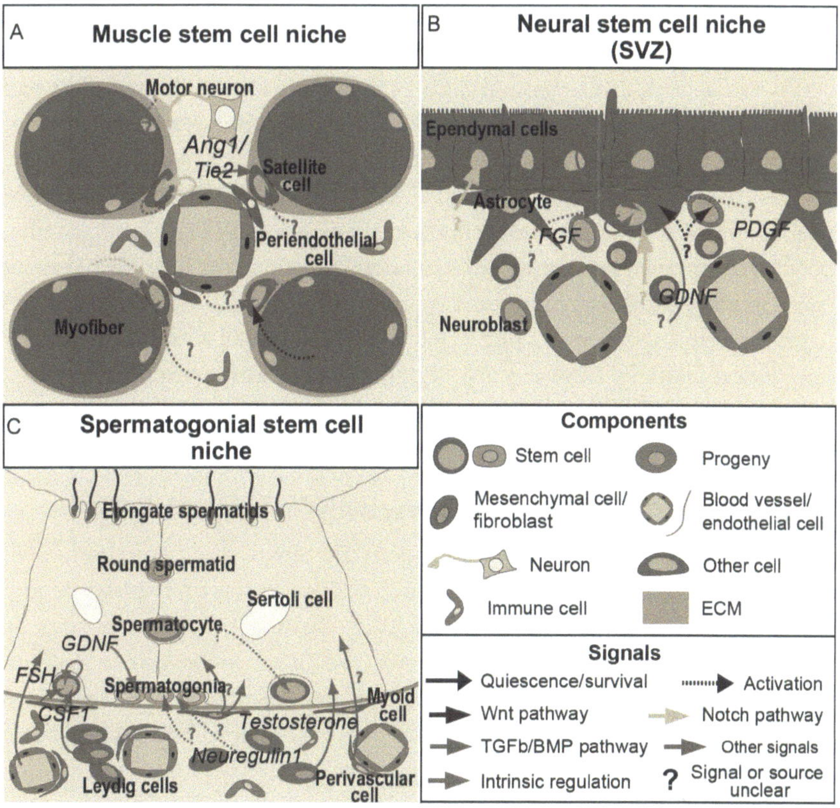

Figure 12.3 Components of the adult muscle, brain, and germ stem cell niches. (A) In the muscle, satellite SCs (blue) are present along muscle fiber progeny (purple), beneath the basal lamina (grey). The SC niche also includes blood vessels (red), periendothelial cells (brown), neurons (yellow), and immune cells (pink). Satellite SCs autoregulate their quiescence through Notch signaling (yellow arrow) and the miRNA machinery (green arrow). Periendothelial cells secrete Ang1 to regulate SC maintenance (brown arrow). SCs control their activation through the regulation of S1P (dashed green arrow). TGFb signaling also influences the regenerative capacity of satellite SCs (dashed red arrows). Myofibers secrete Wnt ligands that promote activation and regeneration (dashed blue arrow). Neural, endothelial, and immune cells secrete factors to impact SC behavior, but precise mechanisms are unclear (dashed brown arrows). (B) The neural SC niche in the SVZ is composed of ependymal cells (brown) and astrocyte progeny (purple) that both directly interact with SCs. Also present in the close environment are other NSC progeny (round purple cells) and blood vessels (red). NSCs autoregulate their own proliferation and quiescence (green arrow). The BMP pathway is precisely regulated with ligands from NSCs (red arrow) and antagonists from ependymal cells (red blunt arrow). Notch (yellow arrows) and GDNF (brown arrow) signaling are critical for ependymal and SCs maintenance as well, but the ligand source is unknown. The miRNA machinery regulates NSC activation (dashed green arrow). The Wnt pathway (dashed blue arrow) also

(*Continued*)

Sox2-expressing cells generate small clusters of Sox2-expressing and more differentiated progeny cells under homeostatic conditions, but the physiological implications of this observation are not yet clear. The niche system within the SVZ has been best studied and incorporates vascular cells, ependymal cells, astrocytes, and even other differentiated SC progeny (Fig. 12.3B). The ependymal cells of the SVZ line the lateral ventricle to shield neighboring SCs, and use beating cilia to generate gradients of morphogens that specify unique cell fates during differentiation (Sawamoto et al., 2006). Recent papers explored the largely ignored question of niche cell establishment and maintenance using neural SC niche cells (NSC) as a model system, and intriguingly found plasticity between ependymal cells and astrocytes (Carlen et al., 2009; Nomura, Goritz, Catchpole, Henkemeyer, & Frisen, 2010).

2.2. Input from neighboring mesenchymal or stromal cells

In numerous tissues, SCs and their progeny are closely associated with mesenchymal or stromal cell neighbors, which have been shown to play a central role in SC regulation (green cells in Fig. 12.1). Stromal cells in the bone marrow are the best studied component of the HSC niche (green cells in Fig. 12.2A). Cells of the osteoblastic lineage have classically been considered the dominant cell type of the quiescent niche, crucial for maintaining and sequestering HSCs through a combination of secreted cytokines, morphogens, and cell–cell adhesion molecules. *In vivo* experiments that depleted osteoblasts caused a corresponding drop in HSC production, suggesting these niche cells are important for maintaining SC numbers, and experimental conditions that expanded osteoblast numbers also increased the size of the HSC pool (Calvi et al., 2003; Kiel, Radice, & Morrison, 2007; Visnjic et al., 2004; Wilson & Trumpp, 2006; Zhang et al., 2003; Zhu et al., 2007). Many *in vitro* experiments also demonstrated the utility of coculturing osteoblasts

Figure 12.3—Cont'd activates proliferation. FGF and PDGF signals from unknown sources promote NSC proliferation (dashed brown arrows). (C) The germ SC niche in the testis is composed of Sertoli cells (light green), myoid cells (dark brown), and Leydig cells (green). Blood vessels (red) surrounded by perivascular cells (brown) and immune cells (pink) are present in the interstitial tissue. SSC maintenance and activation is dependent on Sertoli cell maintenance, which is regulated by testosterone from Leydig cells, FSH from the pituitary gland, and other signals from perivascular cells (brown arrows). SSC survival and maintenance depends on integrin beta1 expression in SSCs (green arrow), GDNF from Sertoli cells (brown arrows), or CSF1 from Leydig and myoid cells (brown arrows). SSCs can be activated by a wide range of signals such as BMP (dashed red arrow). (See color plate.)

with HSCs to improve survival and sustain functionality (Chitteti et al., 2010; Taichman, Reilly, & Emerson, 1996). Other important stromal cells contribute to the bone marrow niche as well; most notably, endothelial cells and CXCL12-abundant reticular (CAR) cells release soluble factors and establish cell–cell contacts to regulate HSC activation and dispersion. Experiments targeting CAR cells for ablation resulted in decreased numbers of activated HSCs with increased expression of progeny genes, suggesting CXCL12-producing cells mediate the survival, retention, and differentiation of primed HSCs in the marrow (Omatsu et al., 2010). Other recent work has provided substantial evidence implicating nestin+ mesenchymal stem cells (MSCs) as potent regulators of HSC activity as well (Mendez-Ferrer et al., 2010). Importantly, in very recent studies, expression of the essential niche signal CXCL12 was disrupted within several niche cell types, that is, osteoblasts, osteoprecursors, endothelial cells, MSCs, and other specific stromal subpopulations to determine differential cellular contributions to HSC maintenance (Ding & Morrison, 2013; Greenbaum et al., 2013). Remarkably, significant reductions in HSC numbers were only observed when CXCL12 was ablated from MSCs, indicating a subset of CXCL12-producing perivascular stromal cells is most instrumental in maintaining HSC survival and retention in the bone marrow, together with some input from endothelial cells. Meanwhile, osteolineage niche cells appear to support more differentiated HSC progeny through CXCL12 production.

Neighboring mesenchymal niche cells are thought to play a central role in activating HFSCs in the transition between the destruction and regeneration phases of the hair cycle. After a period of rest, when HFSCs are first induced to regrow a new follicle, activating signals are thought to emanate from the dermal papilla (DP) (dark green cells in Fig. 12.2B). In the mature follicle the DP exists as a pocket of mesenchymal cells normally localized to the lower bulb where it can regulate hair growth (Clavel et al., 2012), but right before regeneration this compartment approaches the bulge as most of the follicle base undergoes apoptosis. After exposure to stimuli provided by the activating DP niche (Greco et al., 2009; Oshimori & Fuchs, 2012), few fast-cycling HF germ SCs generate transit-amplifying progeny that subsequently multiply and differentiate in order to regenerate a new follicle (Greco et al., 2009; Lee & Tumbar, 2012; Sennett & Rendl, 2012). Recent studies found that the DP is absolutely essential for new HF regeneration (Rompolas et al., 2012) and that the number of cells in the DP determines successful activation of SC during the hair cycle. Below a certain threshold, the DP lose its ability to signal to SCs (Chi, Wu, & Morgan, 2013).

However, a complete understanding of the signals involved in this process remains unknown.

In the intestine, mesenchymal cells neighboring the crypt base have also been implicated as part of the niche (green cells in Fig. 12.2C), although the precise identity of important cells and their corresponding signaling modes are not yet clear (reviewed in Smith, Davies, Silk, & Wong, 2012). In humans, it was shown that pericryptic fibroblasts and underlying smooth muscle cells secrete factors influencing crypt cells (Kosinski et al., 2007). Although the ability of Lgr5+ SCs to form organoids in culture without the influence of neighboring mesenchymal cells challenges this theory (Sato et al., 2009), necessary survival factors of the *in vitro* system are known to be expressed by surrounding cells *in vivo* (Farin et al., 2012; Kosinski et al., 2007).

In the testes, SCs are supported by neighboring cells from another lineage. Testes consist of interstitial tissue and seminiferous tubules in which postmitotic Sertoli cells associate with differentiated germ cells to form the seminiferous epithelium. The basal compartment of the epithelium contains spermatogonial SCs (SSCs) and spermatogonia, whereas more differentiated germ cells are present in the adluminal compartment (reviewed in Oatley & Brinster, 2012). The seminiferous epithelium is bound by a basement membrane that separates it from the interstitial tissue, populated by myoid cells, Leydig cells, blood vessels, and different immune cells (Fig. 12.3C). Sertoli cells in the seminiferous tubules are the most important cellular component of the immediate SSC niche. These cells form tight junctions with each other, creating the blood–testes barrier and physically constraining signals that emanate from the underlying interstitial tissue to the domain of undifferentiated SSCs associated with the basement membrane. More differentiated progeny still associate with the Sertoli cells, but outside of this tight junction barrier. Increasing the number of Sertoli cells in the seminiferous epithelium creates more niches for SSC seeding, and inhibiting the production of supportive signals from these cells depletes SSC numbers (Meng et al., 2000; Oatley, Racicot, & Oatley, 2011).

2.3. SC regulation by ECM and adhesion molecules

The influence of cellular components within the SC niche can extend beyond the production of secreted morphogens and chemokine gradients to include direct cell–cell contact (reviewed in Chen, Lewallen, & Xie, 2013). Because adhesion-mediated interactions are generally static, this kind of niche input typically influences quiescence, maintenance, or retention of SCs no matter the system. Additionally, SCs interact with the ECM (gray area in Fig. 12.1) within the niche as it can act as a reservoir of secreted signals or potentially, as demonstrated largely by *in vitro* work, modulate SC activity in its own right (Engler, Sen, Sweeney, & Discher, 2006; Gilbert et al., 2010).

In the HF, different adhesion molecules were shown to have an important role in epidermal and HF development and maintenance. Complete ablation of integrin beta1 in epithelial cells results in severe skin blistering and impaired HF formation (Raghavan, Bauer, Mundschau, Li, & Fuchs, 2000). Similarly, ablation of the linker protein alpha-catenin impairs hair development and causes major defects in the epidermis (Vasioukhin, Bauer, Degenstein, Wise, & Fuchs, 2001). Also, E-cadherin ablation in adult epidermis leads to a severe differentiation defect, and to decreased proliferation of HF progenitors (Young et al., 2003), suggesting that E-cadherin is important for HF maintenance. However, the exact function of these proteins specifically in adult HFSCs remains unknown.

In the brain, astrocytes are present in both the SVZ and SGZ, producing morphogens and providing direct cell–cell contacts with NSCs to regulate proliferation, differentiation, migration, and synapse formation (Barkho et al., 2006). Vascular cells similarly disperse throughout neural tissue and can act via direct contact to influence progenitor cell fate choices by bringing blood-derived signals into close proximity with target cells, and can additionally provide paths that progenitors migrate along to reach different zones of the brain (Kojima et al., 2010). More specifically, targeted ablation of E-cadherin within all central nervous system (CNS) tissues revealed its functional role in mediating NSC self-renewal *in vivo*, while parallel studies confirmed E-cadherin expression was needed for NSCs to form colonies in culture (Karpowicz et al., 2009).

Both cadherins and integrins are expressed in a polarized orientation by muscle SCs, but the functional role of their localization is not yet clear (Kuang et al., 2008). In the muscle, where SCs are entirely surrounded by matrix, the ECM was shown to localize numerous factors that activate satellite cells, such as HGF, FGF, or IGF, especially in cases of injury (Yin, Price, & Rudnicki, 2013). It was also recently shown that Collagen VI regulates satellite cell self-renewal (Urciuolo et al., 2013).

Reports of direct cell–cell adhesions that regulate HSC activity have been conflicting. Although *N*-cadherin expression was first appreciated in HSCs and the osteoblastic niche a decade ago (Zhang et al., 2003), subsequent studies have both confirmed and negated its role in mediating HSC maintenance (Bromberg et al., 2012; Haug et al., 2008; Hosokawa et al., 2010; Kiel, Acar, Radice, & Morrison, 2009). Separate functional investigations reveal integrins are also expressed by HSC and niche cells and have a role in mediating SC homing to the bone marrow, especially in transplantation experiments (Potocnik, Brakebusch, & Fassler, 2000; Qian, Tryggvason, Jacobsen, & Ekblom, 2006), and *in vitro* studies have hinted at a role in long-term maintenance via interactions with osteoblasts (Schreiber et al., 2009).

Cell adhesion is also believed to influence germ SSC activity, especially with the confirmed expression of several integrins on SSCs (Shinohara, Avarbock, & Brinster, 1999). Undifferentiated SSCs are closely associated with both the basement membrane lining the seminiferous epithelium and next to somatic Sertoli cells within the tubules. Specific studies have shown that SSC expression of integrin beta1 is important for sustaining SC activity and crucial for directing SC homing to the appropriate niche during transplantation studies (Kanatsu-Shinohara et al., 2008). SSCs also directly interact with neighboring Sertoli cells, potentially through cadherin-mediated interactions (Tokuda, Kadokawa, Kurahashi, & Marunouchi, 2007).

2.4. Long-range contributors and macroenvironment

Although the original concept of the SC niche embodied only neighboring elements capable of influencing SC self-renewal and differentiation, we now know that SCs also receive signals from cells more distant in the tissue or even outside the tissue (red and orange cells in Fig. 12.1).

The activation and quiescence of SCs within the HF is known to be influenced by signals that originate globally from the dermis, which helps to coordinate hair cycling between many follicles throughout the skin (Plikus et al., 2011). A recent study further elucidated a role for nascent and mature adipocytes in signaling to the bulge compartment to initiate hair cycling, possibly via signaling through the DP compartment (Festa et al., 2011). Another study also showed the importance of neural input in regulating HFSCs capacity to act as epidermal SCs (Brownell, Guevara, Bai, Loomis, & Joyner, 2011).

Long-range neural input is indisputably important for regulating HSC release into the peripheral bloodstream within the hematopoietic system.

G-CSF stimulation drives SCs out of the bone marrow by affecting norepinephrine signaling, which in turn downregulates osteoblast production of CXCL12, and ultimately allows HSCs to escape into the circulation (Katayama et al., 2006). Remarkably, even light stimuli can influence the sympathetic nervous system with downstream effects on HSC mobilization, which occurs in a circadian rhythm pattern during normal homeostasis (Mendez-Ferrer, Lucas, Battista, & Frenette, 2008). Sympathetic tone has similarly been shown to influence the activity of other components of the cellular niche including nestin+ MSCs, promoting the egress of HSCs from the bone marrow through additional avenues (Mendez-Ferrer et al., 2010). In the case of the hematopoietic system, far-reaching neural input acts on multiple targets to achieve SC mobilization into the circulation.

Within the muscle, satellite SCs are consistently localized near vasculature, ensuring ready access to systemic signals that can regulate SC activity (Christov et al., 2007; Fukada et al., 2007). Signals from the interstitial space outside the muscle basal lamina, such as from immune cells or neurons, can also reach underlying satellite cells and influence SC behavior (Girgenrath et al., 2006). Additional evidence that systemic factors can modulate satellite cell potency comes from experiments characterizing the effects of parabiosis or calorie restriction on the ability of muscle SCs to self-renew and repair degenerating tissue (Cerletti, Jang, Finley, Haigis, & Wagers, 2012; Conboy et al., 2005).

Humoral signals also appear to be crucial for maintenance and maturation of germ SCs, as SSCs preferentially localize near vessels in the interstitium and migrate away as they differentiate (Chiarini-Garcia, Hornick, Griswold, & Russell, 2001; Chiarini-Garcia, Raymer, & Russell, 2003; Yoshida, Sukeno, & Nabeshima, 2007). Signals that reach the SSCs through the interstitial vasculature are prevented from spreading throughout the circulatory system by Sertoli cell tight junctions. Interstitial Leydig cells produce cytokines to support SSC self-renewal, and also secrete testosterone, which is similarly important to ensure proper functioning of Sertoli cells (Davidoff et al., 2004; Oatley, Oatley, Avarbock, Tobias, & Brinster, 2009). Signals from interstitial myoid cells have also been implicated in supporting SSC self-renewal (Nurmio et al., 2012; Qian et al., 2013), while perivascular cells are thought to secrete supporting signals for Sertoli cells (Verhoeven & Cailleau, 1988). Important roles for systemic factors have been demonstrated, but more in the context of supporting the cellular components of the SSC niche (Oatley & Brinster, 2012).

3. MOLECULAR REGULATORS OF ADULT SC NICHES

Adult SC niches can be populated by many different cell types and structural components, each providing molecular and/or physical support to regulate SCs, as described in the previous section. In this part, we discuss signals that have been identified in different adult SC niches. These signals can be organized into two main categories: (1) quiescent or survival signals and (2) activating signals. Quiescent signals are emitted by niches maintaining dormant SCs, whereas activating signals are present after injury or in rapidly renewing tissues during homeostasis.

3.1. Signals maintaining quiescence, survival, and self-renewal

In all organs that contain SCs specific signals promote the survival of these cells. In slowly renewing tissues that contain slow-cycling SC populations, niche signals also allow maintenance of quiescence. Some molecular regulation comes from within the SCs themselves, but most originates from different components of the niche.

3.1.1 Cell autonomous regulation of SC quiescence and survival

The main intrinsic process by which SCs balance quiescence and self-renewal is through regulation of cell cycle entry. This cell-autonomous mechanism regulates HSC quiescence by preventing HSC exhaustion (green arrow in Fig. 12.2A) (Pietras, Warr, & Passegue, 2011). Several studies have highlighted the critical role of the Rb protein family in controlling HSC cell cycle entry, thereby maintaining HSC self-renewal capacity (Viatour et al., 2008). Concomitant ablation of Rb, p130 and p107 activated HSCs proliferation, limiting their reconstitution capacity in long-term transplantation assays. This phenotype was not observed in single deleted mice (Cobrinik et al., 1996; LeCouter et al., 1998; Walkley & Orkin, 2006), suggesting an important functional redundancy within the Rb protein family. Similarly, mice deficient for a single D-cyclin or associated kinase have only minimal hematopoietic phenotypes, whereas concomitant deletion of several D-cyclins or Cyclin-dependent kinases (CDKs) results in embryonic lethality with severe hematopoietic defects (Kozar et al., 2004; Malumbres et al., 2004). The potential importance of D-cyclins/CDKs in adult HSCs is still unknown. The Ink4 protein family of D-cyclin–Cdk4/6 antagonists also regulates HSC cell cycle entry by balancing quiescence and proliferation (reviewed in Pietras et al., 2011). CIP/KIP family proteins p21, p27, and p57, but also master transcriptional regulator p53, are key controls of HSC quiescence as well. Additionally, suppression of PI3K signaling by tumor suppressor PTEN is critical to control D-cyclins for HSC quiescence, illustrating again the central role of cell cycle regulation in the maintenance of HSCs. Similarly in HFs, HFSCs autonomously regulate quiescence through CDKi (green arrow in Fig. 12.2B). The Runx1/p21 complex controls HFSC quiescence during the hair cycle (Lee et al., 2013). During the destruction/catagen phase, downregulation of Runx1 results in p21 upregulation maintaining HFSCs in a quiescent state. Upon regeneration (anagen phase), Runx1 represses p21 expression and then promotes HFSC self-renewal. Cell cycle inhibitors are also crucial for maintenance of adult NSC quiescence (green arrow in Fig. 12.3B). P21 represses NSC proliferation (Kippin, Martens, & van der Kooy, 2005) specifically during development (Molofsky, He, Bydon, Morrison, & Pardal, 2005). P73 also inhibits premature senescence of NSC (Talos et al., 2010).

Other transcription factors have similarly been implicated in SC-autonomous survival. In HFs, HFSCs deficient for Tbx1 were unable to replenish their niche, suggesting a key role in long-term SC maintenance (Chen et al., 2012). NFATc1 intrinsically maintains HFSC quiescence as well (Horsley, Aliprantis, Polak, Glimcher, & Fuchs, 2008). Interestingly, both studies mechanistically link SC survival to bone morphogenetic protein (BMP) signaling. Lhx2 is also involved in HF formation and HFSC maintenance (Rhee, Polak, & Fuchs, 2006), whereas Sox9 is dispensable for hair induction but critical for the formation of the HFSC compartment (Nowak, Polak, Pasolli, & Fuchs, 2008; Vidal et al., 2005). In the muscle, the Transforming Growth Factor beta (TGFb) signaling factor Myostatin is expressed by satellite SCs to maintain their own quiescence (green arrow in Fig. 12.3A) (McCroskery, Thomas, Maxwell, Sharma, & Kambadur, 2003). Genetic ablation of the Notch signaling effector Rbpj also causes satellite SCs to lose quiescence in resting muscle (Mourikis et al., 2012). In this case, the source of the Notch ligand Delta is still unclear, although satellite SCs upregulate its expression after injury (Conboy, Conboy, Smythe, & Rando, 2003). In the intestine, the Wnt target gene Ascl2 is important for CBC survival/maintenance (green arrow in Fig. 12.2C), although the status of Ascl2 in the +4 ISCs is not yet characterized (van der Flier et al., 2009). The pan-ErbB negative regulator Lrig1 marks a quiescent ISC population in the colon and maintains intestinal homeostasis likely by regulating quiescence. Upon deletion of Lrig1, this subpopulation of SCs starts to proliferate and eventually forms tumors (Powell et al., 2012).

In the CNS, the nuclear orphan receptor Tlx (Sun, Yu, Evans, & Shi, 2007), the transcription factor Sox2 (Favaro et al., 2009; Hu et al., 2010), and the longevity-associated factor Foxo3 (Renault et al., 2009) have all been identified as regulators of NSC quiescence. Interestingly, a recent study suggested that GSK3, a downstream effector of numerous signaling pathways, may be a central regulator of NSC homeostasis (Kim et al., 2009). Recently, even components of the miRNA pathway have been shown to regulate SC quiescence. miRNA-489 suppresses satellite cell expression of a protein promoting progenitor expansion in muscle (Cheung et al., 2012). SCs in the skin similarly express miRNA-125b to preferentially self-renew instead of differentiating (Zhang, Stokes, Polak, & Fuchs, 2011). Finally, it is worth noting that some of the autocrine regulators mentioned earlier may nevertheless be dependent on external input, such as from the basement membrane or secreted ligands.

3.1.2 Extrinsic signals that regulate SC quiescence and self-renewal

As described in the first section, many different cell types can constitute the niche and signal to SCs as extrinsic regulators. The TGFb superfamily encompasses multiple signaling pathways known to contribute to SC quiescence. In the bone marrow niche, TGFb maintains HSC hibernation (Yamazaki et al., 2011, 2009). This process involves neighboring glial cells which activate latent TGFb (red arrow in Fig. 12.2A) (Yamazaki et al., 2011). In the intestinal epithelium, it was shown that quiescence of ISC can be promoted by oral administration of TGFb1 (Puolakkainen et al., 1994), although data demonstrating that this process happens endogenously is lacking.

The BMP family, part of the TGFb superfamily, was shown to be important for SC quiescence as well. Inactivation of BMPRIa in epithelial cells causes HFSC activation and premature anagen (Kobielak, Stokes, de la Cruz, Polak, & Fuchs, 2007). BMP also seems important for maintenance of HFSC niche integrity as DP cells deficient for BMPRIa lose hair inducing capacity (Rendl, Polak, & Fuchs, 2008). Moreover, the resting phase of the hair cycle is characterized by a high concentration of BMP ligands in the dermis, generated from subcutaneous fat, which instructs HFSCs to stay quiescent (red arrow in Fig. 12.2B; Plikus et al., 2008). Outside of the skin, the BMP pathway is also involved in preventing proliferation in the intestine. BMP4 is expressed in the intravillus mesenchyme to activate the BMP pathway in differentiated epithelial cells in the villi (Haramis et al., 2004). Another study showed that inactivation of BMPRIa in the intestinal epithelium leads to an expansion of the stem compartment through activation of the canonical Wnt pathway (He et al., 2004). Interestingly, similar observations were made in the colon (Singbrant et al., 2010). Also, it was shown that human pericryptic fibroblasts and smooth muscle cells express BMP antagonists that repress the BMP pathway in ISCs (Kosinski et al., 2007). Taken together these data suggest that a gradient of BMP ligands and antagonists exists along the crypt–villus axis, repressing the BMP pathway in ISC to limit proliferation and thus maintain quiescence (red gradient and blunt arrows in Fig. 12.2C). Also, Wnt5a regulated TGFb has been shown to have a crucial role in *de novo* crypt formation after injury (Miyoshi, Ajima, Luo, Yamaguchi, & Stappenbeck, 2012). Interestingly, a separate study reported that intestinal mesenchymal cells express Wnt5a (Farin et al., 2012), but its link to TGFb and wound repair was not mentioned. Similarly in the hippocampus, the BMP pathway, through BMPRIa, contributes to adult NSCs quiescence maintenance (Mira et al., 2010). A specific balance of BMP signaling occurs in the SVZ to maintain NSC quiescence, involving the production of BMP ligands by NSCs and expression of the BMP antagonist Noggin by ependymal niche cells (red arrow and blunt arrows in Fig. 12.3B) (Lim et al., 2000). Finally, the BMP pathway regulates the niche size for HSCs in the bone marrow through its action on osteoblasts (red arrow in Fig. 12.2A) (Zhang et al., 2003).

Another signal pathway commonly involved in SC homeostasis is canonical Wnt signaling. For years, the Wnt/β-catenin pathway has been associated with intestinal tumorigenesis and specifically the proliferation of intestinal progenitors (Reya & Clevers, 2005). Several studies have implicated this pathway in regulating ISC maintenance, as ablation of Tcf4 (Korinek et al., 1998; van Es et al., 2012) and β-catenin (Fevr, Robine, Louvard, & Huelsken, 2007) leads to reduced ISC survival. Interestingly, a recent study demonstrated numerous and redundant sources of Wnt ligands in the ISC microenvironment (Farin et al., 2012). Indeed, abrogated Wnt3 secretion by Paneth cells is compensated by mesenchymal cell production of Wnt ligands (blue arrows in Fig. 12.2C). Similarly, canonical Wnt signaling is necessary to maintain HFSCs (Huelsken, Vogel, Erdmann, Cotsarelis, & Birchmeier, 2001). Ablating β-catenin in epithelial cells results in loss of SC markers and active proliferation (Lowry et al., 2005). Interestingly, the source of essential Wnt ligands remains unclear and likely does not involve epithelial cells, as the ablation of Wntless (Wls), an essential factor in Wnt secretion, in these cells does not alter HFSC maintenance. (blue arrow in Fig. 12.2B) (Myung, Takeo, Ito, & Atit, 2013). On the other hand, the role of Wnt signaling in the hematopoietic system is under ongoing debate, as several studies reported Wnt ligands maintain HSC self-renewal *in vitro* (Reya et al., 2003; Willert et al., 2003), but *in vivo* data contradict this conclusion (Cobas et al., 2004; Kirstetter, Anderson, Porse, Jacobsen, & Nerlov, 2006; Reya et al., 2003).

The Notch pathway, known to be important for cell fate decisions, has also been identified as a key regulator in SC biology. Notch1 signaling is particularly crucial for the generation of definitive HSCs (Kumano et al., 2003). In adults, Notch is thought to inhibit cytokine-induced differentiation through regulation of GATA2 and Hes1, thus maintaining HSCs (Kumano et al., 2001; Kunisato et al., 2003). Interestingly, the Notch ligand Jagged1 is highly expressed in osteoblasts (yellow arrows in Fig. 12.2A) (Calvi et al., 2003). Notch also controls cell proliferation in the intestine in a Wnt-dependent manner (Fre et al., 2009; Riccio et al., 2008; Rodilla et al., 2009), although its specific mode of action on SCs is unclear. Regardless, several Notch ligands are expressed by surrounding intestinal mesenchymal cells (yellow arrows in Fig. 12.2C) (Fre et al., 2005). In the brain, several studies have identified Notch signaling in maintenance of adult NSCs. In the hippocampus, Notch1 ablation increases proliferation and leads to loss of NSCs (Breunig, Silbereis, Vaccarino, Sestan, & Rakic, 2007). A similar observation was made in the whole brain upon deletion of the downstream transcription factor Rbpjk (Imayoshi, Sakamoto, Yamaguchi, Mori, & Kageyama, 2010). Ependymal cells, which give rise to neuroblasts and astrocytes after a stroke, maintain quiescence through Notch signaling (Carlen et al., 2009), and Notch and PEDF signaling cooperate to regulate self-renewal in the brain (Andreu-Agullo, Morante-Redolat, Delgado, & Farinas, 2009). Although these studies do not identify Notch ligand-expressing cells, these should be in close distance to the NSCs (yellow arrows in Fig. 12.3B).

Several other signaling pathways are involved in regulating SC quiescence. Receptor tyrosine kinase Tie2 signaling, initiated by its ligand Angiopoietin1, promotes HSC quiescence in the bone marrow (Arai et al., 2004), and satellite SC quiescence in the muscle (Abou-Khalil et al., 2009). Osteoblasts produce Angiopoietin1 in the hematopoietic system (brown arrow in Fig. 12.2A) (Arai, Ohneda, Miyamoto, Zhang, & Suda, 2002), and periendothelial cells in the muscle (brown arrow in Fig. 12.3A) (Abou-Khalil et al., 2009). The growth factor GDNF plays an essential role in SC quiescence in the germline and the brain. GDNF-deficient mice have seminiferous tubules that lack germ cells, most likely due to the inability of Sertoli cells to sustain undifferentiated SSCs

(Kubota, Avarbock, & Brinster, 2004; Meng et al., 2000). GDNF is expressed by Sertoli cells upon activation by FSH (brown arrow in Fig. 12.3C) (Tadokoro, Yomogida, Ohta, Tohda, & Nishimune, 2002), and its receptor, GFRalpha1, is expressed by spermatogonia (brown arrow in Fig. 12.3C) (Grisanti et al., 2009). Interestingly, the effect of GDNF signaling on SSC maintenance is promoted *in vitro* by FGF2, EGF (Kanatsu-Shinohara et al., 2005), and IGF (Kubota et al., 2004). In the brain, GDNF is widely expressed and important for neuronal precursor survival (brown arrow in Fig. 12.3B) (Arenas, Trupp, Akerud, & Ibanez, 1995). A number of additional secreted factors are important for maintaining SCs. Stem cell factor (SCF), also known as KITL, is an important promoter of HSC maintenance and is expressed by perivascular cells in the bone marrow (Fig. 12.2A) (Ding, Saunders, Enikolopov, & Morrison, 2012). Interactions between HSCs and osteoblasts involving the adhesion proteins *N*-cadherin and integrins also support HSCs (Schreiber et al., 2009; Zhang et al., 2003). Two recent studies demonstrated a central role of CXCL12 from MSCs in HSC self-renewal and maintenance (brown arrow in Fig. 12.2A) (Ding & Morrison, 2013; Greenbaum et al., 2013). Interestingly, long-range CXCL12 signaling is also important for homing SSC precursors in the embryo, and *in vitro* studies suggest an importance for maintaining adult SSCs as well (Ara et al., 2003; Kanatsu-Shinohara et al., 2012). In the adult testes, CSF1 is expressed in Leydig cell clusters and peritubular myoid cells as a component of the SSC niche that controls self-renewal of the germline (brown arrows in Fig. 12.3C) (Oatley et al., 2009). Finally, in the HF, nerve-secreted Shh is important for the maintenance of bulge HFSCs (brown arrow in Fig. 12.2B) (Brownell et al., 2011).

3.2. SC activating signals for tissue regeneration and repair

As described in the previous section, SC quiescence depends on cell-autonomous factors as well as external inputs in both fast and slow-cycling tissues. Interestingly, only few intrinsic activating SC signals have been described. In the muscle, sphingolipid signaling, through its soluble form S1P, acts in an autocrine/paracrine manner to promote satellite cell proliferation and muscle regeneration (Nagata, Partridge, Matsuda, & Zammit, 2006; Sassoli et al., 2011). In the brain, the miRNA machinery was recently implicated in SC regulation (Szulwach et al., 2010).

In general, SC activation to regenerate tissues during homeostasis or after injury depends almost exclusively on external signals. While TGFb signaling maintains quiescence in several organs, it is also important for SC activation. In the HF, TGFb2 expressed by DP cells transiently activates Smad2/3 and antagonizes refractory BMP signals to allow HF regeneration (dashed red arrow in Fig. 12.2B) (Oshimori & Fuchs, 2012). In the muscle, TGFb is upregulated in satellite cells of aged mice, impairing the regeneration capacity of aging muscle (Carlson, Hsu, & Conboy, 2008). Interestingly, high TGFb levels were also detected in the ECM, but its source remains unclear (dashed red arrow in Fig. 12.3A). In the germline, BMP4 seems to regulate SSC differentiation by acting on spermatogonia, as exposure to BMP4 decreases SSC numbers *in vitro* (dashed red arrow in Fig. 12.3C) (Nagano, Ryu, Brinster, Avarbock, & Brinster, 2003; Pellegrini, Grimaldi, Rossi, Geremia, & Dolci, 2003), but this effect has not been confirmed *in vivo*.

Similarly, Wnt/β-catenin signaling activates SCs in different organs, in addition to promoting survival, self-renewal, and maintenance as previously described. Canonical Wnt signaling activates HSC proliferation in the bone marrow (Reya et al., 2003; Willert et al., 2003) via ligands most likely expressed by osteoblasts (dashed blue arrow in Fig. 12.2A). In the SVZ of the brain, this pathway promotes progenitor cell proliferation. Pharmacological inhibition of GSK3 stabilizes β-catenin in SVZ cells and activates their proliferation (Adachi et al., 2007), although the specific source of Wnt ligands in homeostatic conditions remains unclear (dashed blue arrows in Fig. 12.3B). In the hippocampus, Wnt pathway activation is dispensable for maintenance of the SC compartment, but important for the survival of neural progenitors and neuronal maturation (Kuwabara et al., 2009). Here again, the source of Wnt ligands in the hippocampus was not explored. In addition, the *in vitro* sphere formation capacity of NSCs is Wnt dependent, as well as NSC activation after injury (Wang et al., 2011). The central role of Wnt/β-catenin signaling in intestinal homeostasis was specifically linked to its ability, when overactivated, to promote SC proliferation and adenoma formation. Interestingly, the putative ISC marker Musashi1 is a potent activator of Wnt and Notch pathways, leading to tumor formation (Rezza et al., 2010). More strikingly, specific activation of the Wnt pathway in ISC through stabilization of β-catenin (Sangiorgi & Capecchi, 2008) or Adenomatous Polyposis Coli (APC) deletion (Barker et al., 2009) activates SC proliferation to eventually form tumors. *In vivo*, different redundant sources of Wnt ligands have been identified (dashed blue arrows in Fig. 12.2C) (Farin et al., 2012). In HFs, the canonical Wnt pathway is also important for HFSC activation. Sustained β-catenin stabilization in epithelial cells leads to proliferation and precocious activation of HFSC (Lowry et al., 2005). Interestingly, ablation of Wls specifically in epithelial cells results in decreased proliferation and defective regeneration (Myung et al., 2013), suggesting that Wnt ligands involved in HFSC activation are secreted by epithelial cells (dashed blue arrow in Fig. 12.2B). Two studies have highlighted a role of Wnt signaling in myogenic differentiation and muscle regeneration, as it is required for proper activation and differentiation of satellite cells and hence normal regeneration in an *in vitro* model of cultured myofibers (Brack, Conboy, Conboy, Shen, & Rando, 2008; Otto et al., 2008). Moreover, Wnt ligands are expressed in muscle fibers and canonical Wnt signaling is activated during muscle regeneration after injury (dashed blue arrow in Fig. 12.3A) (Otto et al., 2008).

The Notch pathway is also important for SC regulation and cell fate decisions in different organs. In the intestine as in the muscle, it is necessary for proper differentiation of certain cell types (dashed yellow arrows in Figs. 12.2C and 12.3A) (Conboy & Rando, 2002; Fre et al., 2005; van Es et al., 2005). In one study, Notch signaling promoted satellite cell proliferation during muscle regeneration and was shown to be impaired in aged muscle (Conboy et al., 2003). Other important pathways in SC activation include FGF and PDGF signaling. In HFs, FGF7 secreted by DP cells is upregulated during the transition from telogen to anagen, when HFSCs are activated (dashed brown arrow in Fig. 12.2B). Subcutaneous injection of FGF7 induces HFSC activation and precocious follicle regeneration (Greco et al., 2009). In the brain, FGF2 associated with EGF promotes NSC proliferation *in vitro*, although its role *in vivo* is unclear (dashed brown arrow in Fig. 12.3B) (Gritti et al., 1999; Kuhn, Winkler, Kempermann, Thal, & Gage, 1997). PDGF signaling promotes HFSC activation as well, as PDGF-A ligands generated by subcutaneous fat is required for hair regeneration in the hair cycle, most likely through activation of the PDGFR pathway in DP cells (dashed brown arrow in Fig. 12.2B) (Festa et al., 2011). In the brain, this pathway activates adult NSCs (dashed brown arrow in Fig. 12.3B). In the SVZ, PDGF signaling is necessary for oligodendrogenesis, although it is dispensable for neurogenesis in general. Additionally, PDGF can be a potent mitogen of NSCs, leading to tumor-like hyperplasia (Jackson et al., 2006; Lachapelle, Avellana-Adalid, Nait-Oumesmar, & Baron-Van Evercooren, 2002).

A number of other secreted factors have been linked to SC activation in distinct model systems. G-CSF is an activator of HSCs and is now therapeutically utilized to harvest cells from the bone marrow for transplants (dashed brown arrow in Fig. 12.2A) (Katayama et al., 2006). For germ SCs,

Neuregulin1 has a potent differentiation effect on spermatogonia *in vitro* (Hamra, Chapman, Nguyen, & Garbers, 2007), although this regulation was not yet confirmed *in vivo* (dashed brown arrow in Fig. 12.3C). In the muscle, several long-range signals have been suggested to activate satellite cells, but precise factors and mechanisms are unclear. Endothelial cells express growth factors, such as VEGF, that promote satellite cell proliferation, while neurons secrete neurotrophins, NGF and BDNF, which are thought to influence satellite cell behavior (reviewed in Yin et al., 2013). Similarly, immune cells seem to promote satellite cell proliferation by secreting a range of diffusible factors such as MCP-1 (dashed brown arrows in Fig. 12.3A) (Chazaud et al., 2003; Yin et al., 2013). Androgens are also suggested to affect satellite cells, which express the androgen receptor, but direct evidence of this regulation is lacking (Yin et al., 2013).

4. SC NICHE DYSFUNCTION IN AGING AND CANCER

In addition to studying normal adult SC and niche interactions, it is important to consider the implications of a malfunctioning SC environment during aging and in disease. Aging tissues typically display a diminished capacity for repair, resulting in progressively decreasing integrity. It seems logical that declining SC function could contribute to global tissue aging, but only few studies have explored the role of the aging niche in this context. Some of the most convincing evidence comes from parabiotic studies, in which aged and young mice are joined together with a shared blood circulation (Conboy et al., 2005; Mayack, Shadrach, Kim, & Wagers, 2010). Long-range signaling factors present in the young serum revitalized aged SCs in the older mice, leading to increased long-term HSC numbers and differentiation capacity in the bone marrow, enhancing proliferation of muscle satellite SCs for tissue repair, and even promoting regeneration of aged hepatocytes.

A recent study directly examined signaling changes in aged myofibers, with interesting results on satellite SC maintenance (Chakkalakal, Jones, Basson, & Brack, 2012). With age, myofibers emit increasing levels of FGF2 that negatively affect satellite SC quiescence. While additional FGF ligands are being produced in the niche, aging satellite SCs start to produce fewer FGF pathway inhibitors, resulting in increased proliferation and depletion of reserve SCs. In this case the myofiber niche is important for keeping SCs in a quiescent state through the limited release of soluble factors, a protective mechanism that starts to fail in aging tissue. Convincing studies have similarly demonstrated a central role for the niche in contributing to declining germ SC function and decreased fertility with age. Testes generally atrophy in aged mice; overall weight of testes and number of SSCs decreases, and these SCs have diminished functionality in serial transplantation assays (Zhang, Ebata, Robaire, & Nagano, 2006). However, young SSCs can repopulate old testes stroma, and SSC transplantation into a young environment can rejuvenate their activity, suggesting that both germ SCs and niche undergo decline during aging, but still maintain the potential to produce functional gonads. In another study serial transplantation of adult SSCs through young testes stroma maintains self-renewal and sperm-generating capacity for up to 3 years (Ryu, Orwig, Oatley, Avarbock, & Brinster, 2006), highlighting the powerful role of a young niche in supporting a SC lineage even longer than the lifespan of the animal.

An understanding of HFSC and niche cell maintenance during iterative hair cycles and aging is generally lacking. In mice, the first hair cycle takes only a few days to complete while subsequent cycles are increasingly lengthy (Sennett & Rendl, 2012). In humans, aging follicles similarly decline in productivity as cycling turnover slows, and shorter, smaller HFs become suspended in the resting telogen phase (Kligman, 1988; Trueb, 2006). Very little is known about the cellular dynamics behind these morphological changes, although correlative data has suggested a link between DP niche integrity and continued HF cycling (Chi et al., 2013), and when DP cells are physically ablated in mice, HF regeneration fails (Rompolas et al., 2012). It follows that loss of the mesenchymal compartment and/or activating niche signals during successive hair cycles might contribute to a decline in HF productivity, but the relevance to physiological aging has yet to be explored. Alternatively, an increase in quiescence-promoting signals from the aging skin macroenvironment could negatively impact follicle productivity (Chen et al., 2012).

As SC proliferation and differentiation into the appropriate progeny can be influenced by stimuli in the immediate environment, it also implies that dysregulation within the niche could promote tumorigenesis. The concept of cancer stem cells (CSCs) is still evolving, and so too is the distinct idea of a cancer SC niche, in which normal SCs are prompted to proliferate excessively or a specific lineage overexpands because of aberrant signals from the environment. Evidence for aberrant niche signaling driving tumorigenesis has been described in only a few cancer models. Studies of stromal cells derived from human basal cell carcinomas found increased expression of BMP antagonist Gremlin1 compared to normal fibroblasts, and subsequent work in culture suggested sustained repression of BMP signaling is permissive for CSC expansion (Sneddon et al., 2006). *In vitro* work found enhanced Wnt signaling within colorectal CSCs, which could be enforced by secreted signals generated by myofibroblasts that normally reside in the crypt microenvironment (Vermeulen et al., 2010). Experimentally targeting niche cells for gene deletion in the bone marrow can induce myelodysplasia in mice, presenting the provocative idea that dysregulation in the niche can drive cancer development (Raaijmakers et al., 2010). Also, studies of glioblastoma have found evidence supporting the existence of CSCs that depend on interactions with nearby perivascular and immune niche cells to specifically promote cancer cell survival and proliferation (Charles et al., 2010; Filatova, Acker, & Garvalov, 2013; Heddleston, Li, McLendon, Hjelmeland, & Rich, 2009; Zhu et al., 2011). As powerful signals generated by glioblastoma cells can dramatically alter the gene expression and migration of surrounding vasculature cells, and cancer cells even transdifferentiate into tumor endothelial cells (Wang et al., 2010) it seems that CSCs can mold their own defective niche. Similarly, in the case of cutaneous squamous cell carcinomas, CSCs use secreted signals to simultaneously promote self-renewal and reshape the vasculature in their microenvironment (Beck et al., 2011).

5. CONCLUDING REMARKS

Accumulating data implicates the microenvironment of SCs in governing their behavior and capacity for tissue regeneration. Adult SC niches are composed of cells from different developmental origins, including the SCs themselves and their progeny, mesenchymal neighbors, and other more distant cells. The ECM, basement membrane, and other adhesion molecules are also key players in many adult SC niches (Fig. 12.1). From the molecular data collected in different model systems, three main signaling pathways have been identified as executors of quiescence, survival/maintenance, and/or activation of SCs: the TGFb superfamily signals, the Wnt pathway, and Notch signaling, although many other tissue specific regulators have also been described (Figs. 12.2 and 12.3).

By understanding how SC niches function in physiological conditions and disease we can create better systems to study and manipulate SCs *in vitro*, which will ultimately be crucial for future clinical applications. A small number of niche-centric therapeutic approaches are already being applied in clinic, the most notable example being advances in bone marrow transplantation (Thomas, Stein, Gentile, & Shah, 2010). Recent experimental evidence suggests that agents directly targeting the niche could be useful in the future to enhance transplanted HSC viability (Naveiras et al., 2009). In the future, quiescence-promoting proteins or small molecule treatments could be applied to counteract cancer-driving signals from a diseased niche, or activating signals could be exogenously supplied to jump start tissue regeneration.

ACKNOWLEDGMENTS

We apologize to all colleagues whose relevant work we could not discuss due to space limitations. R. S. was supported by training grant T32GM008553 from NIH/NIGMS. M. R. was supported by a Dermatology Foundation Research Career Development Award, by NYSTEM contracts C026410 and C026411, and by grants from the NIH/NIAMS (R01AR059143; R01AR063151).

REFERENCES

Abou-Khalil, R., Le Grand, F., Pallafacchina, G., Valable, S., Authier, F. J., Rudnicki, M. A., et al. (2009). Autocrine and paracrine angiopoietin 1/Tie-2 signaling promotes muscle satellite cell self-renewal. *Cell Stem Cell, 5*, 298–309.

Adachi, K., Mirzadeh, Z., Sakaguchi, M., Yamashita, T., Nikolcheva, T., Gotoh, Y., et al. (2007). Beta-catenin signaling promotes proliferation of progenitor cells in the adult mouse subventricular zone. *Stem Cells, 25*, 2827–2836.

Andreu-Agullo, C., Morante-Redolat, J. M., Delgado, A. C., & Farinas, I. (2009). Vascular niche factor PEDF modulates Notch-dependent stemness in the adult subependymal zone. *Nature Neuroscience, 12*, 1514–1523.

Ara, T., Itoi, M., Kawabata, K., Egawa, T., Tokoyoda, K., Sugiyama, T., et al. (2003). A role of CXC chemokine ligand 12/stromal cell-derived factor-1/pre-B cell growth stimulating factor and its receptor CXCR4 in fetal and adult T cell development in vivo. *Journal of Immunology, 170*, 4649–4655.

Arai, F., Hirao, A., Ohmura, M., Sato, H., Matsuoka, S., Takubo, K., et al. (2004). Tie2/angiopoietin-1 signaling regulates hematopoietic stem cell quiescence in the bone marrow niche. *Cell, 118*, 149–161.

Arai, F., Ohneda, O., Miyamoto, T., Zhang, X. Q., & Suda, T. (2002). Mesenchymal stem cells in perichondrium express activated leukocyte cell adhesion molecule and participate in bone marrow formation. *Journal of Experimental Medicine, 195*, 1549–1563.

Arenas, E., Trupp, M., Akerud, P., & Ibanez, C. F. (1995). GDNF prevents degeneration and promotes the phenotype of brain noradrenergic neurons in vivo. *Neuron, 15*, 1465–1473.

Barker, N., Ridgway, R. A., van Es, J. H., van de Wetering, M., Begthel, H., van den Born, M., et al. (2009). Crypt stem cells as the cells-of-origin of intestinal cancer. *Nature, 457*, 608–611.

Barker, N., van Es, J. H., Kuipers, J., Kujala, P., van den Born, M., Cozijnsen, M., et al. (2007). Identification of stem cells in small intestine and colon by marker gene Lgr5. *Nature, 449*, 1003–1007.

Barkho, B. Z., Song, H., Aimone, J. B., Smrt, R. D., Kuwabara, T., Nakashima, K., et al. (2006). Identification of astrocyte-expressed factors that modulate neural stem/progenitor cell differentiation. *Stem Cells and Development, 15*, 407–421.

Beck, B., Driessens, G., Goossens, S., Youssef, K. K., Kuchnio, A., Caauwe, A., et al. (2011). A vascular niche and a VEGF-Nrp1 loop regulate the initiation and stemness of skin tumours. *Nature, 478*, 399–403.

Beckervordersandforth, R., Tripathi, P., Ninkovic, J., Bayam, E., Lepier, A., Stempfhuber, B., et al. (2010). In vivo fate mapping and expression analysis reveals molecular hallmarks of prospectively isolated adult neural stem cells. *Cell Stem Cell*, *7*, 744–758.

Bjerknes, M., & Cheng, H. (2006). Intestinal epithelial stem cells and progenitors. *Methods in Enzymology*, *419*, 337–383.

Blanpain, C., & Fuchs, E. (2009). Epidermal homeostasis: A balancing act of stem cells in the skin. *Nature Reviews Molecular Cell Biology*, *10*, 207–217.

Blanpain, C., Lowry, W. E., Geoghegan, A., Polak, L., & Fuchs, E. (2004). Self-renewal, multipotency, and the existence of two cell populations within an epithelial stem cell niche. *Cell*, *118*, 635–648.

Brack, A. S., Conboy, I. M., Conboy, M. J., Shen, J., & Rando, T. A. (2008). A temporal switch from notch to Wnt signaling in muscle stem cells is necessary for normal adult myogenesis. *Cell Stem Cell*, *2*, 50–59.

Breunig, J. J., Silbereis, J., Vaccarino, F. M., Sestan, N., & Rakic, P. (2007). Notch regulates cell fate and dendrite morphology of newborn neurons in the postnatal dentate gyrus. *Proceedings of the National Academy of Sciences of the United States of America*, *104*, 20558–20563.

Bromberg, O., Frisch, B. J., Weber, J. M., Porter, R. L., Civitelli, R., & Calvi, L. M. (2012). Osteoblastic N-cadherin is not required for microenvironmental support and regulation of hematopoietic stem and progenitor cells. *Blood*, *120*, 303–313.

Brownell, I., Guevara, E., Bai, C. B., Loomis, C. A., & Joyner, A. L. (2011). Nerve-derived sonic hedgehog defines a niche for hair follicle stem cells capable of becoming epidermal stem cells. *Cell Stem Cell*, *8*, 552–565.

Buczacki, S. J., Zecchini, H. I., Nicholson, A. M., Russell, R., Vermeulen, L., Kemp, R., et al. (2013). Intestinal label-retaining cells are secretory precursors expressing Lgr5. *Nature*, *495*, 65–69.

Byrd, D. T., & Kimble, J. (2009). Scratching the niche that controls Caenorhabditis elegans germline stem cells. *Seminars in Cell and Developmental Biology*, *20*, 1107–1113.

Calvi, L. M., Adams, G. B., Weibrecht, K. W., Weber, J. M., Olson, D. P., Knight, M. C., et al. (2003). Osteoblastic cells regulate the haematopoietic stem cell niche. *Nature*, *425*, 841–846.

Cambuli, F. M., Rezza, A., Nadjar, J., & Plateroti, M. (2013). Musashi1-Egfp mice, a new tool for differential isolation of the intestinal stem cell populations. *Stem Cells*, *31*(10), 2273–2278.

Carlen, M., Meletis, K., Goritz, C., Darsalia, V., Evergren, E., Tanigaki, K., et al. (2009). Forebrain ependymal cells are Notch-dependent and generate neuroblasts and astrocytes after stroke. *Nature Neuroscience*, *12*, 259–267.

Carlson, M. E., Hsu, M., & Conboy, I. M. (2008). Imbalance between pSmad3 and Notch induces CDK inhibitors in old muscle stem cells. *Nature*, *454*, 528–532.

Casanova-Acebes, M., Pitaval, C., Weiss, L. A., Nombela-Arrieta, C., Chevre, R., A-Gonzalez, N., et al. (2013). Rhythmic modulation of the hematopoietic niche through neutrophil clearance. *Cell*, *153*, 1025–1035.

Cerletti, M., Jang, Y. C., Finley, L. W., Haigis, M. C., & Wagers, A. J. (2012). Short-term calorie restriction enhances skeletal muscle stem cell function. *Cell Stem Cell*, *10*, 515–519.

Chakkalakal, J. V., Jones, K. M., Basson, M. A., & Brack, A. S. (2012). The aged niche disrupts muscle stem cell quiescence. *Nature*, *490*, 355–360.

Chang, C. Y., Pasolli, H. A., Giannopoulou, E. G., Guasch, G., Gronostajski, R. M., Elemento, O., et al. (2013). NFIB is a governor of epithelial-melanocyte stem cell behaviour in a shared niche. *Nature*, *495*, 98–102.

Charles, N., Ozawa, T., Squatrito, M., Bleau, A. M., Brennan, C. W., Hambardzumyan, D., et al. (2010). Perivascular nitric oxide activates notch signaling and promotes stem-like character in PDGF-induced glioma cells. *Cell Stem Cell*, *6*, 141–152.

Chazaud, B., Sonnet, C., Lafuste, P., Bassez, G., Rimaniol, A. C., Poron, F., et al. (2003). Satellite cells attract monocytes and use macrophages as a support to escape apoptosis and enhance muscle growth. *Journal of Cell Biology*, *163*, 1133–1143.
Chen, T., Heller, E., Beronja, S., Oshimori, N., Stokes, N., & Fuchs, E. (2012). An RNA interference screen uncovers a new molecule in stem cell self-renewal and long-term regeneration. *Nature*, *485*, 104–108.
Chen, S., Lewallen, M., & Xie, T. (2013). Adhesion in the stem cell niche: Biological roles and regulation. *Development*, *140*, 255–265.
Cheung, T. H., Quach, N. L., Charville, G. W., Liu, L., Park, L., Edalati, A., et al. (2012). Maintenance of muscle stem-cell quiescence by microRNA-489. *Nature*, *482*, 524–528.
Chi, W., Wu, E., & Morgan, B. A. (2013). Dermal papilla cell number specifies hair size, shape and cycling and its reduction causes follicular decline. *Development*, *140*, 1676–1683.
Chiarini-Garcia, H., Hornick, J. R., Griswold, M. D., & Russell, L. D. (2001). Distribution of type A spermatogonia in the mouse is not random. *Biology of Reproduction*, *65*, 1179–1185.
Chiarini-Garcia, H., Raymer, A. M., & Russell, L. D. (2003). Non-random distribution of spermatogonia in rats: Evidence of niches in the seminiferous tubules. *Reproduction*, *126*, 669–680.
Chitteti, B. R., Cheng, Y. H., Poteat, B., Rodriguez-Rodriguez, S., Goebel, W. S., Carlesso, N., et al. (2010). Impact of interactions of cellular components of the bone marrow microenvironment on hematopoietic stem and progenitor cell function. *Blood*, *115*, 3239–3248.
Chow, A., Lucas, D., Hidalgo, A., Mendez-Ferrer, S., Hashimoto, D., Scheiermann, C., et al. (2011). Bone marrow CD169+ macrophages promote the retention of hematopoietic stem and progenitor cells in the mesenchymal stem cell niche. *Journal of Experimental Medicine*, *208*, 261–271.
Christov, C., Chretien, F., Abou-Khalil, R., Bassez, G., Vallet, G., Authier, F. J., et al. (2007). Muscle satellite cells and endothelial cells: Close neighbors and privileged partners. *Molecular Biology of the Cell*, *18*, 1397–1409.
Chwalinski, S., Potten, C. S., & Evans, G. (1988). Double labelling with bromodeoxyuridine and [3H]-thymidine of proliferative cells in small intestinal epithelium in steady state and after irradiation. *Cell and Tissue Kinetics*, *21*, 317–329.
Clavel, C., Grisanti, L., Zemla, R., Rezza, A., Barros, R., Sennett, R., et al. (2012). Sox2 in the dermal papilla niche controls hair growth by fine-tuning BMP signaling in differentiating hair shaft progenitors. *Developmental Cell*, *23*, 981–994.
Clevers, H. (2013). The intestinal crypt, a prototype stem cell compartment. *Cell*, *154*, 274–284.
Cobas, M., Wilson, A., Ernst, B., Mancini, S. J., MacDonald, H. R., Kemler, R., et al. (2004). Beta-catenin is dispensable for hematopoiesis and lymphopoiesis. *Journal of Experimental Medicine*, *199*, 221–229.
Cobrinik, D., Lee, M. H., Hannon, G., Mulligan, G., Bronson, R. T., Dyson, N., et al. (1996). Shared role of the pRB-related p130 and p107 proteins in limb development. *Genes and Development*, *10*, 1633–1644.
Conboy, I. M., Conboy, M. J., Smythe, G. M., & Rando, T. A. (2003). Notch-mediated restoration of regenerative potential to aged muscle. *Science*, *302*, 1575–1577.
Conboy, I. M., Conboy, M. J., Wagers, A. J., Girma, E. R., Weissman, I. L., & Rando, T. A. (2005). Rejuvenation of aged progenitor cells by exposure to a young systemic environment. *Nature*, *433*, 760–764.
Conboy, I. M., & Rando, T. A. (2002). The regulation of Notch signaling controls satellite cell activation and cell fate determination in postnatal myogenesis. *Developmental Cell*, *3*, 397–409.

Coskun, V., Wu, H., Blanchi, B., Tsao, S., Kim, K., Zhao, J., et al. (2008). CD133+ neural stem cells in the ependyma of mammalian postnatal forebrain. *Proceedings of the National Academy of Sciences of the United States of America*, *105*, 1026–1031.

Davidoff, M. S., Middendorff, R., Enikolopov, G., Riethmacher, D., Holstein, A. F., & Muller, D. (2004). Progenitor cells of the testosterone-producing Leydig cells revealed. *Journal of Cell Biology*, *167*, 935–944.

de Cuevas, M., & Matunis, E. L. (2011). The stem cell niche: Lessons from the Drosophila testis. *Development*, *138*, 2861–2869.

Ding, L., & Morrison, S. J. (2013). Haematopoietic stem cells and early lymphoid progenitors occupy distinct bone marrow niches. *Nature*, *495*, 231–235.

Ding, L., Saunders, T. L., Enikolopov, G., & Morrison, S. J. (2012). Endothelial and perivascular cells maintain haematopoietic stem cells. *Nature*, *481*, 457–462.

Engler, A. J., Sen, S., Sweeney, H. L., & Discher, D. E. (2006). Matrix elasticity directs stem cell lineage specification. *Cell*, *126*, 677–689.

Farin, H. F., Van Es, J. H., & Clevers, H. (2012). Redundant sources of Wnt regulate intestinal stem cells and promote formation of Paneth cells. *Gastroenterology*, *143*(1518–1529), e7.

Favaro, R., Valotta, M., Ferri, A. L., Latorre, E., Mariani, J., Giachino, C., et al. (2009). Hippocampal development and neural stem cell maintenance require Sox2-dependent regulation of Shh. *Nature Neuroscience*, *12*, 1248–1256.

Festa, E., Fretz, J., Berry, R., Schmidt, B., Rodeheffer, M., Horowitz, M., et al. (2011). Adipocyte lineage cells contribute to the skin stem cell niche to drive hair cycling. *Cell*, *146*, 761–771.

Fevr, T., Robine, S., Louvard, D., & Huelsken, J. (2007). Wnt/beta-catenin is essential for intestinal homeostasis and maintenance of intestinal stem cells. *Molecular and Cellular Biology*, *27*, 7551–7559.

Filatova, A., Acker, T., & Garvalov, B. K. (2013). The cancer stem cell niche(s): The crosstalk between glioma stem cells and their microenvironment. *Biochimica et Biophysica Acta*, *1830*, 2496–2508.

Fre, S., Huyghe, M., Mourikis, P., Robine, S., Louvard, D., & Artavanis-Tsakonas, S. (2005). Notch signals control the fate of immature progenitor cells in the intestine. *Nature*, *435*, 964–968.

Fre, S., Pallavi, S. K., Huyghe, M., Lae, M., Janssen, K. P., Robine, S., et al. (2009). Notch and Wnt signals cooperatively control cell proliferation and tumorigenesis in the intestine. *Proceedings of the National Academy of Sciences of the United States of America*, *106*, 6309–6314.

Fujisaki, J., Wu, J., Carlson, A. L., Silberstein, L., Putheti, P., Larocca, R., et al. (2011). In vivo imaging of Treg cells providing immune privilege to the haematopoietic stem-cell niche. *Nature*, *474*, 216–219.

Fukada, S., Uezumi, A., Ikemoto, M., Masuda, S., Segawa, M., Tanimura, N., et al. (2007). Molecular signature of quiescent satellite cells in adult skeletal muscle. *Stem Cells*, *25*, 2448–2459.

Furuyama, K., Kawaguchi, Y., Akiyama, H., Horiguchi, M., Kodama, S., Kuhara, T., et al. (2011). Continuous cell supply from a Sox9-expressing progenitor zone in adult liver, exocrine pancreas and intestine. *Nature Genetics*, *43*, 34–41.

Gilbert, P. M., Havenstrite, K. L., Magnusson, K. E., Sacco, A., Leonardi, N. A., Kraft, P., et al. (2010). Substrate elasticity regulates skeletal muscle stem cell self-renewal in culture. *Science*, *329*, 1078–1081.

Girgenrath, M., Weng, S., Kostek, C. A., Browning, B., Wang, M., Brown, S. A., et al. (2006). TWEAK, via its receptor Fn14, is a novel regulator of mesenchymal progenitor cells and skeletal muscle regeneration. *The EMBO Journal*, *25*, 5826–5839.

Greco, V., Chen, T., Rendl, M., Schober, M., Pasolli, H. A., Stokes, N., et al. (2009). A two-step mechanism for stem cell activation during hair regeneration. *Cell Stem Cell*, *4*, 155–169.

Greenbaum, A., Hsu, Y. M., Day, R. B., Schuettpelz, L. G., Christopher, M. J., Borgerding, J. N., et al. (2013). CXCL12 in early mesenchymal progenitors is required for haematopoietic stem-cell maintenance. *Nature*, *495*, 227–230.

Grisanti, L., Falciatori, I., Grasso, M., Dovere, L., Fera, S., Muciaccia, B., et al. (2009). Identification of spermatogonial stem cell subsets by morphological analysis and prospective isolation. *Stem Cells*, *27*, 3043–3052.

Gritti, A., Frolichsthal-Schoeller, P., Galli, R., Parati, E. A., Cova, L., Pagano, S. F., et al. (1999). Epidermal and fibroblast growth factors behave as mitogenic regulators for a single multipotent stem cell-like population from the subventricular region of the adult mouse forebrain. *The Journal of Neuroscience*, *19*, 3287–3297.

Hamra, F. K., Chapman, K. M., Nguyen, D., & Garbers, D. L. (2007). Identification of neuregulin as a factor required for formation of aligned spermatogonia. *The Journal of Biological Chemistry*, *282*, 721–730.

Haramis, A. P., Begthel, H., van den Born, M., van Es, J., Jonkheer, S., Offerhaus, G. J., et al. (2004). De novo crypt formation and juvenile polyposis on BMP inhibition in mouse intestine. *Science*, *303*, 1684–1686.

Haug, J. S., He, X. C., Grindley, J. C., Wunderlich, J. P., Gaudenz, K., Ross, J. T., et al. (2008). N-cadherin expression level distinguishes reserved versus primed states of hematopoietic stem cells. *Cell Stem Cell*, *2*, 367–379.

He, X. C., Zhang, J., Tong, W. G., Tawfik, O., Ross, J., Scoville, D. H., et al. (2004). BMP signaling inhibits intestinal stem cell self-renewal through suppression of Wnt-beta-catenin signaling. *Nature Genetics*, *36*, 1117–1121.

Heddleston, J. M., Li, Z., McLendon, R. E., Hjelmeland, A. B., & Rich, J. N. (2009). The hypoxic microenvironment maintains glioblastoma stem cells and promotes reprogramming towards a cancer stem cell phenotype. *Cell Cycle*, *8*, 3274–3284.

Horsley, V., Aliprantis, A. O., Polak, L., Glimcher, L. H., & Fuchs, E. (2008). NFATc1 balances quiescence and proliferation of skin stem cells. *Cell*, *132*, 299–310.

Hosokawa, K., Arai, F., Yoshihara, H., Iwasaki, H., Hembree, M., Yin, T., et al. (2010). Cadherin-based adhesion is a potential target for niche manipulation to protect hematopoietic stem cells in adult bone marrow. *Cell Stem Cell*, *6*, 194–198.

Hsu, Y. C., & Fuchs, E. (2012). A family business: Stem cell progeny join the niche to regulate homeostasis. *Nature Reviews Molecular Cell Biology*, *13*, 103–114.

Hsu, Y. C., Pasolli, H. A., & Fuchs, E. (2011). Dynamics between stem cells, niche, and progeny in the hair follicle. *Cell*, *144*, 92–105.

Hu, Q., Zhang, L., Wen, J., Wang, S., Li, M., Feng, R., et al. (2010). The EGF receptor-sox2-EGF receptor feedback loop positively regulates the self-renewal of neural precursor cells. *Stem Cells*, *28*, 279–286.

Huelsken, J., Vogel, R., Erdmann, B., Cotsarelis, G., & Birchmeier, W. (2001). beta-Catenin controls hair follicle morphogenesis and stem cell differentiation in the skin. *Cell*, *105*, 533–545.

Imayoshi, I., Sakamoto, M., Yamaguchi, M., Mori, K., & Kageyama, R. (2010). Essential roles of Notch signaling in maintenance of neural stem cells in developing and adult brains. *The Journal of Neuroscience*, *30*, 3489–3498.

Jackson, E. L., Garcia-Verdugo, J. M., Gil-Perotin, S., Roy, M., Quinones-Hinojosa, A., VandenBerg, S., et al. (2006). PDGFR alpha-positive B cells are neural stem cells in the adult SVZ that form glioma-like growths in response to increased PDGF signaling. *Neuron*, *51*, 187–199.

Kanatsu-Shinohara, M., Inoue, K., Takashima, S., Takehashi, M., Ogonuki, N., Morimoto, H., et al. (2012). Reconstitution of mouse spermatogonial stem cell niches in culture. *Cell Stem Cell*, *11*, 567–578.

Kanatsu-Shinohara, M., Miki, H., Inoue, K., Ogonuki, N., Toyokuni, S., Ogura, A., et al. (2005). Long-term culture of mouse male germline stem cells under serum-or feeder-free conditions. *Biology of Reproduction, 72*, 985–991.

Kanatsu-Shinohara, M., Takehashi, M., Takashima, S., Lee, J., Morimoto, H., Chuma, S., et al. (2008). Homing of mouse spermatogonial stem cells to germline niche depends on beta1-integrin. *Cell Stem Cell, 3*, 533–542.

Karpowicz, P., Willaime-Morawek, S., Balenci, L., DeVeale, B., Inoue, T., & van der Kooy, D. (2009). E-cadherin regulates neural stem cell self-renewal. *The Journal of Neuroscience, 29*, 3885–3896.

Katayama, Y., Battista, M., Kao, W. M., Hidalgo, A., Peired, A. J., Thomas, S. A., et al. (2006). Signals from the sympathetic nervous system regulate hematopoietic stem cell egress from bone marrow. *Cell, 124*, 407–421.

Kiel, M. J., Acar, M., Radice, G. L., & Morrison, S. J. (2009). Hematopoietic stem cells do not depend on N-cadherin to regulate their maintenance. *Cell Stem Cell, 4*, 170–179.

Kiel, M. J., Radice, G. L., & Morrison, S. J. (2007). Lack of evidence that hematopoietic stem cells depend on N-cadherin-mediated adhesion to osteoblasts for their maintenance. *Cell Stem Cell, 1*, 204–217.

Kim, T. H., Escudero, S., & Shivdasani, R. A. (2012). Intact function of Lgr5 receptor-expressing intestinal stem cells in the absence of Paneth cells. *Proceedings of the National Academy of Sciences of the United States of America, 109*, 3932–3937.

Kim, W. Y., Wang, X., Wu, Y., Doble, B. W., Patel, S., Woodgett, J. R., et al. (2009). GSK-3 is a master regulator of neural progenitor homeostasis. *Nature Neuroscience, 12*, 1390–1397.

Kippin, T. E., Martens, D. J., & van der Kooy, D. (2005). p21 loss compromises the relative quiescence of forebrain stem cell proliferation leading to exhaustion of their proliferation capacity. *Genes and Development, 19*, 756–767.

Kirstetter, P., Anderson, K., Porse, B. T., Jacobsen, S. E., & Nerlov, C. (2006). Activation of the canonical Wnt pathway leads to loss of hematopoietic stem cell repopulation and multilineage differentiation block. *Nature Immunology, 7*, 1048–1056.

Kligman, A. M. (1988). The comparative histopathology of male-pattern baldness and senescent baldness. *Clinics in Dermatology, 6*, 108–118.

Kobielak, K., Stokes, N., de la Cruz, J., Polak, L., & Fuchs, E. (2007). Loss of a quiescent niche but not follicle stem cells in the absence of bone morphogenetic protein signaling. *Proceedings of the National Academy of Sciences of the United States of America, 104*, 10063–10068.

Kojima, T., Hirota, Y., Ema, M., Takahashi, S., Miyoshi, I., Okano, H., et al. (2010). Subventricular zone-derived neural progenitor cells migrate along a blood vessel scaffold toward the post-stroke striatum. *Stem Cells, 28*, 545–554.

Korinek, V., Barker, N., Moerer, P., van Donselaar, E., Huls, G., Peters, P. J., et al. (1998). Depletion of epithelial stem-cell compartments in the small intestine of mice lacking Tcf-4. *Nature Genetics, 19*, 379–383.

Kosinski, C., Li, V. S., Chan, A. S., Zhang, J., Ho, C., Tsui, W. Y., et al. (2007). Gene expression patterns of human colon tops and basal crypts and BMP antagonists as intestinal stem cell niche factors. *Proceedings of the National Academy of Sciences of the United States of America, 104*, 15418–15423.

Kozar, K., Ciemerych, M. A., Rebel, V. I., Shigematsu, H., Zagozdzon, A., Sicinska, E., et al. (2004). Mouse development and cell proliferation in the absence of D-cyclins. *Cell, 118*, 477–491.

Kuang, S., Gillespie, M. A., & Rudnicki, M. A. (2008). Niche regulation of muscle satellite cell self-renewal and differentiation. *Cell Stem Cell, 2*, 22–31.

Kubota, H., Avarbock, M. R., & Brinster, R. L. (2004). Growth factors essential for self-renewal and expansion of mouse spermatogonial stem cells. *Proceedings of the National Academy of Sciences of the United States of America*, *101*, 16489–16494.

Kuhn, H. G., Winkler, J., Kempermann, G., Thal, L. J., & Gage, F. H. (1997). Epidermal growth factor and fibroblast growth factor-2 have different effects on neural progenitors in the adult rat brain. *The Journal of Neuroscience*, *17*, 5820–5829.

Kumano, K., Chiba, S., Kunisato, A., Sata, M., Saito, T., Nakagami-Yamaguchi, E., et al. (2003). Notch1 but not Notch2 is essential for generating hematopoietic stem cells from endothelial cells. *Immunity*, *18*, 699–711.

Kumano, K., Chiba, S., Shimizu, K., Yamagata, T., Hosoya, N., Saito, T., et al. (2001). Notch1 inhibits differentiation of hematopoietic cells by sustaining GATA-2 expression. *Blood*, *98*, 3283–3289.

Kunisato, A., Chiba, S., Nakagami-Yamaguchi, E., Kumano, K., Saito, T., Masuda, S., et al. (2003). HES-1 preserves purified hematopoietic stem cells ex vivo and accumulates side population cells in vivo. *Blood*, *101*, 1777–1783.

Kuwabara, T., Hsieh, J., Muotri, A., Yeo, G., Warashina, M., Lie, D. C., et al. (2009). Wnt-mediated activation of NeuroD1 and retro-elements during adult neurogenesis. *Nature Neuroscience*, *12*, 1097–1105.

Lachapelle, F., Avellana-Adalid, V., Nait-Oumesmar, B., & Baron-Van Evercooren, A. (2002). Fibroblast growth factor-2 (FGF-2) and platelet-derived growth factor AB (PDGF AB) promote adult SVZ-derived oligodendrogenesis in vivo. *Molecular and Cellular Neuroscience*, *20*, 390–403.

LeCouter, J. E., Kablar, B., Hardy, W. R., Ying, C., Megeney, L. A., May, L. L., et al. (1998). Strain-dependent myeloid hyperplasia, growth deficiency, and accelerated cell cycle in mice lacking the Rb-related p107 gene. *Molecular and Cellular Biology*, *18*, 7455–7465.

Lee, J., Hoi, C. S., Lilja, K. C., White, B. S., Lee, S. E., Shalloway, D., et al. (2013). Runx1 and p21 synergistically limit the extent of hair follicle stem cell quiescence in vivo. *Proceedings of the National Academy of Sciences of the United States of America*, *110*, 4634–4639.

Lee, J., & Tumbar, T. (2012). Hairy tale of signaling in hair follicle development and cycling. *Seminars in Cell and Developmental Biology*, *23*(8), 906–916.

Lim, D. A., Tramontin, A. D., Trevejo, J. M., Herrera, D. G., Garcia-Verdugo, J. M., & Alvarez-Buylla, A. (2000). Noggin antagonizes BMP signaling to create a niche for adult neurogenesis. *Neuron*, *28*, 713–726.

Lo Celso, C., Fleming, H. E., Wu, J. W., Zhao, C. X., Miake-Lye, S., Fujisaki, J., et al. (2009). Live-animal tracking of individual haematopoietic stem/progenitor cells in their niche. *Nature*, *457*, 92–96.

Lowry, W. E., Blanpain, C., Nowak, J. A., Guasch, G., Lewis, L., & Fuchs, E. (2005). Defining the impact of beta-catenin/Tcf transactivation on epithelial stem cells. *Genes and Development*, *19*, 1596–1611.

Lugert, S., Basak, O., Knuckles, P., Haussler, U., Fabel, K., Gotz, M., et al. (2010). Quiescent and active hippocampal neural stem cells with distinct morphologies respond selectively to physiological and pathological stimuli and aging. *Cell Stem Cell*, *6*, 445–456.

Malumbres, M., Sotillo, R., Santamaria, D., Galan, J., Cerezo, A., Ortega, S., et al. (2004). Mammalian cells cycle without the D-type cyclin-dependent kinases Cdk4 and Cdk6. *Cell*, *118*, 493–504.

Mauro, A. (1961). Satellite cell of skeletal muscle fibers. *Journal of Biophysical and Biochemical Cytology*, *9*, 493–495.

Mayack, S. R., Shadrach, J. L., Kim, F. S., & Wagers, A. J. (2010). Systemic signals regulate ageing and rejuvenation of blood stem cell niches. *Nature*, *463*, 495–500.

McCroskery, S., Thomas, M., Maxwell, L., Sharma, M., & Kambadur, R. (2003). Myostatin negatively regulates satellite cell activation and self-renewal. *Journal of Cell Biology, 162*, 1135–1147.
Mendez-Ferrer, S., Lucas, D., Battista, M., & Frenette, P. S. (2008). Haematopoietic stem cell release is regulated by circadian oscillations. *Nature, 452*, 442–447.
Mendez-Ferrer, S., Michurina, T. V., Ferraro, F., Mazloom, A. R., Macarthur, B. D., Lira, S. A., et al. (2010). Mesenchymal and haematopoietic stem cells form a unique bone marrow niche. *Nature, 466*, 829–834.
Meng, X., Lindahl, M., Hyvonen, M. E., Parvinen, M., de Rooij, D. G., Hess, M. W., et al. (2000). Regulation of cell fate decision of undifferentiated spermatogonia by GDNF. *Science, 287*, 1489–1493.
Ming, G. L., & Song, H. (2011). Adult neurogenesis in the mammalian brain: Significant answers and significant questions. *Neuron, 70*, 687–702.
Mira, H., Andreu, Z., Suh, H., Lie, D. C., Jessberger, S., Consiglio, A., et al. (2010). Signaling through BMPR-IA regulates quiescence and long-term activity of neural stem cells in the adult hippocampus. *Cell Stem Cell, 7*, 78–89.
Miyoshi, H., Ajima, R., Luo, C. T., Yamaguchi, T. P., & Stappenbeck, T. S. (2012). Wnt5a potentiates TGF-beta signaling to promote colonic crypt regeneration after tissue injury. *Science, 338*, 108–113.
Molofsky, A. V., He, S., Bydon, M., Morrison, S. J., & Pardal, R. (2005). Bmi-1 promotes neural stem cell self-renewal and neural development but not mouse growth and survival by repressing the p16Ink4a and p19Arf senescence pathways. *Genes and Development, 19*, 1432–1437.
Mourikis, P., Sambasivan, R., Castel, D., Rocheteau, P., Bizzarro, V., & Tajbakhsh, S. (2012). A critical requirement for notch signaling in maintenance of the quiescent skeletal muscle stem cell state. *Stem Cells, 30*, 243–252.
Myung, P. S., Takeo, M., Ito, M., & Atit, R. P. (2013). Epithelial Wnt ligand secretion is required for adult hair follicle growth and regeneration. *Journal of Investigative Dermatology, 133*, 31–41.
Nagano, M., Ryu, B. Y., Brinster, C. J., Avarbock, M. R., & Brinster, R. L. (2003). Maintenance of mouse male germ line stem cells in vitro. *Biology of Reproduction, 68*, 2207–2214.
Nagata, Y., Partridge, T. A., Matsuda, R., & Zammit, P. S. (2006). Entry of muscle satellite cells into the cell cycle requires sphingolipid signaling. *Journal of Cell Biology, 174*, 245–253.
Nam, H. S., & Benezra, R. (2009). High levels of Id1 expression define B1 type adult neural stem cells. *Cell Stem Cell, 5*, 515–526.
Naveiras, O., Nardi, V., Wenzel, P. L., Hauschka, P. V., Fahey, F., & Daley, G. Q. (2009). Bone-marrow adipocytes as negative regulators of the haematopoietic microenvironment. *Nature, 460*, 259–263.
Nichols, J., & Smith, A. (2012). Pluripotency in the embryo and in culture. *Cold Spring Harbor Perspectives in Biology, 4*, a008128.
Nomura, T., Goritz, C., Catchpole, T., Henkemeyer, M., & Frisen, J. (2010). EphB signaling controls lineage plasticity of adult neural stem cell niche cells. *Cell Stem Cell, 7*, 730–743.
Nowak, J. A., Polak, L., Pasolli, H. A., & Fuchs, E. (2008). Hair follicle stem cells are specified and function in early skin morphogenesis. *Cell Stem Cell, 3*, 33–43.
Nurmio, M., Kallio, J., Adam, M., Mayerhofer, A., Toppari, J., & Jahnukainen, K. (2012). Peritubular myoid cells have a role in postnatal testicular growth. *Spermatogenesis, 2*, 79–87.
Oatley, J. M., & Brinster, R. L. (2012). The germline stem cell niche unit in mammalian testes. *Physiological Reviews, 92*, 577–595.

Oatley, J. M., Oatley, M. J., Avarbock, M. R., Tobias, J. W., & Brinster, R. L. (2009). Colony stimulating factor 1 is an extrinsic stimulator of mouse spermatogonial stem cell self-renewal. *Development, 136*, 1191–1199.

Oatley, M. J., Racicot, K. E., & Oatley, J. M. (2011). Sertoli cells dictate spermatogonial stem cell niches in the mouse testis. *Biology of Reproduction, 84*, 639–645.

Omatsu, Y., Sugiyama, T., Kohara, H., Kondoh, G., Fujii, N., Kohno, K., et al. (2010). The essential functions of adipo-osteogenic progenitors as the hematopoietic stem and progenitor cell niche. *Immunity, 33*, 387–399.

Ono, Y., Boldrin, L., Knopp, P., Morgan, J. E., & Zammit, P. S. (2010). Muscle satellite cells are a functionally heterogeneous population in both somite-derived and branchiomeric muscles. *Developmental Biology, 337*, 29–41.

Ono, Y., Masuda, S., Nam, H. S., Benezra, R., Miyagoe-Suzuki, Y., & Takeda, S. (2012). Slow-dividing satellite cells retain long-term self-renewal ability in adult muscle. *Journal of Cell Science, 125*, 1309–1317.

Oshimori, N., & Fuchs, E. (2012). Paracrine TGF-beta signaling counterbalances BMP-mediated repression in hair follicle stem cell activation. *Cell Stem Cell, 10*, 63–75.

Otto, A., Schmidt, C., Luke, G., Allen, S., Valasek, P., Muntoni, F., et al. (2008). Canonical Wnt signalling induces satellite-cell proliferation during adult skeletal muscle regeneration. *Journal of Cell Science, 121*, 2939–2950.

Pannerec, A., Marazzi, G., & Sassoon, D. (2012). Stem cells in the hood: The skeletal muscle niche. *Trends in Molecular Medicine, 18*, 599–606.

Pellegrini, M., Grimaldi, P., Rossi, P., Geremia, R., & Dolci, S. (2003). Developmental expression of BMP4/ALK3/SMAD5 signaling pathway in the mouse testis: A potential role of BMP4 in spermatogonia differentiation. *Journal of Cell Science, 116*, 3363–3372.

Pietras, E. M., Warr, M. R., & Passegue, E. (2011). Cell cycle regulation in hematopoietic stem cells. *Journal of Cell Biology, 195*, 709–720.

Plikus, M. V., Baker, R. E., Chen, C. C., Fare, C., de la Cruz, D., Andl, T., et al. (2011). Self-organizing and stochastic behaviors during the regeneration of hair stem cells. *Science, 332*, 586–589.

Plikus, M. V., Mayer, J. A., de la Cruz, D., Baker, R. E., Maini, P. K., Maxson, R., et al. (2008). Cyclic dermal BMP signalling regulates stem cell activation during hair regeneration. *Nature, 451*, 340–344.

Potocnik, A. J., Brakebusch, C., & Fassler, R. (2000). Fetal and adult hematopoietic stem cells require beta1 integrin function for colonizing fetal liver, spleen, and bone marrow. *Immunity, 12*, 653–663.

Powell, A. E., Wang, Y., Li, Y., Poulin, E. J., Means, A. L., Washington, M. K., et al. (2012). The pan-ErbB negative regulator Lrig1 is an intestinal stem cell marker that functions as a tumor suppressor. *Cell, 149*, 146–158.

Puolakkainen, P. A., Ranchalis, J. E., Gombotz, W. R., Hoffman, A. S., Mumper, R. J., & Twardzik, D. R. (1994). Novel delivery system for inducing quiescence in intestinal stem cells in rats by transforming growth factor beta 1. *Gastroenterology, 107*, 1319–1326.

Qian, Y., Liu, S., Guan, Y., Pan, H., Guan, X., Qiu, Z., et al. (2013). Lgr4-mediated Wnt/beta-catenin signaling in peritubular myoid cells is essential for spermatogenesis. *Development, 140*, 1751–1761.

Qian, H., Tryggvason, K., Jacobsen, S. E., & Ekblom, M. (2006). Contribution of alpha6 integrins to hematopoietic stem and progenitor cell homing to bone marrow and collaboration with alpha4 integrins. *Blood, 107*, 3503–3510.

Raaijmakers, M. H., Mukherjee, S., Guo, S., Zhang, S., Kobayashi, T., Schoonmaker, J. A., et al. (2010). Bone progenitor dysfunction induces myelodysplasia and secondary leukaemia. *Nature, 464*, 852–857.

Raghavan, S., Bauer, C., Mundschau, G., Li, Q., & Fuchs, E. (2000). Conditional ablation of beta1 integrin in skin. Severe defects in epidermal proliferation, basement membrane formation, and hair follicle invagination. *Journal of Cell Biology, 150*, 1149–1160.

Ratajczak, M. Z., Majka, M., Kucia, M., Drukala, J., Pietrzkowski, Z., Peiper, S., et al. (2003). Expression of functional CXCR4 by muscle satellite cells and secretion of SDF-1 by muscle-derived fibroblasts is associated with the presence of both muscle progenitors in bone marrow and hematopoietic stem/progenitor cells in muscles. *Stem Cells, 21*, 363–371.

Relaix, F., & Zammit, P. S. (2012). Satellite cells are essential for skeletal muscle regeneration: The cell on the edge returns centre stage. *Development, 139*, 2845–2856.

Renault, V. M., Rafalski, V. A., Morgan, A. A., Salih, D. A., Brett, J. O., Webb, A. E., et al. (2009). FoxO3 regulates neural stem cell homeostasis. *Cell Stem Cell, 5*, 527–539.

Rendl, M., Polak, L., & Fuchs, E. (2008). BMP signaling in dermal papilla cells is required for their hair follicle-inductive properties. *Genes & Development, 22*, 543–557.

Reya, T., & Clevers, H. (2005). Wnt signalling in stem cells and cancer. *Nature, 434*, 843–850.

Reya, T., Duncan, A. W., Ailles, L., Domen, J., Scherer, D. C., Willert, K., et al. (2003). A role for Wnt signalling in self-renewal of haematopoietic stem cells. *Nature, 423*, 409–414.

Rezza, A., Skah, S., Roche, C., Nadjar, J., Samarut, J., & Plateroti, M. (2010). The overexpression of the putative gut stem cell marker Musashi-1 induces tumorigenesis through Wnt and Notch activation. *Journal of Cell Science, 123*, 3256–3265.

Rhee, H., Polak, L., & Fuchs, E. (2006). Lhx2 maintains stem cell character in hair follicles. *Science, 312*, 1946–1949.

Riccio, O., van Gijn, M. E., Bezdek, A. C., Pellegrinet, L., van Es, J. H., Zimber-Strobl, U., et al. (2008). Loss of intestinal crypt progenitor cells owing to inactivation of both Notch1 and Notch2 is accompanied by derepression of CDK inhibitors p27Kip1 and p57Kip2. *EMBO Reports, 9*, 377–383.

Rizk, P., & Barker, N. (2012). Gut stem cells in tissue renewal and disease: Methods, markers, and myths. *Wiley Interdisciplinary Reviews Systems Biology and Medicine, 4*, 475–496.

Rodilla, V., Villanueva, A., Obrador-Hevia, A., Robert-Moreno, A., Fernandez-Majada, V., Grilli, A., et al. (2009). Jagged1 is the pathological link between Wnt and Notch pathways in colorectal cancer. *Proceedings of the National Academy of Sciences of the United States of America, 106*, 6315–6320.

Rompolas, P., Deschene, E. R., Zito, G., Gonzalez, D. G., Saotome, I., Haberman, A. M., et al. (2012). Live imaging of stem cell and progeny behaviour in physiological hair-follicle regeneration. *Nature, 487*, 496–499.

Roobrouck, V. D., Vanuytsel, K., & Verfaillie, C. M. (2011). Concise review: Culture mediated changes in fate and/or potency of stem cells. *Stem Cells, 29*, 583–589.

Ryu, B. Y., Orwig, K. E., Oatley, J. M., Avarbock, M. R., & Brinster, R. L. (2006). Effects of aging and niche microenvironment on spermatogonial stem cell self-renewal. *Stem Cells, 24*, 1505–1511.

Sakamori, R., Das, S., Yu, S., Feng, S., Stypulkowski, E., Guan, Y., et al. (2012). Cdc42 and Rab8a are critical for intestinal stem cell division, survival, and differentiation in mice. *The Journal of Clinical Investigation, 122*, 1052–1065.

Sangiorgi, E., & Capecchi, M. R. (2008). Bmi1 is expressed in vivo in intestinal stem cells. *Nature Genetics, 40*, 915–920.

Sassoli, C., Formigli, L., Bini, F., Tani, A., Squecco, R., Battistini, C., et al. (2011). Effects of S1P on skeletal muscle repair/regeneration during eccentric contraction. *Journal of Cellular and Molecular Medicine, 15*, 2498–2511.

Sato, T., van Es, J. H., Snippert, H. J., Stange, D. E., Vries, R. G., van den Born, M., et al. (2011). Paneth cells constitute the niche for Lgr5 stem cells in intestinal crypts. *Nature, 469*, 415–418.

Sato, T., Vries, R. G., Snippert, H. J., van de Wetering, M., Barker, N., Stange, D. E., et al. (2009). Single Lgr5 stem cells build crypt-villus structures in vitro without a mesenchymal niche. *Nature, 459*, 262–265.

Sawamoto, K., Wichterle, H., Gonzalez-Perez, O., Cholfin, J. A., Yamada, M., Spassky, N., et al. (2006). New neurons follow the flow of cerebrospinal fluid in the adult brain. *Science, 311*, 629–632.

Schofield, R. (1978). The relationship between the spleen colony-forming cell and the haemopoietic stem cell. *Blood Cells, 4*, 7–25.

Schreiber, T. D., Steinl, C., Essl, M., Abele, H., Geiger, K., Muller, C. A., et al. (2009). The integrin alpha9beta1 on hematopoietic stem and progenitor cells: Involvement in cell adhesion, proliferation and differentiation. *Haematologica, 94*, 1493–1501.

Sennett, R., & Rendl, M. (2012). Mesenchymal-epithelial interactions during hair follicle morphogenesis and cycling. *Seminars in Cell and Developmental Biology, 23*(8), 917–927.

Sherwood, R. I., Christensen, J. L., Conboy, I. M., Conboy, M. J., Rando, T. A., Weissman, I. L., et al. (2004). Isolation of adult mouse myogenic progenitors: Functional heterogeneity of cells within and engrafting skeletal muscle. *Cell, 119*, 543–554.

Shinohara, T., Avarbock, M. R., & Brinster, R. L. (1999). beta1- and alpha6-integrin are surface markers on mouse spermatogonial stem cells. *Proceedings of the National Academy of Sciences of the United States of America, 96*, 5504–5509.

Singbrant, S., Karlsson, G., Ehinger, M., Olsson, K., Jaako, P., Miharada, K., et al. (2010). Canonical BMP signaling is dispensable for hematopoietic stem cell function in both adult and fetal liver hematopoiesis, but essential to preserve colon architecture. *Blood, 115*, 4689–4698.

Smith, N. R., Davies, P. S., Silk, A. D., & Wong, M. H. (2012). Epithelial and mesenchymal contribution to the niche: A safeguard for intestinal stem cell homeostasis. *Gastroenterology, 143*, 1426–1430.

Sneddon, J. B., Zhen, H. H., Montgomery, K., van de Rijn, M., Tward, A. D., West, R., et al. (2006). Bone morphogenetic protein antagonist gremlin 1 is widely expressed by cancer-associated stromal cells and can promote tumor cell proliferation. *Proceedings of the National Academy of Sciences of the United States of America, 103*, 14842–14847.

Suh, H., Consiglio, A., Ray, J., Sawai, T., D'Amour, K. A., & Gage, F. H. (2007). In vivo fate analysis reveals the multipotent and self-renewal capacities of Sox2+ neural stem cells in the adult hippocampus. *Cell Stem Cell, 1*, 515–528.

Sun, G., Yu, R. T., Evans, R. M., & Shi, Y. (2007). Orphan nuclear receptor TLX recruits histone deacetylases to repress transcription and regulate neural stem cell proliferation. *Proceedings of the National Academy of Sciences of the United States of America, 104*, 15282–15287.

Szulwach, K. E., Li, X., Smrt, R. D., Li, Y., Luo, Y., Lin, L., et al. (2010). Cross talk between microRNA and epigenetic regulation in adult neurogenesis. *Journal of Cell Biology, 189*, 127 141.

Tadokoro, Y., Yomogida, K., Ohta, H., Tohda, A., & Nishimune, Y. (2002). Homeostatic regulation of germinal stem cell proliferation by the GDNF/FSH pathway. *Mechanisms of Development, 113*, 29–39.

Taichman, R. S., Reilly, M. J., & Emerson, S. G. (1996). Human osteoblasts support human hematopoietic progenitor cells in vitro bone marrow cultures. *Blood, 87*, 518–524.

Takeda, N., Jain, R., Leboeuf, M. R., Padmanabhan, A., Wang, Q., Li, L., et al. (2013). Hopx expression defines a subset of multipotent hair follicle stem cells and a progenitor population primed to give rise to K6+ niche cells. *Development, 140*, 1655–1664.

Takeda, N., Jain, R., LeBoeuf, M. R., Wang, Q., Lu, M. M., & Epstein, J. A. (2011). Interconversion between intestinal stem cell populations in distinct niches. *Science, 334*, 1420–1424.

Takizawa, H., Regoes, R. R., Boddupalli, C. S., Bonhoeffer, S., & Manz, M. G. (2011). Dynamic variation in cycling of hematopoietic stem cells in steady state and inflammation. *Journal of Experimental Medicine, 208*, 273–284.

Talos, F., Abraham, A., Vaseva, A. V., Holembowski, L., Tsirka, S. E., Scheel, A., et al. (2010). p73 is an essential regulator of neural stem cell maintenance in embryonal and adult CNS neurogenesis. *Cell Death and Differentiation, 17*, 1816–1829.

Tanaka, M., & Miyajima, A. (2012). Identification and isolation of adult liver stem/progenitor cells. *Methods in Molecular Biology, 826*, 25–32.

Tanimura, S., Tadokoro, Y., Inomata, K., Binh, N. T., Nishie, W., Yamazaki, S., et al. (2011). Hair follicle stem cells provide a functional niche for melanocyte stem cells. *Cell Stem Cell, 8*, 177–187.

Tatsumi, R., Liu, X., Pulido, A., Morales, M., Sakata, T., Dial, S., et al. (2006). Satellite cell activation in stretched skeletal muscle and the role of nitric oxide and hepatocyte growth factor. *American Journal of Physiology Cell Physiology, 290*, C1487–C1494.

Thomas, A., Stein, C. K., Gentile, T. C., & Shah, C. M. (2010). Isolated CNS relapse of CML after bone marrow transplantation. *Leukemia Research, 34*, e113–e114.

Tian, H., Biehs, B., Warming, S., Leong, K. G., Rangell, L., Klein, O. D., et al. (2011). A reserve stem cell population in small intestine renders Lgr5-positive cells dispensable. *Nature, 478*, 255–259.

Tokuda, M., Kadokawa, Y., Kurahashi, H., & Marunouchi, T. (2007). CDH1 is a specific marker for undifferentiated spermatogonia in mouse testes. *Biology of Reproduction, 76*, 130–141.

Trueb, R. M. (2006). Pharmacologic interventions in aging hair. *Clinical Interventions in Aging, 1*, 121–129.

Tumbar, T., Guasch, G., Greco, V., Blanpain, C., Lowry, W. E., Rendl, M., et al. (2004). Defining the epithelial stem cell niche in skin. *Science, 303*, 359–363.

Urciuolo, A., Quarta, M., Morbidoni, V., Gattazzo, F., Molon, S., Grumati, P., et al. (2013). Collagen VI regulates satellite cell self-renewal and muscle regeneration. *Nature Communications, 4*, 1964.

van der Flier, L. G., van Gijn, M. E., Hatzis, P., Kujala, P., Haegebarth, A., Stange, D. E., et al. (2009). Transcription factor achaete scute-like 2 controls intestinal stem cell fate. *Cell, 136*, 903–912.

van Es, J. H., Haegebarth, A., Kujala, P., Itzkovitz, S., Koo, B. K., Boj, S. F., et al. (2012). A critical role for the Wnt effector Tcf4 in adult intestinal homeostatic self-renewal. *Molecular and Cellular Biology, 32*, 1918–1927.

van Es, J. H., van Gijn, M. E., Riccio, O., van den Born, M., Vooijs, M., Begthel, H., et al. (2005). Notch/gamma-secretase inhibition turns proliferative cells in intestinal crypts and adenomas into goblet cells. *Nature, 435*, 959–963.

Vasioukhin, V., Bauer, C., Degenstein, L., Wise, B., & Fuchs, E. (2001). Hyperproliferation and defects in epithelial polarity upon conditional ablation of alpha-catenin in skin. *Cell, 104*, 605–617.

Verhoeven, G., & Cailleau, J. (1988). Testicular peritubular cells secrete a protein under androgen control that inhibits induction of aromatase activity in Sertoli cells. *Endocrinology, 123*, 2100–2110.

Vermeulen, L., De Sousa, E. M. F., van der Heijden, M., Cameron, K., de Jong, J. H., Borovski, T., et al. (2010). Wnt activity defines colon cancer stem cells and is regulated by the microenvironment. *Nature Cell Biology, 12*, 468–476.

Viatour, P., Somervaille, T. C., Venkatasubrahmanyam, S., Kogan, S., McLaughlin, M. E., Weissman, I. L., et al. (2008). Hematopoietic stem cell quiescence is maintained by compound contributions of the retinoblastoma gene family. *Cell Stem Cell, 3*, 416–428.

Vidal, V. P., Chaboissier, M. C., Lutzkendorf, S., Cotsarelis, G., Mill, P., Hui, C. C., et al. (2005). Sox9 is essential for outer root sheath differentiation and the formation of the hair stem cell compartment. *Current Biology, 15*, 1340–1351.

Visnjic, D., Kalajzic, Z., Rowe, D. W., Katavic, V., Lorenzo, J., & Aguila, H. L. (2004). Hematopoiesis is severely altered in mice with an induced osteoblast deficiency. *Blood, 103*, 3258–3264.

Wagers, A. J., & Conboy, I. M. (2005). Cellular and molecular signatures of muscle regeneration: Current concepts and controversies in adult myogenesis. *Cell, 122*, 659–667.

Walkley, C. R., & Orkin, S. H. (2006). Rb is dispensable for self-renewal and multilineage differentiation of adult hematopoietic stem cells. *Proceedings of the National Academy of Sciences of the United States of America, 103*, 9057–9062.

Wang, R., Chadalavada, K., Wilshire, J., Kowalik, U., Hovinga, K. E., Geber, A., et al. (2010). Glioblastoma stem-like cells give rise to tumour endothelium. *Nature, 468*, 829–833.

Wang, Y. Z., Yamagami, T., Gan, Q., Wang, Y., Zhao, T., Hamad, S., et al. (2011). Canonical Wnt signaling promotes the proliferation and neurogenesis of peripheral olfactory stem cells during postnatal development and adult regeneration. *Journal of Cell Science, 124*, 1553–1563.

Willert, K., Brown, J. D., Danenberg, E., Duncan, A. W., Weissman, I. L., Reya, T., et al. (2003). Wnt proteins are lipid-modified and can act as stem cell growth factors. *Nature, 423*, 448–452.

Wilson, A., Laurenti, E., Oser, G., van der Wath, R. C., Blanco-Bose, W., Jaworski, M., et al. (2008). Hematopoietic stem cells reversibly switch from dormancy to self-renewal during homeostasis and repair. *Cell, 135*, 1118–1129.

Wilson, A., & Trumpp, A. (2006). Bone-marrow haematopoietic-stem-cell niches. *Nature Reviews Immunology, 6*, 93–106.

Winkler, I. G., Sims, N. A., Pettit, A. R., Barbier, V., Nowlan, B., Helwani, F., et al. (2010). Bone marrow macrophages maintain hematopoietic stem cell (HSC) niches and their depletion mobilizes HSCs. *Blood, 116*, 4815–4828.

Wozniak, A. C., & Anderson, J. E. (2007). Nitric oxide-dependence of satellite stem cell activation and quiescence on normal skeletal muscle fibers. *Developmental Dynamics, 236*, 240–250.

Xie, Y., Yin, T., Wiegraebe, W., He, X. C., Miller, D., Stark, D., et al. (2009). Detection of functional haematopoietic stem cell niche using real-time imaging. *Nature, 457*, 97–101.

Yamazaki, S., Ema, H., Karlsson, G., Yamaguchi, T., Miyoshi, H., Shioda, S., et al. (2011). Nonmyelinating Schwann cells maintain hematopoietic stem cell hibernation in the bone marrow niche. *Cell, 147*, 1146–1158.

Yamazaki, S., Iwama, A., Takayanagi, S., Eto, K., Ema, H., & Nakauchi, H. (2009). TGF-beta as a candidate bone marrow niche signal to induce hematopoietic stem cell hibernation. *Blood, 113*, 1250–1256.

Yin, H., Price, F., & Rudnicki, M. A. (2013). Satellite cells and the muscle stem cell niche. *Physiological Reviews, 93*, 23–67.

Yoshida, S., Sukeno, M., & Nabeshima, Y. (2007). A vasculature-associated niche for undifferentiated spermatogonia in the mouse testis. *Science, 317*, 1722–1726.

Young, P., Boussadia, O., Halfter, H., Grose, R., Berger, P., Leone, D. P., et al. (2003). E-cadherin controls adherens junctions in the epidermis and the renewal of hair follicles. *The EMBO Journal, 22*, 5723–5733.

Zhang, Y. V., Cheong, J., Ciapurin, N., McDermitt, D. J., & Tumbar, T. (2009). Distinct self-renewal and differentiation phases in the niche of infrequently dividing hair follicle stem cells. *Cell Stem Cell, 5*, 267–278.

Zhang, X., Ebata, K. T., Robaire, B., & Nagano, M. C. (2006). Aging of male germ line stem cells in mice. *Biology of Reproduction, 74*, 119–124.
Zhang, J., Niu, C., Ye, L., Huang, H., He, X., Tong, W. G., et al. (2003). Identification of the haematopoietic stem cell niche and control of the niche size. *Nature, 425*, 836–841.
Zhang, L., Stokes, N., Polak, L., & Fuchs, E. (2011). Specific microRNAs are preferentially expressed by skin stem cells to balance self-renewal and early lineage commitment. *Cell Stem Cell, 8*.
Zhu, T. S., Costello, M. A., Talsma, C. E., Flack, C. G., Crowley, J. G., Hamm, L. L., et al. (2011). Endothelial cells create a stem cell niche in glioblastoma by providing NOTCH ligands that nurture self-renewal of cancer stem-like cells. *Cancer Research, 71*, 6061–6072.
Zhu, J., Garrett, R., Jung, Y., Zhang, Y., Kim, N., Wang, J., et al. (2007). Osteoblasts support B-lymphocyte commitment and differentiation from hematopoietic stem cells. *Blood, 109*, 3706–3712.

CHAPTER THIRTEEN

Cancer Stem Cells, Pluripotency, and Cellular Heterogeneity: A WNTer Perspective

Yaser Atlasi, Leendert Looijenga, Riccardo Fodde[1]
Department of Pathology, Josephine Nefkens Institute, Erasmus MC, Rotterdam, The Netherlands
[1]Corresponding author: e-mail address: r.fodde@erasmusmc.nl

Contents

Abstract

Cancer stem cells (CSCs) are thought to represent the "beating heart" of malignant growth as they continuously fuel tumors through their ability to self-renew and differentiate. Moreover, they are also believed to underlie malignant behavior, local invasion, and metastasis in distal organ sites upon reversible epithelial-to-mesenchymal transitions (EMTs). Nevertheless, the CSC concept has been the object of controversy, mainly due to the absence of robust operational definitions and to the lack of consistency in the use of the often incorrect nomenclature employed to refer to these cells. Notwithstanding the controversies, it is now generally accepted that primary cancers are organized in hierarchical fashion with neoplastic stem-like cells able to give rise to new CSCs and to more committed malignant cells. Notably, these hierarchical structures are not unidirectional, but are rather characterized by a more dynamic equilibrium where stem-like and

Current Topics in Developmental Biology, Volume 107
ISSN 0070-2153
http://dx.doi.org/10.1016/B978-0-12-416022-4.00013-5

more committed cancer cells transit from one meta-state to the other partly because of cues from the microenvironment (niche), but also because of intrinsic and yet incompletely understood characteristics in the activation/silencing of specific signal transduction pathways. Here, we will focus on the Wnt/β-catenin signaling pathway as one of the major regulator of stemness in homeostasis and cancer, and on germ cell tumors as the type of malignancy that most closely mimics normal embryonic development and as such serve as a unique model to study the role of stem cells in neoplasia.

1. CANCER STEM CELLS: AN OVERVIEW

1.1. Historical perspective

The concept of cancer stem cells (CSCs) is based on the assumption that malignancies are organized in hierarchical fashion with the CSCs representing the tumorigenic subpopulation endowed with self-renewal to preserve their identity and with differentiation capacity to give rise to more differentiated bulk cells with diminished tumor-propagating potential. From a different perspective, the CSC model implies that the intratumor morphological and phenotypic heterogeneity observed throughout a very broad spectrum of cancer types reflects the functional diversity among malignant cells with only a subset of them capable of fuelling tumor growth by giving rise to more committed bulk cells (through differentiation) and, at the same time, by maintaining the CSC population (through self-renewal).

Historically, the CSC concept is not a novel one as already in the 1960s it became clear that in some germ cell tumors (GCTs) multipotent embryonic cancer cells can clonally give rise to teratocarcinomas encompassing all three germ layers (Kleinsmith & Pierce, 1964; Pierce & Dixon, 1959). In fact, studies on teratocarcinomas (Jackson & Brues, 1941) and medulloblastomas (Eisenhardt & Cushing, 1930) dating back to 1930–1940 observed histological similarities between embryonic and tumor tissues and formed the basis of the "embryonal rest" theory that postulates that malignancies are originated from cells with features of those of the developing embryo. These and other studies led Pierce and Speers to propose a view of tumors as "*caricatures*" of the process of self-renewal and differentiation (Pierce & Speers, 1988). However, the formal proof of the existence of neoplastic stem-like cells only came by combining flow cytometry with xenotransplantation of tumor cells in immune-incompetent hosts, thus allowing the physical separation of phenotypically distinct subpopulations of live cancer cells and their *in vivo* functional analysis, namely, their capacity to propagate and

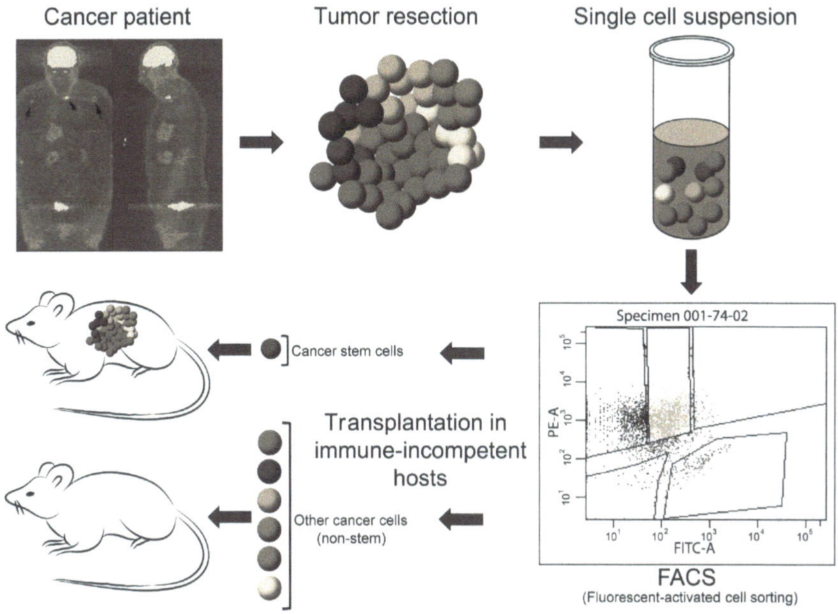

Figure 13.1 Schematic representation of the limited dilution transplantation assays coupled with fluorescent-activated cell sorting (FACS) applied to enrich the CSC subpopulation (red cells) from within a heterogeneous tumor composed of different more committed (i.e., nonstem) lineages. Upon transplantation at limited cellular dilutions, only the CSCs are capable to recapitulate the primary lesion derived from the patient. (See color plate.)

recapitulate the malignancy they are derived from upon transplantation into host immune-incompetent animals (Fig. 13.1). This was first demonstrated by John Dick and colleagues with human acute myeloid leukemia (AML) (Bonnet & Dick, 1997; Lapidot et al., 1994), and subsequently in solid malignancies such as breast (Al-Hajj, Wicha, Benito-Hernandez, Morrison, & Clarke, 2003) and brain cancers (Singh et al., 2003, 2004), and many others (for a comprehensive table see Visvader & Lindeman, 2008). For a more complete and detailed timeline of the scientific milestones leading to the current CSC concept, we refer the reader to the historical perspective by Nguyen, Vanner, Dirks, and Eaves (2012).

More definitive and conclusive experimental evidence for the CSC model has been provided only in 2012 by lineage tracing analysis of mouse models for glioma (Chen et al., 2012), skin tumors (Driessens, Beck, Caauwe, Simons, & Blanpain, 2012), and intestinal adenomas

(Schepers et al., 2012). Here, specific tumor cell types are genetically labeled which allows the "tracing" of the fate of their descendants in their natural *in vivo* context, thus bypassing the need for transplantation assays in immunocompromised hosts. These important studies confirmed that specific cancer cells can give rise to all types of differentiated progenies which constitute the tumor mass.

1.2. CSC definitions and assays

Notwithstanding the above-mentioned studies that led to its delineation and experimental confirmation, the CSC concept has been the object of many controversies and animated discussions on its validity and applicability to a broad spectrum of cancer types. Much of these controversies however, seem to arise from the lack of clear-cut conceptual and operational definitions, and of consistency in the terms used for these cells (Valent et al., 2012). In general, the most simplistic and comprehensive way to descriptively define the CSC concept is that, just like normal tissues, malignant growths are organized in a hierarchical fashion with neoplastic stem cells endowed with self-renewal and differentiation and capable of giving rise to more committed progenitors, but also to new CSCs thus continuously fuelling tumor growth. As normal stem cells are also defined by their capacity to self-renew and differentiate, it is plausible to think that the finely tuned homeostatic equilibrium between self-renewal and differentiation that characterize normal adult stem cell niches is altered within the tumor mass by subtle changes in the rate of symmetric versus asymmetric cell divisions, cell migration, differentiation, and apoptosis. Accordingly, many tumor-initiating mutations affect members of signaling pathways (e.g., Wnt, Notch, Sonic Hedgehog (Shh)) known to play critical roles in the regulation of self-renewal and differentiation in adult stem cell niches (Reya, Morrison, Clarke, & Weissman, 2001) (see below).

When it comes to a more operational definition of CSCs, it is not surprising that the same limiting dilution (xeno)transplantation assay that form the very basis of the CSC concept, also represents a very important tool. However, the large variability in the experimental implementation of this assay (e.g., enzymatic and mechanic dissociation protocols, sorting conditions, use of extracellular matrix and/or cells from the stromal microenvironment, orthotopic vs. subcutaneous transplantation sites, degree of immune deficiency of the recipient hosts, etc.) calls for the establishment of standardized criteria to operationally define CSCs in human malignancies.

Also, the increasing availability of mouse models for human cancers allows performing these transplantation assays in syngeneic hosts thus overcoming the problems linked to the use of partially (NOD/SCID) or highly (NOD/SCID IL2R$\gamma^{-/-}$) immune-deficient recipient mice (Quintana et al., 2008). The development of more "humanized" recipient animals capable of better re-creating the human microenvironment and immune modulation is likely to improve the reliability of xenotransplantation assays and their capacity to mimic the behavior of CSCs in the cancer patient (Lan, Tonomura, Shimizu, Wang, & Yang, 2006; Strowig et al., 2011; Wunderlich et al., 2010). Alternatively, the self-renewal capacity of CSCs can be assayed *in vitro* by sphere formation in serum-free and non-adherent culture conditions (Ponti et al., 2005; Singh et al., 2004). Also in this case, the great variability of the culture conditions employed and the unpredictable behavior of stem-like cells outside the "natural" 3-D environment of the tumor from which they are derived, suggest extreme caution when drawing conclusions of any clinical relevance from these assays.

Ideally, genetic lineage tracing should be considered as the "smoking gun" type of evidence for the assessment of the CSC status as previously shown (Chen et al., 2012; Driessens et al., 2012; Schepers et al., 2012). However, this approach is not easily applicable to human cancer cells, is feasible when appropriate murine models of human malignancies are available, and its experimental complexity make it unsuitable as a routine operational assay. In view of the current status quo, limiting dilution and serial (xeno) transplantation and sphere formation in serum-free and non-adherent culture still represent the most feasible, though imperfect, *in vivo* and *in vitro* assays for the operational definition of CSCs. As these protocols mainly assess the capacity of tumor cells to self-renewal and differentiate, the two main features of "normal" stem cells, the term *cancer* (or *neoplastic*) *stem cells* seems appropriate and, in our opinion, should be preferred to other previously employed terms (e.g., tumor-initiating cells or TICs) which are often incorrect and misleading (TICs do not initiate the cancer but rather propagate it and can be confused with the cells of origin of the tumor).

Given the limitations and caveats of the above assays, it is not surprising that much skepticism has been raised on the validity of the CSC model. First, the unidirectional nature of the cellular hierarchy in cancer has been challenged by studies from the Weinberg laboratory showing that fully differentiated breast cancer cells can, upon EMT induction, dedifferentiate, acquire mesenchymal traits, and express stem cell markers (Chaffer et al., 2011; Mani et al., 2008), a phenomenon that apparently also occurs in established (immortalized) cancer cell lines (Gupta et al., 2011). Although these observations have been exploited to criticize the CSC model (there is no hierarchy if different tumorigenic states can reversibly interconvert), the very fact that dedifferentiation also occurs in normal tissues strengthens the parallel drawn between the hierarchical organization of normal and cancer tissues. In view of the increasing body of evidence pointing at the unexpected plasticity (Takahashi & Yamanaka, 2006) of committed and even fully differentiated cells in normal physiology (Knapp & Tanaka, 2012), and the role played by EMTs (Thiery, Acloque, Huang, & Nieto, 2009) in these dedifferentiation and reprogramming processes, it is highly conceivable that similar phenomena also occur in cancer (Mani et al., 2008) and that such normal and neoplastic cellular hierarchies are substantially more dynamic (i.e., bidirectional) than originally thought in the dogmatic view of differentiation as *a way of no return.*

A second line of criticism toward the CSC model relies on the relative abundance of stem-like neoplastic cells in specific cancer types such as human melanoma (Held et al., 2010; Quintana et al., 2008), lymphomas and leukemia's of mouse origin (Kelly, Dakic, Adams, Nutt, & Strasser, 2007), and in murine malignant peripheral nerve sheath tumors (Buchstaller, McKeever, & Morrison, 2012). Even though the melanoma studies clearly indicate that limiting dilution transplantation assays in immune-incompetent hosts may overestimate the presence of CSCs within tumors, the very fact that the fraction of tumorigenic cells may in some cases even represent the majority of the bulk cancer cells is *per se* not in contradiction with the hierarchical model. For centuries, pathologists have been classifying tumors as low- versus high-grade with reference to the well- or poorly differentiated status of the neoplasia, respectively. Accordingly, it is plausible to think that, while well-differentiated (i.e., low-grade) tumors encompass a minority of CSCs, poorly differentiated and undifferentiated (i.e., high-grade) malignancies have more abundant subpopulations of tumorigenic cells.

A more comprehensive yet conceptual classification system for neoplastic stem cells has been proposed as part of the *Year 2011 Working Conference on CSCs* based on the same principles historically established by pathologists to stage cancers by including both benign and fully malignant cells (Valent et al., 2012). While *bona fide* CSCs are fully malignant stem cells capable of indefinitely self-renewing and of giving rise to overt cancers, their benign or premalignant counterparts (encompassed by the term *neoplastic* SCs) give rise to progenies that may or may not develop into CSCs over time though with no immediate tumor-propagating potential. The distinction between benign and malignant CSCs helps solving some of the discrepancies arising when comparing transplantation and lineage tracing assays. In agreement with this classification, recent studies from Cedric Blanpain's and our laboratory revealed that benign tumors do not encompass CSCs capable of propagating the lesion in immune-deficient recipient mice (Ghazvini et al., 2013; Lapouge et al., 2012). On the other hand, lineage tracing of *Lgr5*$^+$ adenoma stem cells clearly showed their capacity to fuel the growth of benign tumors in the intestine by generating both additional *Lgr5*$^+$ neoplastic stem cells as well as all other more committed tumor cells (Schepers et al., 2012).

1.3. CSC heterogeneity: Proliferative versus quiescent CSCs

The distinction between benign and fully malignant neoplastic stem cells is also of relevance to introduce the concept of CSC heterogeneity (Magee, Piskounova, & Morrison, 2012). This heterogeneity is likely to reflect different combinations of (epi)genetic alterations but also different exposures to cues from the stromal microenvironment (Dalerba et al., 2011; Gerlinger et al., 2012). Moreover, by following the parallel drawn between the hierarchical structure of normal and tumor tissues, it is conceivable that distinct CSC types exist differing in their proliferation rate and/or their capacity to disseminate and invade distant organs.

Normal adult stem cells constitute a population of long-lived cells that have the capacity to self-renew and differentiate not only while fuelling daily turnover in homeostasis but also to underlie the regenerative response upon tissue injury. The coexistence within adult stem cell niches of frequently cycling and quiescent adult stem cells each with distinct functions in homeostasis and tissue damage has been confirmed in a broad spectrum of tissues and organisms (Fuchs, 2009; Roth & Fodde, 2011) including, more recently, the intestine (Buczacki et al., 2013; Roth et al., 2012). This dichotomy of cycling versus quiescent stem cells also applies for the CSC niche where quiescent cancer stem cells (qCSCs) coexist together with their more actively cycling and tumor-propagating counterparts. Indeed, the existence of quiescent CSCs has been demonstrated in several cancer types (Roth & Fodde, 2011) including ovarian carcinoma (Gao, Choi, Kang, Youn, & Cho, 2010; Kusumbe & Bapat, 2009), pancreas adenocarcinoma (Dembinski & Krauss, 2009), melanoma (Roesch et al., 2010), breast cancer (Pece et al., 2010), and in acute myeloid leukemia (AML) (Ishikawa et al., 2007). Quiescent CSCs may have profound clinical implications as malignant cells with dormant characteristics are more likely to be resistant to conventional cytotoxic therapeutic agents whose mode of action relies on the introduction of DNA damage into dividing cells. Moreover, qCSCs may underlie metastasis in cancer patients even decades after removal of the primary tumor (Aguirre-Ghiso, 2007; Muller, Alix-Panabieres, & Pantel, 2010). Indeed, following chemotherapy, ovarian qCSCs acquire the potential to proliferate in a drug-refractory manner (Kusumbe & Bapat, 2009). Likewise, in a mouse model of AML, patient-derived CSCs have been identified within the bone marrow endosteal region where they are protected from chemotherapy-induced apoptosis (Ishikawa et al., 2007). Subsequent characterization of these leukemia stem cells showed that they are partly quiescent and can be induced to enter cell cycle through a pretreatment with granulocyte colony-stimulating factor (G-CSF). Accordingly, a combined G-CSF/chemotherapy treatment enhances the induction of apoptosis and elimination of AML stem cells thus significantly increasing survival when compared to chemotherapy alone (Saito et al., 2010).

1.4. CSCs, EMT, and metastasis

EMT, that is, the developmental process through which epithelial cells acquire a mesenchymal identity is now recognized as an essential feature of tumor invasion (Friedl & Wolf, 2003; Thiery, 2002). Also in this case, the parallel between cancer and normal stem cell niches holds true as in the adult organism EMT plays pivotal roles during wound healing, tissue regeneration, and organ fibrosis. In cancer, EMT is thought to confer tumor cells the capacity to detach from the primary mass by losing cell adhesive properties and acquiring more motile features, thus enabling local invasion, intravasation into blood or lymph vessels, extravasation, and the recapitulation of the primary mass at distant sites through the reverse process of mesenchymal-to-epithelial transition (MET) (Thiery, 2002). In colon cancer, nuclear accumulation of β-catenin, the hallmark of Wnt signaling activation and cancer stemness, is preferentially found at the invasive front of the primary lesion where it earmarks EMT-competent cells which detach from the primary mass into the tumor microenvironment (Brabletz, Jung, Spaderna, Hlubek, & Kirchner, 2005; Fodde & Brabletz, 2007). From this perspective, the establishment of a direct molecular link between EMT and cancer stemness (Mani et al., 2008) further supports the existence of a specific CSC type, the migrating cancer stem cells (mCSCs), endowed with EMT features that confer them the plasticity to adapt to different environments and efficiently form metastatic colonies in distant tissues (Brabletz et al., 2005). Further experimental evidence for the transient EMT–MET switches in metastasis has been provided for bladder (Chaffer et al., 2006), breast (Korpal et al., 2011), and skin (Tsai, Donaher, Murphy, Chau, & Yang, 2012) cancer. More recently, a novel EMT inducer, "pair-related homeobox transcription factor 1" (Prrx1), was identified whose expression, in contrast to more canonical EMT activators such as Twist (Tsai et al., 2012), suppresses rather than enhances stemness traits (Ocana et al., 2012). These observations, although they further support the central role of EMT in tumor cell dissemination and the requirement to revert the process by MET for metastatic colonization, also uncouple stemness from EMT and favor a model where distinct subtypes of CSCs and EMT-competent cells underlie tumor growth, local invasion, and distant metastases due to their distinct cell proliferative, motile, and plastic features (Brabletz, 2012).

1.5. CSCs and therapy resistance

Cancer treatments have become more efficient and successful over the last decades. However, metastases occurring from tumor cell dissemination still represent the main cause of morbidity and mortality among individual cancer patients.

Tumor dormancy is a widely used term to describe the state in which neoplastic cells persist for long periods of time following treatment and/or surgical removal of the primary malignant lesion. From their dormant state, EMT-driven quiescent cancer cells can subsequently switch to a state of active cycling through MET and underlie tumor relapse. In the clinic, tumors are considered dormant if they recur after at least 5 years from surgical treatment.

In general (quiescent) CSCs are assumed to be resistant to therapy which implies that many currently employed cytotoxic regimens would eventually result in an initial shrinkage of the tumor mass but also in the relative enrichment in CSCs which would then underlie tumor relapse and metastasis (Reya et al., 2001). Apart from their slow cycling rates, additional mechanisms are likely to underlie CSCs' innate resistance to therapy: upregulation of cellular efflux pumps, for example, the multidrug resistance transporter MDR-1 (Knutsen et al., 1998; Mickley, Spengler, Knutsen, Biedler, & Fojo, 1997), activation of DNA damage response (Bao et al., 2006), and lower concentrations of critical mediators of ionizing-radiation-induced cell killing such as reactive oxygen species (Diehn et al., 2009). Moreover, additional mechanisms potentially underlying CSC resistance to DNA damaged-induced apoptosis are likely to exist, as suggested by the observed expression of antiapoptotic proteins such as members of the BCL2 family in stem-like cancer cells (Wang, Yang, & Lippman, 2003), and by the autocrine production of antiapoptotic cytokines, as illustrated by the expression of interleukin-4 in colon CSCs (Todaro et al., 2007). Given the CSCs' resistance to conventional radio- and chemotherapies which sharply contrast with the efficacy of these treatment modalities in initially reducing tumors size, the combination of CSC-tailored agents with more conventional cytotoxic drugs offers new perspectives for a more effective and long-term disease eradication (Rich & Bao, 2007).

Of note, the constitutive activation of well-known signal transduction pathways such as Notch, Wnt, and Hedgehog in CSC niches (Reya et al., 2001) and the rate-limiting role they play in regulating self-renewal in CSCs make them an attractive therapeutic target toward the development of novel anticancer therapies (Takebe, Harris, Warren, & Ivy, 2011). In particular, canonical Wnt/β-catenin signaling has emerged as a critical regulator of stem cells and, accordingly, is frequently aberrantly activated across a vast range of human malignancies including colon, breast, and cutaneous malignancies (Malanchi et al., 2008; Nusse et al., 2008; Reya & Clevers, 2005; Vermeulen et al., 2010; Watt & Collins, 2008; Zeng & Nusse, 2010). In a high-throughput screening carried out on EMT-induced breast epithelial cells to identify CSC-specific inhibitors, the potassium ionophore salinomycin was found to have strong selective toxicity for CSCs (Gupta et al., 2009). This compound specifically inhibited proliferation of mammary CSCs and reduced their proportion by >100-fold relative to commonly used breast cancer chemotherapeutic drugs. Furthermore, salinomycin appears to induce terminal epithelial differentiation accompanied by cell cycle arrest. The mechanisms by which salinomycin interferes with CSC survival are still largely unknown. However, recent studies have shown that salinomycin acts as a potent inhibitor of the Wnt signaling cascade in chronic lymphocytic leukemia (Lu et al., 2011) and in osteosarcoma (Tang et al., 2011). Of note, a different screen for novel agents specifically targeting quiescent pancreatic CSCs led to the identification of the plant-derived flavonoid quercetin (Adikrisna et al., 2012), previously reported to specifically inhibit Wnt signaling (Park et al., 2005).

As mentioned earlier, the identification of CSC-specific drugs allows their implementation in combinatorial treatment protocols together with more conventional cytotoxic drugs aimed at bulk tumor cells. Accordingly, the combined administration of the conventional anticancer drug gemcitabine with quercetin on a xenograft mouse model for pancreas cancer was shown to be more effective both *in vitro* and *in vivo*, when compared with the single agents (Adikrisna et al., 2012) thus opening up novel avenues for long-term disease eradication (Rich & Bao, 2007).

2. CONTROL OF (CANCER) STEMNESS BY SIGNALING PATHWAYS: THE EMBRYONIC/GERM CELL CONNECTION

Self-renewal represents the most important property of both normal and CSCs (Reya et al., 2001). In the adult organism, specific signal transduction pathways including Notch, Shh, and Wnt tightly control self-renewal within stem cell niches during homeostasis and tissue regeneration following injury. (Epi)genetic alterations of these pathways leading to their constitutive activation result in loss of the homeostatic balance between self-renewal and differentiation and underlie cancer onset, progression, and malignant behavior in the same adult tissues (Taipale & Beachy, 2001).

Among the pathways known to regulate self-renewal and oncogenesis in hematopoietic, epidermal, and gut stem cells is Wnt signaling (Nusse et al., 2008). Here, we will focus on Wnt as one of the major cancer stemness pathway and employ it to establish a parallel between its role in embryonic stem (ES) and carcinoma (EC) cells, and in GCTs.

2.1. Wnt/β-catenin signaling

Wnt signaling is considered one of the most ancient signaling pathways, conserved among different multicellular organisms and throughout evolution, that plays a crucial role in embryonic development and tissue homeostasis. The word Wnt is in fact a fusion of two terms, the *Drosophila* gene *Wingless* (*Wg*) (Nusslein-Volhard & Wieschaus, 1980) and its mouse orthologous *integration-1* (*int-1*) (Nusse & Varmus, 1982). Wnt signaling is generally subdivided into canonical and noncanonical pathways with the canonical signaling representing the most studied one usually referred to as Wnt/β-catenin signaling (Fig. 13.2). In short, in the absence of the glycoprotein Wnt ligand, the intracellular pool of β-catenin is partly associated to the cell membrane through its interaction with cadherins to mediate cell–cell adhesion, and is partly localized to the cytosol where it makes integral part of the so-called destruction

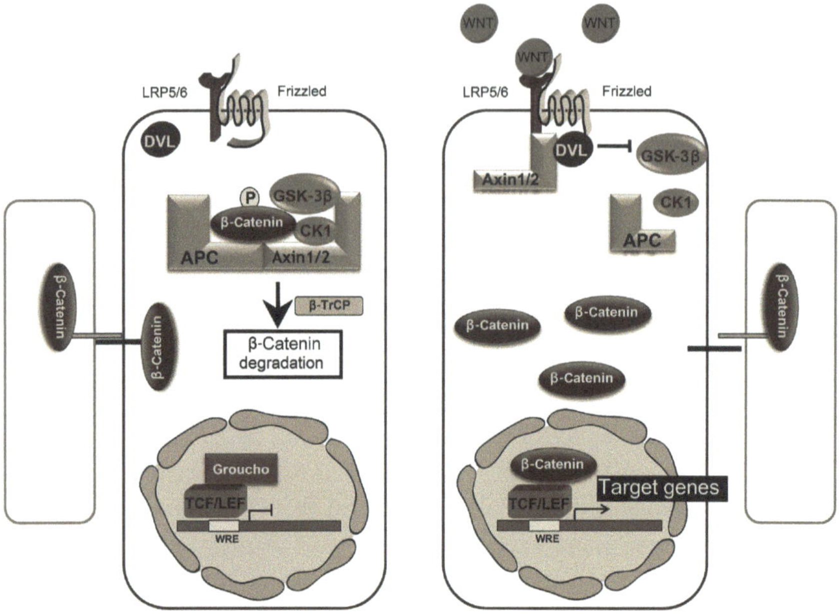

Figure 13.2 Schematic representation of the canonical Wnt/β-catenin signaling pathway. The left panel represents the cell in the absence of the Wnt ligand, while the right panel depicts the actively signaling cell in the presence of the glycoprotein Wnt. See review text for additional details. Legend. DVL, disheveled; GSK-3β, glycogen synthase kinase 3 beta; CK1: casein kinase 1; APC, adenomatous polyposis coli; β-TrCP: beta-transducin repeats-containing protein (substrate recognition subunits for E3 ubiquitin ligases); LRP5/6, LDL receptor-related proteins 5 and 6. (See color plate.)

complex, consisting of the APC (adenomatous polyposis coli) and Axin scaffolding proteins, and the serine–threonine kinases GSK and CK1 (glycogen synthase and casein kinases). In the absence of the Wnt ligands, β-catenin is negatively regulated by Ser-Thr phosphorylation followed by ubiquitination and proteolytic degradation. The presence of Wnt ligands and their interaction with their natural receptors (Frizzled, LRPs) inhibits the destruction complex function thus stabilizing cytoplasmic β-catenin and leading to its translocation to the nucleus. Here, β-catenin acts as a cotranscriptional factor by interacting with different members of the Tcf/Lef family of transcription factors and as such modulating the expression of a broad spectrum of downstream Wnt target genes (http://www.stanford.edu/group/nusselab/cgi-bin/wnt/target_genes).

Wnt/β-catenin signaling plays a central role in regulating self-renewal and differentiation in many adult stem cell niches including the mammary gland, intestine, hair follicle and skin, bone marrow, and brain (Reya & Clevers, 2005). Moreover, lineage tracing experiments using the Wnt target gene *Lgr5* led to the identification of Wnt-responsive stem cells in different organs (Barker, Bartfeld, & Clevers, 2010). Accordingly, mutations in different components of the Wnt pathway have been reported in many cancers in the above-mentioned tissues and organs (Reya & Clevers, 2005). The observations according to which (i) the rate-limiting *Apc* mutation in intestinal cancer must occur in the $Lgr5^+$ stem cell to successfully trigger tumor formation (Barker et al., 2009), and (ii) colon CSCs are earmarked by constitutive Wnt activity (Vermeulen et al., 2010, 2008), further reinforces the great relevance of Wnt signal transduction in both the onset (stem cell of origin) and maintenance (CSC) of cancer.

In the following sections, we will attempt to draw a parallel between embryonic stem cells (ESCs) and CSCs, and the role played by Wnt signaling in regulating self-renewal and differentiation in these two only apparently divergent cell types.

2.2. Embryonic and cancer cells: Two of a kind?

ESCs are derived from the inner cell mass of the preimplantation blastocyst and are earmarked by their pluripotency (i.e., able to differentiate into all derivatives of the three primary germ layers: ectoderm, endoderm, and mesoderm) and by their ability to replicate indefinitely (for a detailed description of ESCs we refer the readers to the chapter by Posfai and Rossant). The golden standard assay to determine the pluripotency potential of ESCs and their capacity to contribute to virtually all tissues in the adult organism is the generation of chimeric mice upon ESC injection into the blastocyst. Similar to ESCs, few studies demonstrated that tumor cells can also contribute to normal embryonic development and lineage differentiation when injected in blastocyst. In the seminal work by Illmensee and Mintz (1976), teratocarcinoma cells cultured for over 8 years by serial passaging were injected as single cells in mouse blastocysts. Surprisingly, tumor cells were still capable of contributing to normal development and generated phenotypically normal chimeric mice with tumor cells present in most adult tissues (Illmensee & Mintz, 1976). Likewise, normal embryonic development has also been observed with tumor cells employed in somatic cell nuclear transfer experiments (Hochedlinger et al., 2004; Li, Connelly, Wetmore, Curran, & Morgan, 2003), and with chicken embryonic cells infected with the Rous sarcoma virus (Dolberg & Bissell, 1984; Rous,

1979; Stoker, Hatier, & Bissell, 1990). Notably, when cultured *in vitro*, these embryonic cells rapidly reveal pronounced tumorigenic features and undergo neoplastic transformation. Hence, the embryonic environment seems to attenuate oncogenesis possibly by inducing differentiation and suppressing cancer stemness.

Overall, these seminal experiments provided the first indications that tumor cells share similarities with ESCs in that they are able to integrate into the developing organism following apparently normal differentiation routes. From this perspective, it is of interest that somatic cell reprogramming is elicited by the combinatorial action of specific transcriptional regulators (i.e., the so-called Yamanaka factors *Oct4*, *Sox2*, *Klf4*, and *c-Myc*) capable of resetting developmental potential and of inducing stem cell characteristics in otherwise terminally differentiated cells. Are the regulatory networks that underlie ESC's self-renewal and pluripotency also active in tumor cells? By comparing expression profiles obtained from different human tumors and from ESCs, it was reported that genes involved in preserving ESC identity are often overexpressed in poorly differentiated tumors (Ben-Porath et al., 2008). In fact, this ESC-specific transcriptional program appears to be active in a broader spectrum of human epithelial malignancies including basal-like, triple-negative breast tumors, gliomas, and bladder cancer (Ben-Porath et al., 2008). Independently from this study, a similar ESC-specific "gene module map" was also identified in diverse human epithelial cancers where it is strongly predictive of metastasis and death (Wong et al., 2008). Notably, c-Myc, on the contrary of other oncogenes, is sufficient for the activation and establishment of the ESC gene expression network in both normal and cancer cells, and underlies a significant increase in the number of CSCs (Wong et al., 2008). Similarly, a c-Myc-centered transcriptional network has more recently been shown to account for the similarities between ES and cancer cell gene signatures by regulating ESC identity. Also in this case, expression of this regulatory gene network is predictive of poor outcome (Kim et al., 2010).

Specific members of the core pluripotency factors including the four Yamanaka reprogramming factors are also expressed in different cancers (Schoenhals et al., 2009). *Oct4* expression has been detected in different somatic tumors (Sotomayor, Godoy, Smith, & Huss, 2009) and its ectopic expression is sufficient to induce dysplastic transformation in adult mice by blocking the differentiation of progenitor cells (Hochedlinger, Yamada, Beard, & Jaenisch, 2005). However, it remains to be proven whether Oct4 protein is expressed in GCTs (Rijlaarsdam et al., 2011). *Sox2* is expressed in different adult stem/progenitor cells and in a broad spectrum of malignancies where accumulating evidence suggests that it can promote tumor growth even in the absence of *Oct4* (Gangemi et al., 2009; Ikushima et al., 2009; Laga et al., 2010; Rodriguez-Pinilla et al., 2007). *KLF4* is expressed in different adult tissues and can function as an oncogene or tumor suppressor gene in different tumor types where its expression is upregulated (McConnell, Ghaleb, Nandan, & Yang, 2007; Rowland & Peeper, 2006).

Hence, substantial body of evidence indicates that ESCs represent a unique *in vitro* and *in vivo* model to study cancer stemness.

2.3. Germ cell tumors

The first similarities between embryonic and tumor cells were observed by studying teratomas. Teratomas encompass a broad spectrum of differentiated cells recapitulating early events of embryonic development. In mice, teratomas occur spontaneously in the testes of approximately 1% of mice on the inbred 129 strain. Similarly, transplantation of genital ridges or early embryos (from the 2-cell stage embryos) from 129 inbred mice into adult mouse testis result in teratocarcinoma formation. These tumors contained various differentiated and undifferentiated embryonic cells and could be serially transplanted into new adult testes (Solter, Skreb, & Damjanov, 1970; Stevens, 1968, 1970). The undifferentiated stem cells within teratocarcinomas are also known as EC cells and represent neoplastic stem cells able to differentiate into different cell lineages. These observations were among the first to suggest the existence of CSCs that can recapitulate the original tumor heterogeneity when serially transplanted into new host.

In man, GCTs represent a heterogeneous group of neoplastic lesions found in the gonads or in extragonadal tissues. Increasing attention is currently being given to the pathogenesis and clinical behavior of GCTs as the type of malignancy that most closely mimics normal embryonic development (Spiller et al., 2012; Westerman et al., 2011). GCTs are predominantly diagnosed in relatively young patients, that is, neonates and infants, as well as adolescents and young adults, with major physical, social, and economic impact. A significant numbers of GCT patients require surgery followed by systemic treatment (irradiation and/or chemotherapy). In neonates and infants, the GCTs mainly mimic the teratomas as found in mice (Oosterhuis & Looijenga, 2005). These are referred to as Type I GCTs, and histologically are composed of either teratoma, yolk sac tumor (YST), or a combination of both.

GCTs of adolescents and adults are generally divided into seminomas and nonseminomas, referred to as Type II GCTs. While seminomas resemble primordial germ cells (PGCs), nonseminoma tumors are classified into different subtypes among which EC represents the malignant stem cell component. Subsequent studies demonstrated that indeed EC cells share many similarities with ESCs suggesting the former as the malignant counterpart of the latter. Notably, seminomas and ECs share the expression of several pluripotency factors such as *NANOG* and *OCT4* though they differ in many other markers and signaling pathways. A number of highly informative diagnostic markers for the earliest developmental stages of GCT pathogenesis have been identified based on knowledge from normal embryogenesis. These include alkaline phosphatase (Roelofs et al., 1999) and the stem cell factor receptor c-KIT (Strohmeyer et al., 1991). However, the main contributions to the spectrum of diagnostic markers for GCTs have been based on the basic knowledge of stem cell regulatory factors which led to the identification of *OCT4* (*POU5F1*) (Looijenga et al., 2003), *SOX2*, and *SOX17* (de Jong et al., 2008). While *OCT4* is consistently found in seminomas and ECs, *SOX17* earmarks the seminomas, whereas *SOX2* is mainly found in ECs.

2.4. Wnt signaling in ESCs, CSCs, and GCTs

Wnt signaling has been studied extensively in mouse embryos and in ESCs. Loss-of-function experiments in mouse embryos have shown that depletion of β-catenin does not have an apparent effect on blastocyst stage embryonic development (Haegel et al., 1995; Huelsken et al., 2000). Similar to the blastocyst, β-catenin-null ESCs can be maintained *in vitro*, suggesting that β-catenin is dispensable in self-renewing ESCs (Lyashenko et al., 2011). However, upon induction of differentiation, when Wnt signaling would otherwise confer differentiation resistance, β-catenin-null ESCs rapidly exit self-renewal and undergo differentiation, suggesting a role for Wnt signaling as a guardian of the self-renewal gate (Wagner, Xu, Yi, Merrill, & Cooney, 2010; Wray et al., 2011). On the other hand, loss-of-function mutations of *Apc* or *Gsk* (both negative regulators of β-catenin) in mouse ESCs confer differentiation resistance and enhance self-renewal (Doble, Patel, Wood, Kockeritz, & Woodgett, 2007; Kielman et al., 2002). These observations led to a new chemically defined ESC medium where inhibition of GSK by small molecule inhibitors or by Wnt3a direct supplementation represents the main component of the culture medium currently employed to maintain ESCs in undifferentiated, self-renewing state (Ying et al., 2008).

In addition to its role in enhancing self-renewal, constitutive activation of Wnt signaling affects cell fate decision and multilineage differentiation of ESCs. In fact, transition from naïve to primed epi-embryonic stem cell stage, one of the very first steps of ESCs differentiation, is prevented by Wnt signaling (ten Berge et al., 2011). Also, constitutive activation of Wnt signaling in mouse ESCs inhibits their capacity to differentiate toward all the three germ layers. Teratomas derived from *Apc*-mutant or GSK-depleted ESCs display a broad spectrum of differentiation defects in endodermal, mesodermal, and (neuro)ectodermal lineages (Kielman et al., 2002; Stoker et al., 1990). On the other hand, Wnt signaling is also required during the course of differentiation for proper lineage commitment. Expression of Wnt ligands and increased Wnt reporter activity were observed during ESC differentiation (Lindsley, Gill, Kyba, Murphy, & Murphy, 2006). These data suggest that during development the downstream effects of Wnt signaling are stage dependent: while Wnt inhibits differentiation in pluripotent ESCs, it is necessary for lineage commitment when differentiation is triggered.

In contrast to mouse ESCs, the role of Wnt signaling in human ESCs is controversial. Different reports indicate that Wnt3a stimulation or GSK inhibition supports short-term though fail to maintain long-term self-renewal of human ESCs (Cai et al., 2007; Dravid et al., 2005; Ullmann et al., 2008; Villa-Diaz et al., 2009). Moreover, long-term culture of hESCs by Wnt3a or GSK inhibition-induced mesodermal, endodermal, and primitive streak gene expression (Davidson et al., 2012). Conversely, inhibition of Wnt does not affect hESC maintenance suggesting that endogenous Wnt signaling does not play a major role in human ESCs self-renewal. OCT4 was also shown to repress β-catenin signaling in human ESCs and targeted knock-down of OCT4 activates Wnt. However, caution should be taken when interpreting these results since the final outcome of Wnt activation in human ESCs is extremely dependent on culture conditions (Blauwkamp, Nigam, Ardehali, Weissman, & Nusse, 2012).

Wnt signaling has also been studied in GCTs. Several components of the canonical pathway are expressed in developing and adult testis in both mouse and man. Activation of Wnt signaling in developing PGCs leads to their impaired development and lack of germ cells (Kimura et al., 2006). Overexpression of a constitutive form of β-catenin in Sertoli cells blocks their differentiation and induced Sertoli cell tumors of the testis (Boyer, Hermo, Paquet, Robaire, & Boerboom, 2008; Boyer, Paquet, Lague,

Hermo, & Boerboom, 2009; Chang, Guillou, Taketo, & Behringer, 2009). In human GCTs, Wnt signaling is active in EC and YSTs (Korkola et al., 2006), and its inhibition in EC cell lines reduces their tumorigenicity (DeAlmeida et al., 2007).

The distinctive expression pattern of the transcription factor *SOX17* in early germ cells, seminoma, and the GCT precursor carcinoma *in situ* (CIS) (de Jong et al., 2008) is of interest in view of its role as a negative regulation factor of β-catenin/TCF transcription activity in the Wnt signal transduction pathway (Zorn et al., 1999). It has also been reported that *SOX4* expression stabilizes β-catenin and enhances Wnt signaling in human colon carcinoma cells (Sinner et al., 2007). A similar role might be replaced by *SOX2* in EC cells. According to this hypothetical scenario, no or little Wnt activity is present in the germ cell lineage of GCC, namely, CIS and seminoma, due to *SOX17* expression. In contrast, Wnt activation characterizes the EC component due to the absence of *SOX17*, whose expression is replaced by *SOX2* in these tumors. This is in line with IHC studies based on primary tumors (Honecker et al., 2004) as well as cell lines (Snow et al., 2009).

2.5. Heterogeneity within embryonic and CSC populations: Role of Wnt signaling

Accumulating evidence suggests that ESCs are not as homogenous as previously thought but instead comprise heterogeneous subpopulations which exist in a finely tuned equilibrium. In mouse ESCs, different reporter lines have been generated to monitor cellular heterogeneity in culture. *Nanog* expression, for example, fluctuates in normal mouse ES culture with approximately 20% of cells being negative (Chambers et al., 2007). Likewise, *Oct4*-CFP/*Rex1*-GFP double knock-in mouse ESCs also revealed both *Rex1*-positive (90%) and -negative (10%) populations within the *Oct4*$^+$ fraction. Although the two populations showed dynamic reversibility, *Rex1*$^+$ cells revealed improved capacity to give rise to chimeras and expressed higher levels of preimplantation embryonic markers (*Esrrb1*, *Klf4*, and *Stella*) with concurrent downregulation of postimplantation markers (e.g., *Fgf5*, *Brachyury*, and *EOMES*) (Toyooka, Shimosato, Murakami, Takahashi, & Niwa, 2008). *Stella* (*Dppa3*) expression also distinguishes interchangeable populations within undifferentiated mouse ESCs with positive cells more closely resembling preimplantation ICM cells (Hayashi, Lopes, Tang, & Surani, 2008). In all *Rex1*, *Nanog*, and *Stella* reporter mouse ES lines, the GFP-negative populations could revert back to the GFP$^+$ status, thus suggesting a dynamic transition between the different substates in ESCs. Similar results were obtained in human ESCs (Bhatia, Pilquil, Roth-Albin, & Draper, 2013).

Although it is possible that the observed cellular heterogeneity of ESCs can result from suboptimal culture conditions which impose coexistence of distinct subpopulations with different differentiation capacities, an alternative view suggests that the different lineage-committed states reflect an intrinsic feature of stem cells, namely, the simultaneous expression of differentiation agonists and antagonists which provide ESCs with a ground state ready for lineage commitment upon external stimuli. Several other studies support the latter scenario with multiple substates coexisting within epiblast and ESC cultures mirroring the embryonic states observed during blastocyst development. Of note, this cellular heterogeneity appears to be required not only for multilineage commitment but also for generating distinct niches, which support further differentiation.

Although the molecular mechanisms underlying the observed ESC heterogeneity are yet largely unknown, the very existence of different metastable states within the epiblast and in ESCs is likely to serve the purpose of providing a window of opportunity for specific lineage differentiation upon the presence of correct environmental cues.

In view of its central role in controlling self-renewal and differentiation in ESCs, Wnt signaling heterogeneity is likely to earmark distinct ES subpopulations with different differentiation states. A basal Wnt signaling activity can be detected in mouse ESCs under self-renewal condition. Using a Wnt reporter ES line, heterogeneous Wnt activity was observed in mouse ESCs cultured on feeder layers. Interestingly, the Wnt^{hi} ESCs showed higher ability to form undifferentiated colonies in colony forming assay. Conversely, inhibition of autocrine and paracrine Wnt signaling in self-renewing ESCs prevented expansion of undifferentiated ESCs (ten Berge et al., 2011). This indicates heterogeneous Wnt activity in mouse ESCs under self-renewal conditions with Wnt positively correlating with their self-renewal capacity. Similarly, heterogeneous Wnt activity was detected in human ESCs where it affects differentiation fate. Wnt signaling was found to correlate with higher expression of mesodermal and endodermal genes. In *in vitro* differentiation assays, Wnt^{hi} ESCs generate endodermal and mesodermal structures whereas their Wnt^{lo} counterparts mainly result in neuroectodermal cells. More importantly, the heterogeneity in GFP (reporter) expression was maintained over time in culture and both GFP-positive and -negative populations re-established the original heterogeneity suggesting that the heterogeneity of Wnt signaling is intrinsically kept in equilibrium (Blauwkamp et al., 2012).

Intratumor Wnt signaling heterogeneity has been well established in colon cancer. Despite the initiating role of *APC* mutations which predicts constitutive Wnt activation in all cells within colon cancers, striking heterogeneity in β-catenin subcellular localization has been observed, with cells clustered along the invasive front showing the highest levels of Wnt signaling (Fodde & Brabletz, 2007). Although different intrinsic and extrinsic factors have been suggested to explain β-catenin heterogeneity, recent studies suggest that Wnt signaling and β-catenin heterogeneity might functionally designate the CSC population within the tumors.

Lentiviral Wnt reporter constructs have been employed by Vermeulen and colleagues to transduce primary spheroid cultures derived from colorectal tumors. These *colospheres* are enriched in CSCs and can reconstitute the original malignancy when injected in immune-deficient mice. Analyzing single cell clones derived from the transduced spheroid cultures showed heterogeneous GFP expression among the secondary colospheres. Wnt^{hi} subclones expressed higher levels of the intestinal stem cell markers *LGR5* and *ASCL2* and showed downregulation of differentiation markers (MUC2 and CK20) when compared to Wnt^{lo} cells. Moreover, the Wnt^{hi} subpopulation showed higher clonogenicity *in vitro* and increased tumorigenicity *in vivo*. Accordingly, Wnt^{hi} spheres reestablished the original Wnt heterogeneity in xenografts, suggesting the existence of a hierarchic equilibrium between Wnt^{hi} and Wnt^{lo} cancer cells. Exposure of Wnt^{lo} cells to stromal secreted factors induced Wnt and CSC markers. This indicates that specific microenvironmental cues can affect the heterogeneity within the tumor cells and reprogram committed cancer cells into CSCs (Vermeulen et al., 2010). Similarly, factors secreted from activated macrophages, a central component of the tumor microenvironment, also activate Wnt signaling in gastric cancer cells (Oguma et al., 2008). Moreover, a gene expression signature derived from Wnt^{hi} colosphere cells has been used to stratify colorectal cancer patients (de Sousa, Vermeulen, Richel, & Medema, 2011). Although the Wnt signature was predictive of tumor recurrence and poor prognosis in CRC, genes upregulated in Wnt^{hi} cells unexpectedly correlated with overall survival. Instead, the Wnt^{lo} signature was found in immature, undifferentiated tumors also in strong correlation with the above-mentioned ESCs gene signature (Ben-Porath et al., 2008). These results suggest that poorly differentiated colon tumors adapt an immature phenotype that resembles the gene expression program employed by ESCs as opposed to the more intestinal stem cell program active in the Wnt^{hi} CSCs.

In a similar study, Horst and colleagues employed a more sensitive Wnt lentiviral reporter construct to transduce established CRC cell lines or primary tumor xenografts. Wnt^{hi} colon cancer cells showed increased expression of intestinal stem cell markers like *LGR5*. Interestingly, Wnt signaling heterogeneity was found to correlate with mitogen-activated protein kinase (MAPK) signaling in CRC cells. Accordingly, MAPK activity inhibition in tumor xenografts reduced Wnt signaling and Wnt heterogeneity within the tumors (Horst et al., 2012). These data are in accordance with previous reports showing that oncogenic *KRAS* mutations increase Wnt signaling and β-catenin nuclear accumulation in CRC cells (Janssen et al., 2006).

Overall, it appears that Wnt signaling heterogeneity in colon tumors functionally define cells with stem cell characteristics that underlie the tumor formation and propagation from those with a more committed phenotype.

3. CONCLUSIONS AND FUTURE PERSPECTIVE

The transplantation of 2-cell stage embryos into mouse testis leading to the formation of teratocarcinomas still represents the archetypal experimental evidence for the CSC concept, encompassing at the same time the very first limiting dilution and lineage tracing experiment (Stevens, 1968). The resulting teratocarcinomas encompassed both differentiated and undifferentiated components and could be serially transplanted, thus demonstrating the self-renewing and differentiation capacity of the EC cells. Forty years later, more sophisticated lineage tracing experiments in mouse models for glioma, intestinal adenomas, and skin tumors provided definitive evidence for the capacity of CSCs to give rise and maintain tumors (Chen et al., 2012; Driessens et al., 2012; Schepers et al., 2012). Of note, while CSCs from somatic tumors are generally thought to be resistant to conventional cytotoxic therapies, teratocarcinomas and EC cells are highly sensitive to cisplatin treatment. This is likely to result from the short G1/S checkpoint in EC and the specific features of DNA damage response in ESCs when compared with adult ones. However, a considerable (~20%) proportion of patients with metastatic GCT present with cisplatin-refractory disease, the underlying cellular and molecular mechanisms of which are yet largely unknown. In view of the role played by Wnt signaling in conferring resistance to chemo- and radiotherapy in cancer (Sun et al., 2012; Woodward et al., 2007), it will be of interest to assess whether the observed Wnt activation in EC and possibly other GCT subtypes plays a role in cisplatin-refractory disease.

Functional heterogeneity may also represent an essential feature of CSCs that underlies their capacity to survive exposure to cytotoxic therapies. Both cycling and quiescent CSCs have been shown to coexist within primary tumors. Likewise, EMT-competent CSCs, likely to detach from the primary lesions and form metastases in distant organ sites, appear distinct from their "tumor-maintaining" counterparts. As such, CSCs from diverse malignancies appear to exist in different, interconvertible "metastates" which makes them true moving targets posing serious challenges for their pharmacological eradication. From this perspective, the simple notion that many CSCs rely on key stem cell regulatory signaling pathways such as Shh, Notch, and Wnt predicts that inhibitors of these pathways may improve therapeutic efficacy of specific cancer types as single agents or, more likely, in combination with current cytotoxic regimens (Takebe et al., 2011). This might be of relevance as these very pathways (as shown here for Wnt signaling) also underlie the heterogeneity observed among stem cells in homeostasis and cancer.

Combinatorial treatments where conventional cytotoxic approaches are combined with specific agents targeting the different CSCs' identities are expected to significantly improve long-term therapeutic efficacy (Fig. 13.3). In colon cancer, for example, CSCs have been shown to resist DNA damage-induced apoptosis because of interleukin-4 expression (Todaro et al., 2007). Therefore, combining anti-IL4 with 5-fluoracil or other first-line colon cancer chemotherapeutic regimens may affect both bulk and stem-like tumor cells. From this perspective, aspirin has been demonstrated to specifically inhibit interleukin-4 expression at the gene level (Cianferoni et al., 2001) and is likely to represent an attractive adjuvant treatment in colon cancer (Bastiaannet et al., 2012; Chan, Ogino, & Fuchs, 2009; Neugut, 2009). Likewise, a number of promising agents aimed at cancer (stem) cells earmarked by constitutive Wnt/β-catenin signaling such as the plant flavonoid quercitin (Park et al., 2005) and the potassium ionophore salinomycin (Gupta et al., 2009) are likely to find therapeutic applications in a broad spectrum of malignancies including breast, osteosarcoma, leukemia, and pancreatic cancer (Adikrisna et al., 2012; Lu et al., 2011; Tang et al., 2011). Interestingly, chronic phase CML represents an example of a tumor type where therapeutic regimens directed against the BCR–ABL translocation were proven extremely effective even though they

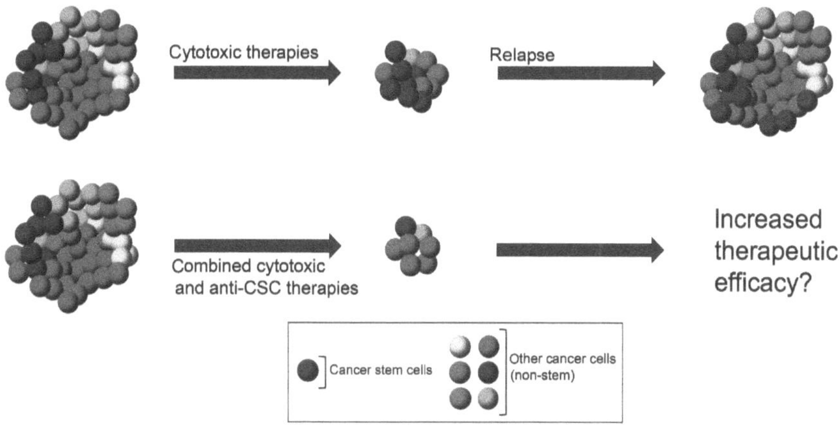

Figure 13.3 Combination of conventional (cytotoxic) and anti-CSC treatment may lead to improved therapeutic efficacy. Assuming that CSC are less sensitive to the cytotoxic effects of conventional chemo- and radiotherapies, these regimens, although initially leading to an effective reduction of the tumor mass, eventually lead to a relative enrichment of the CSCs. Hence, the combined use of conventional and anti-CSC agents is thought to result in improved treatment efficacy. (See color plate.)

fail to eradicate the CSCs (Ross, Hughes, & Melo, 2011). However, this implies that these patients will have to take these drugs permanently, unless novel agents directed against the CML CSCs are identified.

An alternative approach to the therapeutic targeting of CSCs relies on agents capable of inducing their differentiation, the so-called differentiation therapy. This approach is based on the assumption that (terminal) differentiation is accompanied by concomitant loss in proliferative capacity and in programmed cell death. Although the effectiveness of differentiation therapy has been demonstrated in several laboratory studies, its translation in the clinical setting is to date limited to the use of *all trans* retinoic acid in acute promyelocytic leukemia (Petrie, Zelent, & Waxman, 2009).

Overall, notwithstanding the controversies arisen around the CSC concept, it seems more than likely that the improved characterization of the stem-like component of human malignancies and the elucidation of the mechanisms underlying their resistance to conventional cytotoxic regimens will eventually lead to the improved efficacy of anticancer therapies.

ACKNOWLEDGMENTS

This study was supported among others by grants from the Erasmus MC, the Dutch Cancer Society, the BSIK program of the Dutch Government Grant 03038 (www.stemcells.nl), and the EU FP6 and FP7 consortia Migrating Cancer Stem Cells program (MCSCs; www.mcscs.eu) and TuMIC (integrated concept of tumor metastasis, http://itgmv1.fzk.de/www/tumic/tumic_main.htm). The authors are grateful to Mr. Frank van der Panne for his assistance with the artwork.

REFERENCES

Adikrisna, R., Tanaka, S., Muramatsu, S., Aihara, A., Ban, D., Ochiai, T., et al. (2012). Identification of pancreatic cancer stem cells and selective toxicity of chemotherapeutic agents. *Gastroenterology*, *143*, 234–245, e7.

Aguirre-Ghiso, J. A. (2007). Models, mechanisms and clinical evidence for cancer dormancy. *Nature Reviews Cancer*, 7, 834–846.

Al-Hajj, M., Wicha, M. S., Benito-Hernandez, A., Morrison, S. J., & Clarke, M. F. (2003). Prospective identification of tumorigenic breast cancer cells. *Proceedings of the National Academy of Sciences of the United States of America*, *100*, 3983–3988.

Bao, S., Wu, Q., McLendon, R. E., Hao, Y., Shi, Q., Hjelmeland, A. B., et al. (2006). Glioma stem cells promote radioresistance by preferential activation of the DNA damage response. *Nature*, *444*, 756–760.

Barker, N., Bartfeld, S., & Clevers, H. (2010). Tissue-resident adult stem cell populations of rapidly self-renewing organs. *Cell Stem Cell*, 7, 656–670.

Barker, N., Ridgway, R. A., van Es, J. H., van de Wetering, M., Begthel, H., van den Born, M., et al. (2009). Crypt stem cells as the cells-of-origin of intestinal cancer. *Nature*, *457*, 608–611.

Bastiaannet, E., Sampieri, K., Dekkers, O. M., de Craen, A. J., van Herk-Sukel, M. P., Lemmens, V., et al. (2012). Use of aspirin postdiagnosis improves survival for colon cancer patients. *British Journal of Cancer*, *106*, 1564–1570.

Ben Porath, I., Thomson, M. W., Carey, V. J., Ge, R., Bell, G. W., Regev, A., et al. (2008). An embryonic stem cell-like gene expression signature in poorly differentiated aggressive human tumors. *Nature Genetics*, *40*, 499–507.

Bhatia, S., Pilquil, C., Roth-Albin, I., & Draper, J. S. (2013). Demarcation of stable subpopulations within the pluripotent hESC compartment. *PLoS One*, *8*, e57276.

Blauwkamp, T. A., Nigam, S., Ardehali, R., Weissman, I. L., & Nusse, R. (2012). Endogenous Wnt signalling in human embryonic stem cells generates an equilibrium of distinct lineage-specified progenitors. *Nature Communications*, *3*, 1070.

Bonnet, D., & Dick, J. E. (1997). Human acute myeloid leukemia is organized as a hierarchy that originates from a primitive hematopoietic cell. *Nature Medicine*, *3*, 730–737.

Boyer, A., Hermo, L., Paquet, M., Robaire, B., & Boerboom, D. (2008). Seminiferous tubule degeneration and infertility in mice with sustained activation of WNT/CTNNB1 signaling in sertoli cells. *Biology of Reproduction*, *79*, 475–485.

Boyer, A., Paquet, M., Lague, M. N., Hermo, L., & Boerboom, D. (2009). Dysregulation of WNT/CTNNB1 and PI3K/AKT signaling in testicular stromal cells causes granulosa cell tumor of the testis. *Carcinogenesis*, *30*, 869–878.

Brabletz, T. (2012). EMT and MET in metastasis: Where are the cancer stem cells? *Cancer Cell*, *22*, 699–701.

Brabletz, T., Jung, A., Spaderna, S., Hlubek, F., & Kirchner, T. (2005). Opinion: Migrating cancer stem cells—An integrated concept of malignant tumour progression. *Nature Reviews Cancer*, *5*, 744–749.

Buchstaller, J., McKeever, P. E., & Morrison, S. J. (2012). Tumorigenic cells are common in mouse MPNSTs but their frequency depends upon tumor genotype and assay conditions. *Cancer Cell, 21*, 240–252.

Buczacki, S. J., Zecchini, H. I., Nicholson, A. M., Russell, R., Vermeulen, L., Kemp, R., et al. (2013). Intestinal label-retaining cells are secretory precursors expressing Lgr5. *Nature, 495*, 65–69.

Cai, L., Ye, Z., Zhou, B. Y., Mali, P., Zhou, C., & Cheng, L. (2007). Promoting human embryonic stem cell renewal or differentiation by modulating Wnt signal and culture conditions. *Cell Research, 17*, 62–72.

Chaffer, C. L., Brennan, J. P., Slavin, J. L., Blick, T., Thompson, E. W., & Williams, E. D. (2006). Mesenchymal-to-epithelial transition facilitates bladder cancer metastasis: Role of fibroblast growth factor receptor-2. *Cancer Research, 66*, 11271–11278.

Chaffer, C. L., Brueckmann, I., Scheel, C., Kaestli, A. J., Wiggins, P. A., Rodrigues, L. O., et al. (2011). Normal and neoplastic nonstem cells can spontaneously convert to a stem-like state. *Proceedings of the National Academy of Sciences of the United States of America, 108*, 7950–7955.

Chambers, I., Silva, J., Colby, D., Nichols, J., Nijmeijer, B., Robertson, M., et al. (2007). Nanog safeguards pluripotency and mediates germline development. *Nature, 450*, 1230–1234.

Chan, A. T., Ogino, S., & Fuchs, C. S. (2009). Aspirin use and survival after diagnosis of colorectal cancer. *Journal of the American Medical Association, 302*, 649–658.

Chang, H., Guillou, F., Taketo, M. M., & Behringer, R. R. (2009). Overactive beta-catenin signaling causes testicular sertoli cell tumor development in the mouse. *Biology of Reproduction, 81*, 842–849.

Chen, J., Li, Y., Yu, T. S., McKay, R. M., Burns, D. K., Kernie, S. G., et al. (2012). A restricted cell population propagates glioblastoma growth after chemotherapy. *Nature, 488*, 522–526.

Cianferoni, A., Schroeder, J. T., Kim, J., Schmidt, J. W., Lichtenstein, L. M., Georas, S. N., et al. (2001). Selective inhibition of interleukin-4 gene expression in human T cells by aspirin. *Blood, 97*, 1742–1749.

Dalerba, P., Kalisky, T., Sahoo, D., Rajendran, P. S., Rothenberg, M. E., Leyrat, A. A., et al. (2011). Single-cell dissection of transcriptional heterogeneity in human colon tumors. *Nature Biotechnology, 29*, 1120–1127.

Davidson, K. C., Adams, A. M., Goodson, J. M., McDonald, C. E., Potter, J. C., Berndt, J. D., et al. (2012). Wnt/beta-catenin signaling promotes differentiation, not self-renewal, of human embryonic stem cells and is repressed by Oct4. *Proceedings of the National Academy of Sciences of the United States of America, 109*, 4485–4490.

de Jong, J., Stoop, H., Gillis, A. J., van Gurp, R. J., van de Geijn, G. J., Boer, M., et al. (2008). Differential expression of SOX17 and SOX2 in germ cells and stem cells has biological and clinical implications. *The Journal of Pathology, 215*, 21–30.

de Sousa, E. M., Vermeulen, L., Richel, D., & Medema, J. P. (2011). Targeting Wnt signaling in colon cancer stem cells. *Clinical Cancer Research, 17*, 647–653.

DeAlmeida, V. I., Miao, L., Ernst, J. A., Koeppen, H., Polakis, P., & Rubinfeld, B. (2007). The soluble wnt receptor Frizzled8CRD-hFc inhibits the growth of teratocarcinomas in vivo. *Cancer Research, 67*, 5371–5379.

Dembinski, J. L., & Krauss, S. (2009). Characterization and functional analysis of a slow cycling stem cell-like subpopulation in pancreas adenocarcinoma. *Clinical & Experimental Metastasis, 26*, 611–623.

Diehn, M., Cho, R. W., Lobo, N. A., Kalisky, T., Dorie, M. J., Kulp, A. N., et al. (2009). Association of reactive oxygen species levels and radioresistance in cancer stem cells. *Nature, 458*, 780–783.

Doble, B. W., Patel, S., Wood, G. A., Kockeritz, L. K., & Woodgett, J. R. (2007). Functional redundancy of GSK-3alpha and GSK-3beta in Wnt/beta-catenin signaling shown by using an allelic series of embryonic stem cell lines. *Developmental Cell, 12*, 957–971.

Dolberg, D. S., & Bissell, M. J. (1984). Inability of Rous sarcoma virus to cause sarcomas in the avian embryo. *Nature, 309*, 552–556.

Dravid, G., Ye, Z., Hammond, H., Chen, G., Pyle, A., Donovan, P., et al. (2005). Defining the role of Wnt/beta-catenin signaling in the survival, proliferation, and self-renewal of human embryonic stem cells. *Stem Cells, 23*, 1489–1501.

Driessens, G., Beck, B., Caauwe, A., Simons, B. D., & Blanpain, C. (2012). Defining the mode of tumour growth by clonal analysis. *Nature, 488*, 527–530.

Eisenhardt, L., & Cushing, H. (1930). Diagnosis of intracranial tumors by supravital technique. *The American Journal of Pathology, 6*, 541–552, 7.

Fodde, R., & Brabletz, T. (2007). Wnt/beta-catenin signaling in cancer stemness and malignant behavior. *Current Opinion in Cell Biology, 19*, 150–158.

Friedl, P., & Wolf, K. (2003). Tumour-cell invasion and migration: Diversity and escape mechanisms. *Nature Reviews Cancer, 3*, 362–374.

Fuchs, E. (2009). The tortoise and the hair: Slow-cycling cells in the stem cell race. *Cell, 137*, 811–819.

Gangemi, R. M., Griffero, F., Marubbi, D., Perera, M., Capra, M. C., Malatesta, P., et al. (2009). SOX2 silencing in glioblastoma tumor-initiating cells causes stop of proliferation and loss of tumorigenicity. *Stem Cells, 27*, 40–48.

Gao, M. Q., Choi, Y. P., Kang, S., Youn, J. H., & Cho, N. H. (2010). CD24+ cells from hierarchically organized ovarian cancer are enriched in cancer stem cells. *Oncogene, 29*, 2672–2680.

Gerlinger, M., Rowan, A. J., Horswell, S., Larkin, J., Endesfelder, D., Gronroos, E., et al. (2012). Intratumor heterogeneity and branched evolution revealed by multiregion sequencing. *The New England Journal of Medicine, 366*, 883–892.

Ghazvini, M., Sonneveld, P., Kremer, A., Franken, P., Sacchetti, A., Atlasi, Y., et al. (2013). Cancer stemness in Apc- vs. Apc/KRAS-driven intestinal tumorigenesis. *PLoS One, 8*, e73872.

Gupta, P. B., Fillmore, C. M., Jiang, G., Shapira, S. D., Tao, K., Kuperwasser, C., et al. (2011). Stochastic state transitions give rise to phenotypic equilibrium in populations of cancer cells. *Cell, 146*, 633–644.

Gupta, P. B., Onder, T. T., Jiang, G., Tao, K., Kuperwasser, C., Weinberg, R. A., et al. (2009). Identification of selective inhibitors of cancer stem cells by high-throughput screening. *Cell, 138*, 645–659.

Haegel, H., Larue, L., Ohsugi, M., Fedorov, L., Herrenknecht, K., & Kemler, R. (1995). Lack of beta-catenin affects mouse development at gastrulation. *Development, 121*, 3529–3537.

Hayashi, K., Lopes, S. M., Tang, F., & Surani, M. A. (2008). Dynamic equilibrium and heterogeneity of mouse pluripotent stem cells with distinct functional and epigenetic states. *Cell Stem Cell, 3*, 391–401.

Held, M. A., Curley, D. P., Dankort, D., McMahon, M., Muthusamy, V., & Bosenberg, M. W. (2010). Characterization of melanoma cells capable of propagating tumors from a single cell. *Cancer Research, 70*, 388–397.

Hochedlinger, K., Blelloch, R., Brennan, C., Yamada, Y., Kim, M., Chin, L., et al. (2004). Reprogramming of a melanoma genome by nuclear transplantation. *Genes & Development, 18*, 1875–1885.

Hochedlinger, K., Yamada, Y., Beard, C., & Jaenisch, R. (2005). Ectopic expression of Oct-4 blocks progenitor-cell differentiation and causes dysplasia in epithelial tissues. *Cell, 121*, 465–477.

Honecker, F., Kersemaekers, A. M., Molier, M., Van Weeren, P. C., Stoop, H., De Krijger, R. R., et al. (2004). Involvement of E-cadherin and beta-catenin in germ cell tumours and in normal male fetal germ cell development. *The Journal of Pathology, 204*, 167–174.

Horst, D., Chen, J., Morikawa, T., Ogino, S., Kirchner, T., & Shivdasani, R. A. (2012). Differential WNT activity in colorectal cancer confers limited tumorigenic potential and is regulated by MAPK signaling. *Cancer Research, 72*, 1547–1556.

Huelsken, J., Vogel, R., Brinkmann, V., Erdmann, B., Birchmeier, C., & Birchmeier, W. (2000). Requirement for beta-catenin in anterior-posterior axis formation in mice. *The Journal of Cell Biology, 148*, 567–578.

Ikushima, H., Todo, T., Ino, Y., Takahashi, M., Miyazawa, K., & Miyazono, K. (2009). Autocrine TGF-beta signaling maintains tumorigenicity of glioma-initiating cells through Sry-related HMG-box factors. *Cell Stem Cell, 5*, 504–514.

Illmensee, K., & Mintz, B. (1976). Totipotency and normal differentiation of single teratocarcinoma cells cloned by injection into blastocysts. *Proceedings of the National Academy of Sciences of the United States of America, 73*, 549–553.

Ishikawa, F., Yoshida, S., Saito, Y., Hijikata, A., Kitamura, H., Tanaka, S., et al. (2007). Chemotherapy-resistant human AML stem cells home to and engraft within the bone-marrow endosteal region. *Nature Biotechnology, 25*, 1315–1321.

Jackson, E. B., & Brues, A. M. (1941). Studies on a transplantable embryoma of the mouse. *Cancer Research, 1*, 494–498.

Janssen, K. P., Alberici, P., Fsihi, H., Gaspar, C., Breukel, C., Franken, P., et al. (2006). APC and oncogenic KRAS are synergistic in enhancing Wnt signaling in intestinal tumor formation and progression. *Gastroenterology, 131*, 1096–1109.

Kelly, P. N., Dakic, A., Adams, J. M., Nutt, S. L., & Strasser, A. (2007). Tumor growth need not be driven by rare cancer stem cells. *Science, 317*, 337.

Kielman, M. F., Rindapaa, M., Gaspar, C., van Poppel, N., Breukel, C., van Leeuwen, S., et al. (2002). Apc modulates embryonic stem-cell differentiation by controlling the dosage of beta-catenin signaling. *Nature Genetics, 32*, 594–605.

Kim, J., Woo, A. J., Chu, J., Snow, J. W., Fujiwara, Y., Kim, C. G., et al. (2010). A Myc network accounts for similarities between embryonic stem and cancer cell transcription programs. *Cell, 143*, 313–324.

Kimura, T., Nakamura, T., Murayama, K., Umehara, H., Yamano, N., Watanabe, S., et al. (2006). The stabilization of beta-catenin leads to impaired primordial germ cell development via aberrant cell cycle progression. *Developmental Biology, 300*, 545–553.

Kleinsmith, L. J., & Pierce, G. B., Jr. (1964). Multipotentiality of single embryonal carcinoma cells. *Cancer Research, 24*, 1544–1551.

Knapp, D., & Tanaka, E. M. (2012). Regeneration and reprogramming. *Current Opinion in Genetics & Development, 22*, 485–493.

Knutsen, T., Mickley, L. A., Ried, T., Green, E. D., du Manoir, S., Schrock, E., et al. (1998). Cytogenetic and molecular characterization of random chromosomal rearrangements activating the drug resistance gene, MDR1/P-glycoprotein, in drug-selected cell lines and patients with drug refractory ALL. *Genes, Chromosomes & Cancer, 23*, 44–54.

Korkola, J. E., Houldsworth, J., Chadalavada, R. S., Olshen, A. B., Dobrzynski, D., Reuter, V. E., et al. (2006). Down-regulation of stem cell genes, including those in a 200-kb gene cluster at 12p13.31, is associated with in vivo differentiation of human male germ cell tumors. *Cancer Research, 66*, 820–827.

Korpal, M., Ell, B. J., Buffa, F. M., Ibrahim, T., Blanco, M. A., Celia-Terrassa, T., et al. (2011). Direct targeting of Sec23a by miR-200 s influences cancer cell secretome and promotes metastatic colonization. *Nature Medicine, 17*, 1101–1108.

Kusumbe, A. P., & Bapat, S. A. (2009). Cancer stem cells and aneuploid populations within developing tumors are the major determinants of tumor dormancy. *Cancer Research, 69*, 9245–9253.

Laga, A. C., Lai, C. Y., Zhan, Q., Huang, S. J., Velazquez, E. F., Yang, Q., et al. (2010). Expression of the embryonic stem cell transcription factor SOX2 in human skin: Relevance to melanocyte and merkel cell biology. *The American Journal of Pathology, 176*, 903–913.

Lan, P., Tonomura, N., Shimizu, A., Wang, S., & Yang, Y. G. (2006). Reconstitution of a functional human immune system in immunodeficient mice through combined human fetal thymus/liver and CD34+ cell transplantation. *Blood, 108*, 487–492.

Lapidot, T., Sirard, C., Vormoor, J., Murdoch, B., Hoang, T., Caceres-Cortes, J., et al. (1994). A cell initiating human acute myeloid leukaemia after transplantation into SCID mice. *Nature, 367*, 645–648.

Lapouge, G., Beck, B., Nassar, D., Dubois, C., Dekoninck, S., & Blanpain, C. (2012). Skin squamous cell carcinoma propagating cells increase with tumour progression and invasiveness. *The EMBO Journal, 31*, 4563–4575.

Li, L., Connelly, M. C., Wetmore, C., Curran, T., & Morgan, J. I. (2003). Mouse embryos cloned from brain tumors. *Cancer Research, 63*, 2733–2736.

Lindsley, R. C., Gill, J. G., Kyba, M., Murphy, T. L., & Murphy, K. M. (2006). Canonical Wnt signaling is required for development of embryonic stem cell-derived mesoderm. *Development, 133*, 3787–3796.

Looijenga, L. H. J., Stoop, H., De Leeuw, P. J. C., De Gouveia Brazao, C. A., Gillis, A. J. M., Van Roozendaal, K. E. P., et al. (2003). POU5F1 (OCT3/4) identifies cells with pluripotent potential in human germ cell tumors. *Cancer Research, 63*, 2244–2250.

Lu, D., Choi, M. Y., Yu, J., Castro, J. E., Kipps, T. J., & Carson, D. A. (2011). Salinomycin inhibits Wnt signaling and selectively induces apoptosis in chronic lymphocytic leukemia cells. *Proceedings of the National Academy of Sciences of the United States of America, 108*, 13253–13257.

Lyashenko, N., Winter, M., Migliorini, D., Biechele, T., Moon, R. T., & Hartmann, C. (2011). Differential requirement for the dual functions of beta-catenin in embryonic stem cell self-renewal and germ layer formation. *Nature Cell Biology, 13*, 753–761.

Magee, J. A., Piskounova, E., & Morrison, S. J. (2012). Cancer stem cells: Impact, heterogeneity, and uncertainty. *Cancer Cell, 21*, 283–296.

Malanchi, I., Peinado, H., Kassen, D., Hussenet, T., Metzger, D., Chambon, P., et al. (2008). Cutaneous cancer stem cell maintenance is dependent on beta-catenin signalling. *Nature, 452*, 650–653.

Mani, S. A., Guo, W., Liao, M. J., Eaton, F. N., Ayyanan, A., Zhou, A. Y., et al. (2008). The epithelial–mesenchymal transition generates cells with properties of stem cells. *Cell, 133*, 704–715.

McConnell, B. B., Ghaleb, A. M., Nandan, M. O., & Yang, V. W. (2007). The diverse functions of Kruppel-like factors 4 and 5 in epithelial biology and pathobiology. *Bioessays, 29*, 549–557.

Mickley, L. A., Spengler, B. A., Knutsen, T. A., Biedler, J. L., & Fojo, T. (1997). Gene rearrangement: A novel mechanism for MDR-1 gene activation. *The Journal of Clinical Investigation, 99*, 1947–1957.

Muller, V., Alix-Panabieres, C., & Pantel, K. (2010). Insights into minimal residual disease in cancer patients: Implications for anti-cancer therapies. *European Journal of Cancer, 46*, 1189–1197.

Neugut, A. I. (2009). Aspirin as adjuvant therapy for colorectal cancer: A promising new twist for an old drug. *Journal of the American Medical Association, 302*, 688–689.

Nguyen, L. V., Vanner, R., Dirks, P., & Eaves, C. J. (2012). Cancer stem cells: An evolving concept. *Nature Reviews Cancer, 12*, 133–143.

Nusse, R., Fuerer, C., Ching, W., Harnish, K., Logan, C., Zeng, A., et al. (2008). Wnt signaling and stem cell control. *Cold Spring Harbor Symposia on Quantitative Biology*, *73*, 59–66.

Nusse, R., & Varmus, H. E. (1982). Many tumors induced by the mouse mammary tumor virus contain a provirus integrated in the same region of the host genome. *Cell*, *31*, 99–109.

Nusslein-Volhard, C., & Wieschaus, E. (1980). Mutations affecting segment number and polarity in Drosophila. *Nature*, *287*, 795–801.

Ocana, O. H., Corcoles, R., Fabra, A., Moreno-Bueno, G., Acloque, H., Vega, S., et al. (2012). Metastatic colonization requires the repression of the epithelial-mesenchymal transition inducer Prrx1. *Cancer Cell*, *22*, 709–724.

Oguma, K., Oshima, H., Aoki, M., Uchio, R., Naka, K., Nakamura, S., et al. (2008). Activated macrophages promote Wnt signalling through tumour necrosis factor-alpha in gastric tumour cells. *The EMBO Journal*, *27*, 1671–1681.

Oosterhuis, J. W., & Looijenga, L. H. (2005). Testicular germ-cell tumours in a broader perspective. *Nature Reviews Cancer*, *5*, 210–222.

Park, C. H., Chang, J. Y., Hahm, E. R., Park, S., Kim, H. K., & Yang, C. H. (2005). Quercetin, a potent inhibitor against beta-catenin/Tcf signaling in SW480 colon cancer cells. *Biochemical and Biophysical Research Communications*, *328*, 227–234.

Pece, S., Tosoni, D., Confalonieri, S., Mazzarol, G., Vecchi, M., Ronzoni, S., et al. (2010). Biological and molecular heterogeneity of breast cancers correlates with their cancer stem cell content. *Cell*, *140*, 62–73.

Petrie, K., Zelent, A., & Waxman, S. (2009). Differentiation therapy of acute myeloid leukemia: Past, present and future. *Current Opinion in Hematology*, *16*, 84–91.

Pierce, G. B., & Dixon, F. J., Jr. (1959). Testicular teratomas. I. Demonstration of teratogenesis by metamorphosis of multipotential cells. *Cancer*, *12*, 573–583.

Pierce, G. B., & Speers, W. C. (1988). Tumors as caricatures of the process of tissue renewal: Prospects for therapy by directing differentiation. *Cancer Research*, *48*, 1996–2004.

Ponti, D., Costa, A., Zaffaroni, N., Pratesi, G., Petrangolini, G., Coradini, D., et al. (2005). Isolation and in vitro propagation of tumorigenic breast cancer cells with stem/progenitor cell properties. *Cancer Research*, *65*, 5506–5511.

Quintana, E., Shackleton, M., Sabel, M. S., Fullen, D. R., Johnson, T. M., & Morrison, S. J. (2008). Efficient tumour formation by single human melanoma cells. *Nature*, *456*, 593–598.

Reya, T., & Clevers, H. (2005). Wnt signalling in stem cells and cancer. *Nature*, *434*, 843–850.

Reya, T., Morrison, S. J., Clarke, M. F., & Weissman, I. L. (2001). Stem cells, cancer, and cancer stem cells. *Nature*, *414*, 105–111.

Rich, J. N., & Bao, S. (2007). Chemotherapy and cancer stem cells. *Cell Stem Cell*, *1*, 353–355.

Rijlaarsdam, M. A., van Herk, H. A., Gillis, A. J., Stoop, H., Jenster, G., Martens, J., et al. (2011). Specific detection of OCT3/4 isoform A/B/B1 expression in solid (germ cell) tumours and cell lines: Confirmation of OCT3/4 specificity for germ cell tumours. *British Journal of Cancer*, *105*, 854–863.

Rodriguez-Pinilla, S. M., Sarrio, D., Moreno-Bueno, G., Rodriguez-Gil, Y., Martinez, M. A., Hernandez, L., et al. (2007). Sox2: A possible driver of the basal-like phenotype in sporadic breast cancer. *Modern Pathology*, *20*, 474–481.

Roelofs, H., Manes, T., Janszen, T., Millan, J. L., Oosterhuis, J. W., & Looijenga, L. H. (1999). Heterogeneity in alkaline phosphatase isozyme expression in human testicular germ cell tumours: An enzyme-/immunohistochemical and molecular analysis. *The Journal of Pathology*, *189*, 236–244.

Roesch, A., Fukunaga-Kalabis, M., Schmidt, E. C., Zabierowski, S. E., Brafford, P. A., Vultur, A., et al. (2010). A temporarily distinct subpopulation of slow-cycling melanoma cells is required for continuous tumor growth. *Cell, 141*, 583–594.

Ross, D. M., Hughes, T. P., & Melo, J. V. (2011). Do we have to kill the last CML cell? *Leukemia, 25*, 193–200.

Roth, S., & Fodde, R. (2011). Quiescent stem cells in intestinal homeostasis and cancer. *Cell Communication & Adhesion, 18*, 33–44.

Roth, S., Franken, P., Sacchetti, A., Kremer, A., Anderson, K., Sansom, O., et al. (2012). Paneth cells in intestinal homeostasis and tissue injury. *PLoS One, 7*, e38965.

Rous, P. (1979). A transmissible avian neoplasm (Sarcoma of the common fowl) by M. D. Peyton Rous, Experimental Medicine for Sept. 1, 1910, vol. 12, pp. 696–705. *The Journal of Experimental Medicine, 150*, 738–753.

Rowland, B. D., & Peeper, D. S. (2006). KLF4, p21 and context-dependent opposing forces in cancer. *Nature Reviews Cancer, 6*, 11–23.

Saito, Y., Uchida, N., Tanaka, S., Suzuki, N., Tomizawa-Murasawa, M., Sone, A., et al. (2010). Induction of cell cycle entry eliminates human leukemia stem cells in a mouse model of AML. *Nature Biotechnology, 28*, 275–280.

Schepers, A. G., Snippert, H. J., Stange, D. E., van den Born, M., van Es, J. H., van de Wetering, M., et al. (2012). Lineage tracing reveals Lgr5+ stem cell activity in mouse intestinal adenomas. *Science, 337*, 730–735.

Schoenhals, M., Kassambara, A., De Vos, J., Hose, D., Moreaux, J., & Klein, B. (2009). Embryonic stem cell markers expression in cancers. *Biochemical and Biophysical Research Communications, 383*, 157–162.

Singh, S. K., Clarke, I. D., Terasaki, M., Bonn, V. E., Hawkins, C., Squire, J., et al. (2003). Identification of a cancer stem cell in human brain tumors. *Cancer Research, 63*, 5821–5828.

Singh, S. K., Hawkins, C., Clarke, I. D., Squire, J. A., Bayani, J., Hide, T., et al. (2004). Identification of human brain tumour initiating cells. *Nature, 432*, 396–401.

Sinner, D., Kordich, J. J., Spence, J. R., Opoka, R., Rankin, S., Lin, S. C., et al. (2007). Sox17 and Sox4 differentially regulate beta-catenin/T-cell factor activity and proliferation of colon carcinoma cells. *Molecular and Cellular Biology, 27*, 7802–7815.

Snow, G. E., Kasper, A. C., Busch, A. M., Schwarz, E., Ewings, K. E., Bee, T., et al. (2009). Wnt pathway reprogramming during human embryonal carcinoma differentiation and potential for therapeutic targeting. *BMC Cancer, 9*, 383.

Solter, D., Skreb, N., & Damjanov, I. (1970). Extrauterine growth of mouse egg-cylinders results in malignant teratoma. *Nature, 227*, 503–504.

Sotomayor, P., Godoy, A., Smith, G. J., & Huss, W. J. (2009). Oct4A is expressed by a subpopulation of prostate neuroendocrine cells. *Prostate, 69*, 401–410.

Spiller, C. M., Feng, C. W., Jackson, A., Gillis, A. J. M., Rolland, A. D., Looijenga, L. H. J., et al. (2012). Endogenous Nodal signalling regulates germ cell potency during mammalian testis development. *Development, 139*, 4123–4132.

Stevens, L. C. (1968). The development of teratomas from intratesticular grafts of tubal mouse eggs. *Journal of Embryology and Experimental Morphology, 20*, 329–341.

Stevens, L. C. (1970). The development of transplantable teratocarcinomas from intratesticular grafts of pre- and postimplantation mouse embryos. *Developmental Biology, 21*, 364–382.

Stoker, A. W., Hatier, C., & Bissell, M. J. (1990). The embryonic environment strongly attenuates v-src oncogenesis in mesenchymal and epithelial tissues, but not in endothelia. *The Journal of Cell Biology, 111*, 217–228.

Strohmeyer, T., Peter, S., Hartmann, M., Munemitsu, S., Ackermann, R., Ullrich, A., et al. (1991). Expression of the hst-1 and c-kit protooncogenes in human testicular germ cell tumors. *Cancer Research, 51*, 1811–1816.

Strowig, T., Rongvaux, A., Rathinam, C., Takizawa, H., Borsotti, C., Philbrick, W., et al. (2011). Transgenic expression of human signal regulatory protein alpha in Rag2-/-gamma(c)-/- mice improves engraftment of human hematopoietic cells in humanized mice. *Proceedings of the National Academy of Sciences of the United States of America, 108*, 13218–13223.

Sun, Y., Campisi, J., Higano, C., Beer, T. M., Porter, P., Coleman, I., et al. (2012). Treatment-induced damage to the tumor microenvironment promotes prostate cancer therapy resistance through WNT16B. *Nature Medicine, 18*, 1359–1368.

Taipale, J., & Beachy, P. A. (2001). The Hedgehog and Wnt signalling pathways in cancer. *Nature, 411*, 349–354.

Takahashi, K., & Yamanaka, S. (2006). Induction of pluripotent stem cells from mouse embryonic and adult fibroblast cultures by defined factors. *Cell, 126*, 663–676.

Takebe, N., Harris, P. J., Warren, R. Q., & Ivy, S. P. (2011). Targeting cancer stem cells by inhibiting Wnt, Notch, and Hedgehog pathways. *Nature Reviews Clinical Oncology, 8*, 97–106.

Tang, Q. L., Zhao, Z. Q., Li, J. C., Liang, Y., Yin, J. Q., Zou, C. Y., et al. (2011). Salinomycin inhibits osteosarcoma by targeting its tumor stem cells. *Cancer Letters, 311*, 113–121.

ten Berge, D., Kurek, D., Blauwkamp, T., Koole, W., Maas, A., Eroglu, E., et al. (2011). Embryonic stem cells require Wnt proteins to prevent differentiation to epiblast stem cells. *Nature Cell Biology, 13*, 1070–1075.

Thiery, J. P. (2002). Epithelial-mesenchymal transitions in tumour progression. *Nature Reviews Cancer, 2*, 442–454.

Thiery, J. P., Acloque, H., Huang, R. Y., & Nieto, M. A. (2009). Epithelial-mesenchymal transitions in development and disease. *Cell, 139*, 871–890.

Todaro, M., Alea, M. P., Di Stefano, A. B., Cammareri, P., Vermeulen, L., Iovino, F., et al. (2007). Colon cancer stem cells dictate tumor growth and resist cell death by production of interleukin-4. *Cell Stem Cell, 1*, 389–402.

Toyooka, Y., Shimosato, D., Murakami, K., Takahashi, K., & Niwa, H. (2008). Identification and characterization of subpopulations in undifferentiated ES cell culture. *Development, 135*, 909–918.

Tsai, J. H., Donaher, J. L., Murphy, D. A., Chau, S., & Yang, J. (2012). Spatiotemporal regulation of epithelial-mesenchymal transition is essential for squamous cell carcinoma metastasis. *Cancer Cell, 22*, 725–736.

Ullmann, U., Gilles, C., De Rycke, M., Van de Velde, H., Sermon, K., & Liebaers, I. (2008). GSK-3-specific inhibitor-supplemented hESC medium prevents the epithelial-mesenchymal transition process and the up-regulation of matrix metalloproteinases in hESCs cultured in feeder-free conditions. *Molecular Human Reproduction, 14*, 169–179.

Valent, P., Bonnet, D., De Maria, R., Lapidot, T., Copland, M., Melo, J. V., et al. (2012). Cancer stem cell definitions and terminology: The devil is in the details. *Nature Reviews Cancer, 12*, 767–775.

Vermeulen, L., De Sousa, E. M. F., van der Heijden, M., Cameron, K., de Jong, J. H., Borovski, T., et al. (2010). Wnt activity defines colon cancer stem cells and is regulated by the microenvironment. *Nature Cell Biology, 12*, 468–476.

Vermeulen, L., Todaro, M., de Sousa Mello, F., Sprick, M. R., Kemper, K., Perez Alea, M., et al. (2008). Single-cell cloning of colon cancer stem cells reveals a multi-lineage differentiation capacity. *Proceedings of the National Academy of Sciences of the United States of America, 105*, 13427–13432.

Villa-Diaz, L. G., Pacut, C., Slawny, N. A., Ding, J., O'Shea, K. S., & Smith, G. D. (2009). Analysis of the factors that limit the ability of feeder cells to maintain the undifferentiated state of human embryonic stem cells. *Stem Cells and Development, 18*, 641–651.

Visvader, J. E., & Lindeman, G. J. (2008). Cancer stem cells in solid tumours: Accumulating evidence and unresolved questions. *Nature Reviews Cancer, 8*, 755–768.

Wagner, R. T., Xu, X., Yi, F., Merrill, B. J., & Cooney, A. J. (2010). Canonical Wnt/beta-catenin regulation of liver receptor homolog-1 mediates pluripotency gene expression. *Stem Cells, 28*, 1794–1804.

Wang, S., Yang, D., & Lippman, M. E. (2003). Targeting Bcl-2 and Bcl-XL with nonpeptidic small-molecule antagonists. *Seminars in Oncology, 30*, 133–142.

Watt, F. M., & Collins, C. A. (2008). Role of beta-catenin in epidermal stem cell expansion, lineage selection, and cancer. *Cold Spring Harbor Symposia on Quantitative Biology, 73*, 503–512.

Westerman, B. A., Braat, A. K., Taub, N., Potman, M., Vissers, J. H., Blom, M., et al. (2011). A genome-wide RNAi screen in mouse embryonic stem cells identifies Mp1 as a key mediator of differentiation. *The Journal of Experimental Medicine, 208*, 2675–2689.

Wong, D. J., Liu, H., Ridky, T. W., Cassarino, D., Segal, E., & Chang, H. Y. (2008). Module map of stem cell genes guides creation of epithelial cancer stem cells. *Cell Stem Cell, 2*, 333–344.

Woodward, W. A., Chen, M. S., Behbod, F., Alfaro, M. P., Buchholz, T. A., & Rosen, J. M. (2007). WNT/beta-catenin mediates radiation resistance of mouse mammary progenitor cells. *Proceedings of the National Academy of Sciences of the United States of America, 104*, 618–623.

Wray, J., Kalkan, T., Gomez-Lopez, S., Eckardt, D., Cook, A., Kemler, R., et al. (2011). Inhibition of glycogen synthase kinase-3 alleviates Tcf3 repression of the pluripotency network and increases embryonic stem cell resistance to differentiation. *Nature Cell Biology, 13*, 838–845.

Wunderlich, M., Chou, F. S., Link, K. A., Mizukawa, B., Perry, R. L., Carroll, M., et al. (2010). AML xenograft efficiency is significantly improved in NOD/SCID-IL2RG mice constitutively expressing human SCF, GM-CSF and IL-3. *Leukemia, 24*, 1785–1788.

Ying, Q. L., Wray, J., Nichols, J., Batlle-Morera, L., Doble, B., Woodgett, J., et al. (2008). The ground state of embryonic stem cell self-renewal. *Nature, 453*, 519–523.

Zeng, Y. A., & Nusse, R. (2010). Wnt proteins are self-renewal factors for mammary stem cells and promote their long-term expansion in culture. *Cell Stem Cell, 6*, 568–577.

Zorn, A. M., Barish, G. D., Williams, B. O., Lavender, P., Klymkowsky, M. W., & Varmus, H. E. (1999). Regulation of Wnt signaling by Sox proteins: XSox17 alpha/beta and XSox3 physically interact with beta-catenin. *Molecular Cell, 4*, 487–498.

CHAPTER FOURTEEN

Cellular Mechanisms of Somatic Stem Cell Aging

Yunjoon Jung*, Andrew S. Brack*,†,‡,1

*Center for Regenerative Medicine, Massachusetts General Hospital, Boston, Massachusetts, USA
†Harvard Medical School, Boston, Massachusetts, USA
‡Harvard Stem Cell Institute, Boston, Massachusetts, USA
[1]Corresponding author:

Contents

Abstract

Tissue homeostasis and regenerative capacity rely on rare populations of somatic stem cells endowed with the potential to self-renew and differentiate. During aging, many tissues show a decline in regenerative potential coupled with a loss of stem cell function. Cells including somatic stem cells have evolved a series of checks and balances to sense and repair cellular damage to maximize tissue function. However, during aging the mechanisms that protect normal cell function begin to fail. In this review, we will discuss how common cellular mechanisms that maintain tissue fidelity and organismal lifespan impact somatic stem cell function. We will highlight context-dependent changes and commonalities that define aging, by focusing on three age-sensitive stem cell compartments: blood, neural, and muscle. Understanding the interaction between extrinsic regulators and intrinsic effectors that operate within different stem cell compartments is likely to have important implications for identifying strategies to improve health span and treat age-related degenerative diseases.

Current Topics in Developmental Biology, Volume 107
ISSN 0070-2153
http://dx.doi.org/10.1016/B978-0-12-416022-4.00014-7

1. INTRODUCTION

Aging leads to profound effects on many, if not all tissues of the body, including muscle weakness (Lang et al., 2010), graying and loss of hair (Nishimura, Granter, & Fisher, 2005), a decline in cognition (Bishop, Lu, & Yankner, 2010), and impaired immune function (Geiger, de Haan, & Florian, 2013). The regenerative response of tissues after injury is often delayed leading to slower repair of parenchyma that is commonly replaced by accumulation of adipogenic or fibrogenic accumulation (Kapetanaki, Mora, & Rojas, 2013).

Maintenance and repair of many adult tissues rely on stem cells. These cells reside at the top of a cellular hierarchy endowed with the ability to self-renew and differentiate, whereas their downstream progeny is restricted to replenishing the differentiated tissue (Orford & Scadden, 2008; Simons & Clevers, 2011). Stem cells spend relatively long periods of time in a quiescent state compared to their progeny, which proliferate to produce numerous differentiated cells that replace or repair the tissue throughout the lifespan of the organism (Li & Clevers, 2010; Orford & Scadden, 2008). In response to increased demand such as growth or regeneration after injury, stem cells break from quiescence, enter the cell cycle, and divide either symmetrically or asymmetrically to replace the stem cell pool and the committed progenitor pool. To avoid abnormal growth or loss of tissues, the balance between production of stem cells and differentiated progeny needs to be tightly regulated. Multiple levels of cell autonomous and extrinsic factors tightly control fate decisions of stem cells. For example, a specialized microenvironment, also known as the stem cell niche, provides extrinsic signals in the form of paracrine or juxtacrine signaling that is essential for maintenance of stem cell function and restricting stem cell numbers (Li & Clevers, 2010; Morrison & Spradling, 2008). It is possible that extrinsic signals derived from the local niche and systemic environment shape the epigenetic landscape of the stem cell, which influences gene expression to dictate stem cell fate (Pollina & Brunet, 2011).

Recent technological advances in genetic reporters and cell surface marker detection have revealed a greater complexity in stem cell populations than previously anticipated (Grompe, 2012; Simons & Clevers, 2011). Across different niches, stem cells with a restricted proliferative history, termed slow dividing stem cells, are endowed with high self-renewing potential compared with stem cells from the same tissue that have undergone more divisions during their history (Chakkalakal, Jones, Basson, & Brack, 2012; Foudi et al., 2009; Wilson et al., 2008; Zhang, Cheong, Ciapurin, McDermitt, & Tumbar, 2009). That slow dividing cell give rise to frequently dividing cells, but not vice versa, demonstrates a hierarchical relationship that is controlled by or correlated with proliferative output. As the markers to define stem cells increase, the degree of heterogeneity within a population is becoming appreciated. Within the same tissue, subsets of stem cells can be indiscriminately identified that are biased to differentiate into distinct cell types, albeit restricted in the same developmental lineage. Due to this level of complexity, it is possible that changes in function between two points (i.e., adult and aged) are a feature of extrinsic and intrinsic changes in all stem cells or the expansion of biased subsets over others.

Studies on stem cell aging and the molecular regulation of lifespan were pioneered in nonmammalian systems (Jones & Rando, 2011; Kenyon, 2010). In *Drosophila*, the number of stem cells in the testis and ovary declines during aging, due in part to age-dependent changes in the niche (Boyle, Wong, Rocha, & Jones, 2007; Pan et al., 2007). Moreover, deregulation of Notch/JNK (Jun-activated kinase) signaling during aging causes loss of intestinal tissue homeostasis through overproliferation and inappropriate differentiation of intestinal stem cells (ISCs) (Biteau, Hochmuth, & Jasper, 2008). Remarkably, reducing proliferation of ISCs through repression of insulin-like growth factor (IGF) and JNK stress pathways demonstrates an inverse correlation between lifespan and ISC proliferation. Maximal lifespan was achieved when ISC proliferation was reduced, which correlated with improved metabolic homeostasis of aged *Drosophila* (Biteau et al., 2010). This demonstrates a direct link between lifespan and stem cell activity, at least in the intestine. Moreover, stem cell function and lifespan are affected by metabolic and epigenetic factors that change with age (Bratic & Larsson, 2013; Eijkelenboom & Burgering, 2013; Laplante & Sabatini, 2012; Pollina & Brunet, 2011).

At the organismal level, aging is based on a chronological clock. At the cellular level, age can be broken down into two components, replicative and chronological age. Replicative age relates to the proliferative output of stem cells during their history. Chronological aging is linked to the age of the intracellular constituents of stem cells. Hierarchically upstream stem cells are widely considered restricted in their proliferative output compared to downstream progenitors. Therefore, aging of quiescent stem cells occurs primarily on a chronological clock, whereas the age of downstream progeny is based on a replicative clock.

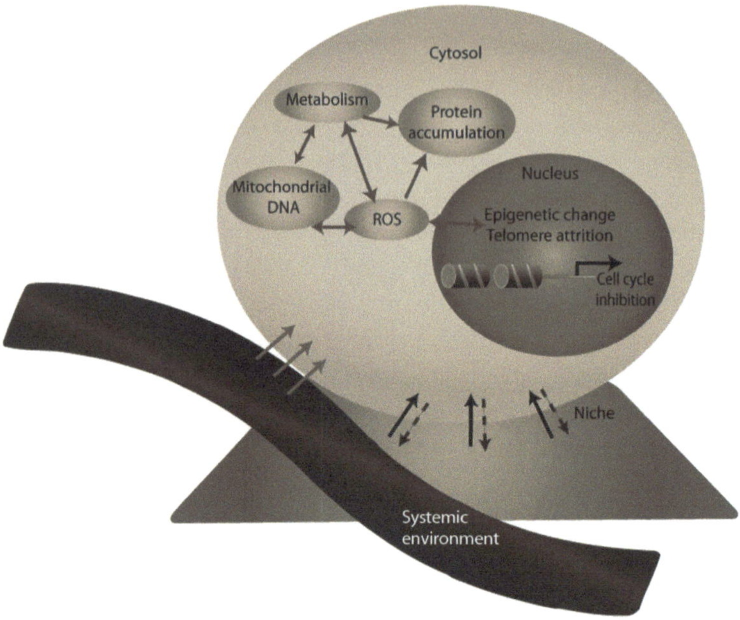

Figure 14.1 Multiple components driving stem cell aging. Stem cell aging occurs through changes in both cell-extrinsic regulators and intrinsic effectors. Cell-external influences such as changes in systemic factors from the bloodstream (blue) or factors from the stem cell niche (gray) can alter stem cell function by inducing signaling pathways and modifying the epigenetic and genetic signature of stem cells. The aged stem cell may also directly alter the niche. Aging is also driven by deregulation of cell-intrinsic effectors that control proteostasis, mitochondrial function, and metabolic control. Many of these modifications are interrelated suggesting one can impact another, leading to stem cell failure at multiple levels. ROS, reactive oxygen species; mtDNA, mitochondrial DNA. (See color plate.)

During each round of division a cell has to faithfully copy its DNA, repair any errors, and transcribe and translate proteins necessary for ensuring the appropriate fate and functionality of the cell. Each of these processes is error prone and demands high-fidelity repair processes and functional checkpoints. In the absence of replication, the cellular constituents may be the same age as the organism itself. In a nonreplicating cell, the cellular constituents must repair faulty DNA and remove damaged and misfolded proteins that are normally cleared through cell division to avoid protein toxicity. Whether a stem cell ages on a chronological or replicative clock will affect the cellular damage and repair processes that are invoked, which ultimately may impact the aging phenotype of the stem cell.

In the mammalian organism, it is unquestionable that tissue homeostasis and regenerative capacity decline during aging. It is clear that age-dependent stem cell dysfunction can manifest in many forms, such as depletion of available stem cells, deregulation of cell fate, that is, loss of self-renewal and/or differentiation, increased apoptosis, and senescence. However age-dependent changes across different niches are not conserved. For example, as we discuss later, the number of skeletal muscle stem cells and neural stem cells (NSCs) decreases, whereas the number of blood forming stem cells is maintained during aging. These differences may reflect the extent of accrued damage, the capacity to repair, or the ability to persist in spite of damage. It is noteworthy that the defects that impact stem cell function with age are unique to stem cells, such as impaired self-renewal potential; however, some changes such as apoptosis or senescence may be generalizable to all types of somatic cells during aging and organismal lifespan.

In this review, we discuss the mechanisms and consequences of cellular and organismal aging that are used reiteratively in the regulation of mammalian somatic stem cells. We also highlight the common and context-specific age-dependent changes of stem cell function by focusing on three paradigmatic stem cell populations: blood, neural, and skeletal muscle, due to their previously characterized aged phenotypes.

2. MOLECULAR PLAYERS OF CELLULAR AGING

In this section, we outline the cell-intrinsic mechanisms and extrinsic modifiers that have been demonstrated to control cellular aging (Fig. 14.1).

2.1. Genome

DNA is relatively unstable and prone to DNA mutations through by-products of cellular metabolism such as reactive oxygen species (ROS) and environmentally induced lesions such as exposure to ultraviolet (UV) or irradiation (IR). Accumulation of irreversible genomic DNA damage has been implicated as a prominent cause of aging (Eijkelenboom & Burgering, 2013; Kenyon, 2010; Sperka, Wang, & Rudolph, 2012). The WRN (Werner syndrome ATP-dependent) helicase and ATM (Ataxia Telangiectasia Mutated) kinase are essential for DNA repair. Deletion of either leads to a premature aging phenotype in mice (Lombard et al., 2000; Wong et al., 2003).

Maintenance of genomic integrity and fidelity is dependent on protective DNA repair mechanisms. ROS-induced damage is repaired by base excision repair, whereas repair of double strand breaks (DSBs) after IR is achieved by nucleotide excision repair (Blanpain, Mohrin, Sotiropoulou, & Passegue, 2011). Cells invoke distinct methods to repair DSBs depending on the cell cycle status; quiescent cells are repaired by non-homologous end joining (NHEJ), proliferating cells are repaired by homologous recombination (HR). Therefore, due to their quiescent nature, somatic stem cells may rely on NHEJ for repair. NHEJ is a more error-prone repair mechanism, as it does not rely on the other intact DNA strand as a template. Therefore, it is possible that the quiescent state of stem cells may increase their likelihood of accumulating DNA damage and impacting stem cell function; however, it is also possible that when the quiescent stem cell enters the cell cycle any damage will be repaired using HR.

Telomere integrity is cited as a major regulator of lifespan longevity (Flores et al., 2008; Vaziri & Benchimol, 1996). Age-associated telomere reduction has been shown to threaten chromosome integrity of highly proliferative aging tissues (Vaziri & Benchimol, 1996). A gradual decline of telomere length with age has been observed in mouse (Flores et al., 2008) and human tissues (Harley, Futcher, & Greider, 1990). Further evidence supports the notion that age-related decrease in telomere length could occur through loss of telomerase, which maintains telomere length. Telomerase mutations, which are found in patients with *dyskeratosis congenita*, who have shorter telomeres, show a premature aging phenotype (Mitchell, Wood, & Collins, 1999). In a mouse model, loss of telomerase promotes lineage skewedness, which is one of the aging phenotypes in HSCs (hematopoietic stem cells) (Ju et al., 2007). Overexpression of telomerase delays the appearance of age-dependent phenotypes and shows cancer resistance (Tomas-Loba et al., 2008). These studies show that shortened telomeres or a deficit in telomerase function can cause a functional decline of tissues. Therefore, the proliferative output and levels of telomerase will influence the contribution of telomere biology to stem cell decline during aging.

Accumulation of DNA damage results in a cell checkpoint response, involving the upregulation of cell cycle inhibitors, such as $p16^{Ink4a}$, $p19^{Arf}$, and p53 that leads to cell cycle arrest, senescence, apoptosis, or differentiation (Signer & Morrison, 2013; Sperka et al., 2012). $p16^{Ink4a}$, $p19^{Arf}$, and p53 are increased with age (Sahin & DePinho, 2012; Sperka et al., 2012) and their loss is associated with tumor incidence (Kemp, Donehower, Bradley, & Balmain, 1993; Matheu et al., 2007). Intuitively, higher expression of tumor suppressors may be associated with increased incidences of cancers with age as a compensatory mechanism to prevent the prevalence of tumors (Matheu et al., 2007). Insufficiency of BubR1, which is one of the major components of the mitotic checkpoint for spindle assembly, increases expression of both $p16^{Ink4a}$ and $p19^{Arf}$ that causes premature aging by progressively increased aneuploidy and senescence (Baker et al., 2004). As expected, loss of $p16^{Ink4a}$ attenuates age-dependent decline in proliferation and function of stem cells (Janzen et al., 2006; Krishnamurthy et al., 2006; Molofsky et al., 2006). In contrast to differentiated tissue, there is a scarcity of data showing that senescence occurs to stem cells *in vivo*. Interestingly, the influence of senescent cells on tissue homeostasis may be through their influence on neighboring cells via secretion of paracrine factors (Burtner & Kennedy, 2010; Campisi, 2005; Krishnamurthy et al., 2004).

2.2. Mitochondrial DNA mutation

Oxidative phosphorylation in the mitochondrial electron transport chain generates ROS, which is highly reactive and toxic to mitochondrial (mt) DNA, leads to a decline in mitochondrial functions (Kujoth et al., 2005; Trifunovic et al., 2004). With age, increased levels of ROS have been observed along with dysfunctional mitochondria and considered as a cause of aging (Bratic & Larsson, 2013; Trifunovic et al., 2004).

Can mtDNA mutations affect lifespan and aging? Homozygous knock-in mice engineered to express mtDNA polymerase that have defective proof-reading, show increased mtDNA point mutations and deletions, which is associated with reduced lifespan and premature aging phenotypes (Kujoth et al., 2005; Trifunovic et al., 2004). In addition, mice that lack mtDNA polymerase exonuclease display neural and hematopoietic progenitor dysfunction and a progeria phenotype (Ahlqvist et al., 2012). In contrast, introduction of random point mutations in mtDNA to a mouse model was not sufficient to reduce lifespan (Edgar et al., 2009). Therefore, the extent of mtDNA damage or the specific mutation in the mitochondrial genome will influence the impact to the stem cell.

2.3. Epigenome

The epigenetic code enables cells to receive and remember environmentally induced signals to create a more stable state (Cavalli & Paro, 1998; Grewal & Klar, 1996). DNA methylation and histone modifications provide transient regulation of gene expression, but not through the DNA sequence itself.

The identification of specific enzymes that balance these modifications suggests that modulation of the epigenome may influence stem cell fate and the aging process.

Histone acetylases and deacetylases add and remove acetyl groups on histones, respectively. Sirtuins (Sirt) are a family of histone deacetylases that have been shown to regulate organismal lifespan as well as oxidative stress and DNA damages (Kaeberlein, McVey, & Guarente, 1999; Kennedy, Austriaco, Zhang, & Guarente, 1995; Mostoslavsky et al., 2006; Rodgers et al., 2005). Sirt1 and Sirt2 have been implicated in life extension in different model organisms (Kaeberlein et al., 1999; Kennedy et al., 1995). However, moderate overexpression of Sirt1 in mice was not sufficient to increase lifespan (Herranz et al., 2010). In contrast, overexpression of Sirt6 extends lifespan of male mice by regulating the IGF signaling pathway (Kanfi et al., 2012), which has a key role in aging (Kenyon, 2010).

Histone methylation also plays a role in aging. The ASH-2 trithorax complex 9, which trimethylates H3K4, is a lifespan regulator in *Caenorhabditis elegans*. Deficiencies in the ASH-2 complex and the H3K4 methyltransferase SET-2 are shown to extend worm lifespan (Greer et al., 2010). Moreover, deficiencies in the ASH-2 complex only in parents can be inherited to descendants spanning several generations (Greer et al., 2011). Therefore, ancestral chromatin states may be incompletely reprogrammed and influence gene expression of descendants during future generations.

Chromatin modifiers have been shown to control cellular proliferation, metabolism, and even longevity (Florian et al., 2012; Greer et al., 2010; Jacobs, Kieboom, Marino, DePinho, & van Lohuizen, 1999). The polycomb repressive complex (PRC), which consists of PRC1 (core subunit Bmi1, Cbx, Ring1, and Phc) and PRC2 (structural members including Ezh2, Eed, and Suz12) directly methylate specific lysines on histones to control levels of gene expression (Margueron & Reinberg, 2011; Sauvageau & Sauvageau, 2010). Bmi1 has been related to organismal longevity (Greer et al., 2010), self-renewal and differentiation of HSCs (Hidalgo et al., 2012; Lessard & Sauvageau, 2003; Park et al., 2003), NSC (Fasano et al., 2007; Molofsky et al., 2003), lung (Zacharek et al., 2011), prostate (Lukacs, Memarzadeh, Wu, & Witte, 2010), and epidermal stem cells (Ezhkova et al., 2009). Expression of $p16^{Ink4a}$ and $p19^{Arf}$, which are encoded from the *Cdkn2a* locus, increases with age (Krishnamurthy et al., 2004). The *Ink4a/Arf* locus is regulated by PRC1 including Bmi1 (Lessard & Sauvageau, 2003; Lukacs et al., 2010; Molofsky et al., 2003; Park et al., 2003; Zacharek et al., 2011). Indeed, attenuation of $p16^{Ink4a}$ partially rescues self-renewal functions of stem cells lacking Bmi1 (Lessard & Sauvageau, 2003; Molofsky et al., 2003; Park et al., 2003; Sauvageau & Sauvageau, 2010). A recent report shows association of PRC2 with aging phenotypes in HSCs (Beerman et al., 2013). Given the roles of PRC-mediated transcriptional regulation of specific genes critical for self-renewal, differentiation, and DNA damage repair in adult stem cells their contribution would not be surprising.

The addition of methyl groups to specific regions of the DNA sequence suppresses gene transcription. DNA methylation patterns are controlled by at least three DNA methyltransferases, DNMT1, DNMT3A, and DNMT3B that catalyze the transfer of a methyl group to DNA (Smith & Meissner, 2013). Aged tissues show a decrease in DNA methylation (Maegawa et al., 2010; Pollina & Brunet, 2011). However, specific loci tend to become hypermethylated (Maegawa et al., 2010). A conditional loss-of-function approach to delete *Dnmt1* from hair follicle stem cells led to stem cell failure and an exaggerated aged phenotype to the hair follicles and skin (Li et al., 2012). In addition, inhibition of DNMT1 or 3A in a mesenchymal stem cell population increased cell senescence *in vitro*, through the upregulation of cell cycle inhibitors $P16^{Ink4a}$ and $p21^{Cip1/Waf1}$ (So, Jung, Lee, Kim, & Kang, 2011). Understanding the function of the methylated target genes will be important to decipher whether global or specific methylation drives stem cell aging.

2.4. Protein homeostasis

Tight regulation of the proteome is required for normal cell function. The accumulation of misfolded or aggregated proteins can disrupt intracellular signaling cascades and induce toxicity and apoptosis. In response to stress, cellular homeostasis is achieved by removal of the cell via apoptosis or removal of damaged macromolecules via autophagy. Autophagy is a potent repressor of unwanted apoptosis or necrosis possibly through liberation of Bcl-2 (Lee et al., 2009; Pattingre et al., 2005). Removal of unwanted proteins relies on autophagasomes, chaperones, lysosomes, and ubiquitin-proteasome system (UPS) (Ellis & Pinheiro, 2002; Hartl, Bracher, & Hayer-Hartl, 2011). Increased protein accumulation in aged tissues suggests that homeostasis of the proteome is deregulated during aging (Chiti, Stefani, Taddei, Ramponi, & Dobson, 2003). UPS is one of the main proteolytic mechanisms to ensure degradation of damaged proteins (Hartl et al., 2011;

Rubinsztein, Marino, & Kroemer, 2011). UPS selectively tags polyubiquitin chains on damaged proteins and degrades the tagged proteins by proteasome complex machinery. During aging, proteasome activity declines (Conconi, Szweda, Levine, Stadtman, & Friguet, 1996; Shibatani, Nazir, & Ward, 1996).

Genetic inhibition of autophagy induces premature aging phenotypes. In *C. elegans*, loss of members of the autophagy pathway, Atg3 (autophagy-related 3), Atg9, Atg18, and Beclin1, decrease the lifespan (Toth et al., 2008). In addition, the deletion of Sestrin1 (activator the AMP-responsive protein kinase and repressor of mTOR pathway) (Lee, Budanov, et al., 2010), possibly through decreased autophagy, reduces the longevity of *Drosophila*. In contrast, overexpression of Atg8 increased *Drosophila* lifespan (Simonsen et al., 2008). In mice liver, expression levels of Lamp2a decline with age. Overexpression of hepatocyte-specific Lamp2a, a key protein in the chaperone-mediated autophagy (CMA) pathway, prevents the aging-associated defect in CMA, which leads to a reduction in toxic proteins and aggregates, and apoptotic cells within the liver (Zhang & Cuervo, 2008). These studies indicate that aging is correlated with a decline in autophagy function.

2.5. Energy metabolism

Mitochondrial and metabolic activity during development and normal aging may influence life span and rate of aging phenotypes (Sahin & DePinho, 2012; Signer & Morrison, 2013). The FoxO family of transcription factors, as a part of IGF/mTOR (target of rapamycin) pathway has been shown to regulate metabolism and oxidative stress by promoting antioxidant enzymes (Murphy et al., 2003; Wang, Bohmann, & Jasper, 2005). Stimulation of the IGF pathway induces the PI3K/AKT/mTORC2 pathway, which leads to inactivation of FoxO due to phosphorylation (Kenyon, 2010). Caloric restriction (CR) has been shown to increase lifespan in various species from yeast to mammals (Kenyon, 2010; Mair & Dillin, 2008). Although the underlying mechanisms behind lifespan extension are not fully resolved, recent papers show that CR preserves numbers and functions of stem cells (Cerletti, Jang, Finley, Haigis, & Wagers, 2012; Yilmaz et al., 2012). Genetic analysis in rodents shows modulation of TOR pathway is a major effector of CR response. mTOR exists as two complexes, mTORC1 and mTORC2. As mentioned before, mTORC2 is a downstream target of IGF-PI3K pathway that activates the Akt-FoxO signaling pathway to promote cell proliferation and regulate oxidative stress. In contrast, mTORC1 promotes protein translation, ribosome biogenesis, and regulates autophagy (Laplante & Sabatini, 2012). Therefore, mTOR influences many critical cellular processes and will likely be involved in age-associated stem cell decline.

3. EXTRINSIC REGULATION OF AGED CELLULAR AND TISSUE HOMEOSTASIS

In Section 2, the intrinsic effectors of aged cell function were considered. However, it is known that tissue-specific stem cells reside in niches. Across different mammalian stem cell compartments niches have been operationally defined (Morrison & Spradling, 2008). Local signaling from the niche regulates tissue maintenance by preserving the function of stem cells. Studies from *drosophila* have provided direct evidence how local signaling factors from the niche are required for maintenance of stem cell number and function throughout life (Boyle et al., 2007; Pan et al., 2007).

In addition to the niche, tissue-specific stem cell niches lie near blood vessels, containing numerous soluble growth factors and cytokines that can potentially influence stem cell function. Direct evidence of a systemic influence on stem cell function during aging was provided using heterochronic parabiosis, the surgical pairing of two mice to achieve a shared circulation between young and aged mice (Conboy, Conboy, & Rando, 2013). This system has been used to determine whether the decline in tissue-specific stem cell function with age was due to cell-intrinsic irreversible, age-related changes or cell-extrinsic influence by the environment in muscle (Brack et al., 2007; Conboy et al., 2005), liver (Conboy et al., 2005), heart (Loffredo et al., 2013), and the central nervous system (Ruckh et al., 2012; Villeda et al., 2011). The data from these diverse tissue types indicate that the aged systemic environment markedly contributes to aged stem cell phenotypes and nonstem cell phenotypes.

Another example of tissue rejuvenation was provided by aged skin. It is known that Nf-κb increases with age in multiple tissues (Chambers et al., 2007; Helenius, Hanninen, Lehtinen, & Salminen, 1996; Korhonen, Helenius, & Salminen, 1997). Nf-κb is activated by different cell stressors that accelerate aging such as DNA damage and oxidative stress (Pasparakis, 2009). Moreover, signals such as Sirt1 and FoxO3 that positively regulate longevity are found to repress Nf-κb (Lin, Hron, & Peng, 2004; Yeung et al., 2004). Remarkably, blocking Nf-κb activity for 2 weeks in aged skin led to partial rejuvenation of the gene profile, reverting half of the age-dependent genes back to youthful levels (Adler, Kawahara, Segal, & Chang, 2008). Moreover, many of the phenotypes associated with aged skin, such as increased epidermal thickness, increased senescence, and decreased proliferation were all partially restored back to levels seen in younger skin. Together, these results demonstrate the potential of multiple aged tissues to undergo molecular and phenotypic rejuvenation after modulation of the extrinsic environment.

4. AGING IN TISSUE-SPECIFIC STEM CELLS

The general cellular processes that breakdown during aging appear to impact many different cell types including somatic stem cells. The unique functions, intrinsic regulators, and specialized environments of tissue-specific stem cells are likely to impose distinct responses and phenotypic consequences to aging. In this section, we will discuss age-related phenotypes and the mechanisms that lead to stem cell demise in three distinct stem cell niches (Fig. 14.2).

4.1. Hematopoietic stem cells

HSCs reside in the bone marrow, have long-term self-renewal potential, and can differentiate into committed progenitors that are critical for generating downstream progeny of the blood system (Orkin & Zon, 2008).

Aging in the blood system drives changes in HSC numbers, decreases regenerative potential, and skews differentiation potential toward myeloid lineages (Beerman, Maloney, Weissmann, & Rossi, 2010; Geiger et al., 2013). These phenotypes are contributed to by both cell-autonomous and cell-extrinsic factors. Contrary to other stem cell niches, the number of HSCs in the bone marrow increases with age (Beerman, Bhattacharya, et al., 2010; Beerman et al., 2013; Challen, Boles, Chambers, & Goodell, 2010). Intuitively, increased HSCs in aged bone marrow may seem beneficial. However, analyzed on a per-cell basis or in a competitive transplant setting, the gold standard for assessing stem cell activity, aged HSCs show defects in self-renewal potential and long-term reconstitution of the blood (Chambers et al., 2007; Janzen et al., 2006; Rossi et al., 2005). Therefore, HSCs are functionally impaired due in part to cell-autonomous defects. It remains to be answered why the number of HSCs in aged population increases.

One of the hallmarks of aging in HSCs is a skewed differentiation potential toward the myeloid lineage at the expense of the lymphoid and erythroid lineages (Beerman, Bhattacharya, et al., 2010; Benz et al., 2012; Challen et al., 2010; Dykstra, Olthof, Schreuder, Ritsema, & de Haan, 2011; Morita, Ema, & Nakauchi, 2010; Rossi et al., 2005). Such biased lineage potentials in aged HSCs are not from changes in differentiation potential of individual HSCs but rather by changes in composition of the HSC pool (Beerman, Bhattacharya, et al., 2010; Beerman, Maloney, et al., 2010; Challen et al., 2010; Morita et al., 2010). Prevalence of myeloid-biased ($CD150^{high}$) HSCs may be a result of more robust self-renewal potential than of lymphoid-biased ($CD150^{low}$) HSCs (Beerman, Bhattacharya, et al., 2010; Challen et al., 2010). Both aged $CD150^{high}$ and $CD150^{low}$ HSCs show reduced proliferative capacity and homing to the bone marrow. This indicates that the functional heterogeneity within the HSC pool is cell autonomously maintained and has the potential to skew lineage phenotypes over time.

Genomic integrity of any cell is dependent on protective DNA repair mechanisms. With age, DNA damage accumulates in HSCs, indicating a decrease in the function of appropriate repair mechanisms (Rossi, Seita, et al., 2007). The mode of DNA repair differs between quiescent and cycling stem cells. Quiescent HSCs rely on NHEJ for DNA repair, which is a more error-prone mechanism and will impact the accrual of DNA damage throughout life, leading to genomic rearrangements (Mohrin et al., 2010). Whether the accrued damage is cleared when aged quiescent HSCs enter into the cell cycle and take up a more effective mode of DNA repair will determine the extent that DNA damage accumulation impacts aged stem cell function.

Mutation or deletion in DNA repair components such as *Msh2*$^{-/-}$ (mismatch repair) (Reese, Liu, & Gerson, 2003), *Brca2*$^{-/-}$ (HR) (Navarro et al., 2006), Xpd^{TTD} (nucleotide excision repair) (Rossi, Bryder, et al., 2007), *Ku80*$^{-/-}$ (NHEJ) (Rossi, Bryder, et al., 2007), and $Lig4^{Y288C}$ (NHEJ) (Nijnik et al., 2007) all show reduced HSC function, leading to depletion of HSCs in mice. Deletion of DNA damage sensor *Atm* causes elevation of ROS, and loss of HSC quiescence, defects in repopulating capacity, and ultimately a depletion of the HSC pool (Ito et al., 2006). Antioxidant treatment on *Atm*$^{-/-}$ HSCs rescues the reduced repopulating potential when challenged by serial transplantation. This implies that elevated ROS can cause aging phenotypes in HSCs.

ROS can also mutate mtDNA, which is irreversible and deleterious to mitochondrial function (Kujoth et al., 2005; Trifunovic et al., 2004). In HSCs, the skewed differentiation toward myeloid lineage was observed in mutant mtDNA polymerase γ mice, and this phenotype was rescued by antioxidant-mediated ROS inhibition (Trifunovic et al., 2004), which implies mtDNA mutagenesis modulates adult stem cell function, thus leading to aging phenotypes. However, Bryder and colleagues show stark transcriptional differences between the aged mutant and wild-type mice. Moreover, unlike physiological aging that has epigenetic and genetic signature (Chambers et al., 2007), the transcriptional signature of HSCs is not impacted by the mitochondrial mutation (Norddahl et al., 2011). To date, it remains to be investigated whether mtDNA mutation directly contributes to physiological stem cell aging.

Other major players of cell-intrinsic regulation of HSC aging are oxidative damage, metabolic stress, and autophagy. Autophagy is a stress-response mechanism to clear damaged proteins and a major contributor in aging (Rubinsztein et al., 2011). Loss of autophagy by deletion of *Atg7* (Mortensen et al., 2011) and *Fip200* (Liu et al., 2010), members of autophagosome, increases ROS levels that lead to HSC depletion in mice, which indicates that autophagy is essential for HSC homeostasis. Is aging of the stem cell caused by a decline in autophagy? Recently, Passegue and colleagues challenged this notion (Warr et al., 2013). The authors revealed that freshly isolated aged HSCs have basal levels of autophagy, unlike adult HSCs that only when activated and stressed, mount an autophagic response. Prevention of the autophagic response led to HSC apoptosis, suggesting that steady-state autophagy is essential for aged HSC maintenance.

Oxidative stress and its regulation via FoxO are important for aging in HSCs. Conditional deletion of *FoxO1*, *3*, and *4* leads to a loss in HSC quiescence, increased apoptosis, myeloid-biased differentiation, and decreased long-term HSC maintenance (Miyamoto et al., 2007; Tothova et al., 2007). A partial rescue of these phenotypes was observed using the ROS inhibitor *N*-acetyl-L-cysteine (Tothova et al., 2007), indicating that FoxO family members act as an important modulator of oxidative stress and HSC function throughout life.

The Sirtuin (*Sirt*) family is another candidate regulator of stem cell aging. Deletion of *Sirt1* in adult HSCs causes increased proliferation and elevated DNA damage, which leads to loss of long-term HSC populations (Singh et al., 2013). *Sirt3*, which controls the activity of mitochondrial enzyme acetyl coenzyme A synthetase 2 (AceCS2) and electron transport chain complex 1 in mitochondria (Finkel, Deng, & Mostoslavsky, 2009), is critical for adult and aged HSC function under transplantation stress conditions (Brown et al., 2013). Interestingly, overexpression of *Sirt3* can compensate for the

ROS-mediated aged HSC phenotype (Bratic & Larsson, 2013). It remains unclear whether other members of Sirtuin family can also reverse aging-related phenotypes in aged stem cells.

mTORC1 functions as an energy sensor and attenuator of autophagy (Nicklin et al., 2009). Calorie restriction attenuates the mTORC1 signaling pathway, which increases stem cell proliferation and stem cells in the intestine through mTORC1 expressed in the niche (Yilmaz et al., 2012). With age, expression of mTORC1 increases in HSCs and its repression by rapamycin restores self-renewal of aged HSCs (Chen, Liu, Liu, & Zheng, 2009). In contrast, indirect activation of mTOR via conditional deletion of *Pten* in young HSCs induces a progeria phenotype and depletion of HSCs (Kalaitzidis et al., 2012; Lee, Nakada, et al., 2010; Magee et al., 2012). Therefore, HSCs and other stem cells are sensitive to metabolic output via mTORC1 levels. It is unlikely the requirement of mTORC1 is age specific or stem cell specific; however, any age-dependent change in metabolic regulation may increase the sensitivity of aged stem cells to mTORC1 levels.

Using genome-wide DNA methylation analysis of human and mouse HSCs, it was demonstrated that genes associated with myeloid lineage become hypomethylated with age. In contrast, site-specific hypermethylation was observed at genes associated with the PRC2 complex (Beerman et al., 2013; Bocker et al., 2011). Interestingly, age-dependent myeloid-skewedness and DNA methylation are largely dependent on the proliferative history of HSCs. Using proliferative demand as a driver of HSC aging, it was demonstrated that replicative aging is distinct from chronological aging at the DNA methylation level (Beerman et al., 2013). Both aged HSCs and adult HSCs under moderate proliferative demand underwent site-specific hypermethylation at the PRC2 complex, which was accompanied by transcriptional repression, as suggested previously (Chambers et al., 2007). In contrast, adult HSCs with a high level of experimentally enforced proliferation had a distinct methylome. Therefore, site-specific methylome changes appear to correlate with the functional decline of aged HSCs. Moreover, these data confirm at the epigenome level that stem cells with limited turnover throughout life experience the effects of chronological rather than proliferative aging.

The influence of the systemic environment and the local niche on HSC aging is becoming apparent. Recent data show that both systemic and niche-derived factors contribute to lineage skewedness of HSCs. Rante/Ccl5 cytokine, which is highly expressed in the local niche and aged blood, has been shown to increase myeloid progenitor proliferation, whereas knockout of *Ccl5* increases lymphoid lineages during transplantation (Ergen, Boles, & Goodell, 2012). In addition, Challen et al. show lineage skewedness can be enhanced by differential responsiveness of TGFβ (Challen et al., 2010). The authors demonstrate that TGFβ provokes differential responses to myeloid-biased ($CD150^{high}$) HSCs on activation. G-CSF is a well-known cytokine that promotes HSC proliferation, mobilization into the bloodstream, and eventual stem cell depletion (Ju et al., 2007; Song, Zhang, Ju, & Rudolph, 2012). Normal HSCs transplanted into telomerase-deficient mice that possess a hostile microenvironment show limited engraftment that was abrogated by G-CSF blocking antibody (Ju et al., 2007). Moreover, pharmacological strategies can partially reprogram the aged epigenetic HSC signature to that of adult HSCs (Alberts, Geneste, & Treisman, 1998), thus providing another example of molecular and functional reversion to a youthful state (Nebbioso, Carafa, Benedetti, & Altucci, 2012).

4.2. Neural stem cell

Neurons and glial cells are generated from NSCs, which reside in the subventricular zone (SVZ) of the lateral ventricle and the subgranular zone (SGZ) of the dentate gyrus of the hippocampus (Kempermann, Jessberger, Steiner, & Kronenberg, 2004). Astroglial NSCs in the SVZ give rise to transit-amplifying cells, which are responsible for generating neurons that maintain the olfactory bulb (Imayoshi et al., 2008). NSCs from the SGZ generate neural progenitors that generate postmitotic neurons, which are critical for learning, memory, and behavior (Shors, Townsend, Zhao, Kozorovitskiy, & Gould, 2002). A progressive and dramatic decline in neurogenesis has been observed across different species during aging (Kempermann, Kuhn, & Gage, 1998; Kuhn, Dickinson-Anson, & Gage, 1996). This has been correlated with a decline in both the number of NSCs and proliferative expansion of transit-amplifying progenitors (Encinas et al., 2011; Lugert et al., 2010, 2012). Using markers to define distinct NSC subsets, Lugert et al. showed a reduction in the number of active NSCs due to their transition to a quiescent state in aged mice and therefore, preventing their contribution to neurogenesis (Lugert et al., 2010). In addition, a subset of aged cycling NSCs forms astrocytes instead of neurons (Encinas et al., 2011). Together, these changes are consistent with a failure to produce newborn neurons in the aged brain. Moreover, these studies highlight the heterogeneity within the NSC pool and the divergent responses of distinct subsets to respond to aging.

Several cell-intrinsic factors including the cell cycle inhibitors $p16^{Ink4a}$ and $p57^{Kip2}$, mTOR signaling pathway, and FoxO family members have been correlated with aged phenotypes in NSCs (Furutachi, Matsumoto, Nakayama, & Gotoh, 2013; Molofsky et al., 2006; Nishino, Kim, Chada, & Morrison, 2008; Paik et al., 2009; Paliouras et al., 2012; Renault et al., 2009). Expression of $p16^{Ink4a}$ increases with age in neural progenitors (Molofsky et al., 2006). Deletion of the *Ink4a* locus reveals increased proliferation and neurogenesis from neural progenitors in SVZ compared to young counterpart, but not in the SGZ (Molofsky et al., 2006). In adult neural progenitors, $p16^{Ink4a}$ repression occurs through Hmga2, which is a member of the high-motility group A family that has no transcriptional activity but influences chromatin structure (Nishino et al., 2008). In aged neural progenitors, Hmga2 is repressed through upregulation of *let-7b* microRNA, leading to an increase in $p16^{Ink4a}$ expression. Overexpression of *let-7b* is sufficient to upregulate $p16^{Ink4a}$ expression by targeting Hmga2 expression. In contrast, deletion of *Hmga2* leads to an increased expression of $p16^{Ink4a}$, and a loss in the number and self-renewal potential of NSCs in fetal and young mice but not aged mice. This decline was partially rescued by $p16^{Ink4a}$ deletion. Together, these results indicate that changes in *let-7b* and Hmga2 may contribute to NSC aging phenotypes. It will be interesting to uncover why aged NSCs in SGZ of the aged brain were not sensitive to $P16^{Ink4a}$ levels.

Quiescence is critical for stem cell maintenance, in part through regulation of cell cycle inhibitors such as $p57^{Kip2}$ (Matsumoto et al., 2011; Zou et al., 2011). Recently, Furutachi et al. showed a role of $p57^{Kip2}$on SGZ NSC maintenance during aging (Furutachi et al., 2013). Short-term deletion of $p57^{Kip2}$ specifically in Nestin-positive NSCs led to a transient amplification of radial NSCs and neurogenesis. In contrast, chronic deletion of $p57^{Kip2}$ leads to a reduction in number of NSCs and neurogenesis, indicating that $p57^{Kip2}$ is important for aged NSC maintenance. Moreover, these findings suggest that a short-term increase in NSC output is beneficial for neurogenesis, however, long-term proliferation leads to NSC exhaustion.

mTOR has been identified as a regulator of NSC quiescence (Paliouras et al., 2012). In the SVZ of adult mice, mTORC1 expression is absent in quiescent NSCs and present in transit-amplifying progenitors. In the aged SVZ, a decline in mTORC1 activity was observed along with reduced NSC proliferation. Reduction of mTORC1 by administration of rapamycin promotes adult NSC quiescence. Administration of EGF, strong inducer of mTOR, activated quiescent NSCs in aged mice, an effect abolished upon rapamycin treatment. This report indicates that mTOR activity determines the ratio between quiescent and activated NSCs.

Another cell-intrinsic player of aging in NSCs is the FoxO family. Recent studies show the role of FoxO proteins in regulating stem cell homeostasis in NSCs (Paik et al., 2009; Renault et al., 2009). One group utilized *FoxO3*$^{-/-}$ (Renault et al., 2009), and the other studied conditional deletion of *FoxO1/3/4* (Paik et al., 2009). Both studies show that deletion of *FoxO* leads to a transient increase in the neonatal NSC pool *in vivo* but eventual depletion of NSCs in the adult. Reduced self-renewal potential assessed by *in vitro* neurosphere formation assay and differentiation potential was also observed in *FoxO* deleted mice. In the absence of FoxO, NSCs undergo increased oxidative stress, altered glucose metabolism (Renault et al., 2009), and elevated Wnt signaling (Paik et al., 2009). This suggests that the FoxO family regulates homeostasis of NSC through regulation of diverse genes and pathways.

The decline in neurogenesis during aging may also be regulated by changes in systemic factors in the blood circulation. Wyss-Coray and colleagues analyzed the systemic factors that influence neurogenesis using heterochronic parabiosis and identified that elevated levels of chemokine CC-chemokine ligand 11 (Ccl11) in plasma from aged mice and adult mice paired with aged mice (Villeda et al., 2011). After systemic delivery of Ccl11, adult mice developed learning and memory deficits, which were abrogated with coinjection of antibodies neutralizing Ccl11 along with Ccl11. Therefore, aging-related increase in Ccl11 is at least partly responsible for the reduction in aged neurogenesis. How Ccl11 affects neurogenesis remains unclear.

In addition to the systemic environment, changes in the NSC niche may impose age-dependent changes to NSC function. Components of the Wnt signaling pathway regulate adult neurogenesis (Seib et al., 2013). A recent study shows that expression of the soluble Wnt inhibitor, Dkk1 (Seib et al., 2013) increased with age in SGZ NSCs, which decreased adult neurogenesis. Conditional deletion of *Dkk1* in Nestin-positive NSCs upregulated Wnt signaling, which led to increased self-renewal and differentiation into new neurons (Seib et al., 2013). Deletion of *Dkk1* in aged mice showed evidence of enhanced behavioral performance levels, similar to young counterparts. This report emphasizes that changes in niche factors during aging affect NSCs function. Interestingly, exercise provides a positive stimulus for hippocampal neurogenesis in adult and aged mice (van Praag, Christie, Sejnowski, & Gage, 1999; van Praag, Shubert, Zhao, & Gage, 2005).

Another factor enriched in aged niches is TGFβ, which has been shown to be increased in blood (Challen et al., 2010) and muscle (Carlson et al., 2009) and leads to aging phenotypes. Levels of TGFβ increase in endothelial cells of the NSC niche during aging and after exposure to IR (Pineda et al., 2013). Inhibition of TGFβ signaling by delivery of a neutralizing antibody or administration of a pharmacological TGFβ inhibitor restored neurogenesis of irradiated adult mice and aged mice compared to young mice. This study demonstrates that TGFβ from the local niche is one driver of stem cell aging and these changes can be reversed when inhibited by pharmacological intervention.

4.3. Skeletal muscle stem cells

The differentiated muscle fiber functions as a niche cell that provides cues to retain muscle stem cells or satellite cells (MuSCs) in a quiescent and nondifferentiated state (Bischoff, 1986). In aged muscle, the number of MuSCs declines by approximately 50% compared to adult muscle (Brack, Bildsoe, & Hughes, 2005; Cerletti et al., 2012; Chakkalakal et al., 2012). It was recently demonstrated that aged MuSCs break out of their stable quiescent state due to a redeployment of the developmental mitogen, fibroblast growth factor-2 (FGF2) in aged muscle fibers but not by the MuSCs themselves (Chakkalakal et al., 2012). FGF2-mediated proliferation was associated with increased apoptosis, myogenic commitment, and stem cell decline. Therefore, the aged niche is a modulator of stem cell number, similar to that observed in the aged niche of *Drosophila* (Boyle et al., 2007; Pan et al., 2007). Interestingly, although proliferative output increased, a subset of MuSCs that had undergone fewer divisions throughout adult life were able to retain long-term self-renewal potential when transplanted into adult muscle. These data demonstrate the importance of the quiescent state for maintenance of stem cell potential. Using a genetic strategy to increase FGF signaling in MuSCs only via deletion of *Sprouty1*, an intracellular feedback inhibitor of FGF signaling (Shea et al., 2010), elevated FGF signaling in aged MuSCs exacerbated the age-dependent satellite cell decline. Importantly, MuSC loss negatively impacted regenerative capacity (Chakkalakal et al., 2012). In contrast, short-term deletion of *Sprouty1*, to acutely activate FGF signaling, was met with an enhanced regenerative outcome (Chakkalakal et al., 2012), confirming that FGF signaling acts as a potent myogenic factor to prime MuSCs for functional myogenic contribution (Shefer, Van de Mark, Richardson, & Yablonka-Reuveni, 2006). This demonstrates that cell fate decisions depend on the level and duration of signaling. Other ligands including TGFβ2 and Delta1 that increase and decrease, respectively, in aged muscle fibers may also participate in deregulation of aged satellite cell homeostasis (Carlson et al., 2009; Conboy, Conboy, Smythe, & Rando, 2003). It can be concluded that niche-derived changes occurring during homeostasis can impact the success of subsequent regenerative insults. In the future, it will be important to determine what drives induction of growth factors in the aged niche.

Transplantation assays to compare engraftment potential between adult and aged MuSCs have given conflicting results, depending on whether the adult recipients were preirradiated (Chakkalakal et al., 2012; Collins, Zammit, Ruiz, Morgan, & Partridge, 2007). Engraftment potential of aged MuSCs was comparable to adult MuSCs if the adult host was irradiated (Chakkalakal et al., 2012; Collins et al., 2007). It is known that a fraction of MuSCs are depleted in response to IR (Heslop, Morgan, & Partridge, 2000). Therefore, IR decreases the number and function of adult stem cell "competitors" in the recipient muscle and provides a less competitive environment for the engrafted aged MuSCs to reoccupy the muscle.

It is generally accepted that MuSCs lose their regenerative potential during aging, due to intrinsic defects and cell-extrinsic changes in the aged environment (Brack & Rando, 2012). As we discuss next, the mechanisms that cause those changes are beginning to be identified.

To date, age-dependent DNA damage has not been extensively studied in aged MuSCs. However, dysfunction of aged MuSCs and muscle regeneration is exacerbated in the absence of Ku80, a component of the nonhomologous end-joining complex, that participates in DNA repair (Didier, Hourde, Amthor, Marazzi, & Sassoon, 2012). This demonstrates that DNA repair processes facilitate MuSC maintenance; however, determining whether DNA damage is causative to MuSC aging will require the formal demonstration that augmentation of DNA repair pathways in aged MuSCs rescues their function.

In addition, consistent with increased DNA damage response and cell cycle arrest, it has been suggested that aged MuSCs display increased abundance of the Ink4a/Arf family of cell cycle inhibitors (Carlson, Hsu, & Conboy, 2008). This may partially explain the delay in MuSC activation in response to injury. Whether Ink4a/Arf-mediated senescence occurs in aged MuSCs remains unexplored.

Metabolic factors control the level of mitochondrial and nuclear DNA damage induced by ROS, a by-product of oxidative phosphorylation. CR is a known modulator of metabolism and lifespan (Mair & Dillin, 2008). Recently, CR was demonstrated to increase MuSC number and function in adult and aged mice (Cerletti et al., 2012). The reduction in caloric load in adult and aged mice was correlated with increased number of mitochondria and elevated expression of well-known metabolic regulators Sirtuin and FoxO in FACS sorted MuSCs. Dissecting the complex relationship between DNA repair, metabolism, and autophagy and how they contribute to stem cell aging will require sophisticated genetic approaches.

To date, the contribution of the extrinsic environment to MuSC decline during aging is more clearly defined. It has been demonstrated that extrinsic modifications can rejuvenate MuSC function and muscle repair in the context of heterochronic tissue transplants (Carlson & Faulkner, 1989) and heterochronic parabiosis (Brack et al., 2007; Conboy et al., 2005). The extent of rejuvenation will likely depend on a multitude of factors including the age of the organism, the type and severity of injury, and the inflammation invoked by the injury (Shavlakadze, McGeachie, & Grounds, 2010; Smythe et al., 2008).

Using parabiotic pairings, Conboy et al. showed that unidentified factors in the serum of young mice could reverse the proliferative decline of MuSCs and that proliferation of adult MuSCs was impaired when exposed to aged systemic environment (Conboy et al., 2005). This seminal work provided a novel paradigm demonstrating that at least a subset of tissue-specific stem cells in aged organisms are functionally competent if exposed to a favorable milieu.

Early activation of adult MuSCs requires a high Notch/low Wnt state, followed by a low Notch/High Wnt state for activated MuSC to commit to differentiation (Brack, Conboy, Conboy, Shen, & Rando, 2008; Conboy & Rando, 2002). In aged muscle, both signaling cascades are deregulated; early Notch signaling is repressed and Wnt signaling is derepressed, leading to delayed satellite cell entry and a myogenic-to-fibrogenic fate conversion of a subset of aged MuSCs, respectively (Brack et al., 2007; Conboy et al., 2003). Serum factors have been identified that inhibit Notch and upregulate Wnt signaling. Elevated TGFβ2 in aged serum is able to repress Notch signaling and to increase cell cycle inhibitors of the Ink4a/Arf family to repress cell cycle entry of aged MuSCs. The identification of a Frizzled-binding protein that was upregulated in aged serum and when depleted, abrogated the fate change induced by aged serum, suggested a component of the Wnt pathway was involved (Brack et al., 2007). This was later identified as a complement family member (C1q) that activated Wnt signaling in an age-dependent manner (Naito et al., 2012). This series of papers shows that the systemic aged environment modulates satellite cell fate changes, but

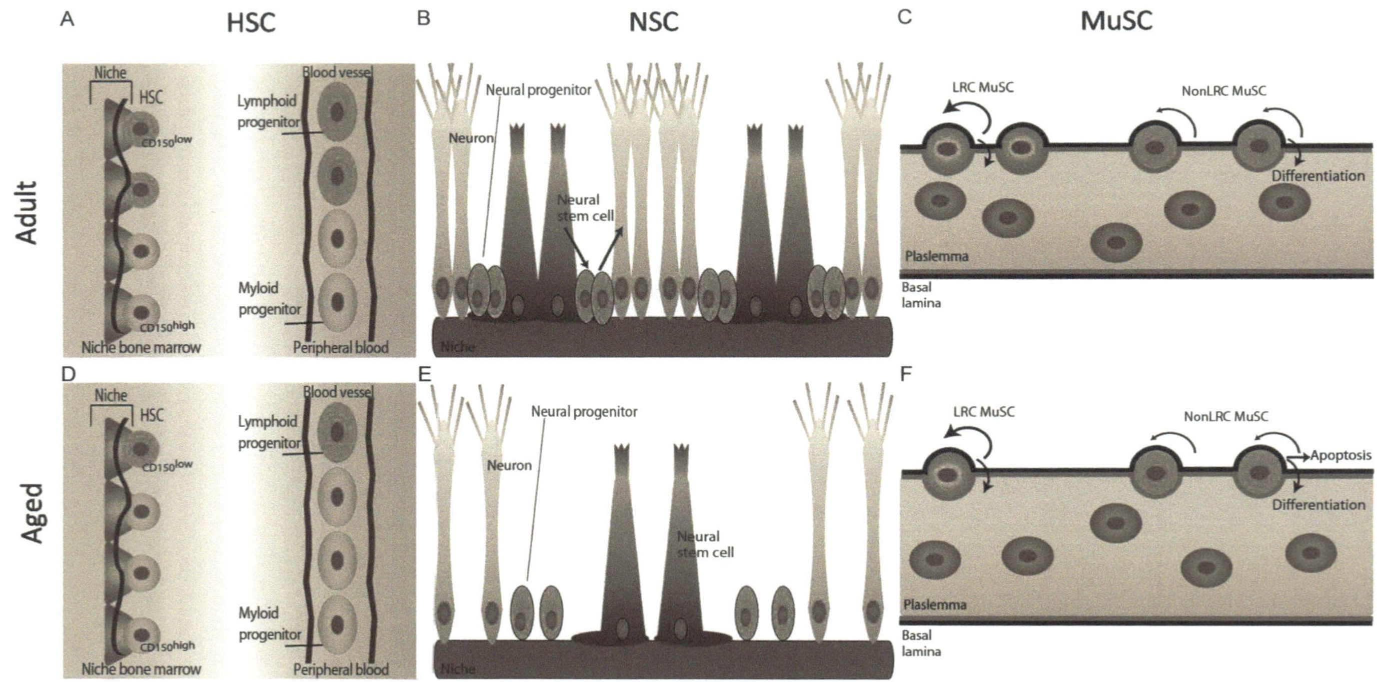

Figure 14.2 Schematic diagrams of adult stem cells and niche in bone marrow, brain, and muscle tissue. (A, D) Quiescent adult hematopoietic stem cells (HSCs) are located in the endosteal region of adult (A) and aged (D) bone marrow (left) residing in a niche composed of endothelial cells, osteoblasts (gray), and perivascular cells (dark blue line). The adult HSC pool consists of lymphoid-biased ($CD150^{low}$) HSCs (green) and myeloid-biased ($CD150^{high}$) HSCs (blue), with high self-renewal potential and balanced differentiation toward lymphoid (green, elongated) and myeloid progenitor cells (blue, elongated) in the peripheral blood (right, A and D). In aged HSC niche, selective expansion/retention of myeloid-biased $CD150^{high}$ HSCs skewing differentiation toward the myeloid lineage due to the robust self-renewal potential of aged $CD150^{high}$ HSCs compared to $CD150^{low}$ HSCs. (B, E) Neural stem cells (NSCs) reside at subgranular zone (SGZ) of the dentate gyrus of the hippocampus in adult (B) and aged (E) brain. NSCs (red) can give rise to neural progenitors (green), which progressively differentiate into neurons (blue). Systemic factors from the blood and local niche factors (purple) control NSC number and function. With age, number of NSCs and adult neurogenesis (making new neurons) declines. (C, F) Muscle satellite stem cells (MuSCs) in adult (C) and aged (F) muscle fibers. MuSCs (green) reside between the plasmalemma (gray line) and the basal lamina (black line) of muscle fibers. Long-term label retaining (LRC) MuSCs (dark green) have a high capacity for self-renewal and differentiation into muscle. nonLRC MuSCs (light green) are capable of differentiation but have limited self-renewal capacity. With age, the number of LRC MuSCs and fusion competent progenitors declines leading to a reduced number of functional stem cells and defective muscle repair. (See color plate.)

whether the subset of aged MuSCs that loses myogenic fate is intrinsically distinct from those that retain a MuSC identity remains unknown. Overall, the changes that occur to the aged niche are likely multifactorial and unlikely controlled through a single factor.

5. DISCUSSION AND CONCLUDING COMMENTS

Our understanding of the molecular control of stem cell function and the mechanisms that go awry during aging is rapidly expanding due to technological advances. For example, the identification of markers that allow unambiguous isolation of specialized stem cells, advances in sequencing technology that enable molecular analysis of rare stem cells, and appropriate assays to interrogate stem cell function *in vivo*. Through these advances, a picture emerges of many overlapping mechanisms that impact stem cell function in different niches during the aging process. The contribution of age-dependent changes to DNA fidelity, the epigenetic landscape, metabolic stress, and extrinsic factors all impact stem cell function during aging. Therefore, it is clear that the major regulators of general cellular fitness and lifespan also govern stem cell aging. This begs the question, what is unique about stem cell aging? Stem cells are unique in their ability to self-renew, a function that is lost during aging. Moreover, stem cells are relatively scarce compared to other cell types; therefore, the consequences of stem cell decline during aging are severe. Hence, identifying the mechanisms responsible for stem cell decline whether specific to stem cells or general regulators of cell aging and lifespan will illustrate the careful balance between tissue homeostasis, aging, and cancer.

In this review, we focused on three well-defined stem cell systems that have been characterized during aging: blood, neural, and muscle. Unfortunately, the specter of aging does not spare a single body part and therefore most stem cell compartments will undergo some form of age-related decline. As we begin to uncover age-dependent changes in other stem cell niches, context-specific differences and commonalities that define aging will become crystallized for the interests of human health.

A common theme across stem cells niches is the observed hierarchical relationship between stem cells of different potential, with upstream stem cells endowed with long-term self-renewal potential and a propensity to maintain a dormant quiescent state, relative to their more committed daughters. Can the loss of self-renewal potential during aging be explained by an increase in downstream stem cells at expense of the hierarchically upstream subset? Both aged HSCs and MuSCs retain a small fraction of label retaining cells endowed with self-renewal potential, however, the relative fraction of the more committed subset increases in the aged MuSC pool. It remains unknown whether this hierarchical skewing occurs in the HSC or NSC compartment during aging. If so, this could explain how the HSC pool is maintained in aged bone marrow at the same time that functionality declines.

It is interesting that unlike the NSC and MuSC compartment, the number of HSCs during aging does not decline. One interpretation is that the HSC system has compensatory mechanisms to prevent stem cell loss. Preservation of the quiescent state is critical for stem cell maintenance. However, the turnover of HSCs, MuSCs, and NSCs is relatively infrequent and therefore unlikely to explain the stark differences. In the future, direct comparisons between stem cell compartments may reveal context-dependent differences in protective or stress pathways that lead to preservation or loss of one stem cell compartment relative to another.

The identification of functionally heterogeneous subsets of adult stem cells under different molecular control dictates that any changes in stem cell function of a whole population maybe through selective loss or expansion of subsets during stress, turnover, and aging. A clonal subset that is intrinsically superior or acquires increased fitness under selective pressure will dominate the aged stem cell pool. Therefore, age-dependent changes in the extrinsic milieu and epigenetic state of the stem cell may cause functional changes to the whole pool of stem cells or allow for selection of one subset at the expense of the other. Determining the consequence of clonal expansion or loss in specific subsets will require a more comprehensive study into stem cell heterogeneity and specific signals that govern the functional properties of such subsets.

The clinical applications of stem cell augmentation are obvious. Several methods of intervention have been discussed that functionally improve aged stem cells. In some instances, the exposure of aged cells to a young environment restores cellular function, suggesting that the aged environment induces a repressive state that can be reversed in a youthful setting. Other promising strategies include the restoration of extrinsic signaling pathways such as Notch, Wnt, and TGFβ to youthful levels; decreasing stress pathways through metabolic reprogramming; or the reduction in inflammatory pathways all provide favorable functional outcome to stem cells and regenerative response of aged organisms. Whether these stem cell modifiers represent a functional compensation by an aged stem cell or true molecular rejuvenation to a young stem cell state remains to be tested. In the absence of full molecular rejuvenation, any functional improvement to aged stem cells is likely to be transient. It is also exciting to consider that stem cells can be modified through lifestyle factors. Exercise and reduced caloric intake appear to enhance tissue-specific stem cell activity in both adult and aged organisms. Understanding how aging of stem cells can be prevented to enhance tissue homeostasis and regeneration potential will be critical for healthy living and the treatment of age-related diseases in the ever expanding human aged population.

ACKNOWLEDGMENTS

We apologize to our colleagues whose work could not be cited due to the space limitation. We would like to thank Brack lab members for critical discussion and reading of the manuscript and Justine Zayhowski for artwork. This work was supported by NIH Grants (R01 AR060868, R01 AR061002) and Ellison Medical Foundation New Scholars Award to ASB.

Disclosure of Potential Conflicts of Interest: The authors declare no competing financial interest.

REFERENCES

Adler, A. S., Kawahara, T. L., Segal, E., & Chang, H. Y. (2008). Reversal of aging by NFkappaB blockade. *Cell Cycle*, *7*, 556–559.

Ahlqvist, K. J., Hamalainen, R. H., Yatsuga, S., Uutela, M., Terzioglu, M., Gotz, A., et al. (2012). Somatic progenitor cell vulnerability to mitochondrial DNA mutagenesis underlies progeroid phenotypes in Polg mutator mice. *Cell Metabolism*, *15*, 100–109.

Alberts, A. S., Geneste, O., & Treisman, R. (1998). Activation of SRF-regulated chromosomal templates by Rho-family GTPases requires a signal that also induces H4 hyperacetylation. *Cell*, *92*, 475–487.

Baker, D. J., Jeganathan, K. B., Cameron, J. D., Thompson, M., Juneja, S., Kopecka, A., et al. (2004). BubR1 insufficiency causes early onset of aging-associated phenotypes and infertility in mice. *Nature Genetics*, *36*, 744–749.

Beerman, I., Bhattacharya, D., Zandi, S., Sigvardsson, M., Weissman, I. L., Bryder, D., et al. (2010). Functionally distinct hematopoietic stem cells modulate hematopoietic lineage potential during aging by a mechanism of clonal expansion. *Proceedings of the National Academy of Sciences of the United States of America*, *107*, 5465–5470.

Beerman, I., Bock, C., Garrison, B. S., Smith, Z. D., Gu, H., Meissner, A., et al. (2013). Proliferation-dependent alterations of the DNA methylation landscape underlie hematopoietic stem cell aging. *Cell Stem Cell*, *12*, 413–425.

Beerman, I., Maloney, W. J., Weissmann, I. L., & Rossi, D. J. (2010). Stem cells and the aging hematopoietic system. *Current Opinion in Immunology*, *22*, 500–506.

Benz, C., Copley, M. R., Kent, D. G., Wohrer, S., Cortes, A., Aghaeepour, N., et al. (2012). Hematopoietic stem cell subtypes expand differentially during development and display distinct lymphopoietic programs. *Cell Stem Cell*, *10*, 273–283.

Bischoff, R. (1986). Proliferation of muscle satellite cells on intact myofibers in culture. *Developmental Biology*, *115*, 129–139.

Bishop, N. A., Lu, T., & Yankner, B. A. (2010). Neural mechanisms of ageing and cognitive decline. *Nature*, *464*, 529–535.

Biteau, B., Hochmuth, C. E., & Jasper, H. (2008). JNK activity in somatic stem cells causes loss of tissue homeostasis in the aging Drosophila gut. *Cell Stem Cell*, *3*, 442–455.

Biteau, B., Karpac, J., Supoyo, S., Degennaro, M., Lehmann, R., & Jasper, H. (2010). Lifespan extension by preserving proliferative homeostasis in Drosophila. *PLoS Genetics*, *6*, e1001159.

Blanpain, C., Mohrin, M., Sotiropoulou, P. A., & Passegue, E. (2011). DNA-damage response in tissue-specific and cancer stem cells. *Cell Stem Cell*, *8*, 16–29.

Bocker, M. T., Hellwig, I., Breiling, A., Eckstein, V., Ho, A. D., & Lyko, F. (2011). Genome-wide promoter DNA methylation dynamics of human hematopoietic progenitor cells during differentiation and aging. *Blood*, *117*, e182–e189.

Boyle, M., Wong, C., Rocha, M., & Jones, D. L. (2007). Decline in self-renewal factors contributes to aging of the stem cell niche in the Drosophila testis. *Cell Stem Cell*, *1*, 470–478.

Brack, A. S., Bildsoe, H., & Hughes, S. M. (2005). Evidence that satellite cell decrement contributes to preferential decline in nuclear number from large fibres during murine age-related muscle atrophy. *Journal of Cell Science*, *118*, 4813–4821.

Brack, A. S., Conboy, I. M., Conboy, M. J., Shen, J., & Rando, T. A. (2008). A temporal switch from notch to Wnt signaling in muscle stem cells is necessary for normal adult myogenesis. *Cell Stem Cell*, *2*, 50–59.

Brack, A. S., Conboy, M. J., Roy, S., Lee, M., Kuo, C. J., Keller, C., et al. (2007). Increased Wnt signaling during aging alters muscle stem cell fate and increases fibrosis. *Science*, *317*, 807–810.

Brack, A. S., & Rando, T. A. (2012). Tissue-specific stem cells: Lessons from the skeletal muscle satellite cell. *Cell Stem Cell*, *10*, 504–514.

Bratic, A., & Larsson, N. G. (2013). The role of mitochondria in aging. *Journal of Clinical Investigation*, *123*, 951–957.

Brown, K., Xie, S., Qiu, X., Mohrin, M., Shin, J., Liu, Y., et al. (2013). SIRT3 reverses aging-associated degeneration. *Cell Reports*, *3*, 319–327.

Burtner, C. R., & Kennedy, B. K. (2010). Progeria syndromes and ageing: What is the connection? *Nature Reviews Molecular Cell Biology*, *11*, 567–578.

Campisi, J. (2005). Senescent cells, tumor suppression, and organismal aging: Good citizens, bad neighbors. *Cell*, *120*, 513–522.

Carlson, M. E., Conboy, M. J., Hsu, M., Barchas, L., Jeong, J., Agrawal, A., et al. (2009). Relative roles of TGF-beta1 and Wnt in the systemic regulation and aging of satellite cell responses. *Aging Cell*, *8*, 676–689.

Carlson, B. M., & Faulkner, J. A. (1989). Muscle transplantation between young and old rats: Age of host determines recovery. *The American Journal of Physiology*, *256*, C1262–C1266.

Carlson, M. E., Hsu, M., & Conboy, I. M. (2008). Imbalance between pSmad3 and Notch induces CDK inhibitors in old muscle stem cells. *Nature*, *454*, 528–532.

Cavalli, G., & Paro, R. (1998). The Drosophila Fab-7 chromosomal element conveys epigenetic inheritance during mitosis and meiosis. *Cell*, *93*, 505–518.

Cerletti, M., Jang, Y. C., Finley, L. W., Haigis, M. C., & Wagers, A. J. (2012). Short-term calorie restriction enhances skeletal muscle stem cell function. *Cell Stem Cell*, *10*, 515–519.

Chakkalakal, J. V., Jones, K. M., Basson, M. A., & Brack, A. S. (2012). The aged niche disrupts muscle stem cell quiescence. *Nature*, *490*, 355–360.

Challen, G. A., Boles, N. C., Chambers, S. M., & Goodell, M. A. (2010). Distinct hematopoietic stem cell subtypes are differentially regulated by TGF-beta1. *Cell Stem Cell*, *6*, 265–278.

Chambers, S. M., Shaw, C. A., Gatza, C., Fisk, C. J., Donehower, L. A., & Goodell, M. A. (2007). Aging hematopoietic stem cells decline in function and exhibit epigenetic dysregulation. *PLoS Biology*, *5*, e201.

Chen, C., Liu, Y., Liu, Y., & Zheng, P. (2009). mTOR regulation and therapeutic rejuvenation of aging hematopoietic stem cells. *Science Signaling*, *2*, ra75.

Chiti, F., Stefani, M., Taddei, N., Ramponi, G., & Dobson, C. M. (2003). Rationalization of the effects of mutations on peptide and protein aggregation rates. *Nature*, *424*, 805–808.

Collins, C. A., Zammit, P. S., Ruiz, A. P., Morgan, J. E., & Partridge, T. A. (2007). A population of myogenic stem cells that survives skeletal muscle aging. *Stem Cells*, *25*, 885–894.

Conboy, M. J., Conboy, I. M., & Rando, T. A. (2013). Heterochronic parabiosis: Historical perspective and methodological considerations for studies of aging and longevity. *Aging Cell*, *12*, 525–530.

Conboy, I. M., Conboy, M. J., Smythe, G. M., & Rando, T. A. (2003). Notch-mediated restoration of regenerative potential to aged muscle. *Science*, *302*, 1575–1577.

Conboy, I. M., Conboy, M. J., Wagers, A. J., Girma, E. R., Weissman, I. L., & Rando, T. A. (2005). Rejuvenation of aged progenitor cells by exposure to a young systemic environment. *Nature*, *433*, 760–764.

Conboy, I. M., & Rando, T. A. (2002). The regulation of notch signaling controls satellite cell activation and cell fate determination in postnatal myogenesis. *Developmental Cell*, *3*, 397–409.

Conconi, M., Szweda, L. I., Levine, R. L., Stadtman, E. R., & Friguet, B. (1996). Age-related decline of rat liver multicatalytic proteinase activity and protection from oxidative inactivation by heat-shock protein 90. *Archives of Biochemistry and Biophysics*, *331*, 232–240.

Didier, N., Hourde, C., Amthor, H., Marazzi, G., & Sassoon, D. (2012). Loss of a single allele for Ku80 leads to progenitor dysfunction and accelerated aging in skeletal muscle. *EMBO Molecular Medicine*, *4*, 910–923.

Dykstra, B., Olthof, S., Schreuder, J., Ritsema, M., & de Haan, G. (2011). Clonal analysis reveals multiple functional defects of aged murine hematopoietic stem cells. *The Journal of Experimental Medicine*, *208*, 2691–2703.

Edgar, D., Shabalina, I., Camara, Y., Wredenberg, A., Calvaruso, M. A., Nijtmans, L., et al. (2009). Random point mutations with major effects on protein-coding genes are the driving force behind premature aging in mtDNA mutator mice. *Cell Metabolism*, *10*, 131–138.

Eijkelenboom, A., & Burgering, B. M. (2013). FOXOs: Signalling integrators for homeostasis maintenance. *Nature Reviews Molecular Cell Biology*, *14*, 83–97.

Ellis, R. J., & Pinheiro, T. J. (2002). Medicine: Danger—Misfolding proteins. *Nature*, *416*, 483–484.

Encinas, J. M., Michurina, T. V., Peunova, N., Park, J. H., Tordo, J., Peterson, D. A., et al. (2011). Division-coupled astrocytic differentiation and age-related depletion of neural stem cells in the adult hippocampus. *Cell Stem Cell*, *8*, 566–579.

Ergen, A. V., Boles, N. C., & Goodell, M. A. (2012). Rantes/Ccl5 influences hematopoietic stem cell subtypes and causes myeloid skewing. *Blood*, *119*, 2500–2509.

Ezhkova, E., Pasolli, H. A., Parker, J. S., Stokes, N., Su, I. H., Hannon, G., et al. (2009). Ezh2 orchestrates gene expression for the stepwise differentiation of tissue-specific stem cells. *Cell*, *136*, 1122–1135.

Fasano, C. A., Dimos, J. T., Ivanova, N. B., Lowry, N., Lemischka, I. R., & Temple, S. (2007). shRNA knockdown of Bmi-1 reveals a critical role for p21-Rb pathway in NSC self-renewal during development. *Cell Stem Cell*, *1*, 87–99.

Finkel, T., Deng, C. X., & Mostoslavsky, R. (2009). Recent progress in the biology and physiology of sirtuins. *Nature*, *460*, 587–591.

Flores, I., Canela, A., Vera, E., Tejera, A., Cotsarelis, G., & Blasco, M. A. (2008). The longest telomeres: A general signature of adult stem cell compartments. *Genes & Development*, *22*, 654–667.

Florian, M. C., Dorr, K., Niebel, A., Daria, D., Schrezenmeier, H., Rojewski, M., et al. (2012). Cdc42 activity regulates hematopoietic stem cell aging and rejuvenation. *Cell Stem Cell, 10*, 520–530.

Foudi, A., Hochedlinger, K., Van Buren, D., Schindler, J. W., Jaenisch, R., Carey, V., et al. (2009). Analysis of histone 2B-GFP retention reveals slowly cycling hematopoietic stem cells. *Nature Biotechnology, 27*, 84–90.

Furutachi, S., Matsumoto, A., Nakayama, K. I., & Gotoh, Y. (2013). p57 controls adult neural stem cell quiescence and modulates the pace of lifelong neurogenesis. *The EMBO Journal, 32*, 970–981.

Geiger, H., de Haan, G., & Florian, M. C. (2013). The ageing haematopoietic stem cell compartment. *Nature Reviews Immunology, 13*, 376–389.

Greer, E. L., Maures, T. J., Hauswirth, A. G., Green, E. M., Leeman, D. S., Maro, G. S., et al. (2010). Members of the H3K4 trimethylation complex regulate lifespan in a germline-dependent manner in C. elegans. *Nature, 466*, 383–387.

Greer, E. L., Maures, T. J., Ucar, D., Hauswirth, A. G., Mancini, E., Lim, J. P., et al. (2011). Transgenerational epigenetic inheritance of longevity in Caenorhabditis elegans. *Nature, 479*, 365–371.

Grewal, S. I., & Klar, A. J. (1996). Chromosomal inheritance of epigenetic states in fission yeast during mitosis and meiosis. *Cell, 86*, 95–101.

Grompe, M. (2012). Tissue stem cells: New tools and functional diversity. *Cell Stem Cell, 10*, 685–689.

Harley, C. B., Futcher, A. B., & Greider, C. W. (1990). Telomeres shorten during ageing of human fibroblasts. *Nature, 345*, 458–460.

Hartl, F. U., Bracher, A., & Hayer-Hartl, M. (2011). Molecular chaperones in protein folding and proteostasis. *Nature, 475*, 324–332.

Helenius, M., Hanninen, M., Lehtinen, S. K., & Salminen, A. (1996). Changes associated with aging and replicative senescence in the regulation of transcription factor nuclear factor-kappa B. *The Biochemical Journal, 318*(Pt. 2), 603–608.

Herranz, D., Munoz-Martin, M., Canamero, M., Mulero, F., Martinez-Pastor, B., Fernandez-Capetillo, O., et al. (2010). Sirt1 improves healthy ageing and protects from metabolic syndrome-associated cancer. *Nature Communications, 1*, 3.

Heslop, L., Morgan, J. E., & Partridge, T. A. (2000). Evidence for a myogenic stem cell that is exhausted in dystrophic muscle. *Journal of Cell Science, 113*(Pt. 12), 2299–2308.

Hidalgo, I., Herrera-Merchan, A., Ligos, J. M., Carramolino, L., Nunez, J., Martinez, F., et al. (2012). Ezh1 is required for hematopoietic stem cell maintenance and prevents senescence-like cell cycle arrest. *Cell Stem Cell, 11*, 649–662.

Imayoshi, I., Sakamoto, M., Ohtsuka, T., Takao, K., Miyakawa, T., Yamaguchi, M., et al. (2008). Roles of continuous neurogenesis in the structural and functional integrity of the adult forebrain. *Nature Neuroscience, 11*, 1153–1161.

Ito, K., Hirao, A., Arai, F., Takubo, K., Matsuoka, S., Miyamoto, K., et al. (2006). Reactive oxygen species act through p38 MAPK to limit the lifespan of hematopoietic stem cells. *Nature Medicine, 12*, 446–451.

Jacobs, J. J., Kieboom, K., Marino, S., DePinho, R. A., & van Lohuizen, M. (1999). The oncogene and Polycomb-group gene bmi-1 regulates cell proliferation and senescence through the ink4a locus. *Nature, 397*, 164–168.

Janzen, V., Forkert, R., Fleming, H. E., Saito, Y., Waring, M. T., Dombkowski, D. M., et al. (2006). Stem-cell ageing modified by the cyclin-dependent kinase inhibitor p16INK4a. *Nature, 443*, 421–426.

Jones, D. L., & Rando, T. A. (2011). Emerging models and paradigms for stem cell ageing. *Nature Cell Biology, 13*, 506–512.

Ju, Z., Jiang, H., Jaworski, M., Rathinam, C., Gompf, A., Klein, C., et al. (2007). Telomere dysfunction induces environmental alterations limiting hematopoietic stem cell function and engraftment. *Nature Medicine, 13*, 742–747.

Kaeberlein, M., McVey, M., & Guarente, L. (1999). The SIR2/3/4 complex and SIR2 alone promote longevity in Saccharomyces cerevisiae by two different mechanisms. *Genes & Development, 13*, 2570–2580.

Kalaitzidis, D., Sykes, S. M., Wang, Z., Punt, N., Tang, Y., Ragu, C., et al. (2012). mTOR complex 1 plays critical roles in hematopoiesis and Pten-loss-evoked leukemogenesis. *Cell Stem Cell, 11*, 429–439.

Kanfi, Y., Naiman, S., Amir, G., Peshti, V., Zinman, G., Nahum, L., et al. (2012). The sirtuin SIRT6 regulates lifespan in male mice. *Nature, 483*, 218–221.

Kapetanaki, M. G., Mora, A. L., & Rojas, M. (2013). Influence of age on wound healing and fibrosis. *The Journal of Pathology, 229*, 310–322.

Kemp, C. J., Donehower, L. A., Bradley, A., & Balmain, A. (1993). Reduction of p53 gene dosage does not increase initiation or promotion but enhances malignant progression of chemically induced skin tumors. *Cell, 74*, 813–822.

Kempermann, G., Jessberger, S., Steiner, B., & Kronenberg, G. (2004). Milestones of neuronal development in the adult hippocampus. *Trends in Neurosciences, 27*, 447–452.

Kempermann, G., Kuhn, H. G., & Gage, F. H. (1998). Experience-induced neurogenesis in the senescent dentate gyrus. *The Journal of Neuroscience, 18*, 3206–3212.

Kennedy, B. K., Austriaco, N. R., Jr., Zhang, J., & Guarente, L. (1995). Mutation in the silencing gene SIR4 can delay aging in S. cerevisiae. *Cell, 80*, 485–496.

Kenyon, C. J. (2010). The genetics of ageing. *Nature, 464*, 504–512.

Korhonen, P., Helenius, M., & Salminen, A. (1997). Age-related changes in the regulation of transcription factor NF-kappa B in rat brain. *Neuroscience Letters, 225*, 61–64.

Krishnamurthy, J., Ramsey, M. R., Ligon, K. L., Torrice, C., Koh, A., Bonner-Weir, S., et al. (2006). p16INK4a induces an age-dependent decline in islet regenerative potential. *Nature, 443*, 453–457.

Krishnamurthy, J., Torrice, C., Ramsey, M. R., Kovalev, G. I., Al-Regaiey, K., Su, L., et al. (2004). Ink4a/Arf expression is a biomarker of aging. *Journal of Clinical Investigation, 114*, 1299–1307.

Kuhn, H. G., Dickinson-Anson, H., & Gage, F. H. (1996). Neurogenesis in the dentate gyrus of the adult rat: Age-related decrease of neuronal progenitor proliferation. *The Journal of Neuroscience, 16*, 2027–2033.

Kujoth, G. C., Hiona, A., Pugh, T. D., Someya, S., Panzer, K., Wohlgemuth, S. E., et al. (2005). Mitochondrial DNA mutations, oxidative stress, and apoptosis in mammalian aging. *Science, 309*, 481–484.

Lang, T., Streeper, T., Cawthon, P., Baldwin, K., Taaffe, D. R., & Harris, T. B. (2010). Sarcopenia: Etiology, clinical consequences, intervention, and assessment. *Osteoporosis International, 21*, 543–559.

Laplante, M., & Sabatini, D. M. (2012). mTOR signaling in growth control and disease. *Cell, 149*, 274–293.

Lee, J. H., Budanov, A. V., Park, E. J., Birse, R., Kim, T. E., Perkins, G. A., et al. (2010). Sestrin as a feedback inhibitor of TOR that prevents age-related pathologies. *Science, 327*, 1223–1228.

Lee, J. S., Li, Q., Lee, J. Y., Lee, S. H., Jeong, J. H., Lee, H. R., et al. (2009). FLIP-mediated autophagy regulation in cell death control. *Nature Cell Biology, 11*, 1355–1362.

Lee, J. Y., Nakada, D., Yilmaz, O. H., Tothova, Z., Joseph, N. M., Lim, M. S., et al. (2010). mTOR activation induces tumor suppressors that inhibit leukemogenesis and deplete hematopoietic stem cells after Pten deletion. *Cell Stem Cell, 7*, 593–605.

Lessard, J., & Sauvageau, G. (2003). Bmi-1 determines the proliferative capacity of normal and leukaemic stem cells. *Nature, 423*, 255–260.

Li, L., & Clevers, H. (2010). Coexistence of quiescent and active adult stem cells in mammals. *Science, 327*, 542–545.

Li, J., Jiang, T. X., Hughes, M. W., Wu, P., Yu, J., Widelitz, R. B., et al. (2012). Progressive alopecia reveals decreasing stem cell activation probability during aging of mice with epidermal deletion of DNA methyltransferase 1. *The Journal of Investigative Dermatology, 132*, 2681–2690.

Lin, L., Hron, J. D., & Peng, S. L. (2004). Regulation of NF-kappaB, Th activation, and autoinflammation by the fork head transcription factor Foxo3a. *Immunity, 21*, 203–213.

Liu, F., Lee, J. Y., Wei, H., Tanabe, O., Engel, J. D., Morrison, S. J., et al. (2010). FIP200 is required for the cell-autonomous maintenance of fetal hematopoietic stem cells. *Blood, 116*, 4806–4814.

Loffredo, F. S., Steinhauser, M. L., Jay, S. M., Gannon, J., Pancoast, J. R., Yalamanchi, P., et al. (2013). Growth differentiation factor 11 is a circulating factor that reverses age-related cardiac hypertrophy. *Cell, 153*, 828–839.

Lombard, D. B., Beard, C., Johnson, B., Marciniak, R. A., Dausman, J., Bronson, R., et al. (2000). Mutations in the WRN gene in mice accelerate mortality in a p53-null background. *Molecular and Cellular Biology, 20*, 3286–3291.

Lugert, S., Basak, O., Knuckles, P., Haussler, U., Fabel, K., Gotz, M., et al. (2010). Quiescent and active hippocampal neural stem cells with distinct morphologies respond selectively to physiological and pathological stimuli and aging. *Cell Stem Cell, 6*, 445–456.

Lugert, S., Vogt, M., Tchorz, J. S., Muller, M., Giachino, C., & Taylor, V. (2012). Homeostatic neurogenesis in the adult hippocampus does not involve amplification of Ascl1 (high) intermediate progenitors. *Nature Communications, 3*, 670.

Lukacs, R. U., Memarzadeh, S., Wu, H., & Witte, O. N. (2010). Bmi-1 is a crucial regulator of prostate stem cell self-renewal and malignant transformation. *Cell Stem Cell, 7*, 682–693.

Maegawa, S., Hinkal, G., Kim, H. S., Shen, L., Zhang, L., Zhang, J., et al. (2010). Widespread and tissue specific age-related DNA methylation changes in mice. *Genome Research, 20*, 332–340.

Magee, J. A., Ikenoue, T., Nakada, D., Lee, J. Y., Guan, K. L., & Morrison, S. J. (2012). Temporal changes in PTEN and mTORC2 regulation of hematopoietic stem cell self-renewal and leukemia suppression. *Cell Stem Cell, 11*, 415–428.

Mair, W., & Dillin, A. (2008). Aging and survival: The genetics of life span extension by dietary restriction. *Annual Review of Biochemistry, 77*, 727–754.

Margueron, R., & Reinberg, D. (2011). The Polycomb complex PRC2 and its mark in life. *Nature, 469*, 343–349.

Matheu, A., Maraver, A., Klatt, P., Flores, I., Garcia-Cao, I., Borras, C., et al. (2007). Delayed ageing through damage protection by the Arf/p53 pathway. *Nature, 448*, 375–379.

Matsumoto, A., Takeishi, S., Kanie, T., Susaki, E., Onoyama, I., Tateishi, Y., et al. (2011). p57 is required for quiescence and maintenance of adult hematopoietic stem cells. *Cell Stem Cell, 9*, 262–271.

Mitchell, J. R., Wood, E., & Collins, K. (1999). A telomerase component is defective in the human disease dyskeratosis congenita. *Nature, 402*, 551–555.

Miyamoto, K., Araki, K. Y., Naka, K., Arai, F., Takubo, K., Yamazaki, S., et al. (2007). Foxo3a is essential for maintenance of the hematopoietic stem cell pool. *Cell Stem Cell, 1*, 101–112.

Mohrin, M., Bourke, E., Alexander, D., Warr, M. R., Barry-Holson, K., Le Beau, M. M., et al. (2010). Hematopoietic stem cell quiescence promotes error-prone DNA repair and mutagenesis. *Cell Stem Cell, 7*, 174–185.

Molofsky, A. V., Pardal, R., Iwashita, T., Park, I. K., Clarke, M. F., & Morrison, S. J. (2003). Bmi-1 dependence distinguishes neural stem cell self-renewal from progenitor proliferation. *Nature*, *425*, 962–967.

Molofsky, A. V., Slutsky, S. G., Joseph, N. M., He, S., Pardal, R., Krishnamurthy, J., et al. (2006). Increasing p16INK4a expression decreases forebrain progenitors and neurogenesis during ageing. *Nature*, *443*, 448–452.

Morita, Y., Ema, H., & Nakauchi, H. (2010). Heterogeneity and hierarchy within the most primitive hematopoietic stem cell compartment. *The Journal of Experimental Medicine*, *207*, 1173–1182.

Morrison, S. J., & Spradling, A. C. (2008). Stem cells and niches: Mechanisms that promote stem cell maintenance throughout life. *Cell*, *132*, 598–611.

Mortensen, M., Soilleux, E. J., Djordjevic, G., Tripp, R., Lutteropp, M., Sadighi-Akha, E., et al. (2011). The autophagy protein Atg7 is essential for hematopoietic stem cell maintenance. *The Journal of Experimental Medicine*, *208*, 455–467.

Mostoslavsky, R., Chua, K. F., Lombard, D. B., Pang, W. W., Fischer, M. R., Gellon, L., et al. (2006). Genomic instability and aging-like phenotype in the absence of mammalian SIRT6. *Cell*, *124*, 315–329.

Murphy, C. T., McCarroll, S. A., Bargmann, C. I., Fraser, A., Kamath, R. S., Ahringer, J., et al. (2003). Genes that act downstream of DAF-16 to influence the lifespan of Caenorhabditis elegans. *Nature*, *424*, 277–283.

Naito, A. T., Sumida, T., Nomura, S., Liu, M. L., Higo, T., Nakagawa, A., et al. (2012). Complement C1q activates canonical Wnt signaling and promotes aging-related phenotypes. *Cell*, *149*, 1298–1313.

Navarro, S., Meza, N. W., Quintana-Bustamante, O., Casado, J. A., Jacome, A., McAllister, K., et al. (2006). Hematopoietic dysfunction in a mouse model for Fanconi anemia group D1. *Molecular Therapy*, *14*, 525–535.

Nebbioso, A., Carafa, V., Benedetti, R., & Altucci, L. (2012). Trials with 'epigenetic' drugs: An update. *Molecular Oncology*, *6*, 657–682.

Nicklin, P., Bergman, P., Zhang, B., Triantafellow, E., Wang, H., Nyfeler, B., et al. (2009). Bidirectional transport of amino acids regulates mTOR and autophagy. *Cell*, *136*, 521–534.

Nijnik, A., Woodbine, L., Marchetti, C., Dawson, S., Lambe, T., Liu, C., et al. (2007). DNA repair is limiting for haematopoietic stem cells during ageing. *Nature*, *447*, 686–690.

Nishimura, E. K., Granter, S. R., & Fisher, D. E. (2005). Mechanisms of hair graying: Incomplete melanocyte stem cell maintenance in the niche. *Science*, *307*, 720–724.

Nishino, J., Kim, I., Chada, K., & Morrison, S. J. (2008). Hmga2 promotes neural stem cell self-renewal in young but not old mice by reducing p16Ink4a and p19Arf expression. *Cell*, *135*, 227–239.

Norddahl, G. L., Pronk, C. J., Wahlestedt, M., Sten, G., Nygren, J. M., Ugale, A., et al. (2011). Accumulating mitochondrial DNA mutations drive premature hematopoietic aging phenotypes distinct from physiological stem cell aging. *Cell Stem Cell*, *8*, 499–510.

Orford, K. W., & Scadden, D. T. (2008). Deconstructing stem cell self-renewal: Genetic insights into cell-cycle regulation. *Nature Reviews Genetics*, *9*, 115–128.

Orkin, S. H., & Zon, L. I. (2008). Hematopoiesis: An evolving paradigm for stem cell biology. *Cell*, *132*, 631–644.

Paik, J. H., Ding, Z., Narurkar, R., Ramkissoon, S., Muller, F., Kamoun, W. S., et al. (2009). FoxOs cooperatively regulate diverse pathways governing neural stem cell homeostasis. *Cell Stem Cell*, *5*, 540–553.

Paliouras, G. N., Hamilton, L. K., Aumont, A., Joppe, S. E., Barnabe-Heider, F., & Fernandes, K. J. (2012). Mammalian target of rapamycin signaling is a key regulator of the transit-amplifying progenitor pool in the adult and aging forebrain. *The Journal of Neuroscience*, *32*, 15012–15026.

Pan, L., Chen, S., Weng, C., Call, G., Zhu, D., Tang, H., et al. (2007). Stem cell aging is controlled both intrinsically and extrinsically in the Drosophila ovary. *Cell Stem Cell*, *1*, 458–469.

Park, I. K., Qian, D., Kiel, M., Becker, M. W., Pihalja, M., Weissman, I. L., et al. (2003). Bmi-1 is required for maintenance of adult self-renewing haematopoietic stem cells. *Nature*, *423*, 302–305.

Pasparakis, M. (2009). Regulation of tissue homeostasis by NF-kappaB signalling: Implications for inflammatory diseases. *Nature Reviews Immunology*, *9*, 778–788.

Pattingre, S., Tassa, A., Qu, X., Garuti, R., Liang, X. H., Mizushima, N., et al. (2005). Bcl-2 antiapoptotic proteins inhibit Beclin 1-dependent autophagy. *Cell*, *122*, 927–939.

Pineda, J. R., Daynac, M., Chicheportiche, A., Cebrian-Silla, A., Sii Felice, K., Garcia-Verdugo, J. M., et al. (2013). Vascular-derived TGF-beta increases in the stem cell niche and perturbs neurogenesis during aging and following irradiation in the adult mouse brain. *EMBO Molecular Medicine*, *5*, 548–562.

Pollina, E. A., & Brunet, A. (2011). Epigenetic regulation of aging stem cells. *Oncogene*, *30*, 3105–3126.

Reese, J. S., Liu, L., & Gerson, S. L. (2003). Repopulating defect of mismatch repair-deficient hematopoietic stem cells. *Blood*, *102*, 1626–1633.

Renault, V. M., Rafalski, V. A., Morgan, A. A., Salih, D. A., Brett, J. O., Webb, A. E., et al. (2009). FoxO3 regulates neural stem cell homeostasis. *Cell Stem Cell*, *5*, 527–539.

Rodgers, J. T., Lerin, C., Haas, W., Gygi, S. P., Spiegelman, B. M., & Puigserver, P. (2005). Nutrient control of glucose homeostasis through a complex of PGC-1alpha and SIRT1. *Nature*, *434*, 113–118.

Rossi, D. J., Bryder, D., Seita, J., Nussenzweig, A., Hoeijmakers, J., & Weissman, I. L. (2007). Deficiencies in DNA damage repair limit the function of haematopoietic stem cells with age. *Nature*, *447*, 725–729.

Rossi, D. J., Bryder, D., Zahn, J. M., Ahlenius, H., Sonu, R., Wagers, A. J., et al. (2005). Cell intrinsic alterations underlie hematopoietic stem cell aging. *Proceedings of the National Academy of Sciences of the United States of America*, *102*, 9194–9199.

Rossi, D. J., Seita, J., Czechowicz, A., Bhattacharya, D., Bryder, D., & Weissman, I. L. (2007). Hematopoietic stem cell quiescence attenuates DNA damage response and permits DNA damage accumulation during aging. *Cell Cycle*, *6*, 2371–2376.

Rubinsztein, D. C., Marino, G., & Kroemer, G. (2011). Autophagy and aging. *Cell*, *146*, 682–695.

Ruckh, J. M., Zhao, J. W., Shadrach, J. L., van Wijngaarden, P., Rao, T. N., Wagers, A. J., et al. (2012). Rejuvenation of regeneration in the aging central nervous system. *Cell Stem Cell*, *10*, 96–103.

Sahin, E., & DePinho, R. A. (2012). Axis of ageing: Telomeres, p53 and mitochondria. *Nature Reviews Molecular Cell Biology*, *13*, 397–404.

Sauvageau, M., & Sauvageau, G. (2010). Polycomb group proteins: Multi-faceted regulators of somatic stem cells and cancer. *Cell Stem Cell*, *7*, 299–313.

Seib, D. R., Corsini, N. S., Ellwanger, K., Plaas, C., Mateos, A., Pitzer, C., et al. (2013). Loss of Dickkopf-1 restores neurogenesis in old age and counteracts cognitive decline. *Cell Stem Cell*, *12*, 204–214.

Shavlakadze, T., McGeachie, J., & Grounds, M. D. (2010). Delayed but excellent myogenic stem cell response of regenerating geriatric skeletal muscles in mice. *Biogerontology*, *11*, 363–376.

Shea, K. L., Xiang, W., LaPorta, V. S., Licht, J. D., Keller, C., Basson, M. A., et al. (2010). Sprouty1 regulates reversible quiescence of a self-renewing adult muscle stem cell pool during regeneration. *Cell Stem Cell*, *6*, 117–129.

Shefer, G., Van de Mark, D. P., Richardson, J. B., & Yablonka-Reuveni, Z. (2006). Satellite-cell pool size does matter: Defining the myogenic potency of aging skeletal muscle. *Developmental Biology*, *294*, 50–66.

Shibatani, T., Nazir, M., & Ward, W. F. (1996). Alteration of rat liver 20S proteasome activities by age and food restriction. *The Journals of Gerontology Series A, Biological Sciences and Medical Sciences, 51*, B316–B322.
Shors, T. J., Townsend, D. A., Zhao, M., Kozorovitskiy, Y., & Gould, E. (2002). Neurogenesis may relate to some but not all types of hippocampal-dependent learning. *Hippocampus, 12*, 578–584.
Signer, R. A., & Morrison, S. J. (2013). Mechanisms that regulate stem cell aging and life span. *Cell Stem Cell, 12*, 152–165.
Simons, B. D., & Clevers, H. (2011). Strategies for homeostatic stem cell self-renewal in adult tissues. *Cell, 145*, 851–862.
Simonsen, A., Cumming, R. C., Brech, A., Isakson, P., Schubert, D. R., & Finley, K. D. (2008). Promoting basal levels of autophagy in the nervous system enhances longevity and oxidant resistance in adult Drosophila. *Autophagy, 4*, 176–184.
Singh, S. K., Williams, C. A., Klarmann, K., Burkett, S. S., Keller, J. R., & Oberdoerffer, P. (2013). Sirt1 ablation promotes stress-induced loss of epigenetic and genomic hematopoietic stem and progenitor cell maintenance. *The Journal of Experimental Medicine, 210*, 987–1001.
Smith, Z. D., & Meissner, A. (2013). DNA methylation: Roles in mammalian development. *Nature Reviews Genetics, 14*, 204–220.
Smythe, G. M., Shavlakadze, T., Roberts, P., Davies, M. J., McGeachie, J. K., & Grounds, M. D. (2008). Age influences the early events of skeletal muscle regeneration: Studies of whole muscle grafts transplanted between young (8 weeks) and old (13–21 months) mice. *Experimental Gerontology, 43*, 550–562.
So, A. Y., Jung, J. W., Lee, S., Kim, H. S., & Kang, K. S. (2011). DNA methyltransferase controls stem cell aging by regulating BMI1 and EZH2 through microRNAs. *PLoS One, 6*, e19503.
Song, Z., Zhang, J., Ju, Z., & Rudolph, K. L. (2012). Telomere dysfunctional environment induces loss of quiescence and inherent impairments of hematopoietic stem cell function. *Aging Cell, 11*, 449–455.
Sperka, T., Wang, J., & Rudolph, K. L. (2012). DNA damage checkpoints in stem cells, ageing and cancer. *Nature Reviews Molecular Cell Biology, 13*, 579–590.
Tomas-Loba, A., Flores, I., Fernandez-Marcos, P. J., Cayuela, M. L., Maraver, A., Tejera, A., et al. (2008). Telomerase reverse transcriptase delays aging in cancer-resistant mice. *Cell, 135*, 609–622.
Toth, M. L., Sigmond, T., Borsos, E., Barna, J., Erdelyi, P., Takacs-Vellai, K., et al. (2008). Longevity pathways converge on autophagy genes to regulate life span in Caenorhabditis elegans. *Autophagy, 4*, 330–338.
Tothova, Z., Kollipara, R., Huntly, B. J., Lee, B. H., Castrillon, D. H., Cullen, D. E., et al. (2007). FoxOs are critical mediators of hematopoietic stem cell resistance to physiologic oxidative stress. *Cell, 128*, 325–339.
Trifunovic, A., Wredenberg, A., Falkenberg, M., Spelbrink, J. N., Rovio, A. T., Bruder, C. E., et al. (2004). Premature ageing in mice expressing defective mitochondrial DNA polymerase. *Nature, 429*, 417–423.
van Praag, H., Christie, B. R., Sejnowski, T. J., & Gage, F. H. (1999). Running enhances neurogenesis, learning, and long-term potentiation in mice. *Proceedings of the National Academy of Sciences of the United States of America, 96*, 13427–13431.
van Praag, H., Shubert, T., Zhao, C., & Gage, F. H. (2005). Exercise enhances learning and hippocampal neurogenesis in aged mice. *The Journal of Neuroscience, 25*, 8680–8685.
Vaziri, H., & Benchimol, S. (1996). From telomere loss to p53 induction and activation of a DNA-damage pathway at senescence: The telomere loss/DNA damage model of cell aging. *Experimental Gerontology, 31*, 295–301.

Villeda, S. A., Luo, J., Mosher, K. I., Zou, B., Britschgi, M., Bieri, G., et al. (2011). The ageing systemic milieu negatively regulates neurogenesis and cognitive function. *Nature*, *477*, 90–94.

Wang, M. C., Bohmann, D., & Jasper, H. (2005). JNK extends life span and limits growth by antagonizing cellular and organism-wide responses to insulin signaling. *Cell*, *121*, 115–125.

Warr, M. R., Binnewies, M., Flach, J., Reynaud, D., Garg, T., Malhotra, R., et al. (2013). FOXO3A directs a protective autophagy program in haematopoietic stem cells. *Nature*, *494*, 323–327.

Wilson, A., Laurenti, E., Oser, G., van der Wath, R. C., Blanco-Bose, W., Jaworski, M., et al. (2008). Hematopoietic stem cells reversibly switch from dormancy to self-renewal during homeostasis and repair. *Cell*, *135*, 1118–1129.

Wong, K. K., Maser, R. S., Bachoo, R. M., Menon, J., Carrasco, D. R., Gu, Y., et al. (2003). Telomere dysfunction and Atm deficiency compromises organ homeostasis and accelerates ageing. *Nature*, *421*, 643–648.

Yeung, F., Hoberg, J. E., Ramsey, C. S., Keller, M. D., Jones, D. R., Frye, R. A., et al. (2004). Modulation of NF-kappaB-dependent transcription and cell survival by the SIRT1 deacetylase. *The EMBO Journal*, *23*, 2369–2380.

Yilmaz, O. H., Katajisto, P., Lamming, D. W., Gultekin, Y., Bauer-Rowe, K. E., Sengupta, S., et al. (2012). mTORC1 in the Paneth cell niche couples intestinal stem-cell function to calorie intake. *Nature*, *486*, 490–495.

Zacharek, S. J., Fillmore, C. M., Lau, A. N., Gludish, D. W., Chou, A., Ho, J. W., et al. (2011). Lung stem cell self-renewal relies on BMI1-dependent control of expression at imprinted loci. *Cell Stem Cell*, *9*, 272–281.

Zhang, Y. V., Cheong, J., Ciapurin, N., McDermitt, D. J., & Tumbar, T. (2009). Distinct self-renewal and differentiation phases in the niche of infrequently dividing hair follicle stem cells. *Cell Stem Cell*, *5*, 267–278.

Zhang, C., & Cuervo, A. M. (2008). Restoration of chaperone-mediated autophagy in aging liver improves cellular maintenance and hepatic function. *Nature Medicine*, *14*, 959–965.

Zou, P., Yoshihara, H., Hosokawa, K., Tai, I., Shinmyozu, K., Tsukahara, F., et al. (2011). p57(Kip2) and p27(Kip1) cooperate to maintain hematopoietic stem cell quiescence through interactions with Hsc70. *Cell Stem Cell*, *9*, 247–261.

INDEX

Note: Page numbers followed by "*f*" indicate figures and "*t*" indicate tables.

A

D

F

G

L

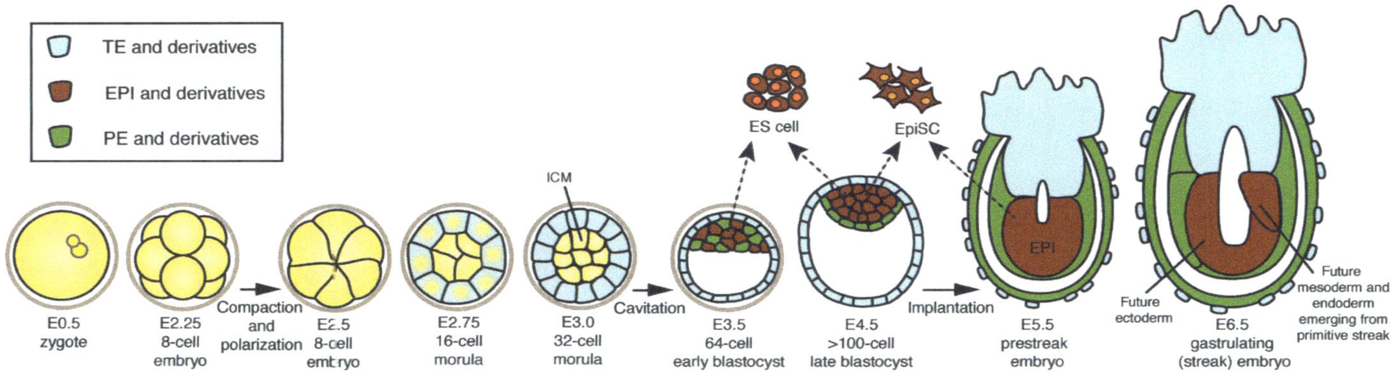

Eszter Posfai *et al.*, Figure 1.1 Schematic overview of early mouse development and origins of pluripotent stem cell types. The totipotent zygote undergoes three rounds of cleavage, producing an 8-cell embryo, which then undergoes compaction and polarization. From the 8-cell stage onward, cell divisions produce two populations of cells: outside cells, which will become the trophectoderm (TE), and inside cells, which will form the inner cell mass (ICM). The ICM further segregates into the primitive endoderm (PE) and the pluripotent epiblast (EPI) by the blastocyst stage. The EPI matures after implantation and gradually loses pluripotency when gastrulation commences at E6.5 (E, embryonic day). During gastrulation, the three germ layers are formed: mesoderm and endoderm precursors migrate through the primitive streak, while nonmigrating EPI cells form ectodermal tissues. Two types of pluripotent stem cells can be derived from the EPI at indicated stages: embryonic stem (ES) cells and epiblast stem cells (EpiSCs).

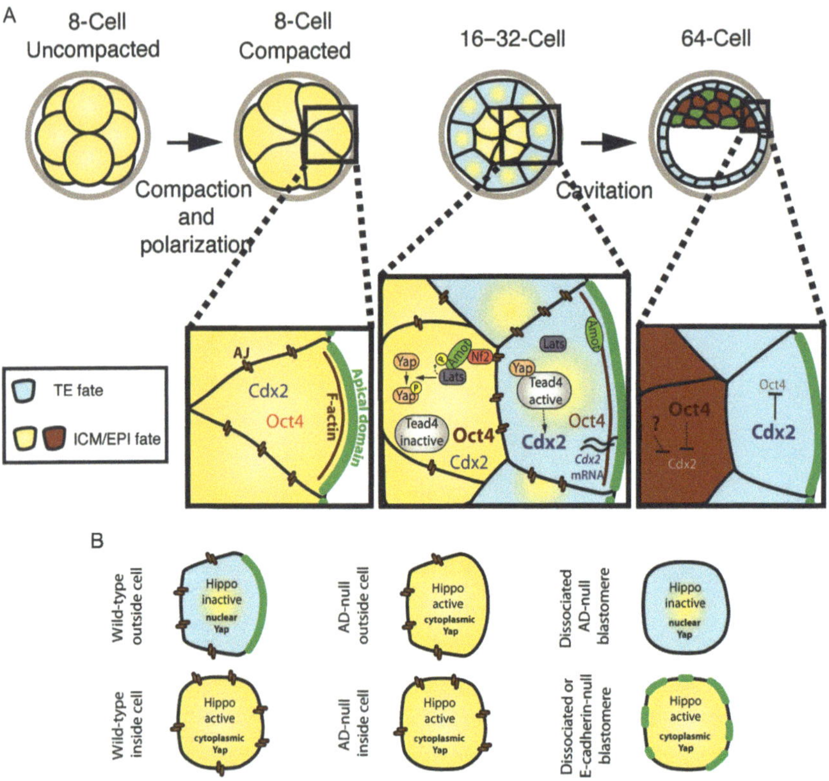

Eszter Posfai *et al.*, Figure 1.2 Model for ICM–TE specification in the mouse embryo. (A) At the 8-cell stage, the embryo undergoes compaction and polarization, forming an apical domain (AD) on outside surfaces and adherens junctions (AJs) on basolateral surfaces. At this stage, ICM and TE markers *Oct4* and *Cdx2* are coexpressed. Further divisions will yield two cell populations: apolar inside and polar outside cells of the embryo. In inside cells, the Hippo pathway is activated through *Amot* binding to AJs via *Nf2* and activating *Lats1/2* kinases. Active *Lats1/2* phosphorylate Yap, preventing its movement into the nucleus. Without Yap, *Tead4* cannot induce the expression of *Cdx2*. In outside cells, *Amot* is sequestered by the AD and does not bind AJs via *Nf2*; thus, it does not activate Hippo signaling. Yap is not phosphorylated and is free to move into the nucleus, activating *Cdx2*. At the blastocyst stage, *Cdx2* inhibits *Oct4* expression in outside cells. *Oct4* may contribute indirectly to *Cdx2* repression in inside cells. (B) Different genetic and/or mechanistic manipulations of blastomeres of the morula stage embryo result in the presence or absence of AJs and AD proteins. Consequently, the Hippo pathway is either active or inactive, resulting in the cytoplasmic or nuclear localization of Yap and the blastomere adopting an ICM or TE fate, respectively.

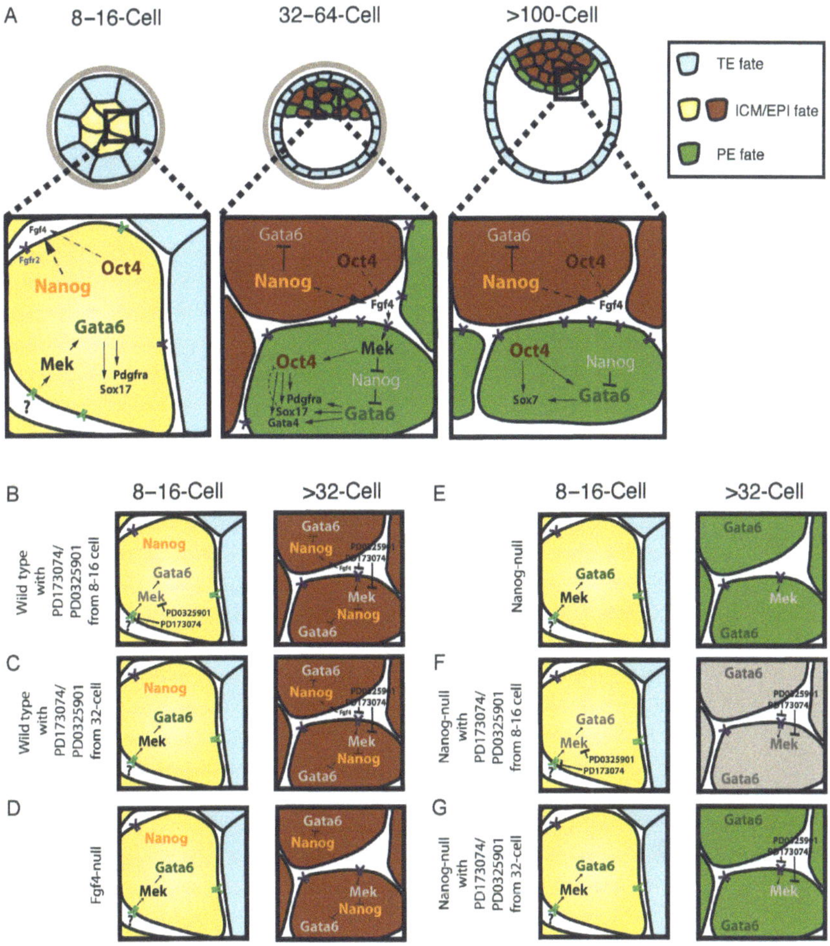

Eszter Posfai *et al.*, Figure 1.3 Model for EPI–PE specification in the mouse embryo. (A) Prior to the 32-cell stage, the EPI marker *Nanog* and the PE marker *Gata6* are coexpressed. *Gata6* expression is FGF4-independent but Mek-dependent at this stage. Segregation of *Nanog*-expressing EPI and *Gata6*-expressing PE precursors occurs from the 32-cell stage onwards: In EPI cells, *Nanog* inhibits *Gata6* expression, while in *Fgfr2*-expressing PE cells, FGF4 (supplied by the EPI cells) via Mek-mediated signaling inhibits *Nanog*. *Oct4* is activated in an Mek-dependent manner in PE cells. In the absence of *Nanog* in PE precursors, *Gata6* is maintained. *Oct4* and *Gata6* are needed for downstream PE-specific TF (*Pdgfra*, *Sox17*, and *Gata4*) expression. In the late blastocyst, PE cells are sorted into a layer lining the blastocoel cavity. PE cells are no longer sensitive to FGF-signaling and express *Sox7* in a *Gata6*- and *Oct4*-dependent manner. (B–G) Different genetic and/or different timing of treatment with small molecule inhibitors (PD173074 or PD0325901) of Mek-dependent signaling results in altered *Nanog* and/or *Gata6* expression and in cells adopting altered EPI or PE fates.

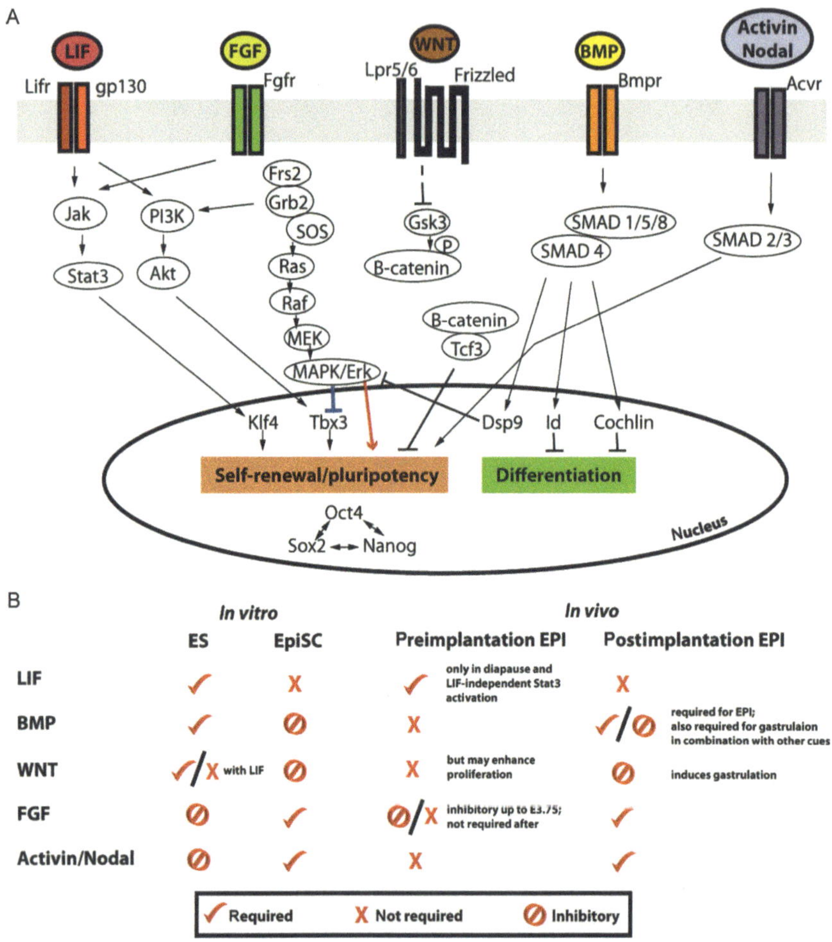

Eszter Posfai *et al.*, Figure 1.4 Signaling pathways involved in regulating self-renewal and pluripotency in the mouse. (A) Cross talk among intracellular signaling cascades involved in regulating self-renewal and pluripotency. Red line ES specific, blue line EpiSC specific. (B) Signaling pathways utilized in the maintenance of self-renewal and pluripotency *in vitro* (ES and EpiSCs) and *in vivo* (pre- and postimplantation EPI).

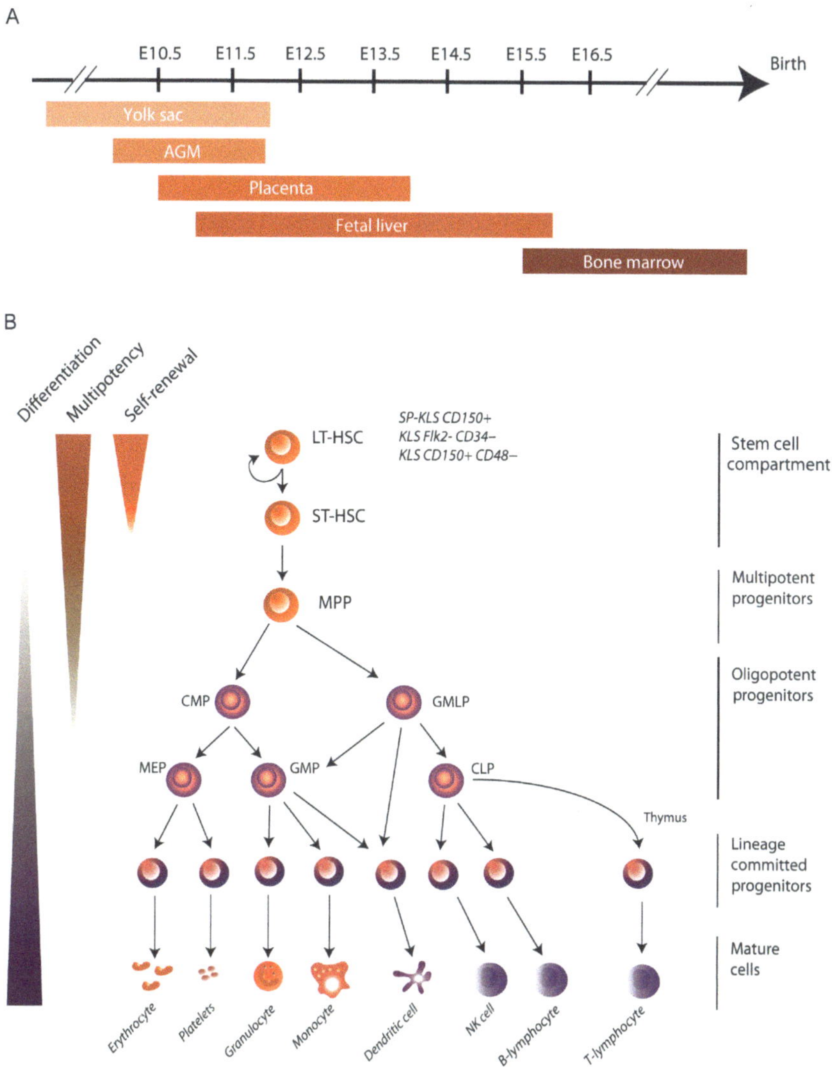

Sean M. Cullen *et al.*, Figure 2.1 (A) Embryonic sites of blood development. The emergence of different sites of hematopoietic activity during development is highly dynamic, involving the coordinated decline of hematopoietic potential in the embryonic site previously occupied. Hematopoietic activity begins in the mesoderm of the extraembryonic yolk sac, followed by the placenta and the aorta-gonad-mesonephros (AGM) region. Hematopoietic cells then migrate to the fetal liver (FL), which becomes the major site of definitive hematopoiesis until the latest stages of embryonic development, when HSCs are recruited to the BM. (B) Overview of the hematopoietic hierarchy. Long-term hematopoietic stem cells (LT-HSCs) produce a continuous supply of cells with progressively decreased multipotency and self-renewal capabilities: short-term HSCs (ST-HSCs), multipotent progenitors (MPPs), and lineage-restricted progenitors. Through rapid proliferation, lineage-restricted progenitors generate mature cells, which are eventually released into the peripheral blood. Phenotypic markers commonly used to identify HSCs are indicated. SP, side population; KSL, c-Kit+Sca-1+Lineage-; CMPs, common myeloid progenitors; CLPs, common lymphoid progenitors; GMLPs, granulocyte–macrophage–lymphocyte progenitors; GMPs, granulocyte–macrophage progenitors; MEPs, megakaryocyte–erythrocyte progenitors; NK, natural killer.

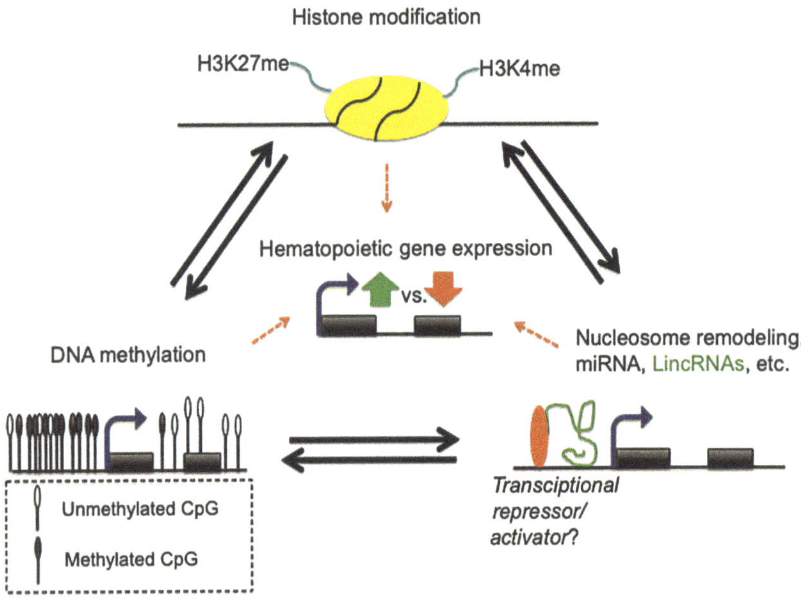

Sean M. Cullen *et al.*, Figure 2.2 Interplay of different classes of epigenetic regulators in hematopoietic stem cells. We propose a model where DNA methylation, histone modification, noncoding RNA, and nucleosome remodeling must all work in concert with one and other in HSCs. Some of these interactions will be of a direct nature, such as histone modifier PRC2, directly interacting with DNA methyltransferases to target specific promoters for DNA methylation. Noncoding RNA molecules, like Mistral in ESC differentiation, could recruit histone modifiers and DNA methyltransferases to sites of activity through direct protein–RNA binding and DNA–RNA interactions. These interconnections will also likely be of a more indirect nature, such as subtle changes in the DNA methylation or histone modification landscape influencing the accessibility and overall effect of other epigenetic regulations. Overall, these interactions will combine to direct changes to hematopoietic gene expression and influence a number of important biological functions in the hematopoietic system, including multipotency maintenance and lineage commitment upon differentiation.

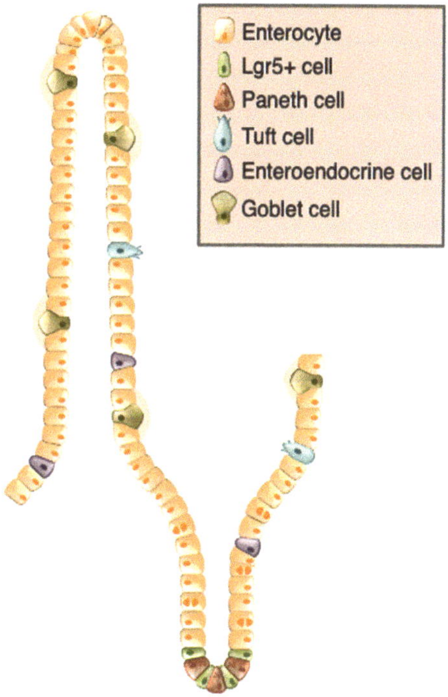

David Wei-Min Tan and Nick Barker, Figure 3.1 Cartoon representation of the intestinal epithelium showing the crypt where stem cells reside, and the villi along which differentiated epithelial cells migrate upward toward the tip.

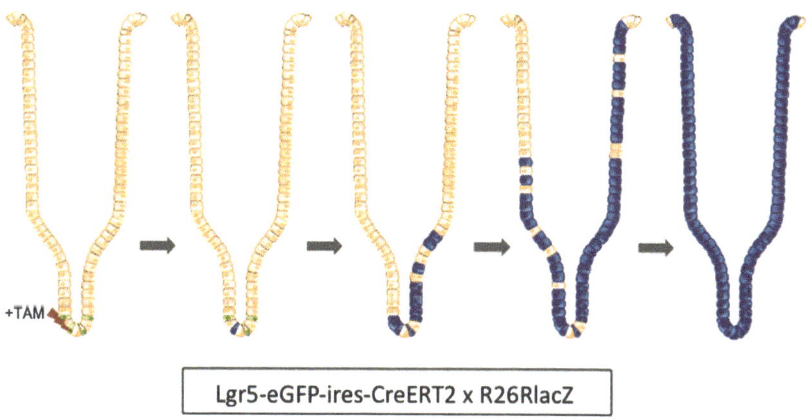

David Wei-Min Tan and Nick Barker, Figure 3.2 How lineage tracing works. By low-frequency genetic marking of cells in a given system, single marked cells transmit the mark to their progeny. From a marked stem cell, either by virtue of its proliferation or by that of its progeny, a labeled clone develops. Intestinal stem cells in which the lacZ transgene has been activated generate progeny that express the β-galactosidase gene. These cells extend from the crypt base to the villus tip and can be visualized as ribbons of blue cells after treatment with X-gal.

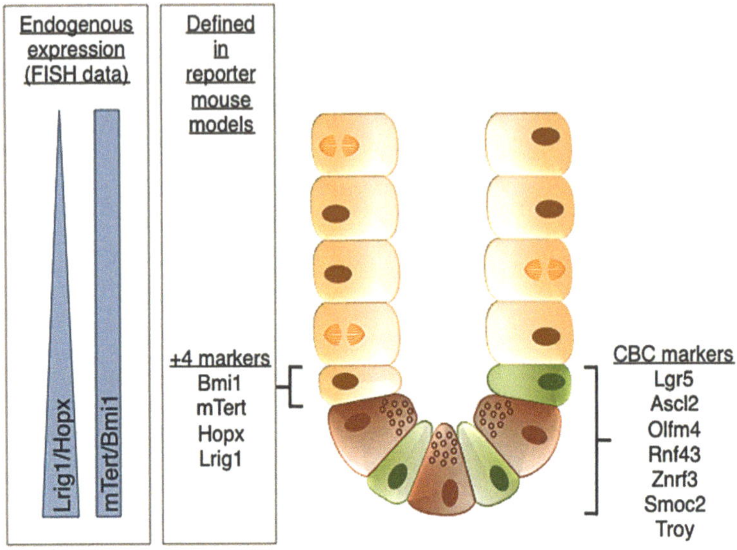

David Wei-Min Tan and Nick Barker, Figure 3.3 Cartoon of an intestinal crypt outlining expression of putative and validated stem cell markers, as well as conflicting data showing broad expression of several markers. CBC markers have been validated and shown to be specifically expressed while data have been reported that show a broader expression of some putative +4 markers. Markers such as Bmi1 and mTert have been observed to be uniformly expressed within the crypt while genes such as Lrig1 and Hopx have been reported to be enriched at the crypt base with a gradient of expression as cells extend toward the crypt/villus border.

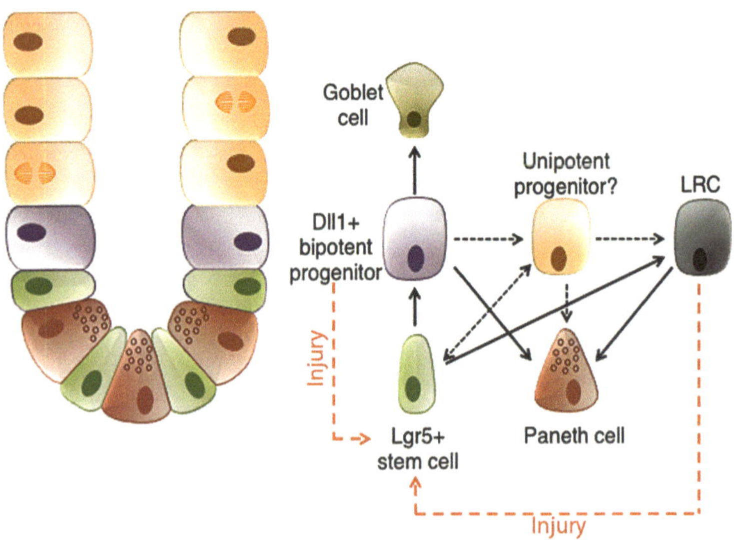

David Wei-Min Tan and Nick Barker, Figure 3.4 Cartoon of the intestinal crypt base showing position and relationships between Lgr5+ CBC cells, Dll1+ progenitor cells, label-retaining cells, and Paneth cells.

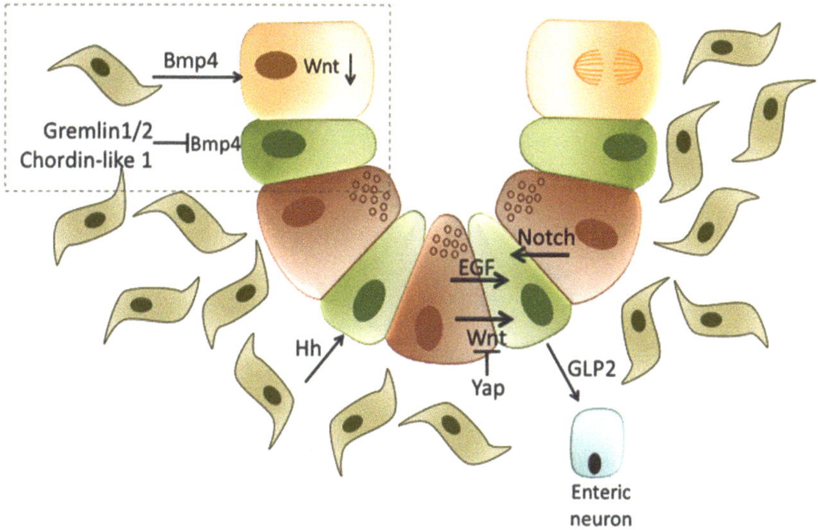

David Wei-Min Tan and Nick Barker, Figure 3.5 Signaling within the stem cell niche. Key niche factors produced by both the epithelial and the stromal compartments mediate cross talk that controls proliferation and differentiation. Dashed box encloses signaling components involved in producing a Wnt gradient along the crypt–villus axis to ensure stem cell maintenance and proliferation in the crypt base while promoting differentiation in cells that migrate upwards.

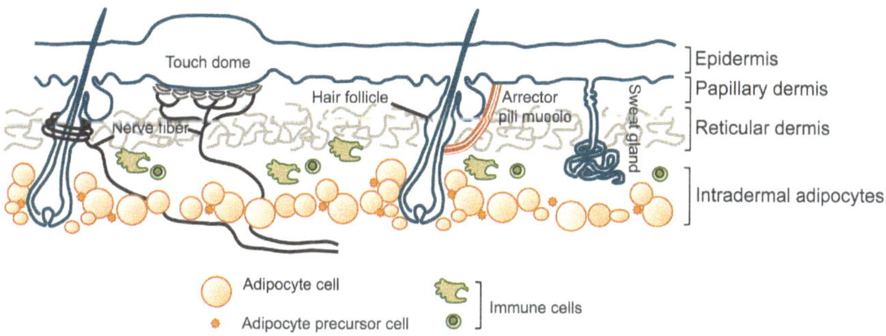

Ana Mafalda Baptista Tadeu and Valerie Horsley, Figure 4.1 Schematic cross-section representation of mammalian skin. The skin is composed of a multitude of cell types and skin appendages that need to interact efficiently and accurately to ensure normal tissue homeostasis.

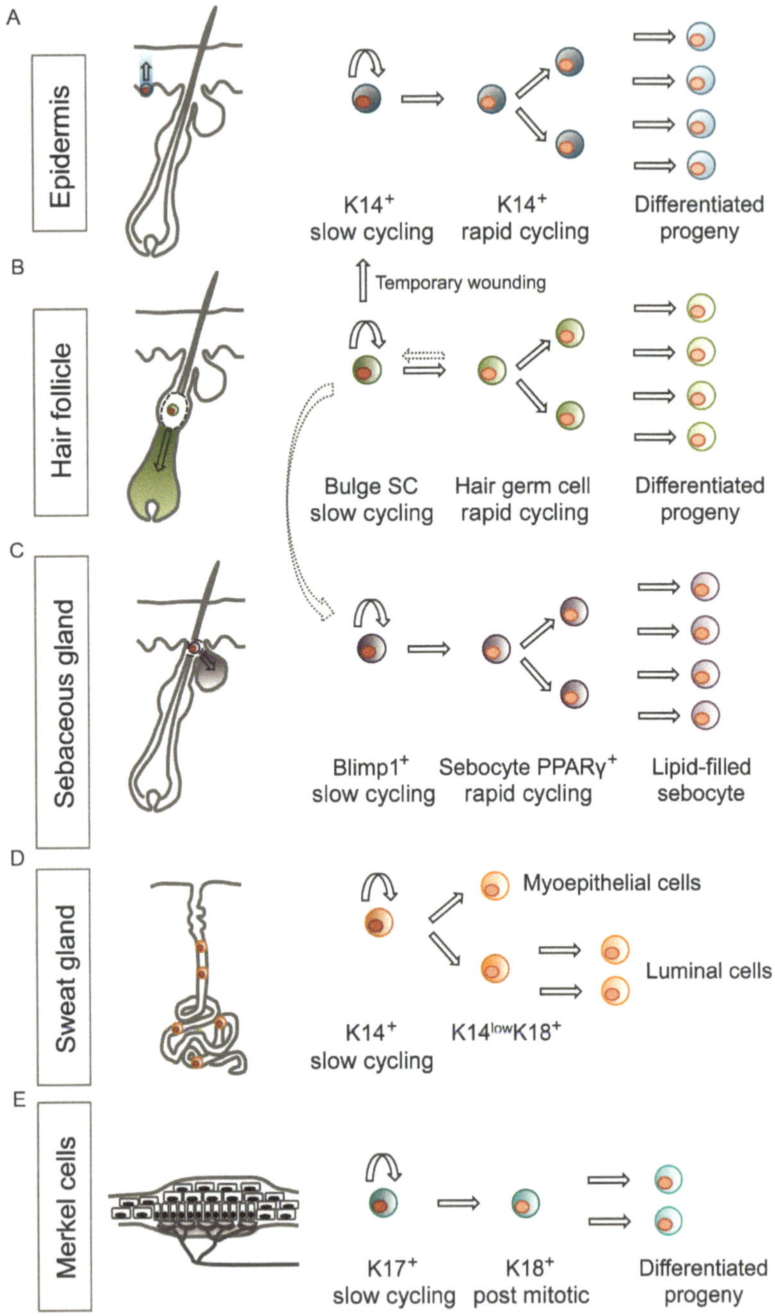

Ana Mafalda Baptista Tadeu and Valerie Horsley, Figure 4.2 Different stem cell populations regulate normal epidermal homeostasis. Schematic representation of the different stem cell population found within different skin compartments. (A) Different lineage tracing experiment have shown that the IFE is maintained by the presence of $K14^{+}$ progenitor basal cells that are able to generate the differentiated lineages that comprise the squamous epithelium of the skin. (B and C) Within the pilosebaceous unit, a slow cycling SC residing in the bulge is able to give rise to all the different HF lineages and regenerate a new hair follicle during anagen. They can also contribute

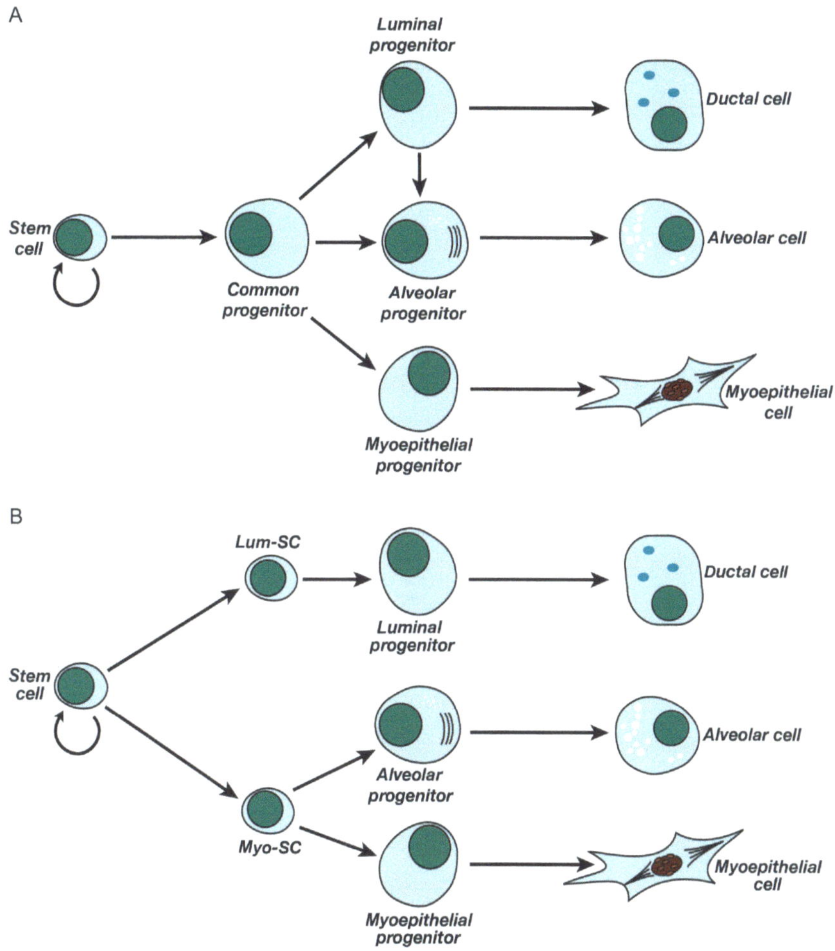

Naiyang Fu *et al.*, Figure 5.1 Schematic diagram showing two potential models of the epithelial hierarchy within the mammary gland. (A) Multipotent stem cells give rise to committed progenitor cells for the myoepithelial and luminal (ductal and alveolar) epithelial lineages but the precise number of progenitor cells is yet to be determined. Luminal progenitors are restricted to either a ductal or alveolar cell fate, and the alveolar-restricted progenitor is thought to be hormone receptor-negative for ER and PR. A bipotent cellular intermediate presumably lies upstream of the lineage-restricted progenitor cells. The prospective isolation of MaSCs, committed luminal progenitors, and mature luminal cells from mouse and human mammary tissue provides strong support for this hierarchical organization. (B) Recent lineage tracing studies suggest that bipotent MaSCs only exist during embryogenesis and that there is a switch to unipotent stem cell populations at birth. Two types of unipotent cells (lum-SC and myo-SC) restricted to either a luminal or myoepithelial fate are postulated to play a major role in the expansion and maintenance of mammary glands in the postnatal gland.

---

to the IFE after wounding and to the SG population that is usually maintained by a resident pool of Blimp1^{+} slow-cycling residing progenitors. (D) In the sweat gland, a K14^{+} progenitor was shown to be able to regenerate a full functioning gland capable of sebum production. Finally, within the touch dome structure (E), K17^{+} slow-cycling cells have been shown to give rise to K18^{+} rapid-cycling progenitor that can further generate differentiated progeny.

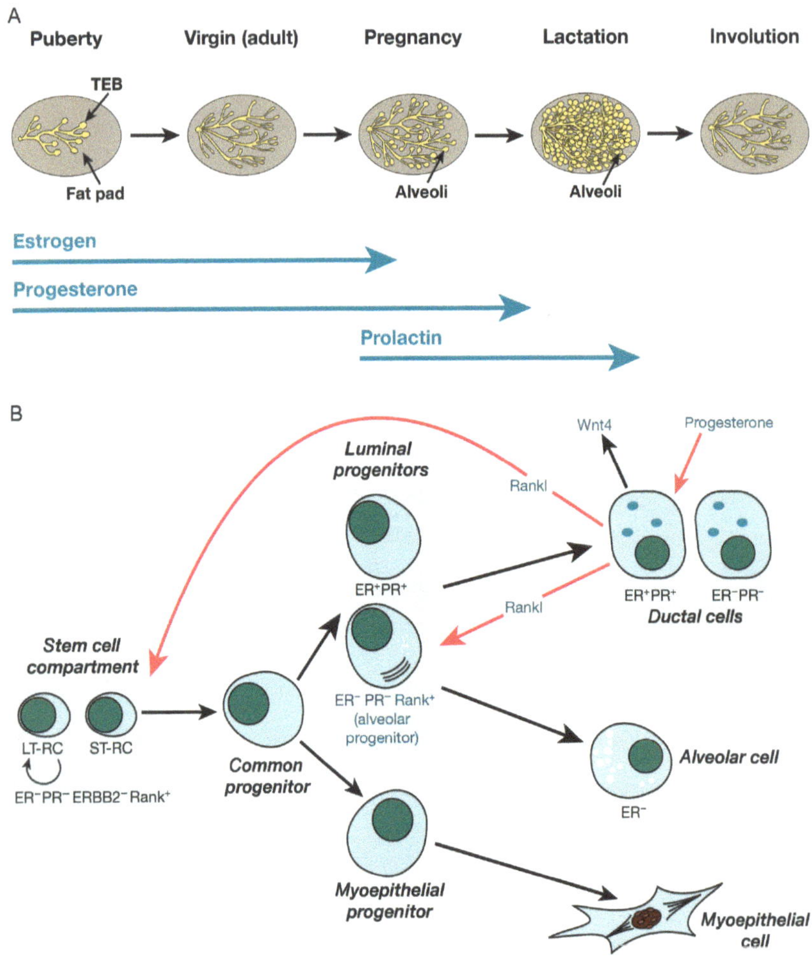

Naiyang Fu *et al.*, Figure 5.2 Hormonal regulation of the epithelial hierarchy. (A) Schematic diagram depicting the primary developmental stages where estrogen and progesterone (steroid hormones), and prolactin (a peptide hormone) act. Estrogen and progesterone are essential for puberty and estrus cycling, while progesterone is important throughout pregnancy. Prolactin plays key roles during pregnancy and lactation. (B) RANKL, an important paracrine mediator induced by progesterone, influences the activity of both mammary stem and progenitor cells. Short-term repopulating cells (ST-RC) are induced in mid-pregnancy. Although PR has many targets including RANKL and Wnt4, RANKL seems to be a likely candidate for transmitting signals from mature luminal cells to MaSCs in response to elevated progesterone levels during pregnancy and diestrus. These short-term MaSCs likely contribute to massive expansion of the gland during pregnancy. Progesterone is also likely to regulate the expansion of alveolar progenitor cells through paracrine signaling from RANKL-producing mature luminal cells to RANK-positive and PR-negative progenitors.

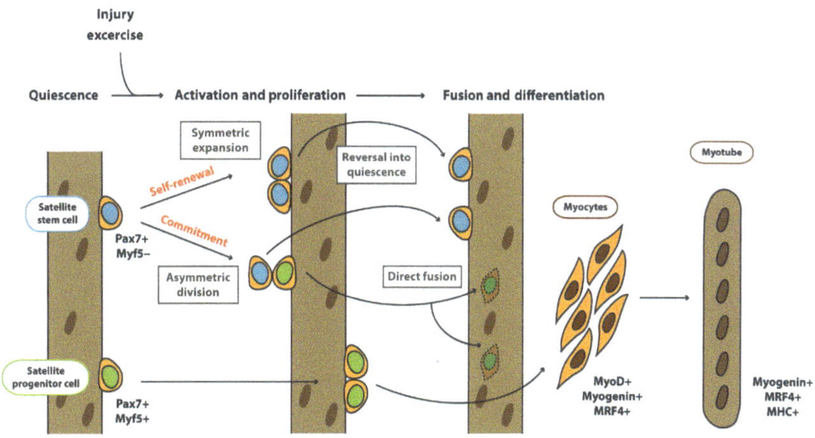

Natasha C. Chang and Michael A. Rudnicki, Figure 6.1 Schematic illustrating the activation and differentiation of satellite cells. Satellite cells are a heterogeneous population of satellite stem cells (Pax7+/Myf5−) and satellite progenitor cells (Pax7+/Myf5+). Upon muscle injury or exercise, satellite cells become activated and exit quiescence. Satellite stem cells (depicted as cells with blue nuclei) can self-renew and undergo either planar symmetric cell division or asymmetric cell division along the apical–basal axis. Expression of *Myf5* indicates myogenic commitment (depicted as cells with green nuclei), and these progenitor cells can contribute to muscle regeneration either by direct fusion into the regenerating myofiber or through the formation of new myotubes, which are characterized by the upregulation of genes that encode for sarcomeric proteins such as myosin heavy chain (MHC).

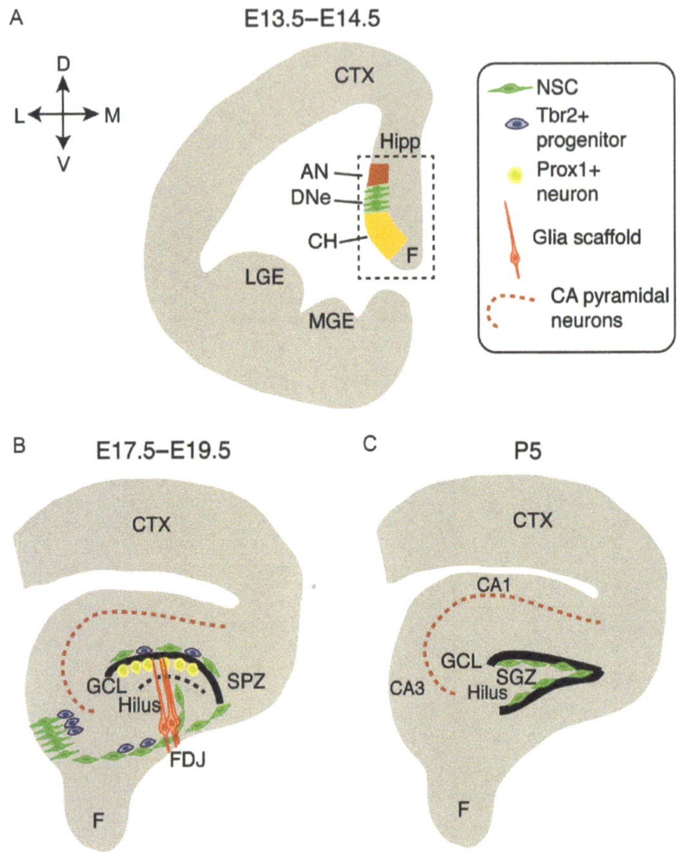

Chiara Rolando and Verdon Taylor, Figure 7.1 Embryonic and early postnatal hippocampal development. (A) Hippocampal primordium develops adjacent to the cortical hem (CH). The dentate neuroepithelium (DNe) contains NSCs that will generate the dentate gyrus. Conversely, the ammonic neuroepithelium (AN) will give rise to pyramidal neurons of the CA3 and CA1. (B) At later developmental stages, NSCs and Tbr2+ progenitors have already started to migrate tangentially and form the SPZ. The trans-hilar radial scaffold forms as progenitors become localized to the FDJ. At the same time, Prox1+ neurons have already populated the GCL. (C) During early postnatal development, progenitor cells that reside in the SPZ undergo redistribution to form the SGZ niche. LGE, lateral ganglionic eminence; MGE, medial ganglionic eminence; CTX, cortex; Hipp, hippocampus; F, fimbria; FJD, fimbriodentate junction; SPZ, subpial zone; GCL, granule cell layer; SGZ, subgranular zone.

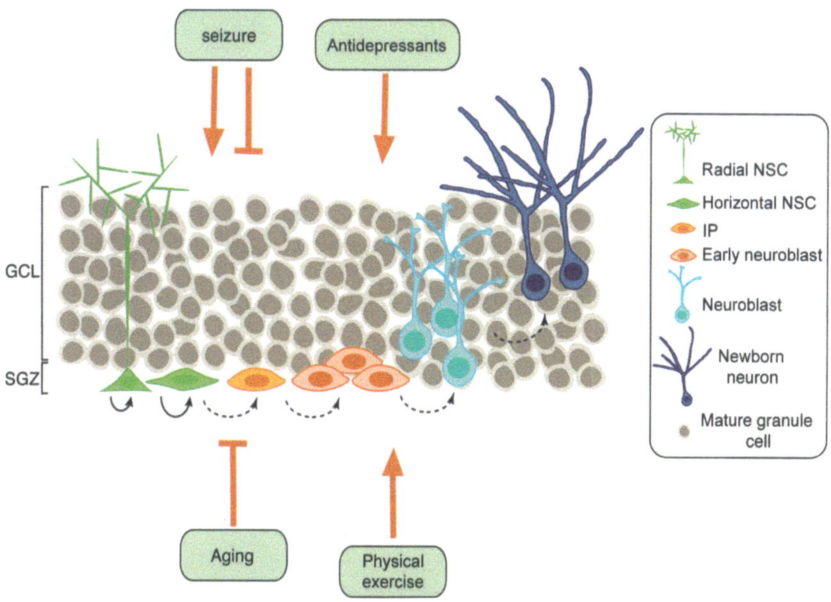

Chiara Rolando and Verdon Taylor, Figure 7.2 Adult neurogenesis in the hippocampal DG. Radial and horizontal NSCs populate the SGZ of the DG. Horizontal NSCs are the active pool that divides multiple times. NSCs originate intermediate progenitors (IPs) that then produce early mitotic neuroblasts. This pool will expand before generating postmitotic neuroblasts and then newborn neurons. In the scheme, different pathological conditions, physical exercise, and therapeutic treatments together with their effect on adult neurogenesis are also represented. GCL, granule cell layer; SGZ, subgranular zone.

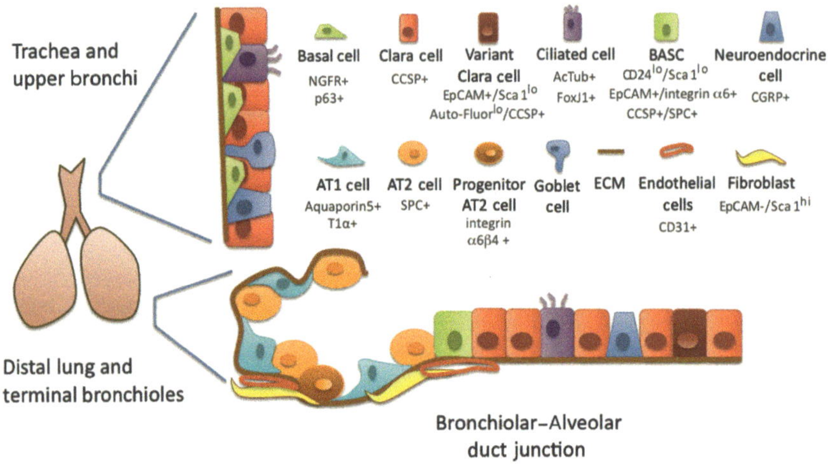

Kristen T. Leeman *et al.*, Figure 8.1 Cell types in the Lung. The proximal region of the murine respiratory system is lined by a pseudostratified epithelium containing secretory CCSP+ Clara cells, mucus-producing goblet cells, and host-defending FoxJ1/Actub+ ciliated cells. Variant Clara cells are thought to give rise to ciliated and Clara cell lineages after injuries such as naphthalene, and are enriched within the $EpCAM^{hi}/Sca1^{lo}/Auto$-$fluorescence^{lo}$ cells. At the basal edge of the epithelium are the NGFR+/p63+ basal cells, which are thought to be able to give rise to Clara and ciliated cells during repair and in culture. In the more distal bronchioles, ciliated and Clara cells are interspersed with CGRP+ neuroendocrine cells. Alveolar epithelial type 1 cells (AT1 cells), which express T1α and Aquaporin5, and SPC+ alveolar epithelial type 2 cells (AT2 cells) line the alveolar space where gas exchange takes place. An alveolar progenitor cell has been identified that can give rise to AT2 and AT1 cells after injuries such as bleomycin, and is termed integrin α6β4+. At the brochioalveolar duct junction, a rare cell population termed the brochioalveolar stem cells (BASC) coexpresses both CCSP and SPC, and is enriched in the $CD24^{lo}/Sca1^{lo}$/EpCAM+/integrin α6+ fraction of lung epithelial cells. BASCs are thought to be able to give rise to both Clara and AT2 cell lineages after injury. Alveolar epithelial cells and BASCs are closely associated with mesenchymal cells such as fibroblasts, the extracellular matrix (ECM), and CD31+ endothelial cells. Useful markers for FACS isolation of cell types are indicated.

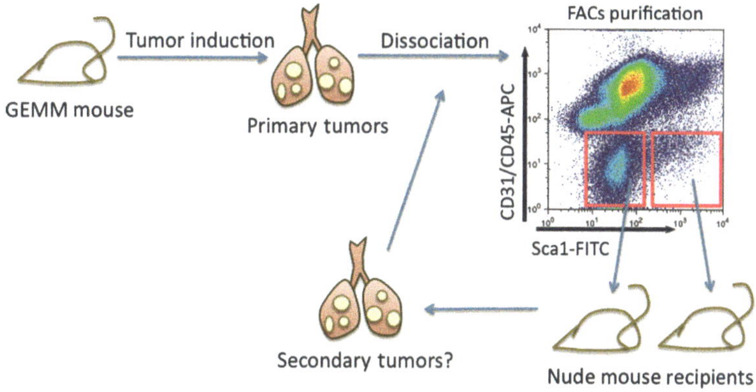

Kristen T. Leeman *et al.*, Figure 8.2 Orthotopic serial transplant assay for lung cancer stem cells. Using genetically engineered mouse models (GEMMs) of lung cancer, tumors with defined genetic characteristics can be induced. These primary tumors can be resected, enzymatically dissociated, and sorted, based on cell surface marker expression. Fluorescence-activated cell sorting (FACS) is typically used to separate tumor cell types. Antibodies against CD31 and CD45 can be used to exclude endothelial cells and hematopoietic cells, respectively, while the remaining tumor cell population can be fractionated using putative cancer stem cell markers such as Sca-1 (*x*-axis of representative FACS plot shown). FACS-isolated populations can then be introduced into the lungs of immunocompromised mice, such as nude mice, and cancer stem cell activity can be measured by secondary tumor growth. As cancer stem cells are thought to have self-renewal ability, a true cancer stem cell pool should be able to propagate tumors in a serial manner.

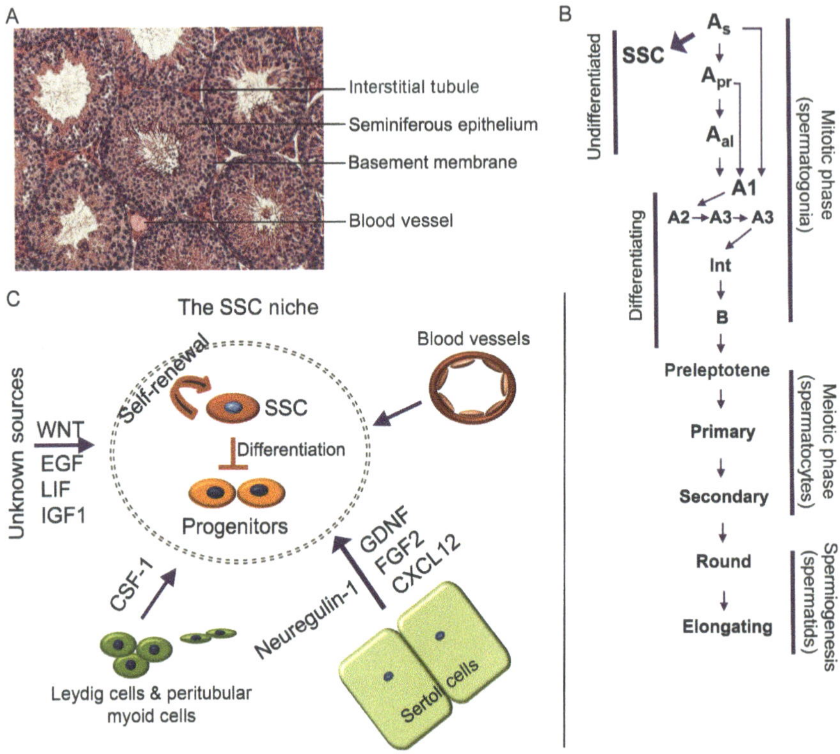

Qi-En Yang and Jon M. Oatley, Figure 9.1 Spermatogenesis and the spermatogonial stem cell niche in mouse testes. (A) Representative image of a hematoxylin and eosin (H&E) stained cross-sections of seminiferous tubules from the testis of an adult mouse. Seminiferous tubules are bound by a basement membrane and lined along the outside by a layer of peritubular myoid cells which provide rigidity. Between tubules is interstitial tissue that is comprised of Leydig cells, blood vessels, and several immune cell types. The seminiferous epithelium consists of an array of germ cell types at different stages of maturation and a single somatic cell population termed Sertoli cells. (B) Schematic of the spermatogenic lineage. The undifferentiated spermatogonial population consists of type A_{single} (A_s), A_{paired} (A_{pr}), and $A_{aligned}$ (A_{al}) subtypes. The A_s spermatogonial population represents a self-renewing reservoir of SSCs from which A_{pr} arise that transiently amplify in number to produce A_{al} progenitors. At periodic intervals, the majority of A_{pr} and A_{al} spermatogonia transition to a differentiating state without a cell division becoming referred to as A1 spermatogonia. Another series of mitotic divisions yields A2, A3, A4, intermediate, and type B spermatogonia that initiate meiosis as preleptotene spermatocytes. Two meiotic divisions ensue with germ cells progressing as primary and then secondary spermatocytes. The end product of the spermatocyte stage is generation of haploid round spermatids that transform into elongated spermatids in a process referred to as spermiogenesis. (C) Current understanding of the SSC niche in mammalian testes. Sertoli cells serve as the major contributor via secreting GDNF, CXCL12, FGF2, and Neuregulin1 to control SSC activities. Leydig cells contribute via secretion of CSF-1 to promote SSC self-renewal. In addition, the vasculature is postulated to play a role but factors emanating from this source that influence SSC activities have not been described.

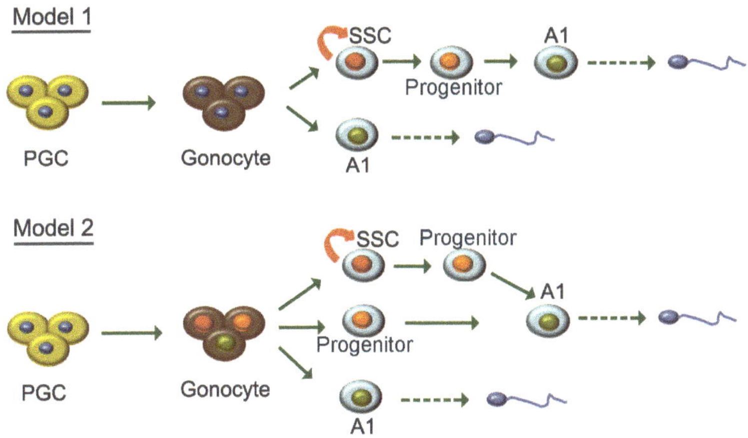

Qi-En Yang and Jon M. Oatley, Figure 9.2 Two proposed models for the formation of the spermatogonial lineage during neonatal development. Model 1 dictates that a subset of prospermatogonia transition to SSCs from which progenitors then arise and another subset gives rise to differentiating spermatogonia directly. Model 2 proposes that the foundational SSC pool, initial progenitor spermatogonial population, and initial differentiating spermatogonial population derive individually from different subsets of gonocytes that are predetermined during embryogenesis.

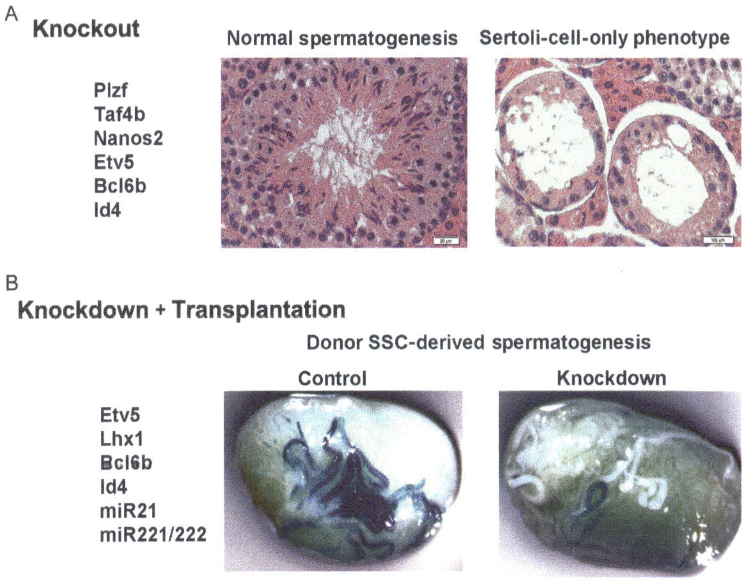

Qi-En Yang and Jon M. Oatley, Figure 9.3 Transcription factors and posttranscriptional regulators that have been identified as important regulators of SSC self-renewal using functional criteria. (A) Transcription factors determined to influence maintenance of the SSC pool based on loss-of-function studies with mutant mice in which the germline is lost after several rounds of normal spermatogenesis resulting in formation of seminiferous tubules with a Sertoli-cell-only phenotype. Testes of mice harboring inactivation of Plzf, Taf4b, Nanos2, Bcl6b, Etv5, and Id4 contain seminiferous tubules with a Sertoli-cell-only phenotype and the percentage increases with advancing age. (B) Transcription factors and miRNAs determined to influence maintenance of the SSC pool based on knockdown of expression in primary cultures of undifferentiated spermatogonia followed by functional transplantation to determine alteration in SSC numbers.

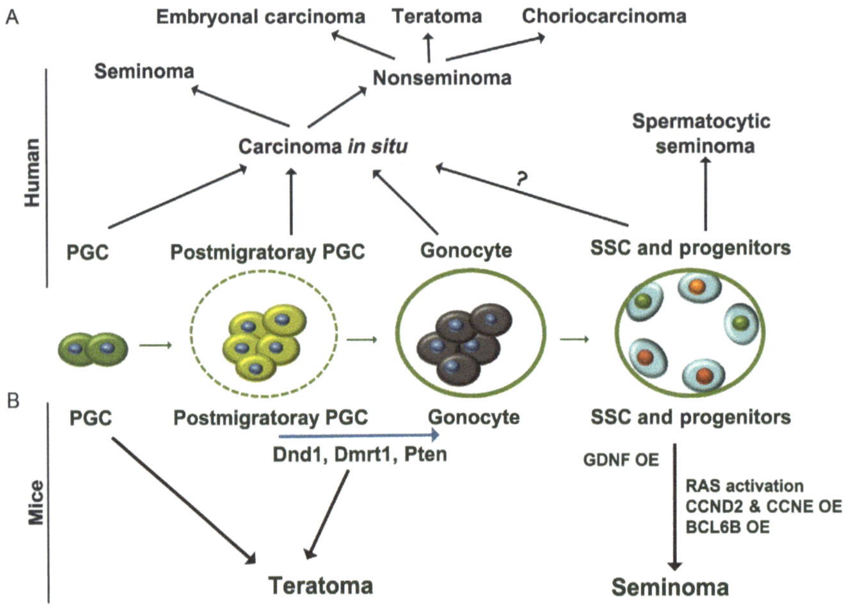

Qi-En Yang and Jon M. Oatley, Figure 9.4 Impact of misregulated molecular mechanisms of spermatogonial lineage development on testicular germ cell tumor formation in humans and mice. (A) Schematic of sources for the variety of testicular germ cell tumors that form in humans. Carcinoma *in situ* (CIS) cells are thought to be the common source of seminoma and nonseminoma tumor types. CIS cells arise in postnatal life as a mass with morphological appearance resembling PGCs or gonocytes. Spermatocytic seminomas are rare benign tumors that share morphological similarities with both spermatogonia and spermatocytes. (B) Impact of misregulated molecular mechanisms influencing spermatogonial stem cell self-renewal on testicular germ cell tumor formation in mice. Teratomas arise from PGCs in which repression of pluripotency and cell cycle arrest is impaired. Seminomas can be induced by GDNF overexpression (OE) in a transgenic mouse line. Primary cultures of mouse undifferentiated spermatogonia engineered to express an activated form of RAS, or overexpress CCND2/CCNE or BCL6B give rise to seminomas following transplantation into recipient mouse seminiferous tubules.

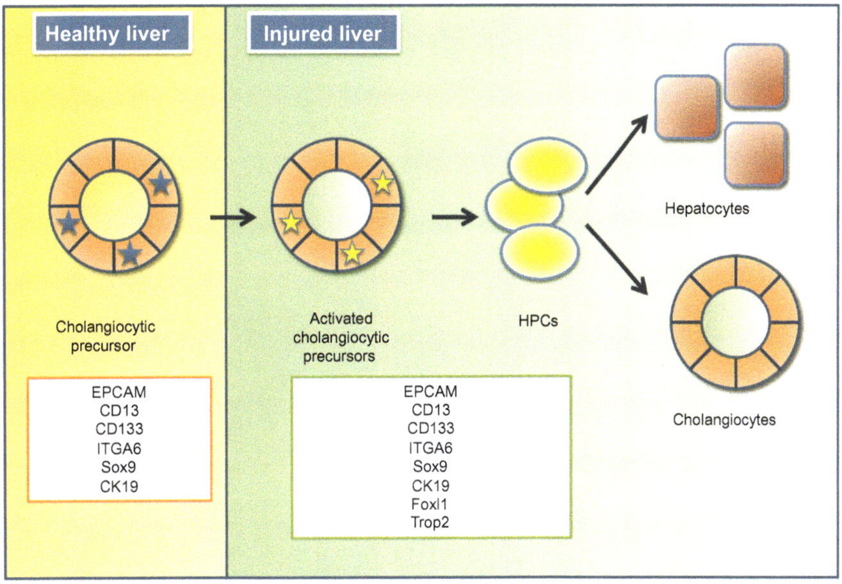

Soona Shin and Klaus H. Kaestner, Figure 10.1 Adult hepatic progenitor cells. A subset of cholangiocytes is activated upon injury and gives rise to HPCs that can differentiate into hepatocytes and cholangiocytes.

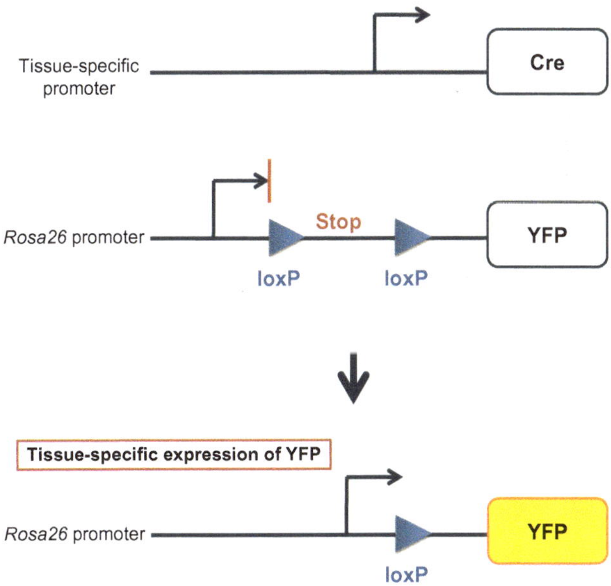

Soona Shin and Klaus H. Kaestner, Figure 10.2 Lineage tracing using the Cre-LoxP approach. *Rosa26* promoter is ubiquitously active in mice. However, transcription of YFP reporter is blocked because of the transcriptional stop sequence (Srinivas et al., 2001). The Cre recombinase excises the sequence located between two loxP sites thus allowing for the expression of YFP in a specific cell type.

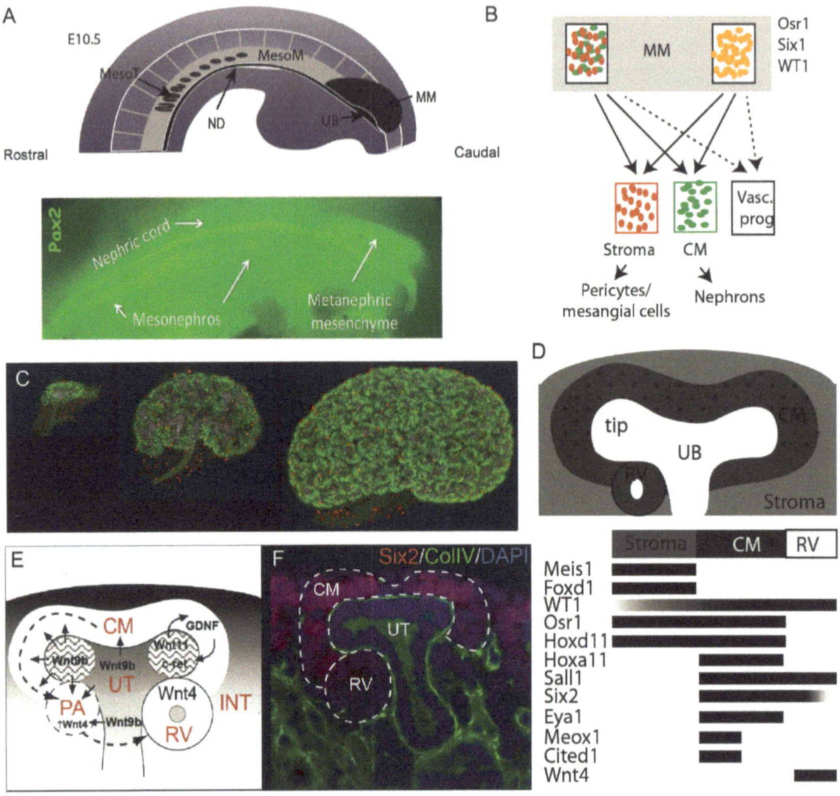

Raphael Kopan *et al.*, Figure 11.1 Development of the mammalian urinary system. (A) Diagram of the developing urinary system in the E10.5 mouse embryo (top) with wholemount Pax2 expression (bottom). ND, nephric duct/nephric cord; MM, metanephric mesenchyme; MesoM, mesonephric mesenchyme; MesoT, mesonephric tubules; UB, point of ureteric bud outgrowth. (B) Two models of the derivation of key kidney progenitor populations from the MM. The MM may represent a heterogeneous field containing several different mesenchymal progenitors (top left), including the stroma, CM and vascular progenitors (vasc.prog.). Alternatively, the MM may represent a homogeneous population of precursors (top right) all co-expressing characteristic MM marker genes and able to differentiate into distinct downstream progenitors. Solid lines indicate what the MM gives rise to. Dotted lines indicate the assumption that this also gives rise to vascular progenitors although renal vasculature is derived from vasculogenesis and angiogenesis. (C) Optical projection series of the developing mouse kidney at E11.5, E13.5 and E15.5 showing ureteric epithelium (white; cytokeratin), cap mesenchyme (green, Six2), and mitosing cells (red, pHH3). (D) Diagram of the nephrogenic niche within the developing metanephros, showing the CM, UB, stroma (SM) and the location of the forming renal vesicles (Stage 1 nephrons; RV). Gene expression within the mesenchymal compartment and derivatives are indicated below this diagram. Shaded regions within CM relate to the assumed position of the uninduced CM based upon expression of *Cited1*. (E) Diagram of the niche indicating the GDNF-c-ret-Wnt11 forward feedback loop driving branching and the Wnt signaling involved in nephron formation and CM survival, (F) Section immunohistochemistry of a single nephrogenic niche at E15.5 showing Six2 in the CM (red), Collagen IV marking all basement membranes (green), and DAPI identifying all nuclei (blue). UT, ureteric tip; CM, cap mesenchyme; RV, renal vesicle. *Panel (A): Adapted from Georgas, Chiu, Lesieur, Rumballe, and Little (2011). Panel (B): From Sequeira Lopez and Gomez (2011). Panel (C): Adapted from Rumballe et al. (2011).*

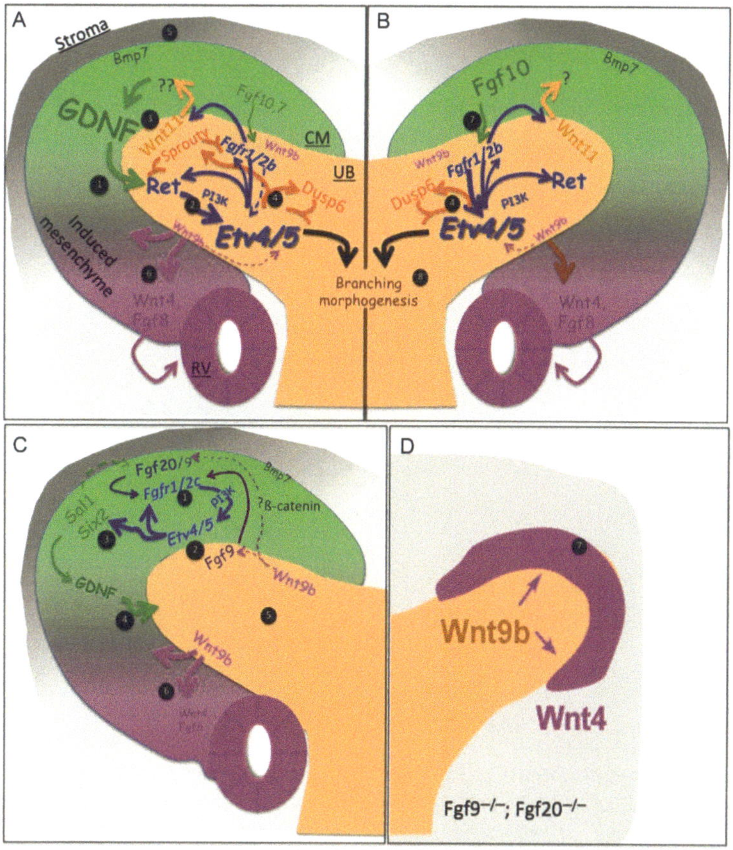

Raphael Kopan *et al.*, Figure 11.2 Genetic network controlling ureteric bud branching and MM maintenance. (A) Ret controls UB tip stem cells (for details see Costantini & Kopan, 2010). Bold arrows mark major effects; dotted arrows, relationships that are week or uncertain. Black arrows summarize complex, multistep processes. (1) GDNF signal to Ret and FgfR in the UB to activate Etv4/5 via PI3K. (2) Etv4/5 activates Ret, FgfR, and Wnt11. (3) Wnt11 signals to the CM to activate GDNF via an undetermined process. (4) At the same time, Etv4/5 also activates Sprouty, an inhibitor of both Ret and Fgfr, and Dusp6, an inhibitor of Ret and Fgfr signaling. This forms an incoherent feed-forward loop. The stroma cells produce (5) BMP7, which promotes aggregation and sorting of CM cells. The UB secretes Wnt9b, the main differentiation-inducing signal in the metanephros (6). Wnt9b-induced mesenchyme undergoes mesenchymal to epithelial transition and produces Wnt4 and FGF8, the latter will induce proliferation in the renal vesicle (RV). (B) Loss of GDNF leads to loss of Ret signaling in the UB tip cells; leading to failure in maintenance of Etv4/5 in the UB and loss of branching/tip cell identity. Mesenchymal Fgf7,10 *can* signal via Fgfr and PI3K to activate Etv4/5 to rescue loss of GDNF but will only do so if the inhibitor *spry* is removed (Costantini & Kopan, 2010) (7). The net output of the Ret/Fgfr signals in the UB is branching morphogenesis, which can be observed in the absence of the CM as long as GDNF is provided. Disruption of this network at any point eliminates progenitor identity in both the MM and the UB tip and renal agenesis. (C) The intra-CM Fgf signaling loop involves Fgf/920 and FgfR1/2 (1) which activate Etv4/5 in the MM via PI3K. (2). Etv4/5 activates Fgfr, and (3) CM maintenance factors Six2 and Sal1 which indirectly ensure GDNF expression. (4) GDNF induces branching. The UB cells produce (5) Wnt9b, which promotes maintenance of CM targets either directly or indirectly, perhaps by elevating Fgf9 and/or Fgf20. Wnt9 also induces differentiation (6) which takes over the entire CM when Fgf9/20 levels are critically reduced (D, (7)).

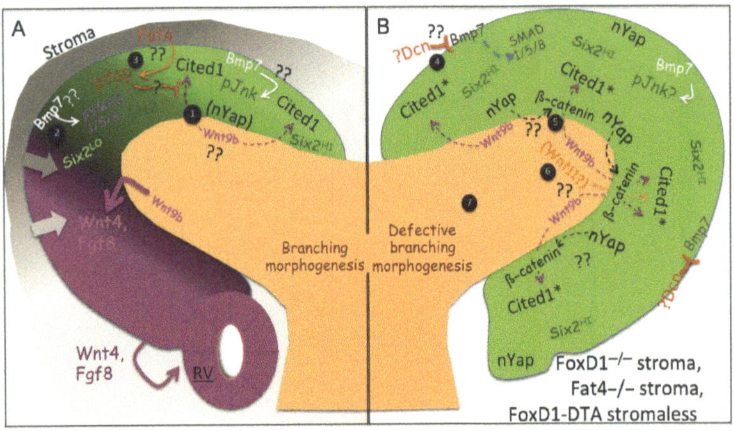

Raphael Kopan *et al.*, Figure 11.3 SM contribution to the niche. (A) The CM is divided into uninduced *Six2*^{HI},*Cited1*$^{+}$ cells (light green) that are refractory to Wnt differentiation signals (1), and *Six2*^{LO}, *Cited1*$^{-}$ cells (dark green) that can respond to Wnt and become competent to differentiate (purple) (2). A Wnt signal activates *Cited1* and additional markers; *Cited1* is not required for renal development and thus labeled in the figure with an asterisk. BMP7 signals through unknown receptor(s) to activate JNK in uninduced CM. Activation of pSMAD1/5/8 is thought to condition the CM and make it competent to respond to Wnt. The stroma also expresses Fat4 (3) that through an unknown kinase phosphorylate Yap. pYap is either unable to cooperate with β-catenin to maintain uninduced CM or interacts with Disheveled to block β-catenin from maintaining uninduced CM. (4). Alternatively, in the absence of FoxD1 or stroma, Decorin (DCN) is overexpressed in the interstitium to either inhibit SMAD phosphorylation or all BMP7 activity, failing to condition CM which remains refractory to Wnt. Without Fat4 Yap is nuclear (nYap). nYap may somehow impact β-catenin to maintain uninduced CM (and the marker *Cited1*) (5) or, in the absence of pYap, weak Wnt ligands like Wnt11 gain the ability to activate β-catenin and maintain uninduced CM and *Cited1* (6). Branching is affected without induced mesenchyme (7).

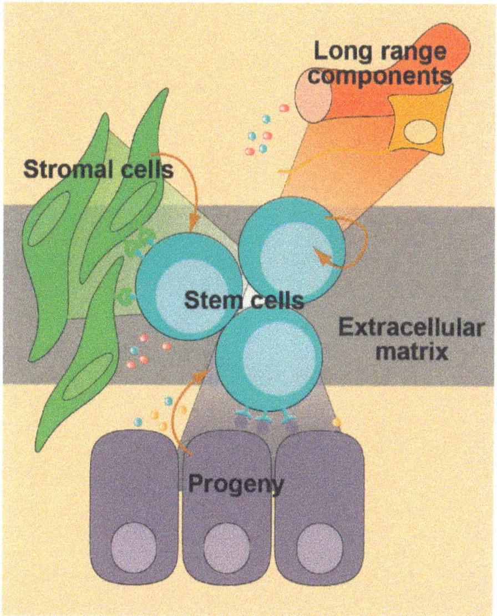

Amélie Rezza *et al.*, Figure 12.1 Common features of adult stem cell niches. Stem cells (SCs) reside in a specific microenvironment called the niche, composed of multiple components: SCs themselves (blue) and their progeny (purple), as well as close surrounding cells from the mesenchyme or stroma (green), more distant cells such as endothelial cells or neurons providing long-range signals (red and orange), and the extracellular matrix/basement membrane (ECM; gray). Each component expresses adhesion molecules, specific receptors, or secreted factors that signal to the SCs (gradients), creating a unique localized niche for SC maintenance and/or activation.

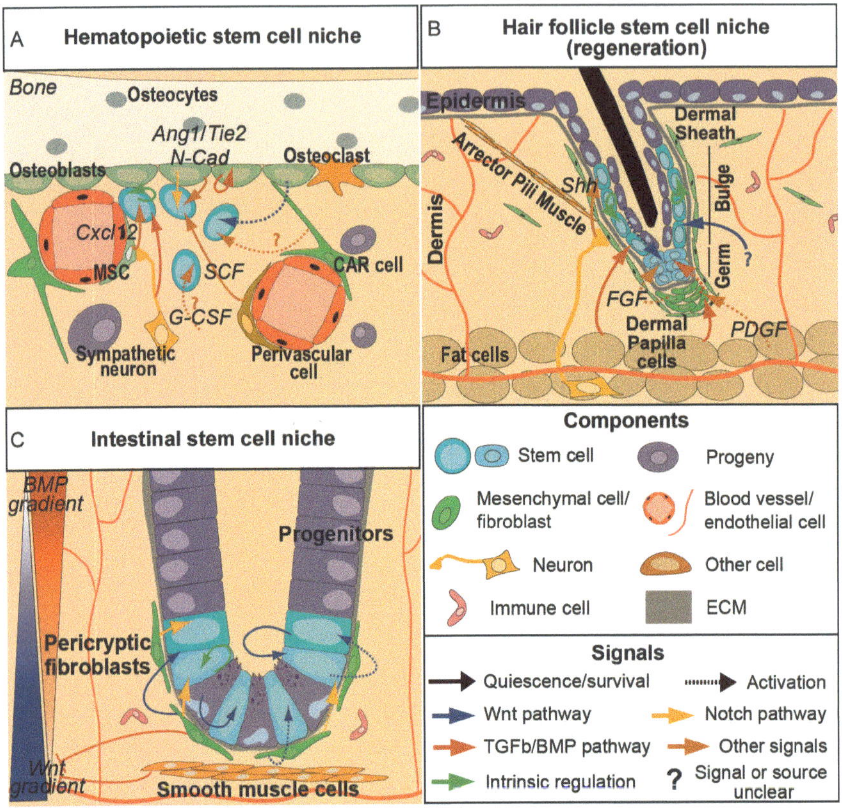

Amélie Rezza *et al.*, Figure 12.2 Cellular and molecular components of three fast-cycling adult stem cell niches: the hematopoietic, the hair follicle, and the intestinal systems. (A) The hematopoietic stem cell (HSC) niche includes osteoblasts (green), osteoclasts (orange), endothelial cells (red), perivascular cells (light brown), HSC progeny (purple), CXCL12-abundant reticular (CAR; dark green) cells, and mesenchymal stem cells (MSC; light green). Many signals regulate HSCs (blue): SCs autoregulate their cell cycle to prevent exhaustion (green arrow). TGFb from neural cells (red arrow), Notch signaling (yellow arrow), Ang1 from osteoblasts, perivascular SCF, and CXCL12 from MSC (brown

(Continued)

Figure 12.2—Cont'd arrows) are important for HSC maintenance. Activating signals are Wnts (still debated; dashed blue arrow), secreted factors from CAR cells and G-CSF (dashed brown arrows). (B) Hair follicle SCs (HFSCs) during regeneration reside in a niche composed of mesenchymal dermal papilla (DP) cells (dark green), fat cells (light brown), and dermis populated by blood vessels (red), immune cells (pink), loose fibroblasts and dermal sheath fibroblasts (light green), arrector pili muscle cells (dark brown), and neurons (yellow). Two populations of SCs are distinguishable during regeneration: bulge SCs (dark blue) and germ SCs (light blue). Maintenance signals to bulge SCs come from different niche components: HFSCs control quiescence through CDKi, Runx1/p21, Tbx1, NFATc1, Lhx2, and Sox9 (green arrow); subcutaneous fat cells express BMP ligands (red arrows); canonical Wnt signals from an unknown source (dark blue arrow); and nerve-secreted Shh (brown arrow). Activating signals include TGFb2 (dashed red arrow) and FGF7 (dashed brown arrow) expressed by DP cells, Wnt ligands expressed by epithelial cells (dashed dark blue arrow), and PDGF from the fat probably acting through DP cells (dashed brown arrow). (C) The intestinal SC niche is composed of ISC progeny Paneth cells (purple), pericryptic fibroblasts (green), and smooth muscle cells (brown). Blood vessels (red) and immune cells (pink) are present in the pericryptic mesenchyme. Two SC pools can be distinguished: Lgr5+ Columnar Basal Cells (light blue) and +4 SCs (dark blue); and many signals regulate their behavior: Ascl2 and Lrig1 control ISC maintenance (green arrow), Wnt signaling regulates SC survival through Wnt ligands from Paneth cells and surrounding fibroblasts (plain and dashed blue arrows), Notch is involved in cell proliferation and differentiation (plain and dashed yellow arrows), BMP signaling is repressed by antagonists from pericryptic fibroblasts and smooth muscle cells (red blunt arrows).

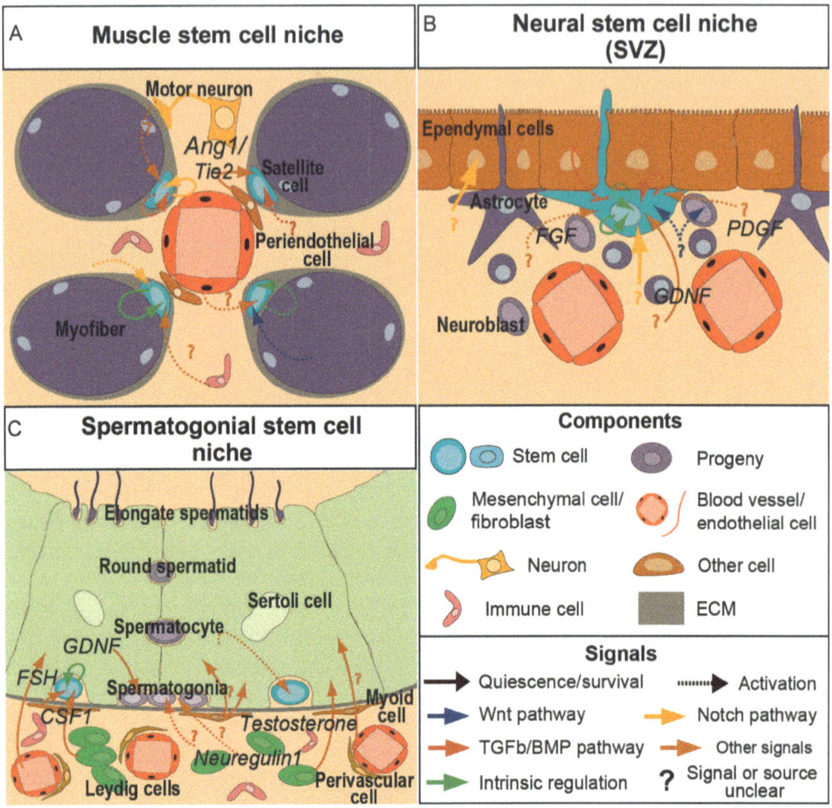

Amélie Rezza *et al.*, Figure 12.3 Components of the adult muscle, brain, and germ stem cell niches. (A) In the muscle, satellite SCs (blue) are present along muscle fiber progeny (purple), beneath the basal lamina (grey). The SC niche also includes blood vessels (red), periendothelial cells (brown), neurons (yellow), and immune cells (pink). Satellite SCs autoregulate their quiescence through Notch signaling (yellow arrow) and the miRNA machinery (green arrow). Periendothelial cells secrete Ang1 to regulate SC maintenance (brown arrow). SCs control their activation through the regulation of S1P (dashed green arrow). TGFb signaling also influences the regenerative capacity of satellite SCs (dashed red arrows). Myofibers secrete Wnt ligands that promote activation and regeneration (dashed blue arrow). Neural, endothelial, and immune cells secrete factors to impact SC behavior, but precise mechanisms are unclear (dashed brown arrows). (B) The neural SC niche in the SVZ is composed of ependymal cells (brown) and astrocyte progeny (purple) that both directly interact with SCs. Also present in the close environment are other NSC progeny (round purple cells) and blood vessels (red). NSCs autoregulate their own proliferation and quiescence (green arrow). The BMP pathway is precisely regulated with ligands from NSCs (red arrow) and antagonists from ependymal cells (red blunt arrow). Notch (yellow arrows) and GDNF (brown arrow) signaling are critical for ependymal and SCs maintenance as well, but the ligand source is unknown. The miRNA machinery regulates NSC activation (dashed green arrow). The Wnt pathway (dashed blue arrow) also activates proliferation. FGF and PDGF signals from unknown sources promote NSC proliferation (dashed brown arrows).

(Continued)

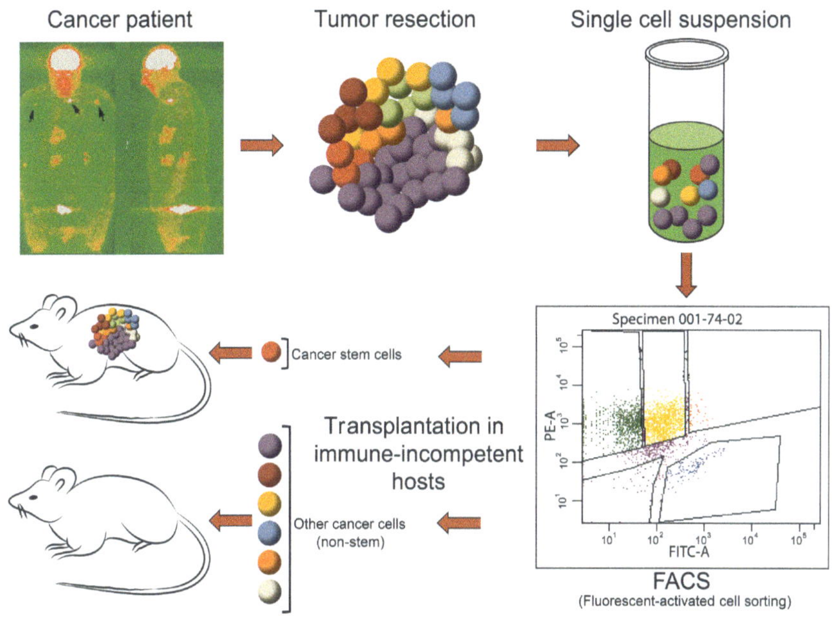

Yaser Atlasi *et al.*, Figure 13.1 Schematic representation of the limited dilution transplantation assays coupled with fluorescent-activated cell sorting (FACS) applied to enrich the CSC subpopulation (red cells) from within a heterogeneous tumor composed of different more committed (i.e., nonstem) lineages. Upon transplantation at limited cellular dilutions, only the CSCs are capable to recapitulate the primary lesion derived from the patient.

Figure 12.3—Cont'd (C) The germ SC niche in the testis is composed of Sertoli cells (light green), myoid cells (dark brown), and Leydig cells (green). Blood vessels (red) surrounded by perivascular cells (brown) and immune cells (pink) are present in the interstitial tissue. SSC maintenance and activation is dependent on Sertoli cell maintenance, which is regulated by testosterone from Leydig cells, FSH from the pituitary gland, and other signals from perivascular cells (brown arrows). SSC survival and maintenance depends on integrin beta1 expression in SSCs (green arrow), GDNF from Sertoli cells (brown arrows), or CSF1 from Leydig and myoid cells (brown arrows). SSCs can be activated by a wide range of signals such as BMP (dashed red arrow).

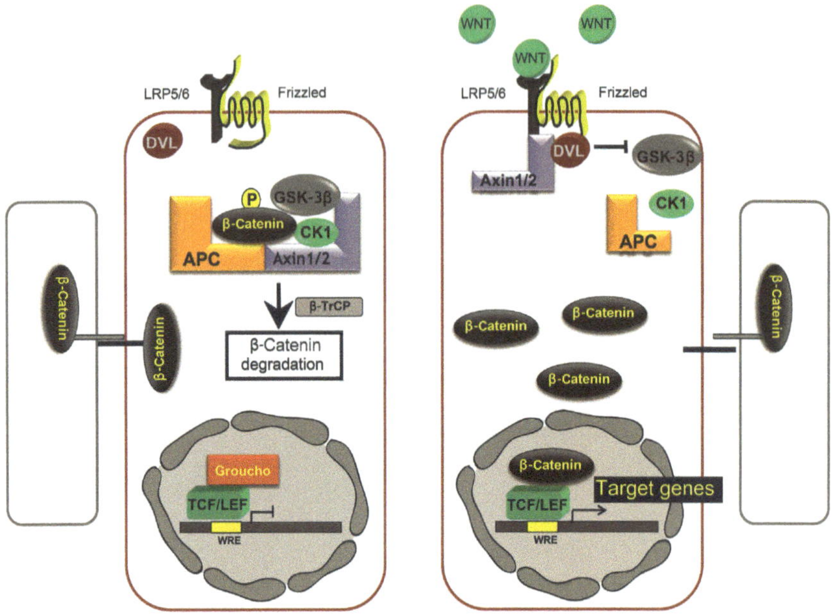

Yaser Atlasi *et al.*, Figure 13.2 Schematic representation of the canonical Wnt/β-catenin signaling pathway. The left panel represents the cell in the absence of the Wnt ligand, while the right panel depicts the actively signaling cell in the presence of the glycoprotein Wnt. See review text for additional details. Legend. DVL, disheveled; GSK-3β, glycogen synthase kinase 3 beta; CK1: casein kinase 1; APC, adenomatous polyposis coli; β-TrCP: beta-transducin repeats-containing protein (substrate recognition subunits for E3 ubiquitin ligases); LRP5/6, LDL receptor-related proteins 5 and 6.

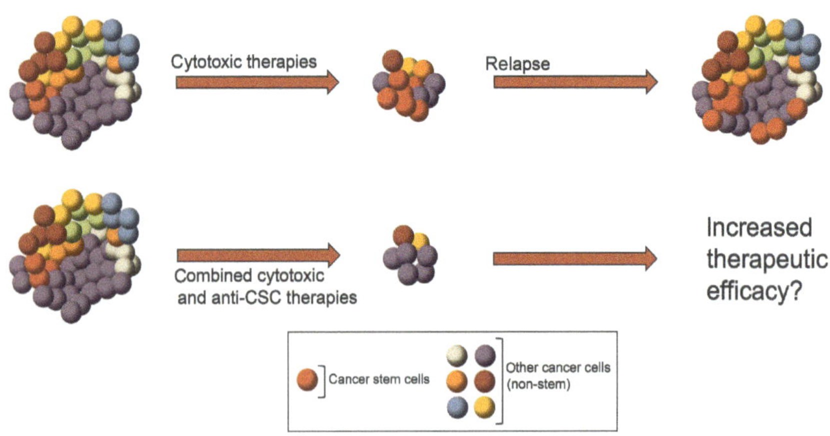

Yaser Atlasi *et al.*, Figure 13.3 Combination of conventional (cytotoxic) and anti-CSC treatment may lead to improved therapeutic efficacy. Assuming that CSC are less sensitive to the cytotoxic effects of conventional chemo- and radiotherapies, these regimens, although initially leading to an effective reduction of the tumor mass, eventually lead to a relative enrichment of the CSCs. Hence, the combined use of conventional and anti-CSC agents is thought to result in improved treatment efficacy.

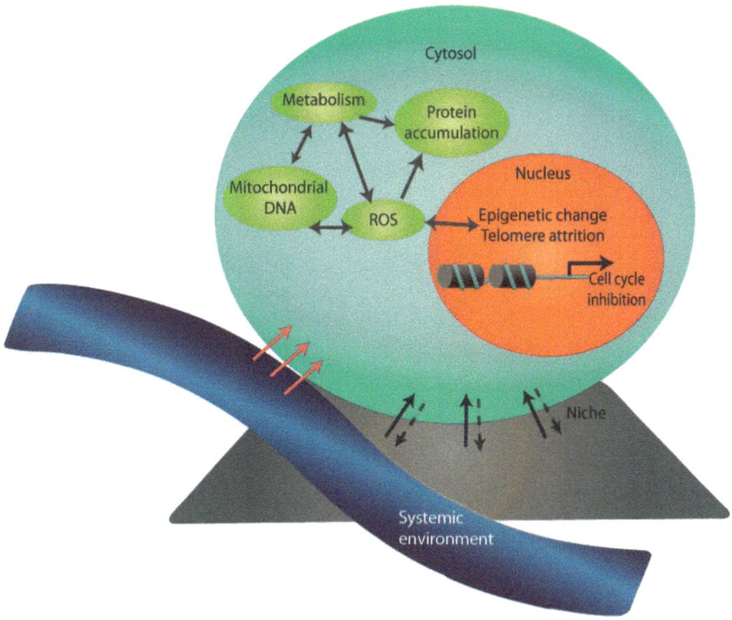

Yunjoon Jung and Andrew S. Brack, Figure 14.1 Multiple components driving stem cell aging. Stem cell aging occurs through changes in both cell-extrinsic regulators and intrinsic effectors. Cell-external influences such as changes in systemic factors from the bloodstream (blue) or factors from the stem cell niche (gray) can alter stem cell function by inducing signaling pathways and modifying the epigenetic and genetic signature of stem cells. The aged stem cell may also directly alter the niche. Aging is also driven by deregulation of cell-intrinsic effectors that control proteostasis, mitochondrial function, and metabolic control. Many of these modifications are interrelated suggesting one can impact another, leading to stem cell failure at multiple levels. ROS, reactive oxygen species; mtDNA, mitochondrial DNA.

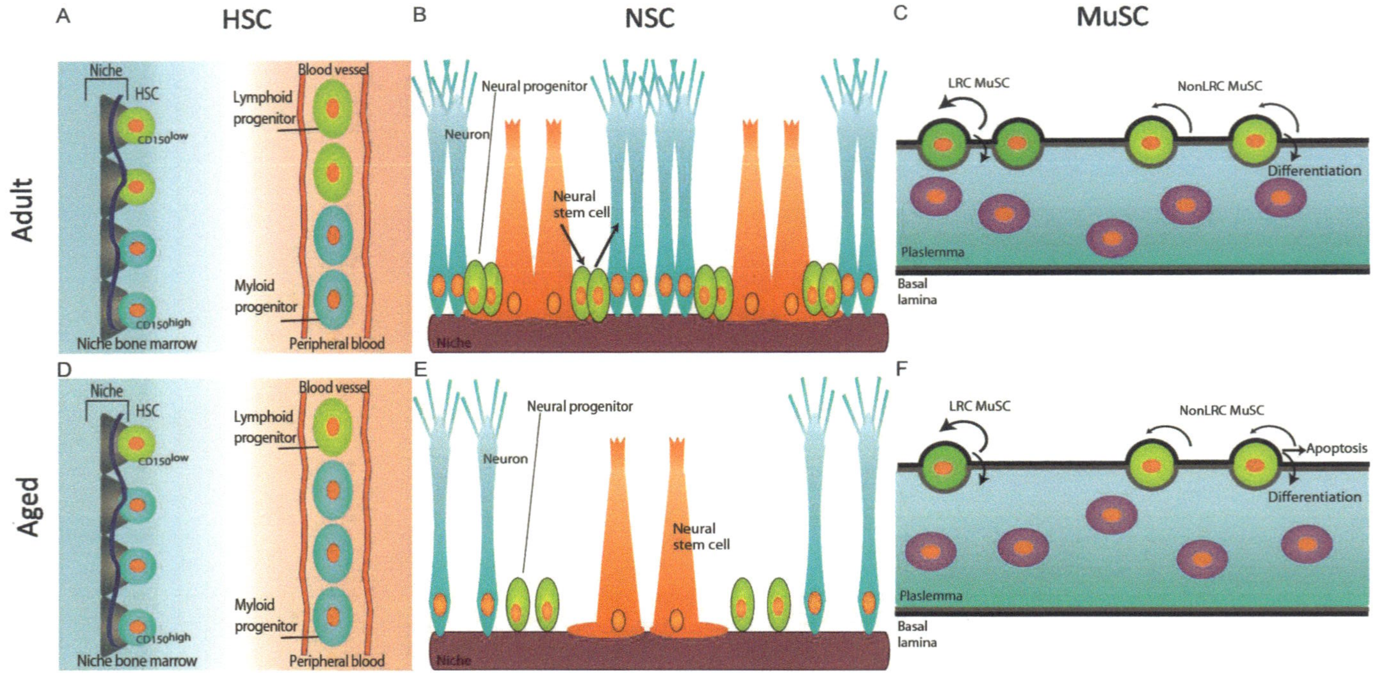

Yunjoon Jung and Andrew S. Brack, Figure 14.2 Schematic diagrams of adult stem cells and niche in bone marrow, brain, and muscle tissue. (A, D) Quiescent adult hematopoietic stem cells (HSCs) are located in the endosteal region of adult (A) and aged (D) bone marrow (left) residing in a niche composed of endothelial cells, osteoblasts (gray), and perivascular cells (dark blue line). The adult HSC pool consists of lymphoid-biased ($CD150^{low}$) HSCs (green) and myeloid-biased ($CD150^{high}$) HSCs (blue), with high self-renewal potential and balanced differentiation toward lymphoid (green, elongated) and myeloid progenitor cells (blue, elongated) in the peripheral blood (right, A and D). In aged HSC niche, selective expansion/retention of myeloid-biased $CD150^{high}$ HSCs skewing differentiation toward the myeloid lineage due to the robust self-renewal potential of aged $CD150^{high}$ HSCs compared to $CD150^{low}$ HSCs. (B, E) Neural stem cells (NSCs) reside at subgranular zone (SGZ) of the dentate gyrus of the hippocampus in adult (B) and aged (E) brain. NSCs (red) can give rise to neural progenitors (green), which progressively differentiate into neurons (blue). Systemic factors from the blood and local niche factors (purple) control NSC number and function. With age, number of NSCs and adult neurogenesis (making new neurons) declines. (C, F) Muscle satellite stem cells (MuSCs) in adult (C) and aged (F) muscle fibers. MuSCs (green) reside between the plasmalemma (gray line) and the basal lamina (black line) of muscle fibers. Long-term label retaining (LRC) MuSCs (dark green) have a high capacity for self-renewal and differentiation into muscle. nonLRC MuSCs (light green) are capable of differentiation but have limited self-renewal capacity. With age, the number of LRC MuSCs and fusion competent progenitors declines leading to a reduced number of functional stem cells and defective muscle repair.

Kolibri 360 GmbH
Ferdinand-Jühlke-Straße 7
99095 Erfurt, Deutschland
produktsicherheit@kolibri360.de